Exercise Immunology

T0300032

Exercise immunology is a discipline at the nexus of exercise physiology and immunology that aims to characterise the effects of exercise on the immune system in health and disease. This new edition of *Exercise Immunology* begins by providing an evidence-based introduction to the effects that individual bouts of exercise and exercise training have on the characteristics and functioning of the immune system.

In addition to introducing the immune system and summarising how different forms of exercise affect the characteristics and functioning of the immune system, this new and fully revised edition will explore exercise immunology in the context of immune ageing, cancer, autoimmune diseases and cardiometabolic disease. In addition, the authors discuss other factors that impact immune health, such as nutrition and environmental stressors, and explain the physiological basis of how exercise changes immune function across the healthspan and lifespan.

This book is written by leading exercise immunologists and is structured to provide a suggested curriculum of an exercise immunology degree component. Every chapter includes summaries of current and up-to-date research and offers practical guidelines to translate laboratory-based information into clinical settings. This textbook is essential for any exercise immunology degree component or advanced exercise physiology degree and will be vital reading for students in exercise and biological sciences and clinicians and researchers interested in the therapeutic applications of exercise.

James E. Turner, PhD is an Associate Professor in the School of Sport, Exercise and Rehabilitation Sciences at the University of Birmingham (United Kingdom). The research James leads explores the interaction between lifestyle and mechanisms of ageing and disease, spanning two themes: overweight/obesity and immune system ageing (with a focus on adipose tissue immunology) and physical activity and cancer immunology (with a focus on factors that influence treatment outcomes).

Guillaume Spielmann, PhD is an Associate Professor of Exercise Immunology at Louisiana State University (United States). Guillaume's research focuses on the impact of exercise on immune ageing and chronic disease risks including cancer and Type 2 diabetes. He is a member of several sport science societies worldwide, including the International Society of Exercise Immunology and the American College of Sports Medicine.

John P. Campbell, PhD is a senior lecturer (Associate Professor) in the Department for Health at the University of Bath (United Kingdom) and John has an adjunct position at the Exercise Medicine Research Institute at Edith Cowan University (Australia). John's principal area of research is understanding how exercise alters anti-cancer immunity, and John also conducts research in cancer immunodiagnostics.

Exercise Immunology

Second Edition

Edited by James E. Turner, Guillaume Spielmann, and John P. Campbell

LONDON AND NEW YORK

Designed cover image: Getty images

Second edition published 2025
by Routledge
4 Park Square, Milton Park, Abingdon, Oxon, OX14 4RN

and by Routledge
605 Third Avenue, New York, NY 10158

Routledge is an imprint of the Taylor & Francis Group, an informa business

First edition published by Routledge 2013

Library of Congress Cataloging-in-Publication Data
A catalog record has been requested for this book

ISBN: 978-1-032-18921-5 (hbk)
ISBN: 978-1-032-18916-1 (pbk)
ISBN: 978-1-003-25699-1 (ebk)

DOI: 10.4324/9781003256991

Typeset in Times New Roman
by codeMantra

Contents

Figures, tables and boxes

Figures

Tables

Boxes

About the editors

James E. Turner, PhD is an Associate Professor in the School of Sport, Exercise and Rehabilitation Sciences at the University of Birmingham (United Kingdom). James started his independent academic career in the Department for Health at the University of Bath between 2013 and 2023. Previously at the University of Birmingham, James undertook postdoctoral training in cancer immunology between 2011 and 2013 and undertook a PhD between 2007 and 2010 examining exercise and immune system ageing. The research James leads explores the interaction between lifestyle and mechanisms of ageing and disease, spanning two themes: overweight/obesity and immune system ageing (with a focus on adipose tissue immunology) and physical activity and cancer immunology (with a focus on factors that influence treatment outcomes).

Guillaume Spielmann, PhD is an Associate Professor of Exercise Immunology at Louisiana State University (United States). Guillaume acquired his PhD at Edinburgh Napier University (United Kingdom) and continued his training at the University of Houston between 2012 and 2015. Guillaume's research focuses on the impact of exercise on immune ageing and chronic disease risks including cancer and Type 2 diabetes. Over the past decade, he has also published numerous manuscripts on the effects of exercise and psychological stress on the immune function of patients, athletes and tactical athletes, such as astronauts. He is a member of several sport science societies worldwide, including the International Society of Exercise Immunology and the American College of Sports Medicine.

John P. Campbell, PhD is a senior lecturer (Associate Professor) in the Department for Health at the University of Bath (United Kingdom). John trained at the Institute of Immunology and Immunotherapy at the University of Birmingham Medical School, after acquiring his PhD from the School of Sport and Exercise Sciences at the University of Birmingham. Prior to that, John obtained a BSc in sport and exercise sciences from Edinburgh Napier University. John's principal area of research is understanding how exercise and other lifestyle factors alter anti-cancer immunity. John also conducts research in cancer immunodiagnostics, and is a co-inventor of numerous blood cancer tests.

About the contributors

Brian J. Andonian, MD, MHSc is an Assistant Professor of Medicine in the Duke Molecular Physiology Institute at the Duke University School of Medicine. He is a rheumatologist who practices lifestyle medicine and uses therapeutic lifestyle interventions (e.g., diet and exercise) to care for patients with autoimmune and rheumatic diseases. He is also a clinician scientist studying the effects of exercise training and lifestyle interventions on immune cell ageing, metabolic health, and molecular function in persons with rheumatic diseases, such as rheumatoid arthritis. The goal of his translational research is to improve outcomes in patients with rheumatic diseases, including disease activity, cardiometabolic risk, and disability, via personalised lifestyle and pharmacologic prescription.

David B. Bartlett, PhD is a senior lecturer of Exercise Immunology at the University of Surrey, UK and an Adjunct Assistant Professor of Medicine at Duke University, USA. He completed his PhD in immunology at the University of Birmingham Medical School under Prof Janet Lord and a post doc in Cancer Metabolism under Dr Dan Tennant before moving to Duke University in North Carolina, where he completed a Marie Curie Fellowship in Exercise Immunology under Prof Bill Kraus. Following this, he started his own research group at the Duke Cancer Institute, investigating the immune-modifying mechanisms of exercise training in patients with cancer. He continues his research at the University of Surrey, examining how exercise can enhance the immune system and promote disease-modifying responses that improve outcomes in clinical populations.

Courtney A. Bouchet, PhD is a postdoctoral fellow in the Department of Biomedical Sciences at Colorado State University. She has a bachelor's degree in Molecular, Cellular, Developmental Biology and Neuroscience from the University of Colorado, Boulder and a master's degree in integrative biology from the University of Colorado, Denver. She earned her PhD in 2022 from Oregon Health and Science University in Portland, Oregon. Through ten publications, she has established her research expertise that spans from the effects of exercise on fear extinction learning, memory, and relapse to the regulation and pain-mediated adaptations of opioid and cannabinoid receptor signalling within the descending pain modulatory pathway. Her research goals are to integrate her diverse interests to investigate synaptic- and circuit-based adaptations that shape the physiological intersections between exercise, stress, and inflammatory pain.

Jesper Frank Christensen, PhD, MSc is a senior researcher at Centre for Physical Activity Research (CFAS), Copenhagen University Hospital (Rigshospitalet), and Associate Professor in Exercise as Medicine with the University of Southern Denmark. He has been the principal investigator for clinical exercise trials across a wide range of malignancies including testicular, prostate, colorectal, and gastro-oesophageal cancers. His exercise oncology research spans the entire translational 'bench-bedside-practice' spectrum of biomedical research to elucidate and integrate information on the biological mechanisms, clinical effects and practice-based implications of exercise training in the oncology setting.

Glen Davison, PhD is an experienced research-active lecturer and chair in sport & exercise sciences, specialising in exercise immunology. Professor Davison is also BASES accredited sport and exercise scientist (physiology) and chartered scientist (CSci). He has worked with amateur, elite and professional athletes from a range of sports, including football, rugby, hockey, athletics, triathlon and cycling. His research interests include nutrition and exercise immunology; interval training; and sport & exercise science support of athletes (i.e. maintaining optimal health and performance). Glen is also a keen middle distance runner.

Kate M. Edwards is an Associate Professor and head of Exercise and Sports Sciences in the Sydney School of Health Sciences, University of Sydney (Australia). She trained at the University of California, San Diego (USA), having obtained her PhD from the University of Birmingham, UK. Kate's research on the adjuvant effects of exercise on immune function, in vaccination responses and efficacy of oncology treatment, is recognised internationally. She holds a senior fellowship to the Higher Education Academy, is a course accreditation reviewer and is elected board member for the Council of Heads of Exercise, Sport and Movement Sciences (Australia).

Monika Fleshner, PhD is a Professor in the Department of Integrative Physiology, member of the Center for Neuroscience, and Director of the Stress Physiology laboratory. She won the international Norman Cousins Award from the Psychoneuroimmunology Research Society and the national Guyton Distinguished Lectureship Award from the Association of Chairs of Departments of Physiology. She served as President of the International Society for Exercise Immunology (2011–2013) and President (2011–2012) and Secretary/Treasurer (2004–2006) of the Psychoneuroimmunology Research Society (PNIRS). She teaches undergraduate and graduate immunology and stress physiology. Professor Fleshner has trained over 50 MS/PhD/postdoctoral students. She has published over 200 scientific papers and book chapters. Her integrative research programme focuses on understanding (1) the impact of acute and chronic stressor exposure (mental and physical) on behaviour, neural, hormonal, and immunological function; (2) how such systems interact to affect the whole organism; and (3) the mechanisms of increased stress robustness (resistance/resilience) produced by exercise, prebiotics, and nutraceuticals.

Erik D. Hanson, PhD is a Kulynych/story fellow and an Associate Professor of Exercise Physiology at the University of North Carolina. His primary research focus includes exercise testing and training in clinical populations, with an emphasis on cancer survivors. He is interested in the role of exercise in improving skeletal muscle function during anti-cancer therapies, along with examining the immune system response to acute and chronic exercise, both as a potential therapy and a means of reducing disease recurrence.

Bruno Gualano, PhD is an Associate Professor at the School of Medicine of the University of São Paulo. He is also an undergraduate course chair of exercise physiology applied to clinical medicine and Head of the Laboratory of Conditioning and Assessment in Rheumatology, Applied Physiology & Nutritional Research Group and Center of Lifestyle Medicine. He also serves as associate editor of the *British Journal of Sports Medicine*. His research line involves clinical exercise physiology and lifestyle medicine.

Brian A. Irving, PhD is a trained exercise physiologist and Professor at Louisiana State University. Previously, he served as the founding director of the Geisinger Obesity Institute's

Metabolic Phenotyping Laboratory and as Assistant Professor of Medicine at the Mayo Clinic. Dr Irving has advanced training in (i) endocrinology and metabolism, (ii) skeletal muscle physiology and (iii) stable-isotope methodologies. His laboratory is interested in developing a more thorough understanding of the short- and long-term metabolic and proteomic adaptions to exercise, dietary, and medical interventions in young and older adults at risk for or with cardiometabolic diseases. Specifically, he is interested in skeletal muscle adaptations that provide protection against cardiometabolic diseases (e.g., type 2 diabetes, cardiovascular disease and peripheral vascular disease). Dr Irving also investigates the independent and combined effects prolonged sitting and physical inactivity have on skeletal muscle and whole-body physiology. In addition, his laboratory seeks to identify mechanisms of exercise intolerance and functional impairment in older adults with obesity.

Arwel W. Jones, PhD is a mid-career researcher with expertise in the efficacy and mechanisms of non-drug interventions to manage respiratory infections and chronic lung disease. His early career investigated the role of nutritional countermeasures to changes in mucosal and systemic immunity following prolonged strenuous exercise or during endurance training. Since then, much of Arwel's work has focused on the disease-modifying effects of moderate exercise. This has included clinical trials of exercise training interventions in people with chronic obstructive pulmonary disease who are susceptible to frequent respiratory infections and investigations of the anti-inflammatory and anti-viral effects of exercise in this population. Arwel's favourite hobbies are watching and playing sports, particularly football (or soccer depending on where you are from).

J. Philip Karl, PhD, RD is a nutrition physiologist in the Military Nutrition Division at the U.S. Army Research Institute of Environmental Medicine. He has contributed to more than 100 publications on topics including nutrition, exercise and gut microbiology. His recent research has primarily focused on determining how interactions between diet, exercise, stress and the gut microbiome impact human physiology, and to identify strategies for leveraging the gut microbiome to improve health and performance. He serves on the editorial board of multiple international journals and has held leadership positions within the American Society of Nutrition and within multiple scientific communities under the U.S. Department of Defense.

Graeme Koelwyn, PhD is the Dr James Hogg chair and Tier 2 Canada research chair in Public Health 'Omics in Exercise and Disease at St Paul's Hospital and Assistant Professor in the Faculty of Health Sciences at Simon Fraser University in Vancouver, Canada. His research applies a translational approach for understanding how heart, lung and/or oncologic diseases communicate with each other through immune-specific mechanisms, leading to adverse systemic, tissue and cellular responses. It also seeks to demonstrate how exercise can therapeutically improve immune function to protect from these diseases and their deleterious interactions.

Marian L. Kohut is a Professor in the Department of Kinesiology, Program of Immunobiology, and Nanovaccine Institute at Iowa State University. She completed her postdoctoral training in psychoneuroimmunology at the University of Rochester and a PhD in exercise science at the University of South Carolina. Her research investigates the mechanisms by which host factors including ageing, exercise and stress impact innate and adaptive immunity in the context of viral infection and vaccination. Her scientific appointments have included reviewer for NIH study sections and special emphasis panels: expert panel member for NIAID Strategic Plan for Research on Vaccine Adjuvants, and general council member for Autumn Immunology Conference.

Karsten Krüger, PhD is head of the Department of Exercise Physiology and Sports Therapy at the Justus-Liebig-University of Giessen, Germany. He has been active in the field of molecular and cellular adaptation to exercise and training for more than 20 years. His research focuses on the regulation of inflammation through physical activity and on adaptation processes in competitive sports. He is the spokesperson for the in:prove project, one of the largest research projects in elite sport in Germany. He is an editor-in-chief of the journal *Exercise Immunology Review* and board member of the International Society of Exercise and Immunology.

Hawley E. Kunz, PhD is an Assistant Professor of Medicine and associate consultant at Mayo Clinic Rochester. Her graduate school studies in exercise and spaceflight immunology and her postdoctoral training in muscle and adipose tissue physiology and metabolism precipitated an interest in the role of tissue–immune interactions in health, disease and adaptive responses to exercise. Her recent and ongoing work includes projects examining the effects of adipose tissue inflammation on metabolic health in obesity; the mechanistic roles of immune cells in skeletal muscle and adipose tissue responses to acute exercise; and the effects of ageing and inflammation on adipose tissue physiology. Ultimately, she hopes her research will lead to the development of targeted therapeutics for improving metabolic health in ageing and obesity.

Emily C. LaVoy, PhD is an Associate Professor in the Department of Health and Human Performance at the University of Houston. Her research investigates the effects of exercise and physical fitness on the immune system. She is particularly interested in how exercise improves diseases and conditions associated with immune dysregulation, such as cancer. She is also highly committed to mentoring future scientists and runs an active research laboratory employing undergraduate and graduate students. She is a member of the American College of Sports Medicine, the American Physiological Society, the International Society of Exercise and Immunology and the PsychoNeuroImmunology Research Society. Emily enjoys long-distance running and cycling, and credits growing up cross-country skiing in Michigan with her initial interest in exercise science.

Sarah C. Marvin is a PhD candidate within the Sydney School of Health Sciences (Faculty of Medicine and Health) at The University of Sydney (Australia). Her research is primarily in the field of exercise oncology, with a particular interest in the role of exercise during immunotherapy treatment. Sarah is a member of Exercise and Sports Science Australia (ESSA), and holds qualifications as both accredited exercise physiologist and accredited exercise scientist.

David C. Nieman, PhD is a Professor in the Department of Biology, College of Arts and Sciences, at Appalachian State University, and director of the Human Performance Lab at the North Carolina Research Campus (NCRC) in Kannapolis, NC. Dr Nieman is a pioneer in the research area of exercise and nutrition immunology and his current work is centred on investigating novel nutritional products as countermeasures to exercise- and obesity-induced immune dysfunction, inflammation, illness and oxidative stress using a multi-omics approach. Dr Nieman has published more than 400 peer-reviewed publications in journals and books, is editor-in-chief of the sports nutrition section of the journals *Nutrients* and *Frontiers in Nutrition*, and sits on ten journal editorial boards. He is the author of nine books on health, exercise science and nutrition. Dr Nieman served as vice-president of the American College of Sports Medicine (ACSM), president of SEACSM and two terms as president of the International Society of Exercise and Immunology.

Samuel J. Oliver, PhD is a Professor in sport & exercise science at Bangor University. He gained his PhD in 2007, studying the influence of nutritional restriction on human performance and immune health. Subsequently, he has investigated the effects of exercise, sleep, diet, hypoxia and cold on the immune system, the findings of which have influenced athletic, armed and emergency service practice. Sam has mentored more than 25 PhD and MRes researchers; he is associate editor for applied physiology, nutrition and metabolism and environmental physiology lead for the Welsh Institute of Performance Sciences. In his spare time, Sam enjoys climbing mountains and surfing.

Brandt D. Pence, PhD is an Associate Professor and Director of Research in the College of Health Sciences, University of Memphis. His research focuses on ageing and the innate immune system, with specific interests in monocyte and macrophage metabolic and mitochondrial dysfunction that drives ageing biology and ageing-related diseases. He is also interested in the effects of lifestyle factors such as exercise and nutrition on immune cell function and metabolism. He is a Fellow of the American College of Sports Medicine and a member of the International Society for Exercise and Immunology, the American Heart Association, the American Aging Association and the Gerontological Society of America. His research is or has been funded by the National Institutes of Health and the American Heart Association. In his free time, he enjoys hiking, trail running and landscape/nature photography.

Ana J. Pinto, PhD is a postdoctoral fellow in the Division of Endocrinology, Metabolism, and Diabetes at the University of Colorado Anschutz Medical Campus, USA. She completed her PhD in clinical exercise physiology at the University of São Paulo, Brazil. Her research interests include testing and optimising exercise and sedentary behaviour interventions to maximise the therapeutic effects of such interventions on physical capacities, cardiometabolic risk factors, clinical symptoms and immune function in patients with rheumatoid arthritis. Her long-term goal is to improve individualised exercise/physical activity prescription and, therefore, patient care and disease management and progression.

David B. Pyne, PhD is an Adjunct Professor with the Research Institute for Sport and Exercise at the University of Canberra (Australia). Pyne has 35 years' experience as a sports scientist, attended four Olympic Games, and consults with a variety of sports. Pyne's research in exercise and the immune system, environmental physiology and fitness for sports is recognised internationally. Pyne has published over 300 peer-reviewed papers, with an H-Index of 94 (Google Scholar), and supervised 27 PhD students to completion. He was the foundation editor of the *International Journal of Sports Physiology and Performance* from 2004 to 2009 and currently serves as consulting editor.

Heather Quiriarte, MS is a research associate in the School of Kinesiology at Louisiana State University. She has studied the impact of various psychological and physical stressors on the immune system, and her involvement in numerous NASA-funded spaceflight and bed rest studies enabled her to realise how interconnected multiple physiological systems are especially the musculoskeletal and immune systems. She expanded her research to better understand how stress and ageing affect muscle and immune health and how exercise interventions can be designed to improve them. She focused on the effects of exercise on circulating exosomal profiles and understanding the crosstalk between the musculoskeletal and immune systems in older adults. She continues to study the interplay between immunity and skeletal muscle and to develop targeted non-pharmacological interventions that could comprehensively improve overall immune competency in clinical populations.

Mark Ross, PhD is an Associate Professor in Exercise Physiology at Heriot-Watt University, Edinburgh, UK. His research examines the role of ageing and exercise on immune cells. Specifically, Mark is working to understand how ageing and regular exercise influences T cell-mediated inflammation and tissue repair, physiological processes which are affected in several age-related diseases, including diabetes and cardiovascular disease. Mark is research lead for the Sport and Exercise Science research group at Heriot-Watt University.

Tim Schauer, PhD is postdoctoral fellow at the Centre for Physical Activity Research (CFAS) located at the Copenhagen University Hospital (Rigshospitalet). He gained his PhD in 2021, studying the influence of exercise, anti-cancer treatment and inflammation on the immune system. His work involves experimental research in both healthy and clinical populations as well as in vitro and ex vivo studies.

Richard J. Simpson, PhD is Professor in the School of Nutritional Sciences and Wellness (College of Agriculture and Life Sciences) at the University of Arizona and hold joint appointments in Pediatrics (College of Medicine), Immunobiology (College of Medicine), the Arizona Cancer Center and the Bio5 Institute. He also holds the Gouri Bhattacharya Endowed Chair in Pediatric Cancer in the Steele Children's Research Center at the University of Arizona. His research interests are concerned with the effects of exercise, stress and ageing on the immune system, and how exercise can mitigate diseases such as cancer through its immune-modulating effects. He is the current president and executive director of the International Society of Exercise Immunology (ISEI) and fellow of the American College of Sports Medicine (ACSM).

William Trim, PhD is a postdoctoral research fellow at the Department of Systems Biology at the Harvard Medical School, USA, working on cell size regulation. He completed his PhD at the University of Bath, UK, and the European Space Agency in Toulouse, France, under Professor Dylan Thompson and Dr James Turner focused on understanding human adipose tissue immunology and metabolism in the context of ageing and space flight. He next completed a postdoctoral position with Dr Lydia Lynch in Trinity College Dublin, Ireland, and Brigham and Women's Hospital, USA, investigating tissue networks underpinning metabolic disease development in human obesity.

Alex Wadley, PhD is an Assistant Professor in Exercise Metabolism at the University of Birmingham, UK. His research examines how single and regular bouts of exercise impact the functioning of the immune system in different population groups. Most notably, Alex is currently examining how periods of exercise training impact autoimmunity in people with type-1 diabetes and systemic lupus erythematosus, and how single bouts of exercise can act as an adjuvant for the harvest of stem cells for transplantation in individuals with blood cancers. Alex runs the Exercise Immunology & Health research group and is the director of Cellular Health and Metabolism Facility at the university.

Philipp Zimmer, PhD is the head of the division of performance and health (full Professor) and director of the Institute for sport and sport science at TU Dortmund University (Germany). Zimmer gained a PhD in exercise and neuroscience and worked as post doc at the German Cancer Research Center and group leader at the German Sport University Cologne. He is investigating the role of exercise for disease prevention in healthy people and as disease-modifying therapy in clinical populations (e.g., persons with cancer and multiple sclerosis). His main area of expertise is translational exercise immunology. He serves as editor-in-chief for the *International Journal of Sports Medicine*.

Preface

Exercise immunology is a field of study at the nexus of exercise physiology and immunology. It has a specific focus on understanding the relationship between exercise, immune function, health and disease. This discipline of study has been ongoing for around 120 years. The field has expanded considerably in the past 50 years, thanks to the combination of increasing interest in understanding the relationship between exercise and health, and the rapid evolution of research technologies in immunology. The exponential growth of exercise immunology research has been mirrored by an expansion of students interested in studying the discipline, as demonstrated by an ever-increasing number of undergraduate and graduate courses teaching exercise immunology – either as a subcomponent (i.e., a module or a course) of a taught undergraduate (e.g., BSc) or postgraduate degree (e.g., MSc), or as an outright course of study (e.g., PhD). Accordingly, this textbook is designed to be used by students enrolled on undergraduate and postgraduate degrees that contain exercise immunology as an area of study.

The goal of this book is to provide an evidence-based introduction on the effects of individual exercise bouts and exercise training on the characteristics of the immune system. This new edition builds on its predecessor – the first edition of this book, published in 2013 by Michael Gleeson, Nicolette Bishop, and Neil Walsh. The first edition provided an athlete-centred introduction of the immune response to physical stressors and presented the immune challenges observed in athletes and sedentary individuals in response to exercise. This second edition provides necessary updates to core chapters and expands the provision of content relating to the clinical relevance of exercise in ageing and disease. In particular, this new edition contains new chapters with a dedicated focus on cancer, autoimmune and cardiometabolic diseases.

This book is structured to provide a suggested curriculum for an exercise immunology module or course. Every chapter includes summaries of past and up-to-date research and offers practical guidelines to translate laboratory-based information into clinical or applied settings. This book shows students how to evaluate the strengths and limitations of the evidence linking exercise, immune system integrity and health, and explains why exercise is associated with anti-inflammatory effects that are potentially beneficial to long-term health. Each chapter summarises key concepts, and throughout each chapter, readers are directed to comprehensive reviews and seminal experimental studies.

In this book, after a historical overview of the field in Chapter 1, we provide a description of the immune system in Chapter 2, introducing the necessary foundation to understand the modulating effect of exercise on aspects of immunity that influence health and disease. In Chapter 2, our description of how the immune system is organised, its components, and their function in protecting the body from pathogens and tumours, is an update from the *Human Immunology* chapter written by Prof Gleeson and Dr Jos Bosch in the first edition of this book. Chapter 3 provides an overview of common methods in exercise immunology, before subsequent chapters summarise how exercise affects different immune system components, from cells and soluble molecules, to tissues, organs and organ systems. Further chapters summarise research spanning overall immune competency and infection risk, ageing and studies that have been conducted in the clinical settings of cancer, autoimmunity and cardiometabolic disease.

After reading this book, students should be able to:

- Describe the characteristics, organisation and function of the different components of the immune system.
- Summarise how immune function can be assessed, and the advantages and disadvantages of using different methods to examine the immune system in different human populations.
- Describe the physiological basis and mechanisms underlying the relationship between exercise, immune function, health and disease.
- Interpret the methods and results of relevant research studies and evaluate the strengths and limitations of the evidence linking exercise, exercise, immune function, health and disease.

The editors and contributors of this book are all active recognised researchers in exercise immunology or stress immunology, and most have permanent academic posts with a responsibility for teaching exercise physiology and exercise immunology to undergraduate students. The target audience of this book ranges from undergraduate and postgraduate students in exercise, biological sciences and medicine, and also professionals working in sport, exercise and clinical healthcare settings. We hope that you enjoy reading this book.

James E. Turner, Guillaume Spielmann, John P. Campbell

Keywords

Immune System – Cellular and Soluble Immunity – Exercise – Stress – Physical Activity – Ageing – Health and Disease – Exercise Physiology

1   History of exercise immunology

James E. Turner and John P. Campbell

Introduction

The study of exercise immunology has been conducted across three centuries and has culminated in many historic findings which have shaped our understanding of how exercise affects health and human performance. This chapter will provide a chronological overview of this history, starting in the 19th century, through to the modern era.

The 19th century

The first report of exercise influencing the immune system was published in the 19th century when it was found that the total leucocyte count in blood measured minutes after completing a marathon race was close to three times resting values (Schultz, 1893). In addition, it was found that 80–90% of leucocytes in the blood samples collected after exercise were polymorphonuclear cells, when usually these cells account for 50–70% (Schultz, 1893). This immunological phenomenon is referred to as leucocytosis (see Box 1.1) and it is a very commonly reported and reproducible effect that exercise has on the immune system. Although the mechanisms were not known at the time, muscular exertion was assumed to be at least partly responsible, and this knowledge translated to other settings. For example, in 1899, leucocytosis was reported in the context of convulsions and thought to be the result of what was said to be a "double cause" – a temporary mechanical effect of muscular work during convulsions, and what was referred to at the time as "an inflammatory toxic cause" (Burrows, 1899). These hypotheses were remarkably close to the biological mechanisms that were subsequently characterised over the following century (see Box 1.1). The next report of exercise influencing the immune system was published in 1901 as part of an investigation of various factors thought to affect leucocyte counts (Cabot, Blake and Hubbard, 1901). The rationale was that body temperature and leucocyte counts were examined following surgical procedures to diagnose and manage complications such as sepsis. However, it was unknown if other factors around the time of blood sampling could change leucocyte counts. Thus, measurements were made before and after anaesthetic administration and before and after surgery itself among patients needing treatment for a variety of reasons (e.g., fractures, hernias, cancer, and other forms of general surgery). Measurements were also made among other people, who either had an acute infection and fever, or who had exercised. In surgical settings, around half of patients exhibited leucocytosis due to anaesthetic and surgery, but around half did not, or exhibited small fluctuations in leucocyte counts. Overall, leucocytosis in surgical settings seemed to be inconsistent in magnitude and direction, possibly because many of the factors now known to affect leucocyte counts may not have been controlled for adequately (e.g., timing of sampling under resting conditions, inter-individual differences

DOI: 10.4324/9781003256991-1

Box 1.1   Leucocytosis – a commonly reported and reproducible effect of exercise on the immune system

Leucocytosis is an increase in the number of leucocytes, usually measured in peripheral blood. In the absence of disease, this increase in cell number is caused by a transient mobilisation or migration of cells into circulation, rather than an increase in cell division. *Cytosis* – the suffix of the term *leucocytosis* – combines *Cyto* (meaning cells) and *Osis* (meaning pathology or process, such as transport) and is used to describe the frequency of leucocytes in blood. Leucocytosis is an increase of the total leucocyte count above $11 \times 10^9 L^{-1}$ of blood and can be further described according to leucocyte subtypes. For example, lymphocytosis is an increase of lymphocytes above $4.5 \times 10^9 L^{-1}$ of blood. Monocytosis is an increase of monocytes above $0.88 \times 10^9 L^{-1}$ of blood. Granulocytosis is an increase of granulocytes above $7.0 \times 10^9 L^{-1}$ of blood (Riley and Rupert, 2015). A high total leucocyte count, or separate high counts of lymphocytes, monocytes or granulocytes, measured under resting conditions, can be associated with infection, inflammation, trauma or use of some medications and sometimes signposts a haematological malignancy (Chabot-Richards and George, 2014). However, non-pathological challenges to physiology (e.g., exercise in the short-term, such as over seconds, minutes or hours) also stimulates leucocytosis.

In the field of exercise immunology, a focus has been on lymphocytosis – with extensive investigation into subtypes such as natural killer cells, T cells and B cells – whereas monocytosis has received less attention, as with granulocytosis, except for some focus on neutrophils, which compared to eosinophils and basophils, drive granulocytosis. The increase in neutrophils in blood was a focus of many investigations in the 1990s and is referred to as neutrocytosis or neutrophilia, whereas the increase in eosinophils and basophils, referred to as eosinophilia and basophilia respectively, has received very little attention in exercise immunology. Since the first observation of total leucocyte counts being nearly three times higher in the minutes after completing a marathon (Schultz, 1893), it has been established that within seconds of starting exercise, there is a rapid mobilisation of lymphocytes and neutrophils into blood. Depending on the intensity and duration of exercise, lymphocyte numbers can fall below baseline 1–2 hours after an exercise bout has finished, and can remain at this level for up to 24 hours. This decrease in cell number is termed lymphocytopenia (the suffix *penia* means decrease or deficiency). The exercise-induced decrease in lymphocytes after exercise was, for a time, thought to represent an 'open-window' representing a decline in immune competency and heightened susceptibility to infection (Pedersen and Ullum, 1994), but this response is now thought to represent immune-surveillance, where cells leave blood and travel to tissues, where they are more likely to encounter normal cells that have been infected or damaged or cancer cells. Typically, exercise-induced changes to lymphocytes are transient, with cell numbers returning to pre-exercise levels within a few hours. Beyond lymphocytes, neutrophil recovery after exercise has also been studied extensively. Neutrophils tend to leave the circulation after exercise finishes, but it is common for neutrophil counts to exhibit a secondary increase 3 or 4 hours after a particularly vigorous bout of exercise, and they can remain at this level for around 24 hours.

As with exercise, psychological stress also invokes lymphocytosis, and one of the earliest reports was in 1926. Patients with strong emotional feelings before surgery showed a more pronounced anticipatory rise in lymphocyte numbers in resting blood collected in the few days prior to surgery compared with patients who reported being less emotionally affected by the forthcoming operation (Mora et al., 1926). The link between leucocyte numbers in blood and acute stressors is so well established that before methods were commonly available to assay catecholamines, leucocyte counts were used as a proxy measurement of sympathetic activation (Hoagland et al., 1946). Indeed, the two primary mechanisms for leucocytosis are catecholamine release and increased cardiac output, due to increased heart rate and stroke volume, resulting in increased blood flow. The shear forces associated with these cardiovascular changes contribute to lymphocytosis, resulting in a demargination of lymphocytes adhered to blood vessel walls (Atherton and Born, 1972). However, sympathetic activation, including circulating epinephrine and norepinephrine, and direct innervation of immunological organs, is the predominant mechanism, as part of the 'fight or flight' response (Elenkov et al., 2000; Shephard, 2003).

caused by treatment, disease stage or type of disease and adrenergic activity, perhaps due to pain and psychological stress). Leucocytosis was evident among people presenting with fever, but the most dramatic and consistent changes were found in the minutes following a marathon race, characterised by very pronounced changes to cell counts that were two to six times higher than resting values. It was concluded that "Very violent physical exertion produces in the blood a condition which leaves physiological limits, and approaches or is identical with that found in disease". More detailed interpretation of the same results was published separately (Larabee, 1902) with expanded analyses on subtypes of leucocytes. Conclusions confirmed those of Schultz in 1893, such as polymorphonuclear cells being the most predominant cell type in blood post-exercise and exhibiting the largest changes. However, the findings extended prior knowledge, by reporting a differential response among other cell types, showing that rarer cells, such as eosinophils, seemed to disappear altogether, and lymphocytes and monocytes fell below pre-marathon levels among some people. These findings, which were somewhat alarming at the time, are now understood to represent immune-surveillance, rather than a response to a "toxic and inflammatory cause" or cell apoptosis (see Chapter 4). This report in 1902 also highlighted other very important considerations that are now understood, such as the potential influence of blood volume changes due to sweating or dehydration or whether other factors, influence results, such as cold exposure and other environmental factors (see Chapter 13) or psychological stress (see Chapter 15).

The early 20th century

Research in exercise immunology between 1900 and the early 1930s is reviewed by Garrey and Bryan (1935). During this period, there were around 25 exploratory studies published, which continued to characterise the magnitude and timing of leucocytosis following various forms of exercise. Studies examined running exercise, from marathons to very short vigorous bouts lasting from several seconds to less than an hour. Other modes of exercise were also explored, including football, water polo, weightlifting, and gymnastics. Through this research, important observations were made, including that leucocytosis happened within seconds or minutes of beginning exercise, and that fitter individuals, compared with less fit individuals, exhibited smaller leucocytosis in response to the same intensity of exercise. Mechanisms were also explored during this period, and some were ruled out. For example, exercise-associated haemoconcentration due to sweating and dehydration was considered an unlikely mechanism, because blood volume only falls by approximately −10%, but leucocyte counts increase by at least +100% to 200%. However, it became more accepted that leucocytosis was caused by a redistribution of neutrophils from vascular beds and from blood-forming organs. Studies at this time attempted to rule out a primary role for some organs, such as by establishing exercise-induced leucocytosis occurred among people without a spleen – a blood storage organ thought to be responsible for shifts in cell distribution. In addition, studies showed a larger magnitude of leucocytosis with a higher intensity of exercise and that exercise-induced leucocytosis is replicated by adrenaline infusion.

Research between the 1930s and 1940s is reviewed by Sturgis and Bethell (1943) and although only a few additional experimental articles had been published in the previous decade, these authors provided an update and greater clarity on interpretational issues and mechanisms. For example, it was accepted in the field that exercise-induced leucocytosis was due to a redistribution of cells, rather than a formation of new cells, given that changes occurred within seconds, and that only a very small proportion of cells exhibited a 'young' phenotype. Indeed, studies had demonstrated that the polymorphonuclear cells present in blood after exercise were of a mature, adult type, with lobed nuclei (i.e., mature neutrophils). It was also accepted in the field that leucocytosis occurred due to circulatory shifts where cells were liberated from capillaries throughout the body, thought to be in locations such as the spleen, liver, lungs, lymph nodes, bone marrow, and muscles. Other potential mechanisms had been deemed unlikely, such as changes to blood lactate, blood glucose, blood pressure, body temperature and capillary dilation, and there was a greater appreciation that lymphocytes also contribute towards leucocytosis. Studies had reported a rapid increase in lymphocytes and neutrophils in response to short vigorous bouts of exercise, but neutrophils increased most after more prolonged exercise and with a delay of several hours.

The mid-20th century

Research between the 1940s and 1980s is reviewed by McCarthy and Dale (1988). Around 25 additional experimental studies examining exercise-induced leucocytosis were published between the 1940s and 1980. During this time, leucocytosis was systematically examined in controlled laboratory investigations (e.g., with treadmill running or cycle ergometry) and in field-based studies, most commonly running events, but also following a variety of other activities, including marching, swimming, canoeing, wrestling, weightlifting, gymnastics, basketball, and football. The systematic laboratory investigations mapped the time-course of leucocyte kinetics with repeated blood sampling and explored the effect of manipulating exercise intensity, ranging from low to vigorous up until exhaustion, and manipulating duration, ranging from seconds and minutes to several hours. It was confirmed that generally, the magnitude of change in leucocyte counts was proportional to the intensity and duration of exercise. This provided a much better understanding of mechanisms, including roles for adrenaline, cortisol, other hormones, and adhesion molecules, due to advanced study design conditions, such as treating exercising individuals with adrenergic receptor blocking drugs, or infusing adrenaline and cortisol to mimic exercise. Interpreting infusion and exercise studies fully established the time-course of leucocytosis and analysis of leucocyte subtypes provided a much clearer picture. For example, vigorous exercise of up to around 1 hour in duration was associated with an almost immediate increase in lymphocytes and neutrophils, most likely caused by a mobilisation from marginal pools, followed by a decline in counts towards normal values several minutes after exercise ceased. Additionally, a so-called biphasic response was characterised, consisting of a second larger increase of neutrophils around 3 hours after the start of exercise, thought to represent mobilisation of cells from bone marrow. Earlier findings were also confirmed, such as that eosinophils decrease, or even disappear following prolonged vigorous exercise, and although counts are very low in blood, basophils were also shown to decrease in numbers and proportion. Another consistent finding was that monocytes increased slightly in blood following most forms of exercise, but the effects were often masked by the large changes in other leucocyte subtypes.

Whilst advances were made in understanding leucocytosis, most research between 1940 and 1980 that investigated how exercise affects the immune system had an underlying emphasis on understanding the relationship between physical exertion and risk or pathogenesis of infectious disease. Before World War II occurred between 1939 and 1945, treatment of acute viral infections, including viral hepatitis (Wong et al., 2015) comprised rest during the acute phase of disease, but as patients recovered, normal physical activities of daily living were reintroduced. However, during World War II, it was noticed that relapse was associated with reintroducing physical activities too early in the recovery phase, thus prolonged and strict bed rest was advocated (Barker, Capps and Allen, 1945a; Barker, Capps and Allen, 1945b). Indeed, being active exacerbated early phases of disease: exercise tolerance tests were even used to check whether patients had recovered fully by examining if signs and symptoms of disease worsened (e.g., liver swelling, tenderness and a change to bilirubin or liver function enzymes). In support, an early clinical trial showed that when comparing strict bed rest to no restriction of activity, the duration in hospital was around one-third less with rest and liver size returned to normal in around half the time (Hughes, 1945). This research changed clinical practice, and by the 1950s, hospital stays had doubled from 30 to 60 days. Strict bed rest was later questioned by the military, due to strains on personnel given that viral hepatitis was extremely common, and subsequent research demonstrated that physical activity, even in early phases of infection, did not lengthen recovery, so long as it was undertaken when patients felt well, and when their bilirubin levels had returned towards normal (Swift et al., 1950; Chalmers et al., 1955). Indeed, the importance of maintaining physical activity and avoiding bed rest in clinical settings gained traction in treating heart disease around this time – as reviewed more recently (Kavanagh, 2000; Redfern et al., 2022) – and we now know bed rest can have very profound and deleterious effects on aspects of physiology and the immune system (see Chapter 13). However, despite the forward-thinking view that being physically active did not have a harmful effect on the setting of viral hepatitis – a standpoint now becoming better known in the chronic disease setting (see Chapters 10–12) – concerns about exercise were maintained in the setting of acute communicable disease. For example, in the 1940s, it was suspected that exercising during polio virus infection exacerbated disease, due to the observation that paralysis was often localised to parts of the body that had been active, and paralysis appeared to develop most commonly among people who had been vigorously active in the preceding days when symptoms developed (Russell, 1947; Hargreaves, 1948; Russell, 1949). In addition, in the 1970s, it was found that 5 out of 78 sudden unexpected deaths during or immediately after exercise, were linked to recent history of upper respiratory tract infections (Jokl and McClellan, 1971). Links between Coxsackie virus infection and arrhythmias led to research examining whether the combination of vigorous exercise, viral infections, and myocarditis explained sudden unexpected death among young physically active people (Wentworth, Jentz and Croal, 1979; Burch, 1979; Roberts, 1986) stimulating further worry about how exercise interacts with the immune system. At this time, it was assumed that young people who undertook vigorous and prolonged exercise during upper respiratory tract infections had an increased risk of developing viral cardiomyopathy, causing irreversible damage to the heart muscle, and possibly fatal outcomes, and it was advised that strenuous exercise should be avoided for two weeks after an infection had resolved (Burch, 1979). Around this time, there was also concern about exercise increasing the risk of splenic rupture with infectious mononucleosis – caused by Epstein Barr Virus – and it was recommended that exercise should be avoided for between one and six months after the illness had subsided (Roberts, 1986). Although specific worries about exercise and polio virus infection eased from 1988, partly due to disease control with the ongoing Global Polio Eradication Initiative (Barrett, 2023), in later years, concern about exercise and the immune system remained the underlying theme of most exercise immunology research from the 1980s into the new millennium (see Box 1.2).

Box 1.2   Does concern about exercise and the immune system persist in the 2020s?

Despite the gradual change of focus in the field of exercise immunology in the 21st century, research from the 1980s onwards, with an emphasis on risk of infection among athletic populations, has continued to influence the discipline. Guidance from the 1980s on strategies for athletes to avoid infections or adapt training when unwell (Roberts, 1986) has continued to be published in the early 21st century (Friman and Wésslen, 2000). These include position statements from the ISEI (Walsh et al., 2011a; Walsh et al., 2011b), the British Association of Sport and Exercise Sciences (Gleeson and Walsh, 2012), and other authoritative reviews (Bermon et al., 2017; Walsh, 2018, Ruuskanen et al., 2022). More broadly, there has been some interest in exercise-associated allergic reactions, including anaphylaxis, urticaria, asthma and bronchoconstriction (Cooper et al., 2007). In addition, awareness of sudden unexpected death among young physically active people due to viral myocarditis – a concern that began in the 1970s – has continued into the 2020s, with precautionary steps taken in sports medicine (Halle et al., 2021). Much greater attention on viral myocarditis was paid during the global pandemic caused by severe acute respiratory syndrome coronavirus 2 (SARS-CoV-2) resulting in coronavirus disease 2019 (COVID-19). Partly due to greater awareness of how coronaviruses can cause severe disease, but also due to the number of people who became infected worldwide, guidelines over cardiac and respiratory return-to-play considerations were developed (Wilson et al., 2020). However, most concern about exercise during the pandemic, particularly among the general population, was focused on exposure to aerosolised virus particles and risk of infection, which was assumed to be greater if exercising indoors or when in close contact with other people. During the pandemic, if infected and acutely unwell, the general advice was to rest or undertake only very light activity. However, the potential positive influence that regular exercise might have on immune system functioning was highlighted by research showing that high cardiorespiratory fitness was associated with a lower risk of complications and hospitalisation, following infection with SARS-CoV-2 (Brawner et al., 2019). In addition, research showed that high levels of habitual physical activity – assessed mostly using self-report data – were associated with a lower risk of infection and that more physically active individuals were less likely to develop very severe disease (Ezzatvar et al., 2022). Thus, in conclusion, concern about exercise and the immune system persists in the 2020s, including risk of infections among athletic populations, potential exercise-associated allergic reactions, viral myocarditis and risk of becoming infected with the strains and variants of SARS-CoV-2 that continue to circulate.

Other studies in the pre-1980s era of exercise immunology had also begun to explore the impact of exercise on lymphocyte function in vitro, concluding that generally, following both short or prolonged exercise, there is a decreased responsiveness of cells to mitogens such as concanavalin-A, phytohaemagglutinin and pokeweed mitogen. A fundamental interpretational concept was highlighted, at that time, which has not always been considered in more recent research: when using in vitro assays, the proportional composition of lymphocyte subtypes in samples – characterised by fewer responding effector cells in the minutes or hours after exercise – is most likely driving observations of altered cell responsiveness, rather than exercise altering cell function per se. By 1980, many questions remained, and topics required further exploration. For example, a fall in lymphocyte counts 1–2 hours after vigorous exercise (i.e., termed 'lymphocytopenia' or 'lymphopenia' – see Box 1.1) had not yet been robustly documented. In addition, mechanisms underlying the different magnitude of post-exercise change among lymphocyte subtypes had not been defined (i.e., studies of adrenergic receptor expression density came later in the 1980s). Finally, the potential clinical relevance of leucocytosis had not been examined among patient groups in disease settings; indeed, all studies had recruited largely healthy and physically active men and women.

The late 20th century

In the 1980s, the field of exercise immunology expanded and knowledge advanced considerably. This would prove to be a defining era for the discipline. As noted earlier, a large body of research had already characterised exercise-induced leucocytosis, and so new effort began to explore the effects of exercise on subpopulations of those cells previously examined, including different types of lymphocytes. Until the first flow cytometer became commercially available in the 1980s, one of the most advanced techniques available for the enumeration of lymphocyte subsets was the 'rosette' formation assay; it is also known by other names, including erythrocyte rosetting or E-rosetting. The principle of this assay is that upon incubating sheep red cells with human blood leucocytes, a proportion of human blood leucocytes bind to sheep red cells, and under the microscope, the bound cells look like a flower (i.e., a rose), with the red cells forming 'petals' around the leucocytes. At the time of its peak use, the rosette assay was used to quantify T cells, as it was thought that the cells that bound to red cells were exclusively T cells; we now know that rosette formation is due to the binding of the lymphocyte adhesion molecule CD2 with sheep red cells, and CD2 is expressed not only by T cells but by a range of immune cells. The non-binding cells in the rosette assay were presumed to be B cells, and thus the rosette assay was used as a means of simultaneously measuring T cells and B cells in human blood samples. The utilisation of this technique enabled researchers to discover, in a landmark study published in *Nature* (Steel et al., 1974), that blood collected after exercise – in the form of stair running – caused a marked increase in subsets of blood lymphocytes, including rosette-forming cells (presumed T cells), and non-rosette-forming cells exhibited the greatest increase – which was attributed at that time to an increase in presumed B cells. We now know that this latter increase in non-T cells in response to exercise is not driven chiefly by B cells, but is instead caused by a preferential increase of natural killer (NK) cells, particularly CD56[dim] NK cells (Campbell et al., 2009), which tend to have low CD2 expression (Cooper, Fehniger and Caligiuri, 2001). NK cells were not fully identified until 1975 (Herberman et al., 1975; Kiessling, Klein and Wigzell, 1975), a year after the study by Steel and colleagues was published. Clarification regarding the effects of exercise on NK cells would come in a seminal finding made by Edwards and colleagues (1984) at a Medical Research Council funded laboratory in England, using one of the first commercially available flow cytometers (FACS II, Becton Dickinson, USA).

Not only is this one of the most important studies in exercise immunology history because of
the novel methodology it employed, but it is also important because of its findings, which continue to have a lasting impact on the discipline. In their study, blood was taken from healthy
participants before and immediately after stair running exercise. Peripheral blood mononuclear
cells were incubated with fluorescently labelled monoclonal antibodies against the following
cell surface antigens: leu-4 (CD3), leu-3a (CD4), leu-1 (CD5), leu-2a (CD8), leu-11a (CD16),
leu-12 (CD19) and leu-7 (CD57). Readers will notice the original antigen nomenclature used
by Edwards and colleagues and how it compares to the modern day cluster of differentiation
(CD) nomenclature, which was introduced as an internationally agreed 'coding' system in the
late 1980s and updated thereafter. Please refer to Chapter 2 (the human immune system) and
Chapter 3 (methods in exercise immunology) for more information about CD nomenclature and
flow cytometry. Firstly, to some surprise at the time, it was found that leu-12 (CD19) expressing
cells – i.e., B cells – were much less responsive to exercise than T cells than expected. Secondly,
it was shown that leu-11a (CD16) expressing cells were highly responsive to exercise – and this
would be the first strong indication to the field that NK cells mobilise profoundly into the blood
during exercise. Thirdly, it was found that leu-3a (CD4) expressing cells were much less responsive to exercise than their leu-2a (CD8) counterparts. Lastly, they discovered that leu-7 (CD57)
expressing cells – a marker expressed on the most cytotoxic NK cells and CD8+ T cells, and
dubbed 'large granular lymphocytes' at that time – were profoundly exercise-responsive. The
findings of Edwards and colleagues, which were originally made using single-colour flow cytometry, would be confirmed and advanced by exercise immunologists over subsequent decades
using more advanced multi-colour flow cytometry. Remarkably, the findings of Edwards and
colleagues accurately depict what we know now: exercise induces a preferential mobilisation of
mature lymphocytes, which are capable of immediate, cytotoxic effector functions, in a process
led by NK cells and CD8+ T cells. Despite its significance to the field of exercise immunology,
the study by Edwards and colleagues was cited fewer than 150 times in 40 years since its publication. This is likely a result of the rapid advancements that were made in exercise immunology
following the widespread adoption of flow cytometry from the late 1980s onwards. Indeed,
from the early 1990s, published studies citing the use of flow cytometry in exercise science increased considerably (PubMed hits for 'exercise AND flow cytometry': N=93 in 1990s; N=260
in 2000s; N=463 in 2010s). Flow cytometry is still used extensively today (see Chapter 3), and
consequently, a great deal is now known about the effects of acute exercise on immune cell
phenotypes (see Chapter 4).

The expansion of interest in exercise immunology research from the 1980s onwards was
strongly driven by increasing public interest in sport, exercise and physical activity. Academi-
cally, research was instigated by an expansion in interest in exercise physiology, stimulated in
part by the work of the Harvard Fatigue Laboratory – as reviewed elsewhere (Tipton, 2016).
Interest in exercise and health was also sparked by the landmark findings of Jeremy Morris
between the 1950s and 1970s. During that time, it was demonstrated that physically active
conductors compared with sedentary drivers of London double-decker buses were protected
from coronary heart disease, and subsequent studies replicated the findings among postal workers delivering letters compared to postal office workers and linked the greatest protection to
vigorous physical activity (Morris et al., 1953; Morris et al., 1973; Paffenbarger, Blair and Lee,
2001). Subsequently, there was a global "health and fitness boom" from the 1970s, with the
rapid expansion of commercial gyms, fitness and sports centres, with public participation in
body building, aerobics and other exercises (Andreasson and Johansson, 2014). Long-distance
running became very popular in the 1970s (Haberman, 2017) driven by the first 'urban tour'
marathon in New York in 1976, which led to cities around the world hosting similar events

(Burfoot, 2007). It is worth emphasising this period was not the beginning of sport and exercise. Marathon running, with its ancient origins thousands of years ago, developed in modern society over many centuries (Lucas, 1976), and team games and other sports developed over hundreds of years (Ndee, 2010; Giossos et al., 2011; Park, 2011; Heggie, 2016). However, the most important point in relation to exercise immunology is that the 1970s was the beginning of mass-participation in exercise among the general population, and this would ignite research interest investigating the potential health effects of long-distance exercise, with many studies conducted in the context of marathon running. For example, a seminal study from this era showed that one-third of 150 runners participating in the 1982 Two Oceans 56 km ultramarathon in Cape Town South Africa self-reported symptoms of upper respiratory tract infections within two weeks of the race (Peters and Bateman, 1983). The age-matched control group – each of whom lived with a race participant – reported half the number of URTIs. Similar observations were made in a study of the 1987 Los Angeles Marathon (Nieman et al., 1990). Of 2,311 respondents who had completed the marathon and did not report an infection in the week prior to the race, 12.9% reported an infection in the week after the race compared with only 2.2% of individuals who withdrew for reasons other than illness. Around this time, it was also established that heavy exercise training appeared to increase the risk of infection (Linde, 1987), leading to the development of the 'J-curve' hypothesis in the early 1990s (Nieman, 1994), inferring that individuals undertaking a very large volume of exercise training over weeks and months exhibited a greater risk of developing infections, whereas those undertaking moderate volumes of exercise were at lower risk than their sedentary counterparts (see Chapter 9). As such, most interest in the late 1980s was related to the performance of elite athletes (Daniels et al., 1985; Friman et al., 1985) which was supported by earlier experimental studies of fever and performance (Grimby, 1962). Concern about upper respiratory tract infections impairing sports performance stimulated the pharmaceutical industry to develop prophylactic immunoglobulin preparations that could be administered to athletes (Weiss et al., 1985; Weiss 1993), although this form of immunotherapy did not become established in sporting or athletic circles. Similarly, countermeasures – such as nutritional supplements – to bolster immune function and reduce the risk of infection among athletic populations were also investigated considerably in the field of exercise immunology (see Chapter 14 for an overview of nutrition in exercise immunology).

As a result of the considerable expansion of research being undertaken in the 1980s and early 1990s, the burgeoning field of exercise immunology became a fully fledged discipline when a group of 14 scientists began a research society dedicated to its study in 1993 (Shephard, 2010). The *International Society of Exercise Immunology* ('ISEI') was established with the objective of fostering scientific research, education and the dissemination of scientific information relating to exercise and immune function. It purposefully coincided with the first international convention on exercise immunology held in Paderborn, Germany that same year. In creating this new society, and an agreement to hold regular international conventions thereafter, the eminent group of exercise immunologists sought to direct research attention not only to the endeavour of athletes, but towards the improvement of health, and the prevention and treatment of disease. By the 1990s, research in clinical settings was only just beginning in earnest, with the hope of showing benefits of exercise among patients with diseases characterised by immunosuppression, including cancer and HIV (Winningham et al., 1986; Eichner, 1987; Painter and Blackburn, 1988; LaPerierre, 1989). Now, in the present era, disease prevention and treatment comprise the large body of research in exercise immunology. Following the success of the first convention, the society met again two years later in Brussels, Belgium. By that time, interest in the society had grown considerably and 94 scientists from 16 countries were present at the meeting in 1995 (Nieman, 1997). The society has continued to expand, and to this day meets regularly – typically every two years. Given the growing interest in the discipline in the early 1990s and an increasing volume of exercise immunology research being undertaken globally, the society launched a new peer-reviewed journal dedicated to the topic of exercise immunology called *Exercise Immunology Review*, which to this day has one edition published annually. It quickly became a highly cited source, and it typically has one of the highest impact factors among sport and exercise science journals; impact factor is a popular metric used to assess scientific interest in a journal, and is usually calculated by measuring citations of articles in a given year. For more information about the society, the next upcoming convention, and/or the *Exercise Immunology Review* journal, please refer to the ISEI's official website: www.exerciseimmunology.com. Notably, interest in exercise immunology also extended into other societies and journals. For example, the Psycho Neuro Immunology Research Society (PNIRS) annual meeting and the official journal of PNIRS (*Brain, Behavior and Immunity*) regularly features exercise immunology research, and so too does the American College of Sports Medicine (ACSM) annual meeting and the official journal of ACSM (*Medicine and Science in Sport and Exercise*). The study of exercise immunology was also aided by the publication of a series of textbooks in the 1990s by leading exercise immunologists. The first book on the topic *Exercise and Immunology* was published in 1992 by Laurel MacKinnon, followed by another (*Exercise and Immune Function*) in 1996 by Laurie Hoffman-Goetz, and another (*Advances in Exercise Immunology*) in 1999 by Laurel MacKinnon. Further textbooks were then published in the new millennium, including the first edition of *Exercise Immunology* in 2013, edited by Michael Gleeson, Nicolette Bishop, and Neil Walsh.

The formal establishment of exercise immunology as a self-recognised discipline in the 1990s coincided with an expansion in the breadth of research being undertaken in the field. This was supported by technical developments in mainstream immunology, which were adopted rapidly and integrated into exercise science. The measurements being conducted therefore became more sophisticated and this enabled numerous advances. These advances included greater understanding of how exercise affects: the phenotypes of leucocyte subpopulations in blood, leucocyte function (e.g., cytotoxicity, migration), soluble immune factors in human biofluids (e.g., IL-6 in blood, IgA in saliva), systemic responses to immune challenge *in vivo* (e.g. immune responses to viral challenge in animals, antibody responses to vaccination in humans), and immunity in tissues (e.g., tumours) – as well as numerous others that are beyond the scope of this chapter, but are covered elsewhere in this book. Some of these key developments are summarised next.

Naturally, one key focus of research from around the 1990s onwards sought to determine the mechanism(s) underlying the apparent increased risk of infection among athletes undertaking high training volumes or competing at an elite level. As mentioned earlier, evidence had emerged that those participating in prolonged and strenuous exercise, such as marathon running, reported higher incidence of infection symptoms compared with controls who did not exercise (Peters and Bateman, 1983; Nieman et al., 1990). This stimulated exercise immunologists to pursue an array of possible mechanisms and included investigations into changes to cellular and soluble constituents of the immune system. One prominent area of investigation focused on the profound and transient time-dependent decrease to blood lymphocytes that was found to arise in the hours after prolonged strenuous, termed 'lymphocytopenia' or 'lymphopenia' (see Box 1.1). This decline in lymphocyte frequency in the blood in the hours after exercise represented a cause for concern and was subject to much research. Particular focus was given to NK cells after it began to emerge in the late 1980s that leu-11a (CD16) and Leu-7 (CD57) positive cells – considered NK cell associated markers at that time – exhibited a marked decrease in the blood 1 hour after exercise (Mackinnon et al., 1987), and this decrease was proportional to a decline in NK cell function *in vitro*. It was found that cell frequency and functional competency returned to normal 24 hours later. Importantly, however, it was proposed that the temporary decline in NK cell frequency may represent a compromised host immune system and could provide an 'open window' for opportunistic infections (Pedersen and Ullum, 1994). This concept was factored into the evidence base for the generation of the 'J-curve' hypothesis (Nieman, 1994), which, as mentioned earlier, predicted that individuals undertaking a very large volume of exercise were at great risk of developing infections. It was unclear why this decrease to NK cells was arising, but it was speculated by many scientists at the time that NK cells were being redistributed from blood to tissues after exercise – this redistribution theory would be further researched in later decades and, as discussed later, that research would culminate in a change in the narrative from exercise being an 'open window' for infection, to exercise potentially opening the door to enhanced immunosurveillance.

Another line of research inquiry that was opened to explain the apparent increased risk of infection among those undertaking strenuous exercise was the possibility that exercise may suppress the availability of soluble immune factors – principally immunoglobulins (Ig) – in different body tissues, with a main focus on proteins in the oral mucosa, for example via saliva sampling. This body of work was stimulated by a study showing that salivary IgA was reduced by 20% after cross-country skiing (Tomasi et al., 1982). In support, it was found that salivary IgA and IgM were reduced in the hours after vigorous intensity cycling exercise (MacKinnon et al., 1987). As a result of these findings, the apparent suppression of salivary immunoglobulins after exercise was included as further evidence in the formulation of the 'J-Curve' hypothesis (Nieman, 1994). Despite several other studies of similar design reported analogous findings, numerous contradictory findings also emerged showing that exercise did not suppress salivary IgA. For example, cycling to exhaustion at different intensities did not affect salivary IgA production in 18 men of different fitness levels, leading the authors to conclude that exercise may alter the amount of saliva that is produced, but not its 'quality' (Blannin et al., 1998). It is now considered that salivary IgA measurement lacks utility for monitoring immune status in athletes, owing to a lack of reproducible findings in the literature, and because of its low specificity and sensitivity in the detection of respiratory tract infection susceptibility (Turner et al., 2021). The utility of salivary IgA is not helped by its profound variability (Rapson et al., 2020), which is likely dictated by many influencing factors including sleep patterns, oral hygiene, nutrition, psychological stress, and others (Brandtzaeg, 2013) – as discussed in Chapter 5. As a consequence, the measurement of salivary IgA in exercise immunology has declined considerably in recent years.

A key interpretational problem from studies exploring infection risk after marathon races is the likelihood that exercising participants were at higher risk of exposure to opportunistic infections (e.g., from crowds and public transport), thus increasing the likelihood of infection symptoms in the days afterwards. This confounder has made it difficult to understand whether exercise itself – i.e., locomotion – affects the competency of the immune system in humans. To overcome this problem, it has been deemed preferable to control the administration of an infectious agent to humans, and subsequently determine how exercise affects the immune response to that infectious agent (see Chapter 7). However, in most cases, it is not practical – due to safety and ethical considerations – to administer live pathogens (e.g., viruses) to humans. Given this, a series of highly controlled animal studies have been conducted to understand how exercise affects the host immune response to challenge with a pathogen. Such studies, to this day, contain some of the best available evidence that exercise – depending on its intensity and duration – may alter host immune competency. For example, in one seminal study (Davis et al., 1997), mice were randomised to one of the following conditions: moderate intensity exercise for 30 minutes, moderate-to-vigorous intensity exercise until volitional fatigue, or no exercise.

Immediately after completing one of those conditions, mice were intra-nasally administered a strong (i.e., sufficient to cause severe symptoms and potentially fatal) dose of herpes simplex type 1 virus (HSV-1). Results indicated that morbidity scores in fatigued mice were twice that of control mice, and mortality rates in fatigued animals were more than double those in control mice. On the other hand, mice exposed to moderate exercise for 30 minutes prior to infection tended to have lower symptom severity and mortality than controls. Such findings would later be replicated in studies using other models of immune challenge, such as influenza virus administration (Murphy et al., 2008). Studies such as these indicated how exercise might affect the response to *subsequent* infection. As such, other studies were conducted to investigate how exercising *during* infection affects immune competency. One such study from the turn of the millennium investigated three different conditions after influenza administration: moderate exercise for 20–30 minutes per day for three days, prolonged vigorous exercise for 150 minutes per day for three days or no exercise (Lowder, Padgett and Woods, 2005). Fascinatingly, the results indicated that moderate exercise for three days enhanced survival compared with no exercise at all, whereas prolonged exercise increased symptoms of infection and tended to worsen survival. Overall, these collective animal studies tend to support the general idea of the J-curve, whereby exposure to moderate exercise can augment immune responses (compared with no exercise) whereas exposure to arduous exercise can suppress immune responses (compared with no exercise). However, despite their highly controlled nature, it remains unclear whether exercise itself – i.e., locomotion – causes these effects, or whether these outcomes were driven by the presence of confounding factors, such as psychological distress which affects immune competency (see Chapter 15), the environment (see Chapter 13) or another confounding factor that alters immunity. As an example, in a study conducted during World War II to understand how fatigue affects immune competency (Levinson, Milzer and Lewin, 1945), poliovirus was given to monkeys and then the monkeys were assigned to either: (i) forced exhaustive swimming exercise for 2–3 hours; (ii) no exercise, and placed in a cage on land (land controls); or (iii) no exercise, and placed in a straitjacket in water (water controls). The pathology of severe poliovirus infection includes paralysis, and so a main outcome of the study was paralysis of different parts of the body. The study found that exercised animals fared particularly badly compared to land controls, whereby 40% of exercised monkeys developed quadriplegia (paralysis of all four limbs) and only 20% showed no symptoms of paralysis, whereas in the land control group, no monkeys exhibited quadriplegia and 52% in that group had no signs of paralysis whatsoever. This would infer that exercise is immunosuppressive. Importantly however, the water control group fared worst of all, as 50% of monkeys in that group developed quadriplegia and only 15% of animals showed no symptoms of paralysis. Stress of being confined to a straitjacket in water, and cold exposure, were likely contributors to the outcomes reported in the water control group. However, these were also confounding factors that likely played a role in the outcomes reported by the animals forced to exercise exhaustively for hours, which induced near-drowning. As such, confounding factors such as stress are important considerations when interpreting findings in exercise immunology, particularly in animal studies where stress levels are difficult to estimate. In addition, it is also worth remembering these studies do not examine naturalistic exposure to a virus (e.g., airborne droplets) as infection is "guaranteed" due to inoculation.

To better understand how exercise might affect the immune response to challenge in humans, research in the 1990s and thereafter began to explore how exercise affects vaccine efficacy (see Chapter 7). It is generally considered that vaccines offer a safe means of challenging the human immune system, and so researchers have explored how vaccine responses – typically in the form of antibody responses – might be affected by exercise. One key early study from this era investigated whether a half-ironman affected the immune response to tetanus and diphtheritis toxoid and purified pneumococcal polysaccharide (Bruunsgaard et al., 1997). Compared with controls who did not exercise, there were no differences in antibody titres in the exercised participants. This aligned with one of the first studies investigating vaccination responses in humans, which was conducted in the 1970s, and preliminarily showed that marathon running did not appear to impair antibody responses to tetanus toxoid vaccination (Eskola et al., 1978). These such findings indicated that prolonged exercise did not impair antibody responses. As discussed later, there would be a shift in emphasis at the turn of the new millennium, whereby researchers started to explore whether exercise could be used to improve antibody responses to vaccination.

The 21st century

In the early 2000s, and fuelled by a lack of resounding evidence that exercise suppressed immunity – nor a robust mechanism to support that line of thinking – questions were asked about the strength of the evidence linking exercise to immune suppression, particularly in humans. For example, prominent researchers asked questions like: "if there is an increased risk for infections with increased exercise duration and intensity, why do overtrained athletes not display the greatest risk for upper respiratory tract infections?" (Malm, 2004). Other leading researchers in the discipline also concluded that epidemiology evidence linking exercise and immune suppression in humans was "not convincing" (Lancaster and Febbraio, 2016). A critical narrative review of the evidence was published soon after, and it outlined some of the limitations to existing evidence linking exercise with immune suppression (Campbell and Turner, 2018). The main limitations cited included a lack of convincing epidemiology evidence, and a lack of a known mechanism to explain immune suppression from exercise. Nevertheless, a debate took place between proponents of the immune suppression theory, and those against it, and it was published, coincidentally, at the beginning of the COVID-19 pandemic (Simpson et al., 2020). In the debate article, it was generally agreed that exercise can preserve or augment aspects of immune function and regulation, but it was clear that further research was needed to investigate whether a bout of arduous exercise could suppress immune competency. Research in this area has generally waned since the end of the 20th century, but there remains a need to adopt immune challenge studies – for example using live attenuated viruses – in humans to fully understand how exercise affects immune responses to pathogens (see Chapter 9). Indeed, until this research is conducted – and even if it is conducted – concern about exercise and the immune system will likely persist (see Box 1.2).

Whilst the 21st century has seen a decline in research exploring the potential detrimental effects of exercise on immune competency, there has been a dramatic increase in research exploring the potential benefits of exercise on immune function and regulation. This shift was instigated in part by increasing public interest in the health benefits of exercise, the expanding ageing society living with health problems, and also because of more funding available for this type of research, specifically towards research with a focus on preventing and treating disease. Indeed, a large proportion of research in exercise immunology is aligned with the global initiative and viewpoint that "exercise is medicine" which has its origins thousands of years ago. It is beyond the scope of this chapter to review the history of "exercise is medicine" but readers are encouraged to familiarise themselves with the following key reviews (Berryman, 2010; Tipton, 2014). As a consequence, there was renewed focus on exercise immunology in understanding the phenotypic and functional changes that arise to immune cells in response to exercise, and whether this leads to health benefits. This research focus was, in part, ignited by researchers in the sister-field of psychoneuroimmunology, who had proposed at the turn of the millennium that stress hormones released during an acute stressor – such as exercise – might, in a conserved evolutionary response, augment immunosurveillance, and thus lead to enhancement of host immunity. As summarised more recently elsewhere (Dhabhar et al., 2012), using a military analogy, it has been proposed that adrenaline released during an acute stressor, such as exercise, induces a preferential mobilisation of cells (soldiers from the 'barracks') capable of immediate effector function into the bloodstream (the 'boulevards'), which are then preferentially redistributed after cessation of the stressor to peripheral tissues or sites of active inflammation (the 'battlefields'). Using multi-parameter flow cytometry, it was confirmed that exercise does indeed preferentially mobilise these cells via adrenergic mechanisms, including a preferential mobilisation of effector memory CD8+ T cells rather than their naive CD8+ T cell counterparts (Simpson et al., 2007; Krüger et al., 2008; Campbell et al., 2009). It was also partially confirmed where these cells 'go' after exercise cessation. Using fluorescent cell tracking in rodents, it was found that T cells were redeployed in large numbers to peripheral tissues, including the gut and lungs, and to the bone marrow following exercise (Krüger et al., 2008). It has generally been concluded in the discipline that this redistribution reflects heightened immune-surveillance in tissue sites where pathogens are likely to be encountered during and after exercise (i.e., lungs, gut). As discussed later, such findings and others, sparked interest in understanding whether exercise can mobilise immune cells, not only to healthy tissue sites, but to sites of inflammation, such as tumours.

Relatedly, one such area of focus that emerged at the beginning of the 21st century was whether exercise could be used to enhance vaccine responses in humans. This line of research was again stimulated by pioneers in the sister-field of psychoneuroimmunology, where acute stress was found to improve vaccine responses, in a process – that as outlined above – might be due to enhanced immune-surveillance. Naturally, therefore, it was hypothesised that the immune-surveillance response that arises during exercise might culminate in enhanced vaccine responses in humans. Thus, more than a dozen studies have been published with the main focus on investigating the effects of an individual bout of short-duration exercise conducted just before vaccine administration on the antibody response in humans. These studies have most commonly examined influenza vaccines, but other vaccines have been used and include meningococcal, pneumococcal, tetanus, COVID-19 vaccines, and others. Different types of exercise have been explored as a means of augmenting immune function. For example, some studies used vigorous intensity aerobic exercise because of its potential to mobilise immune cells, whereas others focused on using eccentric exercise – which is a type of exercise used to induce muscle damage – with the aim of inducing immune cell recruitment into the sites of vaccination, to in turn enhance immune responses within that tissue. Most of this work has been conducted in younger populations, though some have been conducted in older adults. Whilst there have been studies providing tantalising evidence that an individual bout of exercise can enhance antibody responses against some vaccines, overall, it seems that exercise is not able to meaningfully improve antibody responses to influenza vaccination (Grande et al., 2016), which is the most commonly studied vaccine. These findings mirror the aforementioned early findings in the field (Eskola et al., 1978; Bruunsgaard et al., 1997) which showed no tangible difference in antibody responses between exercisers and controls for other vaccine types. Taken together, the variable efficacy of exercise may be due to different vaccines used, different types of exercise and/ or participant characteristics (e.g., young versus older adults). Researchers have also explored whether longer-term exposure to exercise might impact immune responses to vaccination. This area of research has shown more promise, and numerous studies have shown higher antibody responses to vaccination either following exercise training interventions or among people who self-report having high physical activity levels. Indeed, a meta-analysis of these studies has revealed that exercise training or a high physical activity level is linked with significantly higher antibody responses to influenza vaccination (Dinas et al., 2022), which is again the most heavily studied vaccine in the research studies undertaken to date. Given that antibody responses to vaccination in older adults tend to be poorer than younger people, owing to ageing of the immune system (see Chapter 8), a promising observation is that the benefits of exercise on vaccine responses were not influenced by age (Dinas et al., 2022). Again though, variable effects of exercise were noted for different vaccine types, likely influenced by the types of interventions examined and participants involved. Altogether though, these collective findings indicate that being physically active may augment or preserve immune competency, and in turn enhance vaccine responses. This remains a nascent field in exercise immunology, with numerous exciting avenues of research being explored (see Chapter 7).

Due to methodological advances, and in particular, the availability of multi-parameter flow cytometry, the 21st century has seen a marked increase in research understanding how short (i.e., days and weeks) and long-term exercise training (i.e., months), as well as lifelong physical activity, affects immune cell phenotypes in healthy humans. One area that has received particular focus is the effect of exercise on ageing of the immune system, also known as immunosenescence. Features of immunosenescence include proportional increases in late-differentiated 'senescent' memory T cells with concomitant decreases in the proportion of naive T cells, which is thought to be driven by prior/latent infections (e.g., *Cytomegalovirus*, CMV), thymic involution, and other age-associated factors. Particular attention has been paid to immunosenescence because of global demographic changes resulting in people living longer, and so researchers began to ask whether exercise could ameliorate ageing of the immune system, and thus help prevent and treat diseases that arise more in older adults. For an overview of this research, readers are directed to an authoritative review (Duggal et al., 2019). Briefly, early research in this area focused initially on younger and middle-aged adults and found that a higher cardiorespiratory fitness was associated with altered features of immunosenescence, including reduced senescent T cells and maintenance of naive T cells (Spielmann et al., 2011). Efforts soon shifted to understanding the impact of exercise behaviours on features of immunosenescence in older adults. One renowned study in this area found that older adults who reported a high lifetime level of physical activity exhibited a higher frequency of naive T cells and a proportionally lower frequency of memory T cells than age-matched participants who reported a less active lifestyle (Duggal et al., 2018). Different theories exist to explain how physical activity potentially invokes these effects. For example, some hypothesise that immunoregulatory cytokines such as IL-7 and IL-15 may result in preservation of naive T cell output from the thymus (Duggal et al., 2018). Other theories have been put forward, including one which hypothesises that repeated acute exposure to exercise could drive reductions in the number of senescent T cells via apoptosis (Simpson, 2011). Another hypothesis theorises that physical activity may limit the accumulation of senescent T cells by augmenting immune competency leading to a reduction in CMV reactivation which is common among the general population (Turner and Brum, 2017). Regardless of the mechanism, it has subsequently been proposed that by reducing features of immunosenescence, physical activity in older age may help to reduce the burden of disease. As a result, research in this area remains a key focus for many in the field of exercise immunology (see Chapter 8).

Another area of focus in exercise immunology, that has received considerable attention, is the study of 'exerkines' – defined as signalling molecules released in response to individual bouts of exercise and/or regular exercise training, which exert their effects on different cells and organs throughout the body (Chow et al., 2022). As outlined by Chow and colleagues, it is well established that numerous organs, cells and tissues release these factors, including skeletal muscle (dubbed 'myokines'), the heart ('cardiokines'), liver ('hepatokines'), adipose tissue ('adipokines'), and brown adipose tissue ('batokines') among others. It is widely believed that these exerkines – which are predominantly cytokines – may contribute to the maintenance of cardiovascular, metabolic, immune and neurological health, and, therefore, the secretion of these factors may aid in the prevention and treatment of disease. Of these exerkines, by far the greatest body of work, to date, has focused on the cytokine IL-6, and it was the first myokine discovered in a remarkable series of experiments. In the early 1990s, it was becoming increasingly apparent that IL-6 secretion was increased in response to exercise (Northoff and Berg, 1991); at that time, IL-6 was measured by a 7TD1 assay (i.e., cells that are dependent on IL-6 for growth, thus cell growth can be measured, and it is directly proportional to IL-6 concentration) because sensitive ELISAs for IL-6 were not yet available. It was unclear where this IL-6 was coming from, nor why it was being secreted, though it was felt that this might occur as part of an acute phase response to resolve inflammation from exercise. One early line of thinking was that IL-6 might be secreted by immune cells circulating during exercise, but this was soon discounted, as it was discovered that IL-6 might be produced from within the vicinity of the muscle tissue instead (Ostrowski et al., 1998; Steensberg et al., 2000). To some surprise, IL-6 was not secreted as a consequence of muscle damage (Croisier et al., 1999), but instead the IL-6 produced in muscle during exercise seemed to be predominantly related to muscle glycogen content (Steensberg et al., 2001), and it would later be shown in a landmark finding that exercised skeletal myocytes produce IL-6 upon glycogen depletion (Hiscock et al., 2004). The secretion of IL-6 from muscle, in response to energy depletion, was thus linked to metabolic processes during exercise including maintenance of glucose homeostasis and mediating lipolysis (Pedersen, Steensberg and Schjerling, 2001). Whilst a single bout of prolonged strenuous exercise was found to increase the concentration of cytokines in blood, another important body of exercise immunology research was discovering – somewhat paradoxically – that people with a low physical activity level and/or obesity were found to have higher IL-6 and C-Reactive Protein (CRP) in blood than their more active and lean counterparts; CRP production is stimulated by IL-6, and it is one of the most widely used indicators of active inflammation – including inflammation arising from muscle damage. The source of IL-6 in this case is not due to an energy crisis arising from exercise, or muscle damage, but it is instead associated with too little exercise, and concomitantly, the development of low-grade systemic inflammation, which is common in sedentary and overweight people, as reviewed elsewhere (Gleeson et al., 2011). Clearly, exercise can affect IL-6 in numerous ways, and it is also worth noting that IL-6 is a pleiotropic cytokine, with pro- and anti-inflammatory effects, and its signalling is complex – for example, IL-6 can induce pro-inflammatory effects directly, or via alternative processes such as trans-signalling (Rose-John, 2012). IL-6 is therefore one of the most extensively studied cytokines in the discipline of exercise immunology, and given its diverse functions, it continues to be heavily studied in various contexts including: in muscle and adipose tissue (see Chapter 6), the chronic disease setting (see Chapters 10–12) and immunometabolism and nutrition (see Chapter 14).

Another major discipline of exercise immunology that has emerged over the last decade or so is how exercise affects anti-cancer immunity. Major developments in this field are summarised in a review by Fiuza-Luces et al. (2023). Briefly, some of the first research in this area was published over 100 years ago, when it was theorised that physical activity might reduce the risk of cancer development in humans owing to reduced incidence of cancers in individuals leading more active lifestyles in older age (Siversten and Dahlstrom, 1921). Subsequent studies in rodents – where tumours were induced or transplanted – revealed that exercise often, but not always, reduced cancer development. Speculation began to mount that exercise-induced changes to immune cells might be responsible. Then, research conducted in the 1980s and 1990s began to link exercise, the immune system, and tumour responses. Different methods were adopted, including analysing the effects of individual exercise bouts and regular exercise training on human NK cell tumour cytotoxicity in vitro, as well as measuring tumour development in rodents alongside measurements of NK function. For an authoritative review of work from this period, please refer to Hoffman-Goetz and Husted (1995). A key point raised in the aforementioned review was that the anti-tumour effects of exercise seemed tumour-dependent. That is, not all human cancers appeared to be susceptible to physical activity, and similarly, animal tumours (in vivo*)* and human cell lines (in vitro) were not consistently susceptible to the immune changes that arose in response to exercise. Tying these observations together, it was highlighted by Hoffman-Goetz and colleagues (1998) that "*the majority* of *human cancers are nonimmunogenic and therefore specific antitumor immune-surveillance is not a major factor.*" In other words, many cancers clones are not recognised by the immune system – over the years, this important point would sometimes be overlooked in the discipline, and in the sister-field of exercise oncology. Unfortunately, whilst great strides were taken in exercise immuno-oncology at the end of the last century, research in this topic slowed, and it was not until two decades later that major interest in this area returned when research began to integrate exercise, *in vivo* immune cell function and tumour development. Some of this research has been conducted in the context of "cancer risk" (i.e., understanding mechanisms that might explain the lower risk of developing cancer among physically active people); however, other research has had emphasis on understanding how exercise affects tumour biology *in vivo* – which has greater relevance to people who have been diagnosed with cancer. Indeed, at this juncture, a landmark study in rodents was published, which showed that exercise averted tumour outgrowth – across numerous tumour models – in a process that was dependent on exercise-induced NK cell mobilisation by adrenaline (Pedersen et al., 2016). This prompted researchers to replicate these findings across numerous tumour models in animals. Importantly, however, evidence in support of this theory has remained limited to preclinical or *in vitro* human cell line studies employing tumour cells that were sensitive to NK cells thus rendering those tumours susceptible to the immune stimulus of exercise. Moreover, these findings have also failed in translation to humans across three studies (Djurhuus et al., 2022; Djurhuus et al., 2023; Schenk et al., 2022). Each of these studies investigated early-stage prostate cancer, which is generally considered to be poorly immunogenic and not recognisable by the immune system (i.e., the existing tumour cell clones have not been eliminated). Two of the studies investigated the effects of an exercise bout prior to prostatectomy and both trials showed that exercise did not increase NK cell infiltration to tumours (Djurhuus et al., 2022; Schenk et al., 2022). The third study was a training programme conducted over several weeks, and whilst a modest within-group increase in intra-tumoural NK cell numbers was found in the exercising group, no difference in intra-tumoural NK cell numbers between the intervention and the control group was found overall (Djurhuus et al., 2023). This null finding may be considered unsurprising given the poor immunogenicity of early-stage prostate cancer, and the lack of immune infiltration may also help explain why a physically active lifestyle is not associated with a reduced risk of prostate cancer – as reviewed elsewhere (Emery et al., 2022). These clinical trial findings also corroborate the predictions made by Hoffman-Goetz and colleagues in 1998, whereby a lack of immunogenicity possibly renders the effects of exercise redundant. Nevertheless, to date, it remains unknown whether exercise can increase immune cell numbers in tumours that contain immunogenic tumour cell clones. This will no doubt be a focus of exercise immunology research in the years to come. Another area of research that has emerged is whether exercise is beneficial for patients during cancer therapy. In the broader field of exercise oncology, it is well established that exercise around the time of cancer therapy is associated with several benefits, ranging from improvements in overall wellbeing, fitness, fatigue, and quality of life. However, other evidence has shown that exercise undertaken during chemotherapy is associated with better clinical outcomes. Thus, an emerging research area is understanding whether exercise might augment cancer therapies, including immuno-therapies (Gustafson et al., 2021; Collier-Bain et al., 2023). Given the broad possibilities that exercise holds in the cancer setting, exercise immuno-oncology is arguably now the biggest research focus in exercise immunology (see Chapter 10) and perhaps will become the biggest focus of research in exercise science in the near future.

In summary, the study of exercise and its effects on the immune system has been ongoing for more than 100 years, and, as a result, we now have a broad understanding of how exercise affects the phenotypic and functional competency of the immune system in humans. Whilst much of the 20th century focused on how exercise might impair immune competency, the field has transitioned to having a principle focus on understanding how exercise can be harnessed to prevent and help treat diseases. It is anticipated that the study of exercise, immunology, and disease will continue to be an important area of research well into the 21st century and beyond.

References

Andreasson, J., & Johansson, T. (2014). The fitness revolution: Historical transformations in the global gym and fitness culture. *Sport Sci Rev*, XXIII(3–4), 91–112.

Atherton, A., & Born, G. V. (1972). Quantitative investigations of the adhesiveness of circulating polymor-phonuclear leucocytes to blood vessel walls. *J Physiol*, 222(2), 447–474.

Barker, M. H., Capps, R. B., & Allen, F. W. (1945a). Acute hepatitis in the mediterranean theater. *JAMA*, 128, 997.

Barker, M. H., Capps, R. B., & Allen, F. W. (1945b). Chronic hepatitis in the mediterranean theater. *JAMA*, 129, 653.

Barrett, A. D. T. (2023). Close to the finish of the polio endgame. *Nature*, 619, 36–38.

Bermon, S., Castell, L. M., Calder, P. C., Bishop, N. C., Blomstrand, E., Mooren, F. C., … Nagatomi, R. (2017). Consensus statement. Immunonutrition and exercise. *Exerc Immunol Rev*, 23, 8–50.

Berryman, J. W. (2010). Exercise is medicine: A historical perspective. *Curr Sport Med Rep*, 9(4), 185–201.

Blannin, A. K., Robson, P. J., Walsh, N. P., Clark, A. M., Glennon, L., & Gleeson, M. (1998). The effect of exercising to exhaustion at different intensities on saliva immunoglobulin A, protein and electrolyte secretion. *Int J Sports Med*, 19(8), 547–552.

Brandtzaeg, P. (2013). Secretory immunity with special reference to the oral cavity. *J Oral Microbiol*, 5. 10.3402/jom.v5i0.20401.

Brawner, C. A., Ehrman, J. K., Bole, S., Kerrigan, D. J., Parikh, S. S., Lewis, B. K., … Keteyian, S. J. (2019). Inverse relationship of maximal exercise capacity to hospitalization secondary to coronavirus disease. *Mayo Clin Proc*, 96(1), 32–39.

Bruunsgaard, H., Hartkopp, A., Mohr, T., Konradsen, H., Heron, I., Mordhorst, C. H., & Pedersen, B. K. (1997). In vivo cell-mediated immunity and vaccination response following prolonged, intense exercise. *Med Sci Sports Exerc*, 29(9), 1176–1181.

Burch, G. E. (1979). Viral diseases of the heart. *Acta Cardiologica*, 1, 5–9.

Burfoot, A. (2007). The history of the marathon: 1976-present. *Sports Med*, 37(4–5), 284–287.

Burrows. (1899). *Am J Med Sciences*, May – *further reference details not available*.

Cabot, R. C., Blake, J. B., & Hubbard, J. C. (1901). Studies of the blood in its relation to surgical diagnosis. *Ann Surg*, 34(3), 361–374.

Campbell, J. P., Riddell, N. E., Burns, V. E., Turner, M., van Zanten, J. J., Drayson, M. T., & Bosch, J. A. (2009). Acute exercise mobilises CD8+ T lymphocytes exhibiting an effector-memory phenotype. *Brain Behav Immun*, 23(6), 767–775.

Campbell, J. P., & Turner, J. E. (2018). Debunking the myth of exercise-induced immune suppression: Redefining the impact of exercise on immunological health across the lifespan. *Front Immunol*, 9, 648.

Chabot-Richards, D. S., & George, T. I. (2014). Leukocytosis. *Int J Lab* Hematol, 36(3), 279–288.

Chalmers, T. C., Reynolds, W. E., Eckhardt, R. D., Cigarroa, J. G., Reifenstein, R. W., Smith, C. W., & Davidson, C. S. (1955). Treatment of acute infectious hepatitis in the armed forces advantages of ad lib. Bed rest and early reconditioning. *JAMA*, 159, 1431.

Chow, L. S., Gerszten, R. E., Taylor, J. M., Pedersen, B. K., van Praag, H., Trappe, S., … Snyder, M. P. (2022). Exerkines in health, resilience and disease. *Nat Rev Endocrinol*, 18(5), 273–289.

Collier-Bain, H. D., Brown, F. F., Causer, A. J., Emery, A., Oliver, R., Moore, S., … Campbell, J. P. (2023). Harnessing the immunomodulatory effects of exercise to enhance the efficacy of monoclonal antibody therapies against B-cell haematological cancers: A narrative review. *Front Onc*, 13, 1244090.

Cooper, D. M., Radom-Aizik, S., Schwindt, C., & Zaldivar, F. (2007). Dangerous exercise: Lessons learned from dysregulated inflammatory responses to physical activity. *J Appl Physiol*, 2103(2), 700–709.

Cooper, M. A., Fehniger, T. A., & Caligiuri, M. A. (2001). The biology of human natural killer-cell subsets. *Trends Immunol*, 22(11), 633–640.

Croisier, J. L., Camus, G., Venneman, I., Deby-Dupont, G., Juchmès-Ferir, A., Lamy, M., … Duchateau, J. (1999). Effects of training on exercise-induced muscle damage and interleukin 6 production. *Muscle Nerve*, 22(2), 208–212.

Daniels, W. L., Sharp, D. S., Wright, J. E., Vogel, J. A., Friman, G., Beisel, W. R., & Knapik, J. J. (1985). Effects of virus infection on physical performance in man. *Mil Med*, 150, 8–14.

Davis, J. M., Kohut, M. L., Colbert, L. H., Jackson, D. A., Ghaffar, A., & Mayer, E. P. (1997). Exercise, alveolar macrophage function, and susceptibility to respiratory infection. *J Appl Physiol*, 83(5), 1461–1466.

Dhabhar, F. S., Malarkey, W. B., Neri, E., & McEwen, B. S. (2012). Stress-induced redistribution of immune cells--from barracks to boulevards to battlefields: A tale of three hormones-Curt Richter Award winner. *Psychoneuroendocrinology*, 37(9), 1345–1368.

Dinas, P. C., Koutedakis, Y., Ioannou, L. G., Metsios, G., & Kitas, G. D. (2022). Effects of exercise and physical activity levels on vaccination efficacy: A systematic review and meta-analysis. *Vaccines*, 10(5), 769.

Djurhuus, S. S., Schauer, T., Simonsen, C., Toft, B. G., Jensen, A. R. D., Erler, J. T., … Christensen, J. F. (2022). Effects of acute exercise training on tumor outcomes in men with localized prostate cancer: A randomized controlled trial. *Physiological Reports*, 10(19), e15408.

Djurhuus, S. S., Simonsen, C., Toft, B. G., Thomsen, S. N., Wielsøe, S., Røder, M. A., … Christensen, J. F. (2023). Exercise training to increase tumour natural killer-cell infiltration in men with localised prostate cancer: A randomised controlled trial. *BJU International*, 131(1), 116–124.

Duggal, N. A., Niemiro, G., Harridge, S. D. R., Simpson, R. J., & Lord, J. M. (2019). Can physical activity ameliorate immunosenescence and thereby reduce age-related multi-morbidity? *Nat Rev Immunol*, 19(9), 563–572.

Duggal, N. A., Pollock, R. D., Lazarus, N. R., Harridge, S., & Lord, J. M. (2018). Major features of immunesenescence, including reduced thymic output, are ameliorated by high levels of physical activity in adulthood. *Aging Cell*, 17(2), e12750.

Edwards, A. J., Bacon, T. H., Elms, C. A., Verardi, R., Felder, M., & Knight, S. C. (1984). Changes in the populations of lymphoid cells in human peripheral blood following physical exercise. *Clin Exp Immunol*, 58(2), 420–427.

Eichner, E. R. (1987). Exercise, lymphokines, calories, and cancer. *Phys Sportsmed*, 15(6), 109–116.

Elenkov, I. J., Wilder, R. L., Chrousos, G. P., & Vizi, E. S. (2000). The sympathetic nerve--an integrative interface between two supersystems: The brain and the immune system. *Pharmacol Rev*, 52(4), 595–638.

Emery, A., Moore, S., Turner, J. E., & Campbell, J. P. (2022). Reframing how physical activity reduces the incidence of clinically-diagnosed cancers: Appraising exercise-induced immuno-modulation as an integral mechanism. *Front Onc*, 12, 788113.

Eskola, J., Ruuskanen, O., Soppi, E., Viljanen, M. K., Järvinen, M., Toivonen, H., & Kouvalainen, K. (1978). Effect of sport stress on lymphocyte transformation and antibody formation. *Clin Exp Immunol*, 32(2), 339–345.

Ezzatvar, Y., Ramírez-Vélez, R., Izquierdo, M., & Garcia-Hermoso, A. (2022). Physical activity and risk of infection, severity and mortality of COVID-19: A systematic review and non-linear dose-response meta-analysis of data from 1 853 610 adults. *Br J Sports Med*, 22, bjsports-2022–105733.

Fiuza-Luces, C., Valenzuela, P. L., Gálvez, B. G., Ramírez, M., López-Soto, A., Simpson, R. J., & Lucia, A. (2023). The effect of physical exercise on anticancer immunity. *Nat Rev Immunol.* Epub ahead of print.

Friman, G., & Wésslen, L. (2000). Special feature: Infections and exercise in high-performance athletes. *Immunol Cell Biol*, 78, 510–522.

Friman, G., Wright, J., Ilbäck, N.-G., Beisel, W. R., White, J. D., Sharp, D. S., … Vogel, J. A. (1985). Does fever or myalgia indicate reduced physical performance capacity in viral infections? *Acta Med Scand*, 217, 353–361.

Garrey, W. E., and Bryan, W. R. (1935). Variations in white blood cell counts. *Physiol Rev, 15*(4), 597–638.

Giossos, Y., Sotiropoulos, A., Souglis, A., & Dafopoulou, G. (2011). Reconsidering on the early types of football. *Balt J Health Phys Act*, 3(2), 129–134.

Gleeson, M., Bishop, N. C., Stensel, D. J., Lindley, M. R., Mastana, S. S., & Nimmo, M. A. (2011). The anti-inflammatory effects of exercise: Mechanisms and implications for the prevention and treatment of disease. *Nat Rev Immunol*, 11(9), 607–615.

Gleeson, M., Walsh, N. P., & British Association of Sport and Exercise Sciences. (2012). The BASES expert statement on exercise, immunity and infection. *J Sports Sci*, 30(3), 321–324.

Grande, A. J., Reid, H., Thomas, E. E., Nunan, D., & Foster, C. (2016). Exercise prior to influenza vaccination for limiting influenza incidence and its related complications in adults. *Cochrane Database Systema Rev*, 2016(8), CD011857.

Grimby, G. (1962). Exercise in man during pyrogen-induced fever. *Scand J Lab Clin Invest*, 14 (Suppl. 67), 1–112.

Gustafson, M. P., Wheatley-Guy, C. M., Rosenthal, A. C., Gastineau, D. A., Katsanis, E., Johnson, B. D., & Simpson, R. J. (2021). Exercise and the immune system: Taking steps to improve responses to cancer immunotherapy. *J Immunother Cancer*, 9(7), e001872.

Haberman, A. L. (2017) Thousands of solitary runners come together: Individualism and communitarianism in the 1970s running boom. *J Sport Hist*, 44(1), 35–49.

Halle, M., Binzenhöfer, L., Mahrholdt, H., Johannes Schindler, M., Esefeld, K., & Tschöpe, C., (2021) Myocarditis in athletes: A clinical perspective. *Eur J Prev Cardiol*, 28(10), 1050–1057.

Hargreaves, E. R. (1948). Poliomyelitis: Effect of exertion during the paralytic stage. *BMJ*, 2, 1021–1022.

Heggie V. (2016). Bodies, sport and science in the nineteenth century. *Past Present*, 231(1), 169–200.

Herberman, R. B., Nunn, M. E., Holden, H. T., & Lavrin, D. H. (1975). Natural cytotoxic reactivity of mouse lymphoid cells against syngeneic and allogeneic tumors. II. Characterization of effector cells. *Int J Cancer*, 16(2), 230–239.

Hiscock, N., Chan, M. H., Bisucci, T., Darby, I. A., & Febbraio, M. A. (2004). Skeletal myocytes are a source of interleukin-6 mRNA expression and protein release during contraction: Evidence of fiber type specificity. *FASEB J*, 18(9), 992–994.

Hoagland, H., Elmadjian, F., & Pincus, G. (1946). Stressful psychomotor performance and adrenal cortical function as indicated by the lymphocyte response. *J Clin Endocrinol Metab*, 6, 301–311.

Hoffman-Goetz, L., Apter, D., Demark-Wahnefried, W., Goran, M. I., McTiernan, A., & Reichman, M. E. (1998). Possible mechanisms mediating an association between physical activity and breast cancer. *Cancer*, 83(3 Suppl), 621–628.

Hoffman-Goetz, L., & Husted, J. (1995). Exercise and cancer: Do the biology and epidemiology correspond? *Exerc Immunol Rev*, 1, 81–96.

Hughes, J. D. (1945). The treatment of infectious hepatitis by diet and rest. *Bull US Army M Dept*, 4, 662.

Jokl, E., & McClellan, J. T. (1971). *Exercise and Cardiac Death*. Balitmore, MD, University Park Press.

Kavanagh, T. (2000). Exercise in cardiac rehabilitation. *Br J Sports Med*, 34, 3–6.

Kiessling, R., Klein, E., & Wigzell, H. (1975). "Natural" killer cells in the mouse. I. Cytotoxic cells with specificity for mouse Moloney leukemia cells. Specificity and distribution according to genotype. *Eur J Immunol, 5*(2), 112–117.

Krüger, K., Lechtermann, A., Fobker, M., Völker, K., & Mooren, F. C. (2008). Exercise-induced redistribution of T lymphocytes is regulated by adrenergic mechanisms. *Brain, Behav Immun*, 22(3), 324–338.

Lancaster, G. I., & Febbraio, M. A. (2016). Exercise and the immune system: Implications for elite athletes and the general population. *Immunol Cell Biol*, 94(2), 115–116.

LaPerriere, A., Schneiderman, N., Antoni, M. H., & Fletcher, M. (1989). Aerobic exercise training and psychoneuroimmunology in AIDS research. In Baum, A. & Temoshok, L. (eds) *Psychological Perspectives on AIDS*. Hillsdale, NJ, Erlbaum, 259–286.

Larabee, R. C. (1902). Leucocytosis after violent exercise. *J Med Res*, 7(1), 76–82.

Levinson, S. O., Milzer, A., & Lewin, P. (1945). Effect of fatigue, chilling and mechanical trauma on resistance to experimental poliomyelitis. *Am J Hygiene*, 42, 204–213.

Linde, F. (1987). Running and upper respiratory infections. *Scand J Sports Sci*, 9, 21–23.

Lowder, T., Padgett, D. A., & Woods, J. A. (2005). Moderate exercise protects mice from death due to influenza virus. *Brain Beh Immun*, 19(5), 377–380.

Lucas, J. A. (1976). A history of the Marathon race—490 B.C. to 1975. *J Sport Hist*, 3(2), 120–138.

Mackinnon, L. T., Chick, T. W., van As, A., & Tomasi, T. B. (1987). The effect of exercise on secretory and natural immunity. *Adv Exp Med Biol,* 216A, 869–876.

Malm, C. (2004). Exercise immunology: The current state of man and mouse. *Sports Med,* 9, 555–566.

McCarthy, D. A., & Dale, M. M. (1988). The leucocytosis of exercise. A review and model. *Sports Med*, 6(6), 333–363.

Mora, J. M., Amtmann, L. E., & Hoffmann, S. J. (1926). Effect of mental and emotional states on the leukocyte count. *J Am Med Assoc*, 86, 945–946.

Morris, J. N., Chave, S. P., Adam, C., Sirey, C., Epstein, L., & Sheehan, D. J. (1973). Vigorous exercise in leisure-time and the incidence of coronary heart-disease. *Lancet*, 1(7799), 333–339.

Morris, J. N., Heady, J. A., Raffle, P. A., Roberts, C. G., & Parks, J. W. (1953). Coronary heart-disease and physical activity of work. *Lancet*, 262(6795), 1053–1057.

Murphy, E. A., Davis, J. M., Carmichael, M. D., Gangemi, J. D., Ghaffar, A., & Mayer, E. P. (2008). Exercise stress increases susceptibility to influenza infection. *Brain Behav Immun,* 22(8), 1152–1155.

Ndee, H. S. (2010). Public schools in Britain in the nineteenth century: The emergence of team games and the development of the educational ideology of athleticism. *Int J Hist Sport*, 27(5), 845–871.

Nieman, D. C. (1994). Exercise, infection, and immunity. *Int J Sports Med*, 15(S3), S131–S141.

Nieman, D. C. (1997). Second symposium of the international society of exercise and immunology. *Int J Sports Med*, 18(S1).

Nieman, D. C., Johanssen, L. M., Lee, J. W., & Arabatzis, K. (1990). Infectious episodes in runners before and after the Los Angeles Marathon. *J Sports Med Phys Fitness*, 30(3), 316–328.

Northoff, H., & Berg, A. (1991). Immunologic mediators as parameters of the reaction to strenuous exercise. *Int J Sports Med*, 12(S1), S9–S15.

Ostrowski, K., Rohde, T., Zacho, M., Asp, S., & Pedersen, B. K. (1998). Evidence that Interleukin-6 is produced in human skeletal muscle during prolonged running. *J Physiol*, 508(3), 949–953.

Paffenbarger, R. S. Jr., Blair, S. N., & Lee, I. M. (2001). A history of physical activity, cardiovascular health and longevity: The scientific contributions of Jeremy N Morris, DSc, DPH, FRCP. *Int J Epidemiol*, 30(5), 1184–1192.

Painter, P., & Blackburn, G. (1988). Exercise for patients with chronic disease. *Postgrad Med*, 83(1), 185–196.

Park, R. J. (2011). Physicians, scientists, exercise and athletics in Britain and America from the 1867 Boat race to the four-minute mile. *Sport in History*, 31(1), 1–31.

Pedersen, B. K., Steensberg, A., & Schjerling, P. (2001). Muscle-derived Interleukin-6: Possible biological effects. *J Physiol*, 536, 329–337.

Pedersen, B. K., & Ullum, H. (1994). NK cell response to physical activity: Possible mechanisms of action. *Med Sci Sports Exerc*, 26(2), 140–146.

Pedersen, L., Idorn, M., Olofsson, G. H., Lauenborg, B., Nookaew, I., Hansen, R. H., ... Hojman, P. (2016). Voluntary running suppresses tumor growth through Epinephrine- and IL-6-Dependent NK Cell mobilization and redistribution. *Cell Metabolism*, 23(3), 554–562.

Peters, E. M., & Bateman, E. D. (1983). Ultramarathon running and upper respiratory tract infections. An epidemiological survey. *S Afr Med J,* 64(15), 582–584.

Rapson, A., Collman, E., Faustini, S., Yonel, Z., Chapple, I. L., Drayson, M. T., … Heaney, J. L. J. (2020). Free light chains as an emerging biomarker in saliva: Biological variability and comparisons with salivary IgA and steroid hormones. *Brain Behav Immun*, 83, 78–86.

Redfern, J., Robyn, G., O'Neil, A., Grace, S. L., Bauman, A., Jennings, G., Brieger, D., & Briffa, T. (2022). Historical context of cardiac rehabilitation: Learning from the past to move to the future. *Front Cardiovasc Med*, 9, 842567.

Riley, L. K., & Rupert, J. (2015). Evaluation of patients with leukocytosis. *Am Fam Physician*, 92(11), 1004–1011.

Roberts, J. A. (1986). Viral illnesses and sports performance. *Sports Med*, 3, 296–303.

Rose-John, S. (2012). IL-6 trans-signaling via the soluble IL-6 receptor: Importance for the pro-inflammatory activities of IL-6. *Int J Biol Sci*, 8(9), 1237–1247.

Russell, R. W. (1947). Poliomyelitis. The pre-paralytic stage, and the effects of physical activity on the severity of the paralysis. *BMJ*, 2, 1023–1028.

Russell, R. W. (1949). Paralytic poliomyelitis. The early symptoms and the effect of physical activity on the course of the disease. *BMJ*, 1, 465–471.

Ruuskanen, O., Luoto, R., Valtonen, M., Heinonen, O. J., & Waris, M. (2022). Respiratory viral infections in athletes: Many unanswered questions. *Sports Med*, 52(9), 2013–2021.

Schenk, A., Esser, T., Belen, S., Gunasekara, N., Joisten, N., Winker, M. T., … Zimmer, P. (2022). Distinct distribution patterns of exercise-induced natural killer cell mobilization into the circulation and tumor tissue of patients with prostate cancer. *American Journal of Physiology. Cell Physiology*, 323(3), C879–C884.

Schultz, G. (1893). Experimentelle Untersucliungen uber das Vorkommen und die diagnostische Bedeutune leukocytose. *Deutsch Arch Klin Med*, 51, 234–281.

Shephard, R. J. (2003). Adhesion molecules, catecholamines and leucocyte redistribution during and following exercise. *Sports Med*, 33(4), 261–284.

Shephard, R. J. (2010). Development of the discipline of exercise immunology. *Exerc Immun Rev*, 16, 194–222.

Simpson, R. J. (2011). Aging, persistent viral infections, and immunosenescence: Can exercise "make space"? *Exerc Sport Sci Rev*, 39(1), 23–33.

Simpson, R. J., Campbell, J. P., Gleeson, M., Krüger, K., Nieman, D. C., Pyne, D. B., … Walsh, N. P., (2020). Can exercise affect immune function to increase susceptibility to infection? *Exerc Immunol Rev*, 26, 8–22.

Simpson, R. J., Florida-James, G. D., Cosgrove, C., Whyte, G. P., Macrae, S., Pircher, H., & Guy, K. (2007). High-intensity exercise elicits the mobilization of senescent T lymphocytes into the peripheral blood compartment in human subjects. *J Appl Physiol*, 103(1), 396–401.

Siversten, I., & Dahlstrom, A. W. (1921). Relation of muscular activity to carcinoma: A preliminary report. *J Cancer Res*, 6, 365–378.

Spielmann, G., McFarlin, B. K., O'Connor, D. P., Smith, P. J., Pircher, H., & Simpson, R. J. (2011). Aerobic fitness is associated with lower proportions of senescent blood T-cells in man. *Brain Behav Immun*, 25(8), 1521–1529.

Steel, C. M., Evans, J., & Smith, M. A. (1974). Physiological variation in circulating B cell: T cell ratio in man. *Nature*, 247(5440), 387–389.

Steensberg, A., Febbraio, M. A., Osada, T., Schjerling, P., van Hall, G., Saltin, B., & Pedersen, B. K. (2001). Interleukin-6 production in contracting human skeletal muscle is influenced by pre-exercise muscle glycogen content. *J Physiol*, 537, 633–639.

Steensberg, A., van Hall, G., Osada, T., Sacchetti, M., Saltin, B., & Pedersen, B. K. (2000). Production of interleukin-6 in contracting human skeletal muscles can account for the exercise-induced increase in plasma interleukin-6. *J Physiol*, 529, 237–242.

Sturgis, C. C., & Bethell, F. H. (1943). Quantitative and qualitative variations in normal leukocytes. *Physiolog Rev*, 23, 279–303.

Swift, W. E., Garner, H. T., Moore, D. J., Streitfeld, F. H., & Havens, W. P. (1950). Clinical course of viral hepatitis and the effect of exercise during convalescence. *Am J Med*, 8, 614.

Tipton, C. M. (2014). The history of "exercise is medicine" in ancient civilizations. *Adv Physiol Educ*, 38, 109–117.

Tipton, C. M. (2016). The emergence of Applied Physiology within the discipline of Physiology. *J App Physiol,* 121(2), 401–414.

Tomasi, T. B., Trudeau, F. B., Czerwinski, D., & Erredge, S. (1982). Immune parameters in athletes before and after strenuous exercise. *J Clin Immunol*, 2(3), 173–178.

Turner, J. E., & Brum, P. C. (2017). Does regular exercise counter T Cell immunosenescence reducing the risk of developing cancer and promoting successful treatment of malignancies? *Oxid Med Cell Longev,* 4234765.

Turner, S. E. G., Loosemore, M., Shah, A., Kelleher, P., & Hull, J. H. (2021). Salivary IgA as a potential biomarker in the evaluation of respiratory tract infection risk in athletes. *J Allergy Clin Immunol Pract*, 9(1), 151–159.

Walsh, N. P. (2018). Recommendations to maintain immune health in athletes. *Eur J Sport Sci*, 18(6), 820–831.

Walsh, N. P., Gleeson, M., Shephard, R. J., Gleeson, M., Woods, J. A., Bishop, N. C., … Simon, P. (2011a). Position statement. Part one: Immune function and exercise. *Exerc Immunol Rev*, 17, 6–63.

Walsh, N. P., Gleeson, M., Pyne, D. B., Nieman, D. C., Dhabhar, F. S., Shephard, R. J., … Kajeniene, A. (2011b). Position statement. Part two: Maintaining immune health. *Exerc Immunol Rev*, 17, 64–103.

Weiss, M. (1993). Infektprophylaxe mit polyvalent Immunoglobulinen: Diskussion anlässlich der Anwendung vor der Olympiade 1992 bei der deutschen Box-National- staffel. *Deutsche Z Sportmed,* 33, 466–471.

Weiss, M., Fuhrmansky, J., Lulay, R., & Weicker, H. (1985). Haufigkeit und Ursache von Immunoglobulin-mangeln bei Sportlern. *Deutsche Z Sportmed*, 36, 146–153.

Wentworth, P., Jentz, L. A., & Croal, A. E. (1979). Analysis of sudden unexpected death in Southern Ontario with emphasis on myocarditis. *Can Med Assoc J*, 120, 676–680.

Wilson, M. G., Hull, J. H., Rogers, J., Pollock, N., Dodd, M., Haines, J., … Sharma, S. (2020). Cardiorespiratory considerations for return-to-play in elite athletes after COVID-19 infection: A practical guide for sport and exercise medicine physicians. *Br J Sports Med*, 54(19), 1157–1161.

Winningham, M. L., MacVicar, M. G., & Burke, C. A. (1986). Exercise for cancer patients: Guidelines and precautions. *Phys Sportsmed*, 14(10), 125–134.

Wong, D. T., Mihm, M. C., Boyer, J. L., & Jain, D. (2015). Historical path of discovery of viral hepatitis. *Harvard Medical Student Review*, 3 September, 18–36.

2 The human immune system

Guillaume Spielmann, James E. Turner,
and John P. Campbell

Introduction

The immune system comprises physical barriers to protect against the initial invasion of pathogens, including skin, mucosal membranes, stomach acid, and body temperature. In addition, the immune system has a combination of non-specific and specific defences to recognise, attack, and destroy pathogens that have entered the body, and some of these defences also target tumours (see Chapter 10). Indeed, the immune system does not simply comprise molecules and cells – it includes several immunological organs, such as the thymus and bone marrow and other structures, such as the lymphatic system (see Figure 2.1). An immune response requires the precise coordination of many different cell types and molecular messengers. For a comprehensive overview of major immune system components, including the history of immunology, readers are directed to the following articles: Kaufmann (2019) and Moser & Leo (2010). In humans, the critical role of the immune system becomes clinically apparent when it is impaired or overactive. Inherited and acquired immunodeficiency states are characterised by increased susceptibility to infections, sometimes caused by commensal micro-organisms not normally considered to be pathogenic. However, an overactive immune system can also be damaging rather than protective, as with autoimmune diseases, where the immune system inappropriately attacks tissues or organs (see Chapter 11). Similar effects are apparent with allergy, such as with hypersensitivity reactions, where an excessive response can bring about symptoms, like with infections or diseases, that can be fatal in the most extreme cases. However, even when the immune system appears to be functioning normally, cells of the immune system are mechanistically implicated in ageing (see Chapter 8) and many other chronic diseases, including type 2 diabetes and cardiovascular disease (see Chapter 12). The following sections outline the principal effectors of the immune response, including cells, soluble factors, and later sections provide a summary of related immunological processes.

Innate and adaptive immunity

The immune system can be broadly categorised into innate and adaptive sub-parts, sometimes referred to as 'arms'. The different sub-parts of the immune system are sometimes even described as different systems (i.e., the innate immune system and the adaptive immune system) which is an oversimplification, since both innate and adaptive sub-parts usually work in concert to eliminate threats. Innate immunity is the first line of defence against pathogens. This part of the immune system comprises anatomical (skin and mucus membranes), physiological (temperature and acidic pH), soluble (cytokines and complement), and phagocytic (cellular) defences. Together, these components provide 'non-specific' defences against pathogens by: (1) physically hindering initial pathogen invasion and maximising their clearance via epithelial linings of skin and mucosal

DOI: 10.4324/9781003256991-2

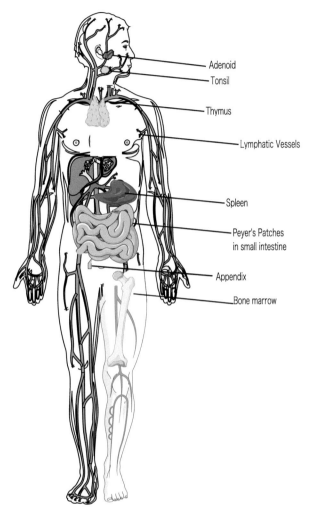

Figure 2.1 Immunological organs and the lymphatic system. The thymus and bone marrow are primary lymphoid tissues, as these are the tissues where maturation of lymphocytes takes place (T cells in thymus, B cells in bone marrow). Through the blood circulation, lymphocytes migrate to other (secondary) lymphoid tissues such as spleen, lymph nodes and the gut-associated lymphoid tissue (e.g. Peyer's patches in the small intestine).

barriers, mucus, ciliary function, and peristalsis; (2) limiting the ability of pathogens to replicate due to chemical factors such as the low pH of stomach fluids, numerous antimicrobial peptides and proteins; (3) engulfing and destroying pathogens via innate immune cells – also called phagocytes, such as monocytes, macrophages, and dendritic cells. These cells can degrade the pathogenic material they have ingested and expose the degraded fragments on their own surface to notify cells of the adaptive immune system of the threat. This process, called antigen presentation enables the adaptive immune cells to recognise foreign material and mount an additional response. Indeed, most challenges that stimulate innate immune processes often lead to activation of adaptive immune processes, which aids substantially in recovery from infection, by contributing to the elimination of pathogens. Adaptive immune cells can recognise and distinguish 'non-self' from 'self' and mount specific responses against pathogens or pathogen-infected cells, without attacking intact healthy body cells. Once cleared, the pathogenic threat is remembered by newly formed memory cells, which facilitate the creation of faster, more efficient immune responses against the pathogen if encountered again with subsequent infections. Adaptive immune cells include B cells which differentiate into plasma cells to produce antibodies as part of the humoral adaptive response and T cells which are activated when presented to antigens by antigen-presenting cells.

Cellular components of the immune system

Most immune system responses are brought about by cells, which can be broadly categorised into three major sub-types (see Table 2.1) – granulocytes, monocytes, and lymphocytes. These cell sub-types all have different characteristics and functions and differ in their appearance, their numbers, and their locations within the body. Collectively, immune cells represent a significant portion of the total cells in the body, with the estimated 1.8×10^{12} immune cells found in an average weight man (73kg) weighing 1.2 kg (Sender et al., 2023). However, it is important to note that even some cells not classically considered to be immune cells have some 'immune' functions, such as secreting soluble mediators known as cytokines (e.g., cells of adipose tissue – adipocytes, cells within the skeletal muscle – myocytes, and other stromal cells – fibroblasts) (see Chapter 6). The remainder of this section focuses on classical immune cells that have most relevance to exercise immunology.

Granulocytes: neutrophils, basophils, and eosinophils

Granulocytes are the most common type of leucocytes, and innate immune cells including neutrophils, basophils, and eosinophils. Granulocytes confer protection against pathogens during the early or acute phase of the immune response. These cells are derived from bone-marrow-residing stem cells and get their name from the vast amount of so-called "azurophilic granules" found in their cytoplasm. These granules, when dyed and observed under a light microscope, exhibit different colours: granules in neutrophils are a neutral pink, granules in basophils appear dark blue, and granules in eosinophils appear bright red. These granules contain potent protein-digesting and bactericidal enzymes and proteins, which facilitate the antibacterial effector functions of granulocytes. Indeed, individuals with abnormally low numbers of granulocytes are more likely to develop frequent and severe bacterial and fungal infections. Granulocytes are also called polymorphonuclear leucocytes because of the varying shapes of the nucleus, which is usually lobed into two to five segments connected by chromatin filaments.

Neutrophils are about 10–12 micrometres (μm) in diameter and contain multi-lobulated nuclei, with variable segmentation depending on neutrophil maturity. The most immature neutrophil, the myeloblast, has a large nucleus with few segments or one lobe, while more mature segmented neutrophils possess two to five lobes in their nuclei. Neutrophils are the most abundant leucocyte in the blood, constituting 50–70% of the total circulating leucocytes. These cells are short-lived, and although their circulating half-life was originally thought to be between 6 and 8 hours (i.e., a lifespan of < 24 hours) (Dancey, Deubelbeiss, Harker, & Finch, 1976), more recent data suggest that their lifespan is closer to 5.4 days (Pillay et al., 2010). The rapid turnover of neutrophils requires continuous release from bone marrow, with around 100,000,000,000 (i.e., 100 billion in short scale, or 100,000 million in long scale) neutrophils produced daily (Rosales, 2018). In the event of pathogen invasion, neutrophils extravasate from the vasculature to reach the site of infection, where they will clear the threat prior to undergoing apoptosis and forming pus. Neutrophils do not normally exit the bone marrow until maturity, but, during an infection, neutrophil precursors called myelocytes and promyelocytes are released. These are

Table 2.1 Key characteristics of human leucocytes and major sub-types that have received focus in exercise immunology

Part A – granulocytes and monocytes				
Leucocyte population	*Leucocyte population sub-type*	*Further sub-types*	*Common identification strategy*	*Characteristics*
Granulocytes				
	Neutrophils		CD66b+	50–70% of all leucocytes, innate phagocytes
		Non-segmented – band	CD66b+CD15++ CD16++	Younger less functionally effective cells
		Segmented	CD66b+CD15+ CD16+++	More mature and more functionally effective cells
	Eosinophils		CD66b+CD49d+ CD125+	2–5% of granulocytes, phagocytose extracellular parasites
		Tissue-resident	CD62L+IL-3Rlow	Reside in tissues, such as the lungs
		Inducible/ inflammatory	CD62LlowIL-3Rhigh	IL-5 producing cells involved in the overall immune response
	Basophils	Mast cells in tissues	CD18+CD31+ CD116+	0–2% of granulocytes, differentiate to mast cells in tissues, produce chemokines, allergy
Monocytes				5–15% of leucocytes
	Classically activated		CD14+CD16−	Phagocytes involved in antimicrobial responses
	Non-classically activated		CD14dimCD16+	Involved in antiviral responses, complement and FcR-mediated phagocytes
	Intermediate		CD14+CD16+	Act as antigen-presenting cells and are strong transendothelial migrators
Macrophages			CD206+	
	M1		HLADR++CD197+	Produce pro-inflammatory cytokines, destroy bacteria and produce ROS and NO
	M2		CD163+CD209+ CCL2+	Produce anti-inflammatory cytokines, involved in tissue repair and wound healing
Dendritic cells			Lin−HLADR+	Tissue-resident professional antigen-presenting cells
	Plasmacytoid		Lin−HLA−DR+ CD303+	Effector sub-type of dendritic cells, produce type-1 interferons
	Myeloid		Lin−HLA−DR+ CD303−	Regulatory sub-type of dendritic cells, produce IL-12

Table 2.1 (Continued)

Part B – lymphocytes				
Leucocyte population	*Leucocyte population sub-type*	*Further sub-types*	*Common identification strategy*	*Characteristics*
Lymphocytes				15–30% of all leucocytes
	Natural killer cells			5–10% of lymphocytes, granular cells, kill virus-infected cells and tumours
		CD56dim effector	CD3−CD56++	Highly cytotoxic, rapid effector function without prior activation
		CD56bright regulatory	CD3−CD56+	Potent cytokine-producing cells
	B cells		CD3−CD19+	5% of lymphocytes, responsible for humoral immunity, can recognise soluble antigens
		Immature B cells	CD3−CD19+CD27−CD10+	Instead of using CD10, IgD− is an alternative. Recent bone marrow emigrants
		Naive B cells	CD3−CD19+CD27−CD10−	Instead of using CD10, IgD+ is an alternative. Have not encountered antigen
		Memory B cells	CD3−CD19+CD27+CD38−	Instead of using CD38, IgD− or IgA+, IgG+, IgM+ alternatives. Long-lived memory cells
		Plasma cells	CD3−CD19+CD27+CD38+	Effector B cells that rapidly produce antibodies
	T cells		CD3+ and CD4+ or CD8+	60% of all lymphocytes (CD4+ helper = 50–80%, CD8+ cytotoxic 20–40%)
		Naive	CD28+CD27+CCR7+CD45RA+	Antigen inexperienced, differentiate into effector/memory cells with novel antigens
		Central memory	CD28+CD27+CCR7+CD45RA−	Long-lived memory cells
		Effector memory	CD28−CD27−CCR7−CD45RA−	Highly cytotoxic effector cells, rapid killing, and cytokine production
		CD45RA+ effector memory	CD28−CD27−CCDR7−CD45RA+KLRG1+CD57+	Terminally differentiated cytotoxic effector cells, rapid killing, and cytokine production
		Regulatory T cells	FoxP3+CD25+CD127+	Maintain peripheral immune tolerance and control inflammation
	Unconventional T cells			Cells spanning innate and adaptive immunity, recognise lipids and small metabolites

(Continued)

Table 2.1 (Continued)

Part B – lymphocytes				
Leucocyte population	*Leucocyte population sub-type*	*Further sub-types*	*Common identification strategy*	*Characteristics*
		Invariant NK T cells	TCRα+T−bet± GATA-3±RORγt±	Between 0.01 and 1% of circulating T cells, multi-functional immunoregulatory cells
		Gamma–delta T cells	TCRγδ+	Recognise a broad range of antigens without the presence of MHC molecules and can present antigens
		GEMT	GEM-TCR+	Possess a highly conserved TCR specific to *Mycobacterium tuberculosis*
		MAIT	TCRαβ+CD3+CD28+ CD27+MDR1+	Possess a semi-invariant TCR which recognises small molecules. MAIT cells play a role in mucosal immunity, with antibacterial and antiviral properties and is involved in tissue repair and homeostasis

Abbreviations: Lin = lineage; **ROS** = Reactive Oxygen Species; **NO** = Nitric Oxide
Abbreviations: Lin = lineage, GEMT = germline-encoded mycolyl lipid-reactive T cells; **MAIT** = Mucosal-associated invariant T cells; **ILCs** = innate lymphoid cells

often referred to as non-segmented 'band cells' and are thought to be not as functionally effective as mature neutrophils. Mature neutrophils are found in the vasculature in two pools: a pool of circulating neutrophils and a marginated pool attached to endothelial cells or in various tissues, such as the liver, spleen, and bone marrow (Hidalgo, Chilvers, Summers, & Koenderman, 2019). Neutrophils have three strategies for directly attacking micro-organisms: phagocytosis (ingestion) followed by intracellular killing; release of soluble antimicrobial compounds (including granule proteins); and generation of neutrophil extracellular traps (NETs).

The first strategy neutrophils can use to attack pathogens is phagocytosis (see Figure 2.2). These cells are capable of ingesting pathogens or other particles coated with antibodies and complement proteins (see the section: soluble components of the immune system) and damaged cells or cellular debris. Indeed, neutrophils can internalise and kill pathogens, with each phagocytic event resulting in the formation of a phagosome (a type of vacuole or membranous sack) into which reactive oxygen species and hydrolytic enzymes are secreted. The consumption of oxygen during the generation of reactive oxygen species (ROS) has been termed the oxidative or respiratory burst, although is unrelated to respiration or energy production. The respiratory burst involves the activation of the enzyme nicotinamide adenine dinucleotide phosphate oxidase, which produces large quantities of superoxide ($O_2 \cdot -$). Superoxide dismutates spontaneously or through catalysis via enzymes known as superoxide dismutases (Cu/Zn-SOD and Mn-SOD), to hydrogen peroxide, which is then converted to hypochlorous acid (HClO) by the heme enzyme myeloperoxidase (MPO) (Mortaz, Alipoor, Adcock, Mumby, & Koenderman, 2018). In addition to HClO, myeloperoxidase also catalyses the production of other hypohalous acids, including hypobromous acid (HOBr) and hypothiocyanous acid (HOSCN), all powerful oxidants. It is thought that the bactericidal properties of these hypohalous acids are enough to

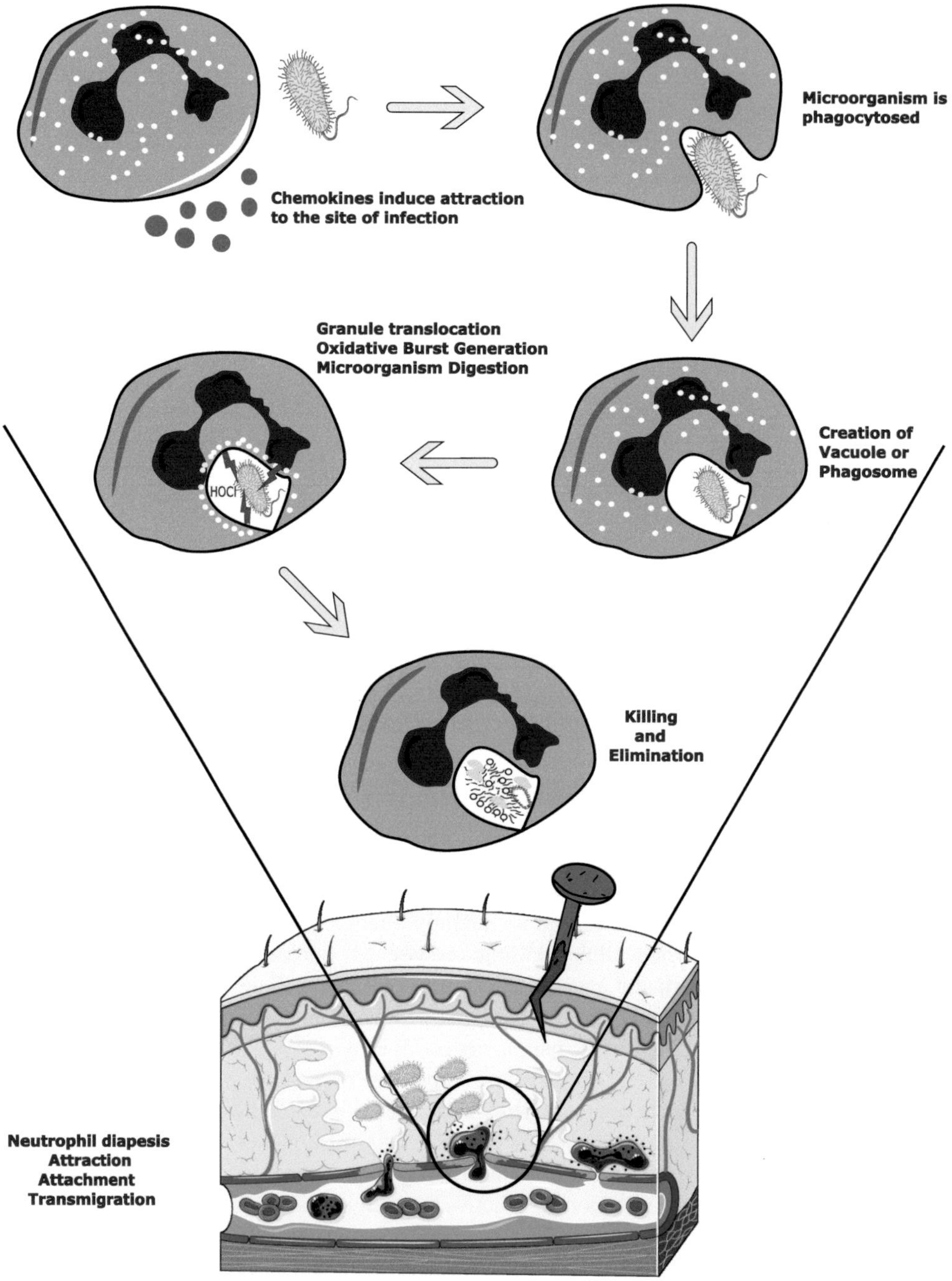

Figure 2.2 Phagocytosis of a pathogen and digestion by activated neutrophils. In response to pathogen entry, for example, by a sharp object breaching the skin, neutrophils can extravasate from the blood stream in response to chemokines, and internalise and kill pathogens. Each phagocytic event results in the formation of a phagosome (a type of vacuole or membranous sack) into which reactive oxygen species and hydrolytic enzymes are secreted – which is often referred to as oxidative burst.

kill bacteria phagocytosed by the neutrophil, but this may instead be a step necessary for the activation of proteases (enzymes such as elastase that break down proteins) (Thomas, 2017).

The second strategy used by neutrophils to combat pathogens is their ability to release granules and products that can destroy bacteria while also stimulating other phagocytes, including monocytes and macrophages. These secretions increase phagocytosis and the formation of reactive oxygen compounds involved in intracellular killing. The release of an assortment of proteins from the cytoplasmic granules is called degranulation. Neutrophils have two types of granules: primary (azurophilic) granules (found in immature neutrophils) and specific granules (which are found in more mature cells). Primary granules contain cationic proteins and defensins that are used to kill bacteria, proteases, and cathepsin G to break down bacterial proteins, lysozyme to break down bacterial cell walls and MPO. In addition, secretions from the azurophilic granules of neutrophils stimulate the phagocytosis of antibody-coated bacteria. The secondary granules contain compounds that are involved in the formation of ROS, lysozyme, and iron chelators, such as lactoferrin, which limits bacterial ability to proliferate.

The third strategy neutrophils can use to eliminate pathogens is by the secretion of extracellular fibrous structures composed of decondensed intracellular chromatin and serine proteases to capture and destroy bacteria. These extracellular traps (NETs) contain high levels of antimicrobial proteins, such as elastase, lactoferrin, calprotectin, myeloperoxidase, and cathepsin G (Estua-Acosta, Zamora-Ortiz, Buentello-Volante, Garcia-Mejia, & Garfias, 2019), which can all kill or inhibit micro-organisms independently of phagocytosis. In addition to their possible antimicrobial properties, NETs may serve as a physical barrier that prevents the further spread of pathogens. During the beginning or acute phase of inflammation, particularly as a result of infection, neutrophils are usually one of the first cells to migrate towards the site of inflammation. They migrate through the blood vessel walls, then through interstitial tissue fluid, following chemical signals, such as interleukin-8 (IL-8), complement fragment C5a, and leukotriene B4 in a process called chemotaxis.

Basophils are the rarest granulocytes in bone marrow and blood, accounting for around 0.5% of peripheral blood leucocytes. Like neutrophils and eosinophils, they have a lobed cell nucleus; however, they have only two lobes and the chromatin filaments that connect them are barely visible. Basophils have receptors that can bind to antibodies, complement proteins, and histamine. The cytoplasm of basophils contains numerous granules that contain histamine, heparin, chondroitin sulphate, peroxidase, platelet-activating factor, and other substances. When an infection occurs, mature basophils will be released from the bone marrow and travel to the site of infection. When basophils are activated, they release interleukin (IL)-4 and IL-13 to promote monocyte differentiation into anti-inflammatory M2 macrophages and assist with tissue repair (Miyake, Ito, & Karasuyama, 2022). In addition, basophils have been implicated in the protection against pathogens and parasites that only infect the superficial layers of the skin, such as ectoparasites, scabies, and helminth infections (Karasuyama, Miyake, Yoshikawa, & Yamanishi, 2018). While the exact mechanisms behind the role of basophils against parasites remain to be fully understood, they have been shown to release extracellular DNA traps to immobilise and kill bacteria, in a similar manner as the neutrophil NET production described above (Yousefi et al., 2015). However, when activated or damaged, basophils can also release histamine and prostaglandin to promote the inflammatory response that helps to kill pathogens. Histamine causes dilation and increased permeability of nearby blood capillaries, while prostaglandins facilitate increased blood flow to the site of infection. Both mechanisms allow blood-clotting elements to be delivered to the infected area. This begins the recovery process and prevents the spread of the pathogen to other tissues.

Eosinophils were first described in 1879 by Paul Ehrlich. They also have a segmented nucleus (two to four lobes) and have a crucial role in killing parasites (e.g., enteric nematodes), but these cells have a limited ability to participate in phagocytosis. They develop and mature in the bone marrow and are recruited into the tissue within 8–12 hours of their release in circulation. Once in contact with invading pathogens, eosinophils, just like neutrophils, can produce oxidative burst through eosinophil peroxidase, superoxide, and H_2O_2 (Svensson & Wenneras, 2005). In addition, eosinophils can also "catapult" DNA to immobilise the foreign material in yet another type of extracellular trap (Yousefi et al., 2008). They also function as antigen-presenting cells, regulate other immune cell functions (e.g. CD4+ T helper cells, dendritic cells, B cells, mast cells, neutrophils, and basophil functions), are involved in the destruction of tumour cells and promote the repair of damaged tissue. However, when eosinophils are found in excess, they can also lead to a variety of allergic reactions, including asthma, allergic airways inflammation, and gastrointestinal diseases (Ramirez et al., 2018). Of note, eosinophils have recently been implicated in adipose-tissue homeostasis, by promoting the differentiation of adipose-tissue resident macrophages into 'alternatively activated' M2 cells which in turn regulate glucose homeostasis and may protect against obesity (Wu et al., 2011). The relationship between immune cells and adipose tissue is described in greater detail in Chapter 6.

Monocytes, macrophages, and dendritic cells

Monocytes are the largest type of leucocytes in the blood, produced by the bone marrow from common myeloid progenitor cells. Monocytes measure between 12 and 20 μm in diameter, with a large smooth bilobed nucleus, often described as 'kidney-shaped'. They circulate in blood for between one and seven days (Patel et al., 2017) depending on their specific sub-type. There are three major sub-types, 'classical' pro-inflammatory cells (CD14+CD16−), 'non-classical' patrolling cells (CD14dimCD16+), and 'intermediate' cells (CD14+/CD16+). Intermediate cells secrete the pro-inflammatory cytokines IL-1β and TNF-α when stimulated or in recognition of pathogens (Gunther & Schultze, 2019). In blood, monocytes normally constitute 5–15% of leucocytes; however, a much greater number of undifferentiated monocytes are stored in the spleen where they can be rapidly deployed to the site of injury when required (Swirski et al., 2009). Once circulating monocytes leave the blood, they will differentiate into macrophages or dendritic cells in the tissues. In response to bacterial lipopolysaccharide (LPS, a component of the cell wall in Gram-negative bacteria), tumour necrosis factor-alpha (TNF-α), or interferon-gamma (IFN-γ) stimulation, tissue-residing macrophages will be polarised into 'classically activated' M1-like macrophages and produce pro-inflammatory cytokines when further stimulated. However, IL-4 and IL-13 stimulation drives macrophages towards more heterogeneous phenotypes of 'alternatively activated' and 'reparative' M2-like macrophages which include M2a, M2b, M2c, and M2d phenotypes.

Monocytes and their macrophage and dendritic-cell progeny serve three main functions: phagocytosis and intracellular killing, antigen presentation to lymphocytes, and cytokine production. Monocytes can perform phagocytosis (in a similar way to neutrophils) using intermediary opsonisation proteins, such as antibody or complement that coat the pathogen, as well as by binding to the microbe directly via pattern-recognition receptors that recognise pathogens. These receptors are referred to as Toll-like receptors (TLRs). Microbial fragments that remain after digestion educate and activate cells of the adaptive immune system. This process of antigen presentation is essential for T-lymphocyte activation, which then mount a specific immune response against the antigen. A secondary role of monocytes is to orchestrate the immune response against pathogens. This is achieved by the initial production of pro-inflammatory cytokines TNF-α, interleukin (IL)-l, and IL-6, along with other soluble mediators such as soluble CD14 by activated monocytes. Upon clearance of the pathogen, monocytes and M2 macrophages will secrete anti-inflammatory cytokines IL-10 and IL-1ra to reduce tissue damage and facilitate tissue repair and remodelling.

Dendritic cells are present in tissues in contact with the external environment, such as the skin (where there is a specialised dendritic-cell type called the Langerhans cell) and the inner lining of the nose, lungs, stomach, and intestine. They can also be found in an immature state in the blood. Once activated, they migrate to the lymph nodes, where they interact with lymphocytes (both T and B cells) to initiate and shape the adaptive immune response. At certain stages of their development, they grow branched projections (dendrites) which give the cell its name *(dendron* being Greek for tree). Immature dendritic cells are also called veiled cells, as they possess large cytoplasmic 'veils' rather than dendrites. Dendritic cells express high levels of Major Histocompatibility Complex class-II (HLA-DR) molecules expressed on their surface and are often referred to as 'professional antigen-presenting cells'. They lack T cell and B cell markers (CD3 and CD19/CD20 respectively), but can be sub-divided into conventional dendritic cells 1 and 2 (cDC1 and cDC2) based on their expression of specific markers such as CD141 and CADM1 (cDC1) and CD169 and CD1c (cDC2) (Cabeza-Cabrerizo, Cardoso, Minutti, Pereira da Costa, & Reis e Sousa, 2021). In addition to playing a key role in presenting antigens to cells of the adaptive immune system, and thus in activating the immune response, it is now clear that dendritic cells also help preserve immune tolerance and reduce the risk of autoimmune reactions.

Lymphocytes: T cells, B cells, Natural Killer cells, innate lymphoid cells, and unconventional T cells

Lymphocytes account for 15–30% of blood leucocytes and vary in size. Large granular lymphocytes include NK cells – innate immune cells – while smaller lymphocytes are mostly T cells and B cells – adaptive immune cells. In healthy lymphocytes, the dense nucleus is approximately the size of a red blood cell (about 7 μm in diameter). Polyribosomes are a prominent feature in the lymphocytes and have an important role in protein synthesis allowing the generation of large quantities of cytokines and immunoglobulins. While T cells and B cells are visually identical, they differ in their function, in their ability to secrete specific proteins and cytokines, and in their expression of various cell surface proteins (CD). Specifically, T cells express CD3, noncovalently associated with the T cell receptor, and either the co-receptors CD4 or CD8. B cells express the molecules CD19 and CD20, and NK cells express CD56 while lacking CD3 (Table 2.1). In practice, many studies rely on the use of fluorescent-labelled (or heavy metal ion-tagged) antibodies targeting these different markers to identify and quantify lymphocyte sub-types via flow cytometry (as described in Chapter 3).

T cells and B cells are the main adaptive immune cells. T cells are named using the beginning letter of the organ in which they develop – the thymus – and B cells are named with the beginning letter of a lymphoid organ only found in birds – the *Bursa of Fabricius* – where B cells develop, but in humans, they develop in the bone marrow. T cells are involved in cell-mediated immunity – directly detecting and eliminating body cells infected with pathogens or tumour cells, whereas B cells are primarily responsible for humoral immunity and the production of antibodies (see the section: soluble immune system components). T cells account for 60–80% of blood lymphocytes and B cells, 5–15%. The function of T cells and B cells is to recognise and respond to specific 'non-self' antigens, following the presentation of degraded fragments of pathogen-derived proteins by antigen-presenting cells, such as dendritic cells and macrophages. B cells and T cells generate specific responses that are tailored to effectively eliminate specific pathogens or pathogen-infected cells. B cells respond to pathogens by producing large quantities of immunoglobulins which neutralise harmful molecules (e.g. toxins) or microbes like bacteria and viruses. In response to pathogens, CD4+ T helper cells produce cytokines that direct the immune response while CD8+ cytotoxic T cells produce toxic granules that contain powerful enzymes released to induce the death of pathogen-infected cells. Following antigen exposure and activation, naive B cells and T cells differentiate into effector and memory cells providing long-lasting immunity to the pathogen. As T cells differentiate, they selectively lose or gain expression of specific co-stimulatory receptors and other surface proteins. Specifically, naive T cells express both CD28 and CD27, two major co-stimulatory molecules that play a crucial role in cellular stimulation, proliferation, and survival, along with the molecules CD45RA and CCR7 which enable them to be activated or migrate to secondary lymphoid organs respectively. It is now well established that naive T cells lose CD45RA expression as they differentiate into central memory cells (CD28+/CD27+/CCR7+/CD45RA−), and finally lose CCR7 when they become effector memory cells (CD28+/−/CD27+/−/CCR7−/CD45RA−). When they enter their terminal stage of differentiation, T cells stop expressing CD28 and CD27, but re-express CD45RA as they gain novel cell surface proteins such as CD57 and KLRG1 (CD28−/CD27−/CCR7−/CD45RA+/KLRG1+/CD57+). More details can be found in Chapter 8). These long-lived cells enable the body to develop a more rapid and effective immune response to a microorganism during subsequent encounters, even years later.

NK cells are lymphocytes that are classically defined as innate immune cells; however, more recent scientific advances have identified NK cell memory-like phenotypes, and thus could be considered as both innate and adaptive immune cells (O'Leary, Goodarzi, Drayton, & von Andrian, 2006). NK cells account for 5–20% of blood lymphocytes and have a major role in defending the host from tumours and virus-infected cells. NK cells distinguish infected cells and tumours from normal cells by recognising changes in the expression levels of a cell surface molecule present on nearly every body cell called Major Histocompatibility Complex class-I (MHC class-I). Once activated, NK cells destroy cells that have a lower cell surface expression of MHC class-I by either releasing cytotoxic enzymes from their intracellular granules, or by inducing death receptor-induced apoptosis of the target cell via the Fas-FasL pathway. However, NK cell activation is tightly regulated by their expression of inhibitory (Killer Inhibition Receptors – KIR) and activatory (Killer Activation Receptors – KAR) receptors which bind to ligands on target cells (Moretta et al., 2004). These receptors provide balancing signals to the NK cell to either induce target cell lysis or inhibit killing. Specifically, when inhibitory receptors KIRs bind to molecules constitutively expressed on uninfected intact 'self' cells, inhibitory signals will be generated to prevent cell death. On the other hand, when activatory receptors KAR bind to molecules expressed on viral-infected cells, in the absence of KIR engagement, the NK cell releases granule contents to kill the infected cell. These granules include cytolysin and perforin (pore-forming proteins which cause the cell membrane to break up, also called lysis) and kill virus-infected cells and tumour cells (Cerwenka & Lanier, 2001).

It is also worth noting that NK cells can also be activated to secrete granules via engagement of CD16 with the Fc region of antibody that is bound to target antigens, thus NK cells bridge innate and adaptive immunity and can, indirectly, induce 'specific' immune responses – a full overview of NK cell activation can be found elsewhere (Morvan & Lanier, 2016).

Other populations of lymphocytes include unconventional T cells and innate lymphoid cells. Unconventional T cells are immune cells that share features of the innate and adaptive immunity, expressing semi-invariant T cell receptors (TCR), but not requiring prior exposure to antigens to exert their effector functions. Among unconventional T cells, invariant NK T cell subsets (iNKT), γδ T cells, germline-encoded mycolyl lipid-reactive (GEMT), and mucosal-associated invariant T cells (MAIT) respond to microbial ligands and cytokines, independently of TCR stimulation (Mayassi, Barreiro, Rossjohn, & Jabri, 2021). Unconventional T cells are a relatively sparse subset of immune cells, which make up between 6% (Singh, Szaraz-Szeles, Mezei, Barath, & Hevessy, 2022) and 30% (Loh et al., 2020) of all T cells. Among these, MAIT cells are the most abundant and make up to 10% of the circulating pool of T cells (Gherardin et al., 2018). Innate lymphoid cells (ILC) are a population of lymphoid cells that do not express T cell or B cell receptors. These recently discovered cell populations are composed of three main groups, categorised by their cytokine secretion profiles and transcriptional factor expression: ILC1s, ILC2s, and ILC3s. The first group, ILC1s, are composed of conventional NK (cNK) cell, and cells that express IFN-γ and the transcription factor "T-bet." ILC2s are composed of cells that express IL-5/IL-13 and the transcription factor GATA3. Finally, ILC3s comprise cells with the natural cytotoxicity receptor (NCR)+, others that do not express NCR, and lymphoid-tissue inducer (LTi) cells that express IL-22, IL17, and the transcription factor RORγt (Spits et al., 2013). Each ILC subset has a different, distinct expression of surface receptors and functions, and it is activated by different molecules. A defining characteristic of unconventional T cells and ILCs is their presence in various tissues, such as mucosal barriers, where they have a fundamental role in maintaining tissue homeostasis and promoting tissue healing. Since unconventional T cells and ILCs do not depend on antigen-presenting cells, they are able to mount rapid responses against pathogens and release cytokines, chemokines, and other defence proteins that stimulate the adaptive immune response (Darrigues, Almeida, Conti, & Ribot, 2022).

Soluble components of the immune system

Soluble molecules provide overall immune system regulation, including influencing leucocyte function and neutralising pathogens. Soluble components of the immune system include cytokines, chemokines, acute-phase proteins, growth factors, complement proteins, and immunoglobulins. Many soluble components can be further divided into sub-types, such as with cytokines, which include interleukins, interferons, tumour necrosis factors, lymphokines, myokines, adipokines, and colony-stimulating factors. Soluble components of the immune system can stimulate the growth, differentiation, and the functional development of leucocytes via specific receptor sites on either secretory cells (autocrine function) or immediately adjacent leucocytes (paracrine function). It is beyond the scope of this chapter to cover all soluble components of the immune system, but readers are directed to Chapter 5 for more coverage and particular relevance to exercise. However, the soluble components that have received most investigation in exercise immunology are included in Table 2.2, and other key concepts are summarised in the text.

There is interaction between the different soluble components of the immune system. For example, acute-phase proteins are produced by the liver in response to cytokines, especially IL-6, which is released from activated monocytes and macrophages when they encounter pathogens. A commonly studied acute-phase protein is C-Reactive Protein, and readers are directed to a brief review of this molecule by Pepys & Hirschfield (2003). Acute-phase proteins have a variety of functions, including stimulating phagocytes, and they can kill bacteria directly or indirectly by reducing the bioavailability of ions essential for the survival of some pathogens, such as haptoglobin, which removes free haemoglobin in plasma, transferrin (together with lactoferrin released from neutrophils) chelate-free iron, and to reduce iron availability and limit iron-dependent bacterial replication. Acute-phase proteins also activate the complement system, consisting of around 40 different proteins that normally circulate in the blood in inactive forms. The presence of certain yeasts, fungi, or bacteria and antibody–antigen complexes activates the complement cascade that results in the breakup of several of the complement proteins into

Table 2.2 Soluble factors of the immune system – origin and function

Soluble factor	Origin	Function
IL-1	Activated macrophages	IL-1α tends to remain cell associated IL-1β acts as a soluble mediator Stimulates IL-2 production from CD4+ Th cells Stimulates B cell proliferation Increases TNF-α, IL-6, and CSF concentration Promote neuroimmune responses (i.e. fever) Excessive production of IL-1β is associated with a variety of inflammation-related pathologies
IL-1ra	Hepatocytes – Adipocytes Monocytes – Macrophages Neutrophils	Anti-inflammatory cytokine by competing with IL-1 Binds to IL-1 receptor without transducing signals
IL-2	Mostly produced by activated CD4+ Th cells	Stimulates IL-2 receptor expression on T cells and B cells Stimulates T cell and B cell proliferation Stimulates release of IFNγ by T cells Stimulates NK cell proliferation and cytotoxic functions
IL-6	Activated CD4+ Th cells, fibroblasts, macrophages, and skeletal muscle	Stimulates B cell differentiation, inflammation, and the acute-phase response Orchestrates both pro- and anti-inflammatory responses depending on its origin and signalling pathway. Muscle-derived IL-6 can be seen as anti-inflammatory, immune-derived IL-6 can be seen as pro-inflammatory
IL-8	Macrophages, various leucocytes	Activates neutrophils and recruits to the site of infection Chemoattractant (chemokine: CXCL8) Promotes angiogenesis
IL-10	Monocytes and macrophages B cells, dendritic cells, NK cells, T cells	Anti-inflammatory cytokine by inhibiting pro-inflammatory cytokine production Protects tissue from immune-related damage High levels have been documented in autoimmune diseases
IL-17	Th17, including CD8+ T cells, γδ T cells, NKT cells, and innate lymphoid cells	Pro-inflammatory cytokine Stimulates neutrophil production and attracts them to the site of infection Is associated with autoimmune diseases such as psoriasis and rheumatoid arthritis
TNF-α	Monocytes T cells, B cells, and NK cell	Enhances tumour cell killing and antiviral activity
IFNγ	Activated T cells	Enhances cell-mediated immunity Activates NK cell and antiviral activity
Acute-Phase Proteins (e.g. C-Reactive Protein)	Hepatocytes	Promotes cell migration to sites of injury and infection Activate the complement Stimulate phagocytosis
Complement	Hepatocytes	Consist of more than 40 proteins Activate leucocytes Neutralise pathogenic organisms
Antimicrobial Proteins	Keratinocytes, sweat and lacrimal glands, neutrophils and monocytes	Over 100 different antimicrobial peptides or proteins (AMP) Destroy bacterial cell wall Inhibits bacterial protein formation Attract immune cells to the site of infection

(*Continued*)

Table 2.2 (Continued)

Soluble factor	Origin	Function
Chemokines	Immune cells, fibroblasts, endothelial, and epithelial cells	Direct immune cells to specific organs, tissue, and other location in the body
		Four main subfamilies of chemokines: CXC, CC, (X)C, and CX3C
		Inflammatory chemokines induce leucocyte recruitment to site of infection/inflammation (e.g. CCL2, CCL8)
		Homeostatic chemokines mediate homeostatic migration of immune cells (e.g.CXCL12, CCL21)
		Dual chemokines exert both roles (e.g. CCL22, CCL20)
Extracellular Vesicles	All cell types	Allows for cell-to-cell communication
		Carry biologically active macromolecules (proteins, DNA, mRNA)
		Regulate immune function

Notes: CD = clusters of differentiation; IL = interleukin; IFN = interferon; CSF = colony-stimulating factor; TNF = tumour necrosis factor; Th = T helper; NK = natural killer

smaller biologically active fragments. The cascade of reactions that activates the complement system leads to the cleavage of a central hepatic-borne protein C3, via the so-called classical, alternative, or lectin pathways. The classical pathway is induced by C1q binding to an antibody attached to a pathogen (either IgG or IgM), while the lectin pathway is initiated by the binding of recognition molecules (specifically, mannose-binding lectins recognition molecules, collections, and ficolins) to oligosaccharides present on the surface of the microorganism. Finally, the alternative pathway is initiated by the spontaneous hydrolysis of C3. Subsequently, the fragments formed from the breakup of complement proteins C3 and CS help neutralise pathogenic threats using a plethora of synergistic pathways: C3b promotes phagocytosis, C3e promotes increased release of leucocytes from the bone marrow, CSa attracts and activates phagocytes, and CSb combines with C6, C7, CS, and C9 to form a membrane attack complex. The latter attaches to bacterial cell membranes, forming pores which allow osmotic influx of water into the bacterium, causing it to swell until it bursts. For a review on complement, readers are directed towards Rothschild-Rodriguez et al. (2022). Other soluble immune components include around 100 types of antimicrobial peptides or proteins (AMP). One of the first proteins identified was salivary lysozyme (Fleming & Allison, 1922). Most AMPs are short cationic molecules, with sizes ranging from ten peptides (neurokinin A) to 149 peptides long (RegIIIα) and are secreted by a variety of cells, ranging from keranocytes in the epidermis, to sebaceous, sweat, and lacrimal glands. In addition, adipocytes, commensal microbiota, and immune cells such as neutrophils, dendritic cells, and monocytes are producers of AMPs in response to infection. A variety of AMPs have been isolated, including lysozyme, lactoferrin, defensins, such as HNP1–3, histatins, dermcidins, hepcidins, and cathelicidins, such as LL-37. AMPs protect the host against infection by (1) creating pores through the bacterial membrane; destroying the bacterial wall in a 'detergent-like' manner (Oren & Shai, 1998), inhibiting protein and/or nucleic acid synthesis; inhibiting cell division, and attracting other immune cells to the site of infection.

Immunoglobulins – also called antibodies – are perhaps the most studied soluble component of the immune system, and the protection they provide is referred to as humoral immunity. Antibodies are produced by terminally differentiated B cells called plasma cells. The primary adaptive humoral immune response occurs when naive B cells encounter a novel antigen for the first time, leading to their activation, proliferation, and differentiation into antibody-producing plasma cells or longer-lived memory B cells. Antibodies, or immunoglobulins (Ig), are glycoproteins composed of two light chains and two heavy chains, which can be divided into five subclasses: IgA, IgD, IgE, IgG, and IgM. Immunoglobulins can either be soluble when released by plasma cell or bound to the membrane of a B cell when it serves as the B cell receptor (BCR). B cells can either be activated by direct contact between the BCR and the antigen it is specific to, or indirectly when B cells and activated CD4+ T helper cells share a cognate interaction. Antigen internalisation or B cell activation by CD4+ T helper cells, induce B cell proliferation and differentiation into memory cells and plasma cells which are capable of secreting large amounts of antibody during their brief life of four to five days. Antibodies circulate in the blood and lymph, binding to antigens and contributing to the destruction of the organism bearing it. Each antibody molecule has the ability to bind to a specific antigen and assist with pathogen destruction. Every antibody has separate regions for each of these two functions (see Figure 2.3). The regions that bind the antigen differ from molecule to molecule and are called variable regions. Conversely, only a few humoral effector mechanisms exist to destroy antigens, so only a limited number of regions are involved; these are called constant regions.

An antibody molecule consists of two pairs of polypeptide chains – two short identical light chains and two longer identical heavy chains. The chains are joined together to form a Y-shaped molecule. The variable regions of heavy and light chains are located at the ends of the arms of

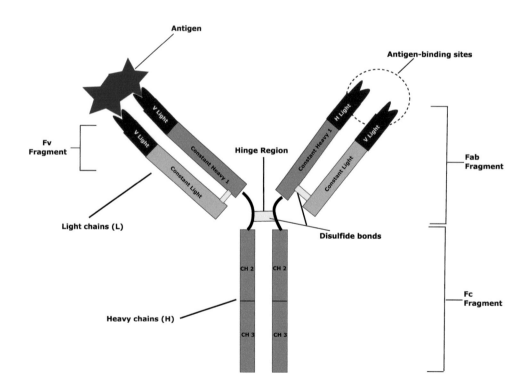

Figure 2.3 Structure of an immunoglobulin. An immunoglobulin consists of two pairs of polypeptide chains – two short identical light chains and two longer identical heavy chains joined together to form a Y-shaped molecule. The variable regions of heavy and light chains are located at the ends of the arms of the Y, where they form the antigen-binding sites. The rest of the antibody molecule, consisting of the constant regions of the heavy and light chains, determines the effector function of the immunoglobulin.

the Y, where they form the antigen-binding sites. Thus, on each antibody molecule, there are two antigen-binding sites, one at each tip of the antibody's two arms (Figure 2.3). The rest of the antibody molecule, consisting of the constant regions of the heavy and light chains, determines the effector function of the immunoglobulin. There are five types of constant region and, hence, five major classes of antibody. Within each class, there will be a multitude of subpopulations of antibodies, each specific to a particular antigen. Whereas IgM and IgG dominate systemic humoral immunity, IgA is the predominant class of immunoglobulin found in mucosal immunity. Antibodies assist with antigen elimination, but do not destroy pathogens directly. Instead, they help identify foreign molecules and cells to facilitate their phagocytosis and destruction by other immune cells. This can be achieved by numerous mechanisms, including (1) binding of specific antigens to limit its toxicity (per example in the case of toxins – a process called neutralisation); (2) linking multiple cell-bound antigens together to create larger antigen complexes (agglutination); (3) precipitating soluble antigens into insoluble, immobile precipitates; (4) inducing the release of histamine from mast cells to increase inflammatory signals in the microenvironment. In addition, antibody–antigen complexes on the surfaces of invading micro-organisms usually cause complement activation, which, once activated, attack the membrane of the invader and, by coating the surface of foreign material, make it even more attractive to phagocytes (a process known as opsonisation).

Mucosal immunity

Although many people assume that blood is the primary location of the immune system, only a very small proportion – around 2% – of all immune cells are found in the circulation (Ganusov & De Boer, 2007). Nearly half of all immune cells are thought to be in the lymph nodes at any one time, with the remaining cells distributed between the spleen, bone marrow, the thymus, and gut. It is important to note that most of these sites are in a state of 'flux' with many immune cell types constantly circulating between blood, tissues, and lymph, as part of immune-surveillance. However, the immune system is far more complex than populations of leucocytes or soluble immune components circulating throughout the body and is a product of many complex interactions between cells, tissues, organs, and organ systems. As described earlier, immune function is a result of both innate and adaptive immune processes, brought about by multiple cell types, soluble immune system components, and complex intracellular, inter-cellular, and inter-tissue and inter-organ communication. Given that the immune system interacts with all parts of the body to defend against pathogens and tumours, it is worth highlighting one of the largest immune defences that provides some of this protection is the mucosal membranes, sometimes referred to as mucosal immunity or even the mucosal immune system.

Mucosal immunity can be considered a first line of protection that reduces the need for systemic immunity, which is principally pro-inflammatory and potentially tissue damaging. The mucosal immune system has two non-inflammatory layers of defence: (1) immune exclusion performed by secretory soluble factors that inhibit surface colonisation of micro-organisms and dampen penetration of potentially dangerous soluble substances; and (2) immunosuppressive mechanisms to avoid local and peripheral hypersensitivity to innocuous antigens. The latter mechanism is referred to as 'oral tolerance' when induced via the gut (Commins, 2015) and likely explains why overt and persistent allergy to food proteins is relatively rare. A similar down-regulatory tone of the immune system normally develops against antigenic components of the commensal microbial flora in the large intestine (see Chapter 6).

The immune system of the gut divides into the physical barrier of the intestine and active immune components, which include both innate and adaptive immune cells. The physical barrier is central to the protection of the body against infections. Acid in the stomach, active peristalsis, mucus secretion, and the tightly connected monolayer of the epithelium each play a major role in preventing micro-organisms from entering the body. Cells of the immune system in the gut are found in the lamina propria. Specialised lymphoid aggregates, called Peyer's patches, reside below specialised epithelial cells whose structure enables sampling of small particles. In addition to the physical and chemical defences conferred by the gut and associated mucosa, the intestinal mucosa also contains at least 80% of the body's activated immunoglobulin-secreting B cells, which are terminally differentiated to Ig-producing plasma cells.

The immune response: a complex coordinated response

Spot the intruder

Pathogens entering the body can be recognised by receptors on innate immune cells called pattern-recognition receptors (PRR). These receptors detect molecular structures present in many types of bacteria, fungi, viruses, and parasites, and include endotoxin or lipopolysaccharide (LPS) (Garcia-Bragado et al., 1979), teichoic acids, and bacterial DNA (Mogensen, 2009). These structures, commonly referred to as pathogen-associated molecular patterns (PAMPs), are recognised as 'danger signals' upon binding by PRRs, which will induce pro-inflammatory and antimicrobial responses to eliminate the pathogen. Two main types of PRRs are found in innate immune cells, membrane-bound receptors called TLRs that recognise extracellular pathogens (Lemaitre, Nicolas, Michaut, Reichhart, & Hoffmann, 1996), and cytosolic PRRs, such as retinoid acid-inducible gene I (RIG-I)-like receptors (RLRs) (Yoneyama et al., 2004) and nucleotide-binding oligomerisation domain (NOD)-like receptors (NLRs) (Kanneganti, Lamkanfi, & Nunez, 2007) that recognise intracellular pathogens. PRRs are expressed mainly by monocytes, macrophages, and dendritic cells but also by a variety of other cell types such as neutrophils, B cells, and epithelial cells (Medzhitov, 2001). A family of ten mammalian TLRs (TLRI-IO) have been identified to date and they recognise conserved PAMPs, including LPS, lipoproteins, peptidoglycan, lipoteichoic acid, zymosan (components of bacterial cell walls), and flagellin (a protein component of the flagellum or 'tail' of motile bacteria, and bacterial DNA and double-stranded RNA found in many viruses). Most TLRs are expressed on the cell surface but some (e.g. TLR9) are located in the cytoplasm. NOD-like receptors are more numerous, and currently 22 different NLRs have been identified (Ting et al., 2008). The NLR family recognise a wide range of ligands from pathogens but also from environmental sources (e.g., asbestos, silica) (Kim, Shin, & Nahm, 2016). As PAMPs are not expressed by host cells, TLR and NOD recognition of PAMPs permits 'self' *versus* 'non-self' discrimination. The binding of these foreign molecules to TLRs and NODs causes activation of immune cells. TLRs, in particular, control both the activation of innate immunity through the induction of antimicrobial activity (e.g. phagocytosis) and the production of inflammatory cytokines.

Alert specific defences against the threat

Innate recognition of PAMPs by PRRs leads to an adaptive immune response through the induction of signalling molecules on the cell surface of macrophages and dendritic cells (collectively known as antigen-presenting cells [APCs]). Thus, the initial activation of the innate immune system is the beginning of a targeted, specific and powerful response from the adaptive immune system. APCs include monocytes, macrophages, B cells, and dendritic cells. The latter are sometimes called 'professional APCs', because presenting fragments of pathogen-derived proteins to naive T cells is their primary function, and it is the crucial step in initiating a primary immune response (Moll, 2003). TLRs on the surface of APCs are activated by binding to

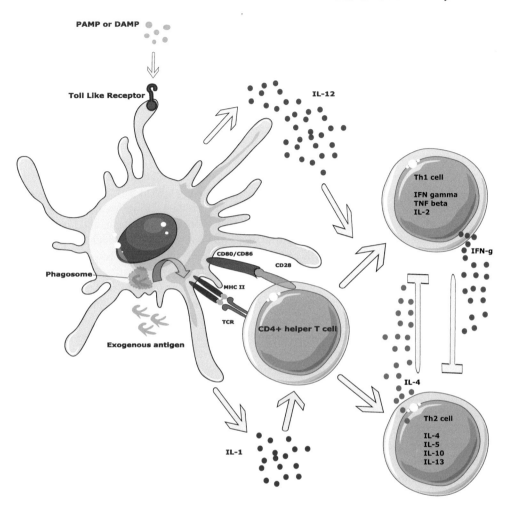

Figure 2.4 Interaction between a dendritic cell and T cell. Binding of pathogen associated molecular patterns (PAMP) and endogenous danger signal molecules such as heat shock proteins to Toll-like receptors leads to the activation of antigen-presenting cells (APC) such as Dendritic cells and subsequent activation of CD4+ helper T cells. APCs take up antigen via endocytic pattern recognition receptors and process (degrade) it to immunogenic peptides, which are displayed to the T cell receptors (TCR) in the polymorphic grove of major histocompatibility complex molecules. This interaction between the APC and the CD4+ helper T cells (via the TCR, and CD28 on the T cell and CD80 or CD86 on the APC) induces T cell activation. Dependent on the types of cytokines released by the APC, the CD4+ helper T cell differentiated into a Th1-cytokine type or Th2-cytokine type secreting cell.

PAMPs followed by subsequent phagocytosis of the pathogen. Lysosomal digestive enzymes and oxidising substances are released into the intracellular vacuole containing the pathogen and pathogen-derived proteins are degraded into short sequences of amino acids (peptides) and incorporated within the polymorphic groove of either MHC class-I or MHC class-II proteins within the cytosol of the APC. Once this process is complete, the newly formed complexes can translocate to the surface of the APC and is presented to other immune cells. Helper CD4+ T cells recognise MHC class-II and is the traditional model of APC and T cell interaction (Figure 2.4).

However, cytotoxic CD8+ T cells can also be activated via their recognition of MHC class-I on the cell surface of professional APCs. It is important to note that only T cells expressing a T cell receptor that is complimentary to the sequence of amino acids within the fragment of the digested pathogen-derived protein will recognise what is being presented by the APC and become activated. The specific structure of the T cell receptor and the sequences of amino acids it recognised are determined by random shuffling of genes during T cell development in the early years of life via processes in the thymus. In the traditional model of APC and T cell interaction, activated CD4+ T helper cells stimulate the overall immune response via cytokine release, including activating B cells – causing them to divide and differentiate into immunoglobulin-producing plasma cells. Following pathogen entry, the initial primary antibody response is relatively slow, and several days are required before enough antigen-specific B cell clones are generated to neutralise the pathogen. It takes two to seven days for neutralising IgM to be found in the blood and two to four weeks before peak IgG antibody levels are attained. However, a second exposure to the same antigen (even years later) produces a much more rapid, stronger, and longer-lasting secondary response. This secondary response relies on memory B cells, which are produced at the same time as effector B cells (i.e., plasma cells) during the primary response. Effector B cells usually only live for a few days, but memory B cells may last for decades. When there is a second exposure to the same antigen, memory B cells rapidly multiply and differentiate into plasma cells to generate large quantities of pathogen-specific antibodies (mainly IgG in the secondary response). Thus, following a first exposure to a specific pathogen, immunity is effectively acquired, such that on a subsequent exposure to the same pathogen – even if this occurs years later – symptoms of illness do not arise, or at least, are less severe.

Formation of effector responses to destroy the invader

It is important to note that B cells also express both MHC class-I and MHC class-II proteins on their cell surface and are able to uptake, process, and present fragments of proteins from pathogens too. As such, antigen-loaded B cells can activate both CD4+ T helper cells and CD8+ cytotoxic T cells (Rastogi et al., 2022) contributing to the cell-mediated immune response, which in combination with an antibody response, is often essential for pathogen clearance. Indeed, many pathogens, including all viruses, can only reproduce within host body cells, which prevents their elimination by antibodies. Consequently, cell-mediated immunity is crucial to ensure the destruction of intracellular pathogens and tumours. Activated CD8+ cytotoxic T cells attack and kill infected cells. This process is either enhanced or suppressed by CD4+ helper T cells, or regulatory T cells (sometimes abbreviated to Treg) respectively, to ensure a swift, yet proportioned immune response occurs against the pathogen. When CD4+ helper T cells bind to specific antigens displayed with MHC class-II proteins on the cell surface of antigen-presenting cells, the cytokine IL-1 is released, which stimulates T cells to clonally divide and proliferate. The activated T cells release another cytokine, IL-2, which further stimulates proliferation and growth of CD4+ helper T cells and CD8+ cytotoxic T cells, which provide immunosurveillance, searching for infected body cells expressing MHC class-I molecules displaying the specific pathogen-derived protein fragment with the sequence of amino acids that they recognise. Upon recognition, CD8+ cytotoxic T cells release lytic granules that contain the cytotoxic effector proteins, perforin, and granzymes. Perforin creates transmembrane pores in the target cells which facilitate the entry of granzymes, a family of serine proteases, that will digest the target cells from within (Hay & Slansky, 2022). Fragments of cell debris and pathogenic material are ingested and digested by phagocytes and other antigen-presenting cells, contributing towards the immune response.

Orchestrate cellular communication and keep the fight targeted

The balance between antibody-mediated and cell-mediated immunity against specific pathogens is achieved by CD4+ helper T cells polarising into so-called Th1 or Th2 cells. Cell-mediated immunity depends on a Th1 profile of cytokines, including IL-12, IFN-γ, and TNF-α. These cytokines activate macrophages and activate CD8+ cytotoxic T cells. Conversely, a Th2 profile of cytokines includes mainly IL-4, IL-5, and IL-13, which are necessary for the promotion of antibody-mediated immunity. Indeed, IL-4 and IL-13 primarily drive B cell differentiation and antibody production, while IL-5 mainly stimulates and primes eosinophils. Interaction of CD28 on CD4+ T helper cells with CD80 on APCs appears to favour Th1 differentiation, whereas interaction with CD28 on the CD4+ helper T cell with CD86 on the APC appears to favour the Th2 phenotype. Certain cytokines secreted by the developed Th1 and Th2 cells act in an autocrine and reciprocally inhibitory fashion: IL-4 promotes Th2 cell expansion and limits the proliferation of Th1 cells, whereas IFN-γ enhances growth of Th1 cells but decreases Th2 cell development. Indeed, the cytokine microenvironment clearly represents a potent determinant of Th1/Th2 polarisation, with IL-4 and IL-12 as the initiating key factors – being derived principally from innate immune responses during T cell priming (Liew, 2002), also known as the "Cytokine Milieu Hypothesis". Considering the self-promoting nature of cytokines produced by either CD4+ helper T cell subset, this leads to a cascade effect that further amplifies either the humoral or cell-mediated immune response. Altogether, exogenous stimuli such as pathogen-derived products, the maturational stage of APCs, as well as genetic factors influence Th1/Th2 differentiation, in addition to complex interactions between antigen dose, T cell receptor engagement, and MHC antigen affinities. High antigen doses appear to favour Th1 development, while low doses favour the Th2 subset (Boonstra et al., 2003). Influential antigenic properties include the nature of the antigen, with bacteria and viruses promoting Th1-cell differentiation and helminths promoting differentiation of the Th2 subset. Th2 differentiation also appears to be promoted by small soluble proteins characteristic of allergens (Liew, 2002). Although it is an oversimplification, the Th1 response can be seen as the major promoter of cell-mediated reactions that provide an effective defence against intracellular pathogens (i.e., viruses and some bacteria that can enter host cells). In contrast, the Th2 response primarily activates humoral immunity and the antibodies produced are usually only effective against pathogens in extracellular fluids. As mentioned previously, Th1- and Th2-cell responses are cross-regulatory and the Th1/Th2 cytokine balance is also influenced by regulatory T cell cells (Piconese & Barnaba, 2015), which may secrete the suppressive cytokines IL-10 and transforming growth factor-ß (TGF-ß). Regulatory T cell cells are important in preventing excessive activity of the immune system and help to bring a stop to immune activation after a pathogen has been eliminated.

Remember the pathogen for future encounters

Following clearance of the pathogen, most of the antigen-specific T and B cells that were generated during the primary response die; however, a small number of B cells, plasma cells, and T cells persist as memory cells. Memory cells are usually defined as quiescent, may be very long-lived and are the basis of immunological memory – a key characteristic of adaptive immunity (Netea, Schlitzer, Placek, Joosten, & Schultze, 2019). Functionally, immunological memory enables a more rapid and effective secondary immune response upon re-exposure to the same pathogen. In contrast to most components of innate immunity, the specificities of adaptive immunity reflect the individual's lifetime exposure to stimuli from pathogens and will consequently differ among people. It is, however, important to note that some cells classically considered to be innate immune cells, such as NK cells can also exhibit memory-like responses.

Conclusion

The human body is frequently exposed to harmful challenges, both from external sources (such as viruses, bacteria, fungi, and parasites: also called pathogens) and internal sources (such as tumours). To achieve immunity throughout the lifespan, the immune system comprises physical barriers to protect against the initial invasion of pathogens, including skin, mucosal membranes, stomach acid, and body temperature. In addition, the immune system has a combination of non-specific and specific defences to recognise, attack, and destroy pathogens that have entered the body, and some of these defences also target tumours. These defences, often dichotomised into innate and adaptive immune system 'arms' or functions, include cellular and soluble components, that work together to detect and eliminate pathogens (and sometimes tumours) providing immunological memory. A wide range of factors affect immune function, including chronological ageing, environmental factors, chronic exposure to psychological stress, sleep disruption, inadequate nutrition, and excessive sedentary behaviours. The remaining chapters of this book summarise how exercise impacts immune function throughout the lifespan and health span. This overview of the immune system and the factors affecting it has been given to facilitate the discussions in the chapters that follow on the measurement of immune system status and the effects of acute and chronic exercise on immune function. In some places, it has been greatly simplified and the complexity of the immune system and its precise coordinated responses should not be underestimated. Readers are directed to other comprehensive sources, such as the following review articles (Kaufmann, 2019; Moser & Leo, 2010) or mainstream immunology textbooks, such as Janeway's *Immunobiology* or Paul's *Fundamental Immunology*).

Key Points

- The immune system recognises and destroys micro-organisms, cells, and cell-parts that are foreign to the body (i.e. non-self) to protect the host. The immune system can broadly be divided into two sub-parts, innate (non-specific, natural) and acquired (adaptive, specific) immunity.
- Innate immunity forms the first line of defence against invading micro-organisms. It consists of three mechanisms that have the common goal of preventing any foreign agent entering the body: (a) physical/structural barriers; (b) chemical barriers; and (c) phagocytic cells (mainly neutrophils and macrophage/monocytes) and other non-specific killer cells (NK cells).
- Neutrophils are the most abundant type of white blood cell or leucocyte. They are the major cell of a subpopulation of leucocytes called granulocytes, so-called because they contain microscopic granules that are released in the killing process. Other types of granulocytes are eosinophils and basophils.
- Other phagocytic cells, monocytes, mature into macrophages in the tissues. Phagocytic cells destroy micro-organisms by engulfing them and releasing toxic substances, including ROS and digestive enzymes, on to the microorganism to kill it and break it up.
- Soluble factors, such as complement, acute-phase proteins, lysozyme, and cytokines, are also important in the innate immune response. Soluble factors help to enhance the innate response, as well as being involved in killing processes directly.
- The adaptive immune response is activated when pathogens escape the innate immune response. Following phagocytosis, macrophages and dendritic cells incorporate parts of foreign proteins (antigen) from the digested microorganism into their own cell surface membrane and present them to T-lymphocytes. Activation of TLRs on the surface of antigen-presenting cells by microbial molecules results in the induction of co-stimulatory molecules and T cell activation.

- There are a number of subpopulations of T cells. The presence of an antigen on a macrophage cell surface stimulates the T cells to divide and proliferate into these subpopulations. CD4+ helper T cells coordinate the cell-mediated adaptive immune response. They activate CD8+ cytotoxic T cells and B cells. CD8+ cytotoxic T cells destroy infected cells and are the main effector cells of cell-mediated immunity.
- B cells proliferate into plasma cells. These secrete vast amounts of antibody (or immunoglobulin) specific to the antigen that triggered the immune response. The B cell response is known as the humoral or fluid adaptive immune response.
- Both B cells and T cells 'exhibit' memory, which means that they can mount a rapid response to that specific antigen upon subsequent exposure. This is the rationale behind immunisation.
- Cell-mediated immunity is promoted by the actions of cytokines secreted by Thl cells, whereas the humoral immune response is activated by cytokines released from Th2 cells.
- Immune function in humans is affected by both genetic and environmental factors. The latter include age, exercise, sex, nutritional status, previous exposure to pathogens, sleep, and psychological stress.

References

Boonstra, A., Asselin-Paturel, C., Gilliet, M., Crain, C., Trinchieri, G., Liu, Y. J., & O'Garra, A. (2003). Flexibility of mouse classical and plasmacytoid-derived dendritic cells in directing T helper type 1 and 2 cell development: Dependency on antigen dose and differential toll-like receptor ligation. *J Exp Med, 197*(1), 101–109.

Cabeza-Cabrerizo, M., Cardoso, A., Minutti, C. M., Pereira da Costa, M., & Reis e Sousa, C. (2021). Dendritic cells revisited. *Annu Rev Immunol, 39*, 131–166.

Cerwenka, A., & Lanier, L. L. (2001). Natural killer cells, viruses and cancer. *Nat Rev Immunol, 1*(1), 41–49.

Commins, S. P. (2015). Mechanisms of oral tolerance. *Pediatr Clin North Am, 62*(6), 1523–1529.

Dancey, J. T., Deubelbeiss, K. A., Harker, L. A., & Finch, C. A. (1976). Neutrophil kinetics in man. *J Clin Invest, 58*(3), 705–715.

Darrigues, J., Almeida, V., Conti, E., & Ribot, J. C. (2022). The multisensory regulation of unconventional T cell homeostasis. *Semin Immunol, 61–64*, 101657.

Estua-Acosta, G. A., Zamora-Ortiz, R., Buentello-Volante, B., Garcia-Mejia, M., & Garfias, Y. (2019). Neutrophil extracellular traps: Current perspectives in the eye. *Cells, 8*(9), 979.

Fleming, A., & Allison, V. D. (1922). Observations on a bacteriolytic substance ("Lysozyme") found in secretions and tissues. *Br J Exp Pathol, 3*(5), 8.

Ganusov, V. V., & De Boer, R. J. (2007). Do most lymphocytes in humans really reside in the gut? *Trends Immunol, 28*(12), 514–518.

Garcia-Bragado, F., Vilardell, M., Caralps, A., Bosch, J. A., Gordo, P., Magrina, N., … Tornos, J. (1979). [Essential mixed IgG-IgM cryoimmunoglobulinemia. Clinical and anatomopathological remission with cyclophosphamide and prednisone treatment]. *Rev Clin Esp, 155*(2), 149–151.

Gherardin, N. A., Souter, M. N., Koay, H. F., Mangas, K. M., Seemann, T., Stinear, T. P., … Godfrey, D. I. (2018). Human blood MAIT cell subsets defined using MR1 tetramers. *Immunol Cell Biol, 96*(5), 507–525.

Gunther, P., & Schultze, J. L. (2019). Mind the map: Technology shapes the myeloid cell space. *Front Immunol, 10*, 2287.

Hay, Z. L. Z., & Slansky, J. E. (2022). Granzymes: The molecular executors of immune-mediated cytotoxicity. *Int J Mol Sci, 23*(3).

Hidalgo, A., Chilvers, E. R., Summers, C., & Koenderman, L. (2019). The neutrophil life cycle. *Trends Immunol, 40*(7), 584–597.

Kanneganti, T. D., Lamkanfi, M., & Nunez, G. (2007). Intracellular NOD-like receptors in host defense and disease. *Immunity, 27*(4), 549–559.

Karasuyama, H., Miyake, K., Yoshikawa, S., & Yamanishi, Y. (2018). Multifaceted roles of basophils in health and disease. *J Allergy Clin Immunol, 142*(2), 370–380.

Kaufmann, S. H. E. (2019). Immunology's coming of age. *Front Immunol, 10*, 684.

Kim, Y. K., Shin, J. S., & Nahm, M. H. (2016). NOD-like receptors in infection, immunity, and diseases. *Yonsei Med J, 57*(1), 5–14.

Lemaitre, B., Nicolas, E., Michaut, L., Reichhart, J. M., & Hoffmann, J. A. (1996). The dorsoventral regulatory gene cassette spatzle/Toll/cactus controls the potent antifungal response in Drosophila adults. *Cell, 86*(6), 973–983.

Liew, F. Y. (2002). T(H)1 and T(H)2 cells: A historical perspective. *Nat Rev Immunol, 2*(1), 55–60.

Loh, L., Gherardin, N. A., Sant, S., Grzelak, L., Crawford, J. C., Bird, N. L., … Kedzierska, K. (2020). Human mucosal-associated invariant T Cells in older individuals display expanded TCRalphabeta Clonotypes with potent antimicrobial responses. *J Immunol, 204*(5), 1119–1133.

Mayassi, T., Barreiro, L. B., Rossjohn, J., & Jabri, B. (2021). A multilayered immune system through the lens of unconventional T cells. *Nature, 595*(7868), 501–510.

Medzhitov, R. (2001). Toll-like receptors and innate immunity. *Nat Rev Immunol, 1*(2), 135–145.

Miyake, K., Ito, J., & Karasuyama, H. (2022). Role of basophils in a broad spectrum of disorders. *Front Immunol, 13*, 902494.

Mogensen, T. H. (2009). Pathogen recognition and inflammatory signaling in innate immune defenses. *Clin Microbiol Rev, 22*(2), 240–273, Table of Contents.

Moll, H. (2003). Dendritic cells and host resistance to infection. *Cell Microbiol, 5*(8), 493–500.

Moretta, L., Bottino, C., Pende, D., Vitale, M., Mingari, M. C., & Moretta, A. (2004). Different checkpoints in human NK-cell activation. *Trends Immunol, 25*(12), 670–676.

Mortaz, E., Alipoor, S. D., Adcock, I. M., Mumby, S., & Koenderman, L. (2018). Update on neutrophil function in severe inflammation. *Front Immunol, 9*, 2171.

Morvan, M. G., & Lanier, L. L. (2016). NK cells and cancer: You can teach innate cells new tricks. *Nat Rev Cancer, 16*(1), 7–19.

Moser, M., & Leo, O. (2010). Key concepts in immunology. *Vaccine, 28*(Suppl 3), C2–13.

Netea, M. G., Schlitzer, A., Placek, K., Joosten, L. A. B., & Schultze, J. L. (2019). Innate and adaptive immune memory: An evolutionary continuum in the host's response to pathogens. *Cell Host Microbe, 25*(1), 13–26.

O'Leary, J. G., Goodarzi, M., Drayton, D. L., & von Andrian, U. H. (2006). T cell- and B cell-independent adaptive immunity mediated by natural killer cells. *Nat Immunol, 7*(5), 507–516.

Oren, Z., & Shai, Y. (1998). Mode of action of linear amphipathic alpha-helical antimicrobial peptides. *Biopolymers, 47*(6), 451–463.

Patel, A. A., Zhang, Y., Fullerton, J. N., Boelen, L., Rongvaux, A., Maini, A. A., … Yona, S. (2017). The fate and lifespan of human monocyte subsets in steady state and systemic inflammation. *J Exp Med, 214*(7), 1913–1923.

Pepys, M. B., & Hirschfield, G. M. (2003). C-reactive protein: A critical update. *J Clin Invest, 111*(12), 1805–1812.

Piconese, S., & Barnaba, V. (2015). Stability of regulatory T Cells undermined or endorsed by different Type-1 cytokines. *Adv Exp Med Biol, 850*, 17–30.

Pillay, J., den Braber, I., Vrisekoop, N., Kwast, L. M., de Boer, R. J., Borghans, J. A., … Koenderman, L. (2010). In vivo labeling with 2H2O reveals a human neutrophil lifespan of 5.4 days. *Blood, 116*(4), 625–627.

Ramirez, G. A., Yacoub, M. R., Ripa, M., Mannina, D., Cariddi, A., Saporiti, N., … Dagna, L. (2018). Eosinophils from physiology to disease: A comprehensive review. *Biomed Res Int, 2018*, 9095275.

Rastogi, I., Jeon, D., Moseman, J. E., Muralidhar, A., Potluri, H. K., & McNeel, D. G. (2022). Role of B cells as antigen presenting cells. *Front Immunol, 13*, 954936.

Rosales, C. (2018). Neutrophil: A cell with many roles in inflammation or several cell types? *Front Physiol, 9*, 113.

Rothschild-Rodriguez, D., Causer, A. J., Brown, F. F., Collier-Bain, H. D., Moore, S., Murray, J., … Campbell, J. P. (2022). The effects of exercise on complement system proteins in humans: A systematic scoping review. *Exerc Immunol Rev, 28*, 1–35.

Sender, R., Weiss, Y., Navon, Y., Milo, I., Azulay, N., Keren, L., … Milo, R. (2023). The total mass, number, and distribution of immune cells in the human body. *Proc Natl Acad Sci U S A, 120*(44), e2308511120.

Singh, P., Szaraz-Szeles, M., Mezei, Z., Barath, S., & Hevessy, Z. (2022). Age-dependent frequency of unconventional T cells in a healthy adult Caucasian population: A combinational study of invariant natural killer T cells, gammadelta T cells, and mucosa-associated invariant T cells. *Geroscience, 44*(4), 2047–2060.

Spits, H., Artis, D., Colonna, M., Diefenbach, A., Di Santo, J. P., Eberl, G., … Mebius, R. E. (2013). Innate lymphoid cells—a proposal for uniform nomenclature. *Nat Rev Immunol, 13*(2), 145–149.

Svensson, L., & Wenneras, C. (2005). Human eosinophils selectively recognize and become activated by bacteria belonging to different taxonomic groups. *Microbes Infect, 7*(4), 720–728.

Swirski, F. K., Nahrendorf, M., Etzrodt, M., Wildgruber, M., Cortez-Retamozo, V., Panizzi, P., … Pittet, M. J. (2009). Identification of splenic reservoir monocytes and their deployment to inflammatory sites. *Science, 325*(5940), 612–616.

Thomas, D. C. (2017). The phagocyte respiratory burst: Historical perspectives and recent advances. *Immunol Lett, 192*, 88–96.

Ting, J. P., Lovering, R. C., Alnemri, E. S., Bertin, J., Boss, J. M., Davis, B. K., … Ward, P. A. (2008). The NLR gene family: A standard nomenclature. *Immunity, 28*(3), 285–287.

Wu, D., Molofsky, A. B., Liang, H. E., Ricardo-Gonzalez, R. R., Jouihan, H. A., Bando, J. K., … Locksley, R. M. (2011). Eosinophils sustain adipose alternatively activated macrophages associated with glucose homeostasis. *Science, 332*(6026), 243–247.

Yoneyama, M., Kikuchi, M., Natsukawa, T., Shinobu, N., Imaizumi, T., Miyagishi, M., … Fujita, T. (2004). The RNA helicase RIG-I has an essential function in double-stranded RNA-induced innate antiviral responses. *Nat Immunol, 5*(7), 730–737.

Yousefi, S., Gold, J. A., Andina, N., Lee, J. J., Kelly, A. M., Kozlowski, E., … Simon, H. U. (2008). Catapult-like release of mitochondrial DNA by eosinophils contributes to antibacterial defense. *Nat Med, 14*(9), 949–953.

Yousefi, S., Morshed, M., Amini, P., Stojkov, D., Simon, D., von Gunten, S., … Simon, H. U. (2015). Basophils exhibit antibacterial activity through extracellular trap formation. *Allergy, 70*(9), 1184–1188.

3 Methods in exercise immunology

Tim Schauer and Jesper Frank Christensen

The stress-induced immune response

Perceived stress can result in a systemic immune response. Stressors inducing this response are varied but can be broadly divided into physical and psychological stressors. As stress is often self-perceived (Segerstrom & Miller, 2004), clear and validated methodology is needed to interpret studies evaluating the immunological response to both physical and psychological stressors. Methodology is primarily chosen based on the underlying research question.

Physical stress

Physical stress comes in many forms such as injury, environmental conditions, or illness but can also be induced by exercise. To quantify the amount of physical stress during exercise, standardised tests, such as a maximum oxygen uptake test, or $\dot{V}O_2$ max test, can be used to determine cardiorespiratory fitness and based on the variables of maximum oxygen uptake, maximal heart rate (HR_{max}) or maximal workload (e.g., watts, or speed), the intensity of exercise can be prescribed. For example, exercise is sometimes prescribed using specific heart rate (HR) zones (e.g., 70–80% of HR_{max},) which can be checked using telemetry (i.e., a chest strap and watch). Exercise can also be prescribed on the basis of percentages of $\dot{V}O_2$ max and checked using indirect calorimetry, which is usually restricted to laboratory-based studies. Using methods such as these, physical stress can be standardised across individuals which subsequently allows for the comparison of the immune response.

$\dot{V}O_2$ max testing

Determining cardiorespiratory fitness using a $\dot{V}O_2$ max test for example is crucial in characterising participants and prescribing standardised exercise which is in turn paramount for the comparison of the exercise-induced immune response between individuals. However, the test may not be feasible, especially in home-based exercise studies. Therefore, variables such as the HR_{max} obtained during an initial $\dot{V}O_2$ max test can be used to design exercise interventions that are delivered outside of laboratory settings. However, the proper execution of a $\dot{V}O_2$ max test is essential and will be explained in the following sections. Maximal oxygen consumption ($\dot{V}O_2$ max) is the maximal rate of oxygen consumed by the body (e.g., during an incremental or graded exercise test). $\dot{V}O_2$ max represents the capacity of the cardiovascular and cardiopulmonary systems to deliver oxygen to the muscle and the consumption of oxygen within the muscle. $\dot{V}O_2$ max can be measured directly and indirectly with direct measurement being the gold standard (American College of Sports Medicine, 2016). Terminology used in $\dot{V}O_2$ max testing is summarised in Table 3.1.

DOI: 10.4324/9781003256991-3

Table 3.1 Terminology during $\dot{V}O_2$ max testing

Term	Description
$\dot{V}O_2$	Volume of oxygen per unit of time
$\dot{V}O_2$ max	Maximal oxygen consumption in the clinical setting is given as absolute units [mL min^{-1}] or relative to body mass [mL min^{-1} kg^{-1}]; other descriptions include allometric scaling of body mass to allow for the size-independent comparison more accurately
$\dot{V}O_2$ peak	Highest attained value during $\dot{V}O_2$ max testing. Can be equal to $\dot{V}O_2$ max and may be used instead of $\dot{V}O_2$ max if no true max is to be expected (i.e., due to advanced age or illnesses)

Table 3.2 Overview of testing modalities to determine $\dot{V}O_2$ max

Modality[a]	Setting	Considerations
Motorised treadmill	Maximal and submaximal protocols; often used for laboratory assessment	*Pros:* variety of protocols available, modifiable for population, monitoring of exercise and health-related outcomes possible *Cons:* expensive equipment, biological samples during exercise difficult to obtain, handrail support should be avoided
Cycle ergometers	Maximal and submaximal protocols; often used for laboratory assessment	*Pros:* variety of protocols available, modifiable for population, biological sampling during exercise easy to obtain, monitoring of exercise and health-related outcomes possible, individuals with limited mobility can be included *Cons:* underestimation of $\dot{V}O_2$ max possible, unfamiliar for some individuals resulting in local muscle fatigue
Field and step tests	Submaximal testing of a large number of individuals	*Pros:* low equipment costs, less labour intensive *Cons:* mostly submaximal, requires estimation, test termination and health-related criteria are difficult to observe, biological sampling often not accessible

Legend: Testing considerations, protocols, and their interpretation as well as details on testing equipment can be found in the ACSMs Guidelines for Exercise Testing and Prescription (American College of Sports Medicine, 2016) and further reviews (Balady et al., 2010; Beltz et al., 2016). [a]Other modalities not presented include swimming/rowing with indirect calorimetry

Direct measurement of $\dot{V}O_2$ max

Direct measurement of $\dot{V}O_2$ max can be achieved by measuring oxygen and carbon dioxide in inspired and expired air. The Douglas Bag method in which expired air is collected in non-permeable bags (Shephard, 2017) is considered a gold standard. However, $\dot{V}O_2$ max is often measured using automated systems with real-time breath-by-breath monitoring using flow devices or mixing chambers. Direct measurement of $\dot{V}O_2$ max includes a systematic and linear (incremental) increase in exercise workload over time until volitional exhaustion. Traditionally, exercises to assess $\dot{V}O_2$ max are limited to cycling or treadmill running but may include other modalities (Table 3.2). In general, $\dot{V}O_2$ max testing reliability depends largely on participant characteristics, such as age, fitness, or disease status, as well as protocol specifics (i.e., overall test duration, increment duration, and increment range). In recent years, individualised protocols have become increasingly popular, especially in the clinical environment where individuals across all ages and fitness levels participate in the same research study. Importantly, the choice of $\dot{V}O_2$ max testing is primarily based on the population. In addition, the same exercise modality should be used during $\dot{V}O_2$ max assessment and the exercise intervention, due to the energy cost and demands of different exercise modes. The occurrence of a $\dot{V}O_2$ plateau validates that

Table 3.3 Criteria commonly used to determine successful $\dot{V}O_2$ max testing

Criteria	Description	Limitation
Primary criteria[a]		
$\dot{V}O_2$ plateau	$\dot{V}O_2$ does not increase proportionally ($\dot{V}O_2$ levelling of) to workload. Can be assessed using modelled vs. actual $\dot{V}O_2$ max (Midgley et al., 2009)	Does not occur in all individuals if incremental testing is used (Poole & Jones, 2017); dependent on age and testing modality
Secondary criteria		
Respiratory Exchange Ratio (RER)	Ratio between produced CO_2 and consumed O_2. Indicates proportion of aerobic and anaerobic energy generation. An RER of >1.10 is generally considered as acceptable (Balady et al., 2010)	Maximal RER can vary from 1.00 to 1.44, depending on factors such as age, workload increment of the protocol, type of population (Poole & Jones, 2017)
Maximal heart rate (HR_{max})	Commonly (*but not recommended*) used cut-off: ≤10 beats min^{-1} or ≤5% of the age-predicted (220-age) maximum (Poole & Jones, 2017)	HR_{max} in the general population can vary with ±12 beats min^{-1}. Dependent on medication, age, training status, geographical origin, workload increment of the protocol (Balady et al., 2010; Poole & Jones, 2017)
Blood lactate	Maximal blood lactate concentration. Lactate accumulates during exercise with oxygen limitation due to energy production from glycolysis. Commonly (*but not recommended*) used cut-off: ≥8mM (Poole & Jones, 2017)	Maximal concentration may vary from 4 to 17 mM (Poole & Jones, 2017) due to biological variation in the population
Other criteria		
Borg Scale	Rating of perceived exertion (RPE) is measured on the Borg Scale (range from 6 to 20). Commonly used criteria: RPE >17	Subjective rating depends largely on the participants understanding of the scale (Beltz et al., 2016). Reliability dependent on fitness status and exercise modality
Assessor judgement	Qualified assessor judged the test as maximal (e.g., by marked hyperventilation)	Subjective rating

Legend: Overview of criteria commonly used for $\dot{V}O_2$ max testing. Commonly used criteria may vary greatly between studies and based on the studied population. Abbreviations: RER (respirator exchange ratio), HR (heart rate), max (maximal). [a] The only unambiguous determinant for $\dot{V}O_2$ max

the $\dot{V}O_2$ max test was maximal (Table 3.3). However, if the $\dot{V}O_2$ plateau is not observed, several secondary criteria are used across the literature, but there is no consensus on which criteria, or combination, is deemed sufficient (Table 3.3). Alternatively, verification protocols may be used with supramaximal workload (i.e., 110% of the workload obtained during the $\dot{V}O_2$ max test) to verify the occurrence of the unambiguous $\dot{V}O_2$ plateau criteria (Poole & Jones, 2017).

Indirect measurement of $\dot{V}O_2$ max

In the absence of specialist equipment (e.g., for indirect calorimetry) or when working with certain populations such as patients with contraindications to exhaustive exercise, indirect (submaximal) measurements of $\dot{V}O_2$ max might be more appropriate. Indirect measurements are based on estimations which often rely on assumptions. Therefore, a variety of different protocols are available which greatly depend on the population and the underlying research question. As an alternative to direct $\dot{V}O_2$ max testing, the test can be undertaken without expired air samples. For example, the watt-max test on cycle ergometers can be used to assess parameters such as HR_{max} as well as maximal sustained watts needed for the design of longitudinal exercise interventions without the need for complex equipment.

Lactate threshold test

The lactate threshold test is another method that can be used to prescribe exercise or characterise populations. However, during the past 50 years, a variety of methods have been used to calculate the lactate threshold, leading to varied nomenclature and controversy (Poole et al., 2021). In principle, (at least) two different thresholds exist. The first lactate threshold is described by the turning point between moderate and heavy exercise (~50% of $\dot{V}O_2$ max), where blood lactate concentration starts to rise above resting levels. The second threshold describes a turning point between heavy and severe exercise (~70% of $\dot{V}O_2$ max), where the rate of appearance of blood lactate continuously exceeds the rate of disappearance (Poole et al., 2021). The lactate threshold is often used in combination with the ventilatory threshold (Ventilation [L min^{-1}] vs. work rate) and the gas exchange threshold (GET; $\dot{V}CO_2$ [L min^{-1}] vs. $\dot{V}O_2$ [L min^{-1}]). Importantly, definitions and terminology may depend largely on the training status of individuals and on the methodological approach used. The lactate threshold can be determined directly by blood lactate or indirectly by the ventilatory and gas exchange response during incremental exercise. Considerations around the lactate threshold nomenclature and methodology have been discussed elsewhere (Balady et al., 2010; Poole et al., 2021).

Resistance exercise

To standardise the workload during resistance exercise, exercise protocols are typically based on the one-repetition maximum (1RM) (i.e., the maximum weight an individual can lift a single time). However, especially in untrained and patient populations, the assessment of 1RM may be of limited use. Therefore, other assessments such as the 5RM can be used to predict the 1RM (Reynolds et al., 2006). As with cardiorespiratory fitness, the choice of the assessment method depends on the population, intervention type, and equipment. Indeed, exercise intensity during resistance exercise may be set using a combination of repetitions and lifting a weight at a percentage of 1RM (e.g., eight repetitions at 50% of 1RM). Protocol specifics such as increments and resting times between sets depend on fitness status and data from comparable studies (Reynolds et al., 2006).

Psychological stress

The relationship between psychological stress and aspects of immunity has influenced the design of exercise studies. Indeed, short-term psychological stress invokes a "fight or flight" response and causes similar changes to physiology as with exercise (e.g., changes to HR). As both psychological and physiological stressors influence the immune system, it is crucial to separate the potential stimuli from another (e.g., blood sampling appointments should include enough time for participants to feel relaxed to avoid measuring adrenergic activation or cardiovascular responses due to unfamiliar and stressful settings). In contrast to physical stress, definitions of psychological stress are varied due to subjective interpretation of stress (Shields & Slavich, 2017). Long-term psychological stress is often quantified via questionnaires such as the Stress and Adversity Inventory (STRAIN) (Shields & Slavich, 2017). Methods used to induce short-term psychological stress can be quantitively assessed by stress-related markers such as HR and cortisol.

The trier social stress test (TSST)

The TSST and its later adaptations describe a group of reliable acute psychological stressors. In the basic form, participants make an interview-style presentation followed by a surprise mental mathematics task in front of an interview panel of people (i.e., subtract the number 13 from 1022 as quick as possible). TSST variations include group-based assessments (TSST-G), virtual reality assessments (TSST-VR) and tests for children (TSST-C). The experimental step-by-step setup including strengths and limitations are discussed elsewhere, but include practicalities, the number of people required to run the experiment due to the audience, the inability to perform repeated exposures (i.e., due to familiarisation) and the methodological variation between studies (Allen et al., 2017; Labuschagne et al., 2019). These tests have been used to examine changes in aspects of soluble and cellular immunity with saliva sampling and blood sampling before, during and in the hours after the stressor.

Public speaking and cold-pressure test

Alternative psychological stress tests have been established, such as public speaking or cold-pressor tests, which solve some of the complications with the TSST. For example, the Leiden Public Speaking Task (Leiden PST) involves delivering a prepared speech in front of a pre-recorded audience as stressor (Westenberg et al., 2009). Cold-pressor tests (CPT) involve submerging the hands of a participant in ice-cold water. In addition, the Maastricht Acute Stress Test (MAST) combines a CPT in predefined time intervals with a mental mathematics task (Smeets et al., 2012) and the socially evaluated cold-pressor test combines a CPT for a maximum of 3 minutes with social evaluation (Schwabe et al., 2008). These tests have been used to examine changes in aspects of soluble and cellular immunity with saliva sampling and blood sampling before, during and in the hours after the stressor.

Physical and psychological stress in animals

Animal models with fish, rodents or primates enable very controlled conditions and procedures to be undertaken – that would not be possible, or at least very complicated, in human studies – to provide detailed understanding of mechanisms. Examples include experimental infection with viruses, implantation of tumour cells and *ex-vivo* labelling of cells that are re-infused as part of cell tracking experiments. While many biochemical and immunological techniques are similar between human and animal studies, in contrast, due to a different biology and anatomy, animal studies can have limited transferability to humans and need to be interpreted carefully. In general, most animal studies within the field of exercise immunology are undertaken with rodents, usually mice and rats. Physical stress models for rodents include low-temperature exposure, electric foot shock (Campos et al., 2013) and exercise (Table 3.4). As with humans, exercise intensity can be determined using $\dot{V}O_2$ max (Kemi et al., 2002; Poole et al., 2020) and a variety of intensities and durations of exercise have been examined. Ethical and methodological considerations over the choice of exercise model in rodents are briefly summarised in Table 3.4 and discussed elsewhere (Fuller & Thyfault, 2021; Omary et al., 2016; Poole et al., 2020). Models of psychological stress in animals sometimes involve both psychological and physical components and may be divided into conditioned and unconditioned designs. Conditioned models include reinforcement, both positive and negative, such as fear conditioning combined with electric shocks (Campos et al., 2013). Unconditioned models include the elevated plus maze, light-dark boxes, and restrained or suppressed feeding models.

Table 3.4 Overview of exercise models to induce physical stress in rodents

Model	Description[a]	Considerations
Wheel running	Running wheel in cage with or without resistance (Pedersen et al., 2016)	*Pros:* Voluntarily and following the circadian rhythm. Not labour intensive as wheels collect data automatically. Takes place in known environment, thereby not inducing psychological stress. Measurable exercise-related improvements *Cons:* Spontaneous usage which cannot be controlled for duration and intensity. Overall usage may vary greatly between animals and from day to day. Running distance is given per cage even if rodents are housed together
Treadmill	External motorised device with multiple lanes (Kurz et al., 2022)	*Pros:* Permits control of exercise intensity, duration, and timing; can be used to determine $\dot{V}O_2$ max, results in similar adaptations as with humans *Cons:* Describes a forced exercise using electric shock or high-pressure air in foreign and stressful environment resulting in psychological stress. Stop-and-go behaviour of rodents during low-to-moderate running speeds. Labour intensive
Swimming	Container filled with water. (Terra et al., 2013)	*Pros:* May be used as continuous and uniform exercise. Low expense for equipment and permits control of duration and timing. Intensity can be set by weights attached to the tail. *Cons:* Animals may resort to diving and bobbing behaviours and swimming induces psychological stress. Labour intensive as investigator needs to be present and control the temperature. Risk of drowning
Resistance exercise[b]	Weighted harnesses, weighted levers to pull for food[c] (Krüger et al., 2013)	*Pros:* Allows for the setting of intensity and duration and results in similar adaptations as seen in humans *Cons:* Describes a forced exercise using negative and positive stimulants such as electric shock or food rewards (after deprivation), resulting in both physical and psychological stress. Labour intensive with often long familiarisation phases, studies are often long-term and often only one rodent at a time can be trained. Foreign and stressful environment induces psychological stress

Legend: Table contents are inspired by Lowe & Alway (2002); Poole et al. (2020). [a]Includes an example publication in which the model was used/described. [b]Can also include electrical stimulation in unconscious animals, compensatory overload by rendering other muscle inactive or chronic stretching models (Lowe & Alway, 2002). [c]Many more exercises exist which include weighted ladder climbing, placing food on top of tall cages and grab/hold exercises.

Sampling of biological material

Studies in exercise immunology rely largely on blood sampling and collecting tissue, including fat and muscle (Figure 3.1). Most considerations regarding biological sampling depend on the research question and methods being used to examine aspects of immunity. A critical consideration is the need for a so-called baseline or reference sample (e.g., a pre-exercise or pre-intervention sample). Sometimes a reference sample may be a different group of people (e.g., a case-control study, where "cases" are people with a disease, and "controls" are healthy people). Another important consideration is that the post-intervention sample obtained at the end of a long-term study (e.g., an exercise training intervention) has to be taken close enough to the intervention (i.e., within days) to reveal the effects of the intervention but not too close to avoid measuring the effects of the last exercise bout. In summary, the following points should also be standardised or at least addressed by the research question and measurements: (i) circadian rhythm, (ii) dietary intake prior to the study and whether samples should be collected fasted, (iii) no alcohol and nicotine use 24–48 hours before, (iv) no physical activity 24–72 hours before, (v) mode of transportation to the laboratory, (vi) resting phase before baseline sampling, (vii) familiarisation to avoid a stress response to new environment, (viii) homogeneity of participant

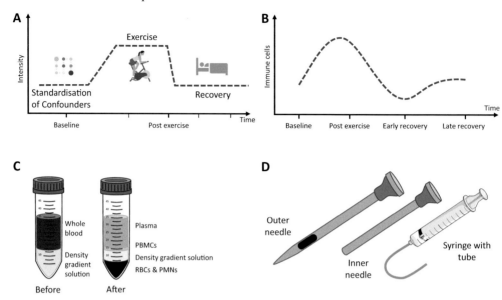

Figure 3.1 Sampling of biological material.

Legend: (A) Overview of an experimental setup for studies examining individual bouts of exercise and measuring immunological parameters in blood with common periods for sampling. (B) Example of immune cell concentration in blood during exercise and suggested sampling timepoints for baseline, post exercise and early as well as late into recovery. (C) PBMC isolation using density gradient centrifugation. Blood is layered on top of gradient media and centrifuged. The PBMC cell layer can be found on top of the gradient media and below the plasma. RBC and PMNs (i.e., neutrophils, eosinophils and basophils), collect below the density layer. In general, density gradient media are available for most cells of interest (e.g., to isolate neutrophils a different separation media can be used compared to lymphocytes and monocytes). (D) Bergström needle set. The technique uses a hollow outer needle and a sharpened and movable hollow inner needle. First, the outer needle is inserted, and suction is applied using the syringe to collect tissue inside of the outer needle. Secondly, the inner hollow needle is used to cut the tissue. Missing on the depiction is a metal rod to push the collected tissue out of the inner needle. This figure was partly generated using Servier Medical Art, provided by Servier, licensed under a Creative Commons Attribution 3.0 unported license. *Abbreviations: Peripheral blood mononuclear cell (PBMC), polymorphonuclear leukocytes (PMNs), Red blood cells (RBC)*

Box 3.1   Practical considerations over blood sampling

Multiple needle insertions may cause a local stress response whereas peripheral venous catheters have higher sheer stress and might increase the risk of haemolysis. A peripheral venous catheter should be placed 15 minutes before the first sample is taken to minimise an acute stress response to the placement itself. Importantly, a peripheral venous catheter should be flushed with saline after samples have been taken to avoid blood coagulation. Approximately 1–3 mL is usually discarded as it may contain trapped blood residues and/or saline and/or air. In general, blood-drawing procedures can alter the concentration of the analyte of interest (Haack et al., 2002).

---

characteristics, (ix) medication (especially anti-inflammatory drugs), and (x) seasonality. For all samples taken during the experimental day, standardised processing is crucial. Especially samples taken during and immediately after exercise are often time-sensitive (i.e., samples should be taken within 1 minute or as quickly as possible after cessation of exercise as many analytes such as immune cells or cytokines quickly egress from the circulation).

Blood

Typically, blood is collected by venepuncture of antecubital veins located close to the forearm or alternatively, resulting in more discomfort, a vein on the back of the hand. Blood from arteries is generally not needed. Other less commonly used sites include fingertips or earlobes. In general, blood should be handled as a potential source of infectious agents. Guidelines for blood sampling (World Health Organization, 2010) vary between countries, but in most cases, a peripheral venous catheter or a cannula is placed at the beginning of an experimental day to enable quick and reliable access and avoid multiple skin punctures.

Immune cell isolation

Immune cells need to be separated from whole blood (i.e., separated from plasma and erythrocytes), and individual immune cell subtypes can also be isolated. In either case, blood samples are collected in tubes containing anti-coagulants (e.g., sodium heparin or ethylenediaminetetraacetic acid; EDTA) to avoid clotting. The most common techniques to separate immune cells are density-gradient centrifugation, immunomagnetic isolation, and flow cytometry. Centrifugation describes a relatively cheap and quick way to isolate immune cells based on physical characteristics such as shape, size, or density as larger (or denser) particles sediment faster. During density-gradient centrifugation (also sometimes called isopycnic centrifugation), cells are centrifuged in the presence of a solution with a dis- or continuous density gradient, resulting in the separation of cells based on their density. The isolation of peripheral blood mononuclear cells (PBMCs) uses a discontinuous gradient (e.g., Ficoll-Paque) to accumulate cells on top of the density layer (Figure 3.1C). In addition to centrifugation, cells can be separated based on cell-surface characteristics. In brief, cells with similar surface characteristics are targeted (e.g., with antibodies linked to magnetic particles placed in a magnetic field). The magnetic field fixates targeted cells to separate them from other cells (Plouffe et al., 2015). Magnetic isolation provides researchers with the opportunity to isolate immune cells in a quick, easy, and more specific way compared with density-gradient centrifugation as specific subsets of (e.g., T cells) can be isolated. Typically, both methods are used together where PBMCs are isolated via density centrifugation followed by magnetic T cell isolation. Commercially, kits are available to isolate all major immune cell subsets. Other more specialised and advanced isolation techniques include: (i) cell-culture separation based on the adhesion potential of cells (commonly used for macrophages), (ii) chromatography for binding capacity and surface interaction with a matrix, (iii) microbubble techniques, (iv) flow cytometry cell sorting, and (v) microfluidic approaches to miniaturise and automate cell and particle isolation.

Immune cell storage

Isolated immune cells can be used immediately for the assessment of relevant outcomes or stored for later use. For storage, isolated immune cells are first washed to remove residuals from

---

Box 3.2   Practical considerations over cell isolation technique

In general, researchers have to decide whether a positive selection (target cells are captured) or a negative selection (everything except the target cells are captured) is needed. Positive selection is generally more sensitive (i.e., less contamination of target cells), but has the potential to activate target cells via (e.g., receptor binding) thereby potentially interfering with downstream processes.

the isolation process. Next, cells are counted and resuspended at a final concentration of 0.5 to 10×10^6 cells mL^{-1} in a so-called "freezing media" (e.g., 10% dimethylsulphoxide and 90% foetal bovine serum). Finally, 1 mL of cells in "freezing media" are placed in cryotubes and stored at −80°C in storage containers allowing for a controlled temperature change (e.g., −1°C per minute). For long-term storage (i.e., more than five days), cryotubes are stored in the vapour phase of liquid nitrogen (i.e., −196°C).

Plasma and serum isolation

The liquid fraction of whole blood can be stored as plasma (i.e., from blood that has not clotted) or serum (i.e., from blood that has clotted). For plasma isolation, blood is collected in anticoagulant-treated tubes (e.g., EDTA) and centrifuged as soon as possible at 1000 to 2000 × g for 10 minutes at 4°C. After centrifugation, the clear top phase (plasma) is collected into multiple aliquots of typically 0.5 mL which are immediately frozen. Long-term storage should be at −80°C. Serum is plasma without clotting factors and is typically used in cell-culture experiments. For serum isolation, blood is collected in tubes coated with clot-activators and left at room temperature for 15–30 minutes until fully clotted followed by centrifugation and storage as described for plasma.

Tissue

In addition to blood sampling, exercise studies often require tissue biopsies, predominantly from muscle and fat. The storage of biopsies is a crucial part in tissue collection processes, and biopsies are either snap frozen or transferred to preservatives (e.g., formalin) to ensure tissue stability.

Adipose tissue

Adipose tissue can be obtained from visceral adipose depots (intra-abdominal fat) in connection with scheduled surgery or from subcutaneous fat depots via needle biopsy from the abdominal

---

Box 3.3   Practical considerations over fresh or frozen cells

Some experiments require freshly isolated immune cells and freezing can change the phenotype and response of immune cells (e.g., typically, responses in functional assays, such as cytokine production, are smaller in magnitude than with fresh cells). Immune cell storage procedures should be tested and individualised depending on the immune cell type and experiments being undertaken.

Box 3.4   Practical considerations over the timing of sample processing

Some exercise-related molecules have very short half-lives. Hence, centrifugation and freezing soon after sample collection is vital for the measurements of some variables (e.g., catecholamines, cytokines, or myokines). Importantly, certain measurements require treatment with stabiliser molecules and protease inhibitors prior to freezing. Haemolysis of samples might interfere with downstream processes and should be noted. For both plasma and serum, repeated freeze–thaw cycles should be avoided.

region under local anaesthesia. Different techniques can be used to obtain biopsies, including punch biopsies, fine needle aspiration or Bergström needle suction (Figure 3.1D) (Alderete et al., 2015). For both the fine needle aspiration and the Bergström technique, manual suction is applied to collect adipose tissue via needles.

Muscle tissue

Muscle biopsies are usually taken from the quadriceps muscle (vastus lateralis) via a 5 mm Bergström needle (Dubowitz et al., 2021) or punch biopsy (gun biopsy) under local anaesthesia. Importantly, the muscle structure should be kept intact when storing biopsies, and biopsies are oriented along the muscle fibres, frozen in isopentane or propane, and cooled by liquid nitrogen or dry ice and cut into thin slices on a cryostat. Slides can be used for histological, histochemical, and immunohistochemical studies (Dubowitz et al., 2021).

Skin tissue

Skin tissue is usually obtained via punch biopsies under local anaesthesia with minimal invasiveness. Punch biopsies can be taken via a pen-like device with a circular blade resulting in 4–6 mm samples, including full-thickness skin with subcutaneous fat (Zuber, 2002). Skin biopsies are often used for blister models or wound-healing studies.

Non-invasive sampling

Many studies in exercise immunology use non-invasive sampling, including saliva, urine, tears, hair, or the gut microbiota, to investigate analytes of interest (Table 3.5).

Blood and tissue collection in animals

Blood and tissue from animals (i.e., mice and rats) is often collected *post-mortem*, but blood and selected other tissues can also be collected from live animals as with human studies (i.e., via blood sampling, biopsies, or surgery). As with human studies, all steps of the tissue collection, from euthanising the animal to tissue isolation and storage depend on the research question and the experimental setup. Blood and subsequent plasma/serum from live animals can be collected during anaesthesia (e.g., from the tail vein or the orbital sinus in the eye) or without

---

Box 3.5   Practical considerations over tissue sampling

To obtain biopsies, the surrounding skin should be cleaned and hair should be cut rather than shaved to lower the risk of infections. Anaesthesia-containing adrenaline should be avoided as it might influence immunological outcomes. The invasive nature of obtaining tissue samples often requires additional consent from participants including information on potential adverse events. Gun or punch biopsies typically result in smaller tissue samples (<100mg) compared to Bergström needles (>400mg) but are less labour intensive. In studies with patients who are undergoing planned surgery, tissue can be collected while under general anaesthesia using more invasive techniques (e.g., sampling visceral adipose tissue) and larger volumes can usually be collected.

Table 3.5 Overview of non-invasive compartments for biological sampling

Type	Description and practical considerations
Saliva	Saliva is collected using swabs or passive drool from under the tongue. Passive drool is considered the gold standard in which saliva is collected directly into collection tubes. Saliva should be collected free from blood and the swab/device should be recommended for the analyte of interest. If other methods than passive drool are used (e.g., stimulated saliva collection), the flow rate of the saliva should be noted as the viscosity may be critical for downstream processes and the comparison between studies (Granger et al., 2012)
Urine	Urine collection techniques depend on downstream applications. Urine should be collected midstream to minimise contaminations and can be collected as (i) 24-hour samples in which individuals collect urine over a period of 24 hours, (ii) first or second morning urine, or (iii) random untimed "spot" urine samples (Thomas et al., 2010). Analytes can be measured using dipsticks (e.g., lateral flow test) or an ELISA
Tears	Tears are collected using microcapillary tubes or absorbing materials. Indirect methods using absorbing materials are easier to implement but might not represent the biochemical composition of tears due to protein retention in the absorbing materials and the potential of tear stimulation. Direct methods include stimulated wash-out techniques using saline or non-stimulated passive tear collection via microcapillary tubes which represent the protein content most accurately but are more labour intensive. Typically, tears are used to determine immunoglobin A (IgA) levels and other proteins (Rentka et al., 2017)
Hair	Around 50–200 mg hair is collected from the posterior vortex region closely to the scalp, placed in aluminium foil and stored at room temperature. Thinning scissors might be used to minimise visibility and the use of dying, bleaching, and shampoo should be standardised. Typically, the collected hair is minced and dissolved in methanol for cortisol determination using ELISA (D'Anna-Hernandez et al., 2011)
Gut microbiota	The immediate freezing of stool samples at −80°C or below is considered the gold standard. Study participants may be provided with at-home collection kits which are optimised for downstream processes, such as microbial signature sequencing, transcriptomic studies, or metabolic studies (Hogue et al., 2019)

Legend: Abbreviations: Enzyme-linked Immunosorbent assay (ELISA)

anaesthesia (e.g., from the dorsal pedal vein located in the foot, the facial and submandibular vein, the saphenous vein, or several tail veins). *Post-mortem* blood is typically collected by cardiac puncture or from the vena cava, and all other organs and bones are accessible for isolation (Knoblaugh & Randolph-Habecker, 2018; Parasuraman et al., 2010). For muscle tissue, the soleus and plantaris muscles of the legs are typically collected *post-mortem* (Gan et al., 2016). For adipose tissue, the subcutaneous and intra-abdominal visceral fat are accessible *post-mortem* (Bagchi & Macdougald, 2019). In many cases, tissues and organs are collected to perform morphological examinations using histology. For tissue histology, samples undergo specific isolation, preservation (fixation), and storage to ensure sample quality (Knoblaugh & Randolph-Habecker, 2018). For considerations on histological experiments, including general microscopy and immunohistochemistry guidelines, see the section Microscopy.

Methods to assess immunity

In the following section, we will introduce the most common techniques to assess immunity in exercise immunology studies. Generally, techniques can be defined based on the environment in which the experiment takes place. First, the term *in vivo* describes methods performed on whole and live organisms (e.g., animal testing or measurements within a human). Second, *ex-vivo* describes experiments outside of an organism as soon as possible after a sample has been taken

Table 3.6 Overview of selected *ex vivo* and *in vitro* methods to assess immunity

Method	Sample type[a]	Application example in exercise immunology[b]
Flow cytometry (FACS)	Cells	Intra- and extracellular immune cell phenotype • Cell isolation (sorting) • Cell counting
Cell culture	Cells, Serum	Co-culture of immune and cancer cells to measure cytotoxicity Exercise serum as growth inhibitor for cancer cells Muscle and fat tissue growth/stimulation
Microscopy	Cells	Muscle cross-sectional area • Fat tissue structure • Cell interactions Immune cell infiltration in tumour tissue
Mitogen stimulation	Cells	Whole blood stimulation with LPS to mimic infection response Specific stimulants used to investigate T or B cell function
ELISA	Plasma, Serum Supernatant	• Pro- and anti-inflammatory cytokines Myokines released during exercise • Cell or tissue explant culture media/supernatant can be assayed
Cell content isolation	Cells	Targeted gene expression analysis of DNA or RNA using qPCR Protein and lipid content for Omics-approach
Blotting techniques	Cells, Plasma, Serum	Western blot to quantify protein content Southern blot to quantify DNA content Northern blot to quantify RNA content

Legend: [a]Major sample types. Most method have versions/applications in which other sample types can readily be used. [b]Not a comprehensive list and many more application examples in the field of exercise immunology exist. Abbreviations: fluorescence activated cell sorting (FACS), lipopolysaccharide (LPS), deoxyribonucleic acid (DNA), ribonucleic acid (RNA), quantitative polymerase chain reaction (qPCR). For further references and methods, see Tables 3.1 and 3.2 in Albers et al. (2013); Table 3.6 in Bermon et al. (2017).

(e.g., blood or tissue samples are used for immediate analysis after collection). Finally, *in vitro* describes techniques using material outside of the normal biological environment (e.g., cell lines for long-term cell culture). Methods used *in vivo* are often used together with *ex vivo* or *in vitro* methods. Table 3.6 provides an overview of the most commonly used *ex vivo* or *in vitro* methods to assess immunity in exercise immunology studies which are often used to compare samples collected pre- and post-exercise. In addition to Table 3.6 and the following sections, the reader is directed to other excellent reviews summarising relevant methodology in detail (Tables 3.1 and 3.2 in Albers et al. [2013] and Table 3.6 in Bermon et al. [2017]). The choice of method to assess immunity always depends on the underlying research question. For example, with the research question "Does a bout of cycling exercise at 80% of $\dot{V}O_2$ max cause a higher natural killer (NK) cell response in blood compared to 60% of $\dot{V}O_2$ max in healthy young individuals?" aims to measure NK cells in blood. The method of choice should be as reliable and accurate as possible considering the expertise and availability of personnel and equipment (i.e., automated cell counter or flow cytometry). Next, the exercise session should be designed to allow for blood sampling before and immediately after the bout of exercise (Figure 3.1) and the samples should be either processed to prepare PBMCs or used as whole blood for the determination of NK cell numbers using flow cytometry.

Whole-body measurements (in vivo)

Besides the laboratory methods summarised in Table 3.6, the immune response can also be monitored in organisms (*in vivo*). These whole-body measurements provide accurate *in-vivo* results but often suffer from variability due to environmental and population-based differences between and within studies. In the field of exercise immunology, some studies have investigated the impact of exercise on vaccine efficacy or infections *in vivo* to gain insights into the immunomodulating capacity of exercise. Indeed, one of the most common measurements in exercise immunology are questionnaires that assess symptoms of upper respiratory tract infections (URTIs) – these can range from daily diaries to weekly or monthly recall. For more details, readers are directed to Table 3.6 in Bermon et al. (2017).

Infectious challenge

One of the most authentic ways of examining immunity and the potential influence of other factors such as exercise is the *in vivo* approach of challenging humans or animals with an infection (see Chapter 7). Although most common in animal models, experimental infection trials are possible in human studies, but are tightly regulated and generally restricted to examining emerging pathogens or vaccine development and are not common in exercise immunology. However, healthy volunteers can in principle be infected with viruses (e.g., influenza) to monitor immune responses and pathology to give insight into disease mechanisms, pathogen detection, and the action of medicines in as few as 10–100 human participants (Roestenberg et al., 2018). Similar procedures are conducted in animal models (Wang et al., 2013) and there are many infectious challenge studies using animal models in exercise immunology (see Chapter 7).

Vaccination

An alternative to using live viruses as an *in-vivo* challenge is to administer vaccines (e.g., against influenza) which assesses the capacity of the adaptive immune system to generate an immune response (see Chapter 7). This adaptive immune response is predominately driven by T and B cells. Vaccine efficacy is most commonly evaluated via antibody titre, but other studies also examine T cell and B cell function, and often incorporate clinical outcomes (e.g., did participants subsequently become infected with the pathogen via naturalistic exposure). Antibody titres can be assessed via enzyme-linked Immunosorbent assay (ELISA) or hemagglutination (inhibition) assays in which serial dilutions of antigens are mixed with whole blood. In the absence of sufficient antigen-specific antibodies, visible clumping of red blood cells occurs (Kaufmann et al., 2017) (see Chapter 7).

Delayed-type hypersensitivity reactions

Delayed-type hypersensitivity (DTH) models, also known as type IV hypersensitivity, assess cell-mediated immune responses that are largely driven by CD4+ and CD8+ T cell activation (see Chapter 7). In exercise immunology, a common approach has been to administer Keyhole limpet haemocyanin (KLH) – a protein found in a specific type of shellfish that elicits an immune response in humans – which can be injected intradermally into the arm and evaluated after 24–72 hours (e.g., via visual inspection of induration and erythema of the skin). Similarly, and often used in conjunction with DTH models, T cell dependent antibody response assays are used to assess antibody formation after KLH challenge, using functional techniques to assess T cell or B cell function (e.g., ELISpot; *enzyme-linked immuno-spot*) or ELISA to assess antibody concentration for up to 28 days after KLH administration. These techniques can be used in human studies (House, 2010) and in rodents (Allen, 2013) (see Chapter 7).

Wounding and blister models

Wound regeneration and repair are complex immunological mechanisms involving a multitude of cells and signalling pathways and have also been used in exercise immunology (see Chapter 7). A common approach in human studies is to take skin biopsies to monitor the effect of interventions such as exercise, on the time course of wound healing. This approach can be combined with *ex-vivo* models to assess epithelial structure (Olivera & Tomic-Canic, 2013). Local inflammatory processes can also be examined using blister models such as the cantharidin-induced skin blisters. Cantharidin forms non-scarring lesions where local inflammatory processes involve neutrophils and macrophages. This model is used to evaluate anti-inflammatory drugs in early-phase trials of humans (Day et al., 2001) and mice (Dawson & Vogelsanger, 2021). Blister models can be evaluated by the blister size, infiltration of immune cells using flow cytometry, inflammatory mediators using ELISA, or histological assessment of tissue (see Chapter 7).

Cell-based measurements (ex vivo–in vitro) and associated laboratory techniques

The cellular composition of blood and tissues are key measurements in exercise immunology. This section will introduce methods which primarily use cell-based approaches (i.e., functional whole-cell experiments). However, some of these methods can be used for the assessment of soluble components too (e.g., for example, flow cytometry can be used in combination with beads, to assess cytokines in plasma).

Flow cytometry

Flow cytometry, also known as FCM, and flow cytometry cell sorting are among the most powerful and widely used methods used across all aspects of immunology. Flow cytometry enables researchers to identify and characterise cells before and after bouts of exercise or examine the composition of tissues. Flow cytometry is often referred to as FACS™ – fluorescence activated cell sorting – which is a trademark of a manufacturer of flow cytometers, Becton Dickinson (BD), and should only be used for BD-related technologies. In summary, cells are labelled with antibodies linked to fluorescent molecules which bind to proteins on the cell surface or within cells. The cells become contained within a flow of liquid within the flow cytometer which passes across a laser beam (Figure 3.2). Light from the lasers reflects off cells in different directions, with signals detected as "forward scatter" (FSC) representing cell size, signals detected as "side scatter" (SSC), representing the internal complexity within cells. Light emitted by the fluorescently tagged antibodies confirms whether a particular protein (which might identify a cell of interest) is present. FSC describes light crossing in a straight line and is proportional to the size of the cell (i.e., the larger the cell, the less light reaches the FSC detector). SSC describes light reflected at an angle from cells (i.e., the more intracellular structures or granularity, the more light is reflected).

There are many different applications of flow cytometry including: (i) the study of whole cells as well as parts like nuclei, chromosomes, or mitochondria, (ii) extracellular vesicles, (iii) cell cycle, DNA, and mRNA, (iv) cell death mechanisms such as apoptosis, (v) phagocytosis and autophagy, (vi) reactive oxygen species, (vii) intracellular signalling pathways and metabolism, (viii) functional cell assays assessing proliferation cytokine production, killing, or migration (Cossarizza et al., 2019). Flow cytometry can also be used to isolate and collect cells for further downstream applications (e.g., cell-culture assays using specific populations of cells).

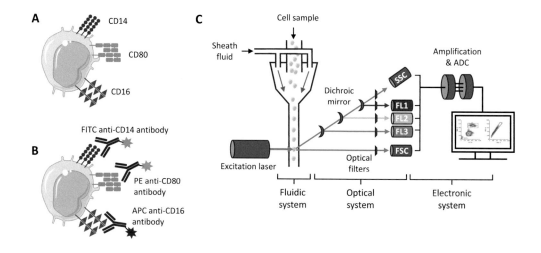

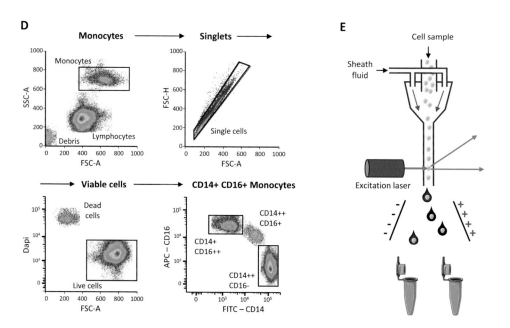

Figure 3.2 Flow cytometry and flow cytometry cell sorting.

Legend: Schematic representation of functions during the flow cytometric assessment of cells. (A) Biological samples are isolated as a single cell suspension to access surface receptors. (B) Fluorochromes such as FITC, APC or PE linked to antibodies are used to target specific surface proteins. The CD nomenclature is used to uniquely identify cell molecules such as surface receptors (i.e., CD14 was the 14th cell molecule to be discovered and quantified). (C) Cells linked to antibodies are introduced into the fluidic system of a flow cytometer in which one cell at a time crosses an array of excitation lasers. Depending on the fluorochromes present on cells different signals are generated, separated and detected in the optical system of a flow cytometer. Finally, signals are amplified, converted and for each single cell, the emission spectrum for each laser and each detector is accessible. (D) Representative gating strategy during data analysis in which cells are selected in gates (i.e., only cells inside of gates are shown in the next plot). First, SSC vs FSC plots are used to roughly identify the population of interest (i.e., monocytes). Second, doublet discrimination (i.e., cells stuck together have more area via FSC-A than height via FSC-H compared to single cells). Third, live-dead discrimination (i.e., cells positive for the vertical-axis marker Dapi are dead and are deselected). Lastly, cells of interest (i.e., single and live cells), are selected based on markers of interest (i.e., monocytes exhibit varying expression of CD14 and CD16 depending on subset). Note that all monocytes express CD14 and are only divided into cells expressing CD14 bright (++) and dim (+). (E) Schematic representation of flow cytometry cell sorting in which cells are separated based on their fluorescent signal into collection tubes. Flow cytometry applications can be used to identify/characterise cells with or without subsequent sorting. This figure was partly generated using Servier Medical Art, provided by Servier, licensed under a Creative Commons Attribution 3.0 unported license. *Abbreviations: ADC (Analog to digital converter), APC (Allophycyanin), CD (cluster of differentiation), DAPI (4',6-diamidino-2-phenylindole), FITC (Fluorescein isothiocyanate), FSC-H (forward scatter height), FL1 (fluorescent signal 1), PE (Phycoerythrin), SSC-A (side scatter area).*

For example, cells can be identified based on light-scattering properties (i.e., fluorescently labelled CD3+ and CD8+ positive cells to identify cytotoxic T cells) (Figure 3.2E). In flow cytometry cell sorting, cells are encapsulated by droplets with one cell per droplet. Droplets are negatively or positively charged depending on pre-selected characteristics (i.e., droplets containing CD4+ cells become negatively charged and droplets with CD8+ positive cells become positively charged, whereas all other droplets do not get a charge). Next, droplets are passed through an electrical field which deflects droplets based on their charge into different collection tubes, so they can then be used in downstream assays examining a specific type of cells (Cossarizza et al., 2019).

Recent advances in flow cytometry include the use of image-based flow cytometry, spectral flow cytometry, and cytometry by time of flight (CyTOF). Image flow cytometry combines the advantages from flow cytometry with digital microscopy to assess cell morphology (Cossarizza et al., 2019). Spectral flow cytometry has expanded the number of fluorescent molecules that can be used in an experiment, allowing cells to be examined in greater detail. With traditional flow cytometry, due to the number of detectors in an instrument, the number of different fluorescently tagged antibodies was limited (e.g., prior to the turn of the millennium, the ability to examine more than six fluorescently tagged antibodies was rare). Although traditional flow cytometry has become more advanced so that up to around 18–20 fluorescently tagged antibodies can be included in an experiment, this can cause complications with compensation or spectral overlap. Thus, spectral flow cytometry uses a greater number of detectors and different detection algorithms so that up to 40 fluorescently tagged antibodies can be included in an experiment, and therefore in principle, 40 different proteins could be assessed on (or within) a single cell. Finally, to overcome the complexities of fluorescence, CyTOF combines the advantages of flow cytometry and mass spectrometry, using antibodies labelled with heavy metal isotopes instead of fluorescent molecules, which allows for the analysis of 40+ targets in a single experiment using weight-dependent time-of-flight measurements.

For more information about flow cytometry, the reader is directed to the basic learning guide published by BD (BD Bioscience, 2002) and comprehensive technical guides (Cossarizza et al., 2019). However, in its basic and most common form (i.e., the assessment of extra- or intracellular proteins on cells), flow cytometry requires several building blocks for a successful experimental setup:

A Antibody panel design: The choice and combination of fluorescent dyes is essential and depends on fluorochrome brightness, target abundance, and, most importantly, the research question. Online tools and manufacturers provide step-by-step guidance. Furthermore, the "*Optimized Multicolor Immunofluorescence Panel (OMIP)*" platform, publicly available in the journal *Cytometry Part A*, provides researchers with optimised multicolour panels for common applications. Alternatively, panels for common human and rodent cells have been summarised by Cossarizza and colleagues (Cossarizza et al., 2019).

B Sample preparation and antibody "staining" procedure: The term "staining" is often used, referring to the incubation of antibodies that target specific cell-surface proteins with either whole blood or cell preparations (e.g., PBMCs). When working with tissue, such as muscle and adipose, a single-cell suspension is crucial, and protocols and their considerations are summarised in the supplementary material of published OMIP articles and by Cossarizza and colleagues (Cossarizza et al., 2019).

C Controls: Several vital controls should be included in every experiment: (1) unstained samples to visualise cell position and autofluorescence on the flow cytometer, (2) Fluorescence

Minus One controls in which the combined expression pattern of all antibodies is compared to the pattern of all antibodies except one (Cossarizza et al., 2019), (3) biological controls with already known expression patterns to control for the experimental setup (Cossarizza et al., 2019), (4) viability controls as dead cells can exhibit high levels of autofluorescence, (5) single-colour controls for compensation to assess spectral overlap of fluorochromes (Cossarizza et al., 2019). Additionally, controls to monitor day-to-day variation in sample-handling and instrument stability are especially useful for longitudinal studies such as exercise interventions (6) control material (i.e., cryopreserved PBMCs) to verify staining pattern and handling, (7) multicolour beads for the standardisation of instrument settings.

D Data acquisition: Maintenance and handling of equipment vary greatly. Therefore, research institutes and universities often have designated core facilities to operate and maintain specialised equipment, and often provide support with data acquisition.

E Data analysis: Data from samples and controls are analysed in software such as FlowJo or FCS Express, which provide researchers with a variety of tools to perform quality control and compensation to assess spectral overlap of fluorochromes. Next, the so-called "gating" begins (i.e., the stepwise selection and exclusion of cells). Before populations of interest can be selected, dead cells are excluded followed by the exclusion of doublets (doublet discrimination) (Figure 3.2D).

F Publication: For comparability across studies and quality assessment, studies involving flow cytometry should adhere to guidelines describing "the minimum information about a Flow Cytometry Experiment (MIFlowCyt)" (Lee et al., 2008) when reporting results.

Cell culture

Cell culture describes the cultivation of cells in an artificial environment. Cell cultures are useful model systems to study the physiology of cells and the influence that exposure to different compounds have on cell viability and growth in both two- and three-dimensional settings (Table 3.7). Cells can be isolated directly from biopsy material as primary cultures, or primary cells and cell lines can be ordered from cell repositories such as the European Collection of Authenticated Cell Cultures (ECACC) or the American Type Culture Collection (ATCC). HeLa

Table 3.7 Frequently used terms in cell culture

Term	Description
Primary cell culture	Cells directly isolated from tissue: limited life span until they undergo senescence
Subculture	Transferring cells from one culture plate to another, also known as passaging or splitting, and done when cells reach a certain concentration or confluency
Confluence	Defined as the percentage of occupied area by cells compared to empty space
Passage number	Identifies how many times a cell line has been sub-cultured. Important indicator to judge cell line authentication if no sequencing has been performed
Cell line	After the first successful subculture, primary cells are known as cell lines. Due to their limited life span, they are sometimes called finite cell line
Continuous cell culture	After immortalisation, cells can be cultured indefinitely. Immortalisation can be achieved by the introduction of mutations allowing for indefinite proliferation
Adherent cell culture	Cell culture of cells attaching to a surface and form 2D or 3D layers. During subculture, cells are detached by trypsin
Suspension cell culture	Cell culture of cells growing in suspension or lightly adhere

cells are one of the most commonly used immortalised cell lines and originated from a cervical cancer specimen collected from Henrietta Lacks in 1951 without consent. The controversy surrounding the ownership of HeLa cells is still ongoing but led to the establishment of guidelines regarding informed consent in clinical studies. Other commonly used cell lines include the mouse skeletal muscle cell line C2C12, the mouse preadipocyte cell line 3T3-L1, the immortalised human embryonic kidney cell line HEK293, or the Jurkat cell line which is used to study T cells. Cell-culture techniques depend on the origin of the cells, but similar techniques apply across all applications (Philippeos et al., 2012), and detailed descriptions are published by ECACC and ATCC. Techniques include (i) aseptic handling, (ii) thawing of cryopreserved cells, (iii) morphological assessment of cells, and (iv) routine subculture of cells. In the field of exercise immunology, common applications include examining the effect of exercise serum on cancer cell growth (Baldelli et al., 2021) as well as the isolation and culturing of fat tissue (Scheele et al., 2022) or muscle tissue (Striedinger et al., 2021).

Microscopy

Microscopy visualises objects too small to be seen by the so-called "naked eye". Optical microscopy is used across all scientific fields, from the bright field microscope for the routine visual examination of cell cultures, to advanced fluorescence-based microscopes, which combine immuno-fluorescent staining (similar to the antibodies uses in flow cytometry) to visualise specific cell types or cell components. Histology describes a field of microscopic anatomy in which tissue sections can be visualised (e.g., the structure of muscle fibres or the amount of immune cell infiltration in tumour tissue). Most histological analysis follows four general steps after the tissue is collected: (i) fixation to preserve tissue structure, (ii) embedding to solidify structures, (iii) sectioning into thin tissue sections, and (iv) staining for cells and structures of interest. Protocols vary greatly on the tissue origin and structures of interest. Arguably, the most common staining used is Haematoxylin and Eosin (often referred to as "H and E"), in which Haematoxylin stains nuclei blue and Eosin stains the cytoplasm and most other structures pink (Kim et al., 2016; Knoblaugh & Randolph-Habecker, 2018). Immunohistochemistry (IHC), a subcategory of histology, describes the staining of tissues using antibodies. Similar to ELISA techniques, specific antibodies are coupled to enzymes which convert substrate to generate a signal (Kim et al., 2016). As an example, in exercise immunology, studies have used histology to assess the cross-sectional area of muscle fibres in order to describe the effect of exercise interventions on muscle function in patients with cancer (Mijwel et al., 2018).

Cytotoxicity assays

Cytotoxicity describes the ability to eliminate targets, such as cancer cells, virus infected cells, or cell infected with bacteria, and often examines the capacity of effector cells of the immune system to kill (i.e., CD8+ T cells and NK cells) (see Chapter 2). In most experiments, target cells are transiently labelled before the addition of effector cells. A variety of assays are available using fluorescent, luminescent, or radioactive compounds for labelling (Zaritskaya et al., 2010). Due to its early implementation, the [51]Cromium release assay is common, but most laboratories have switched to non-radioactive alternatives due to waste management and safety concerns. As NK cells deploy rapid effector functions, NK cytotoxicity assays have shorter incubation times, typically 4 hours, compared with T cell cytotoxicity assays which can incubate for days. NK cytotoxicity assays have the advantage that a mix of cells (e.g., PBMCs) can be used as only NK cells will eliminate target cells in a large quantity during a short incubation. Instead of radioactive reagents, the non-radioactive compound calcein-AM can be used to indirectly determine the amount of dead target cells using a fluorescence plate reader (Neri et al., 2001). Alternatively, flow cytometry allows for the direct determination of dead target cells more accurately. Similarly, T cell cytotoxicity assays using flow cytometric and calcein-AM-based protocols are available (Zaritskaya et al., 2010). As exercise has a profound effect on NK cells, exercise immunology studies have investigated NK cytotoxicity before and after an individual bout of exercise (e.g., in patients with prostate cancer) (Schauer et al., 2022).

Mitogen stimulation

Mitogens describe compounds that induce cell division (i.e., lead to activation and proliferation of cells) (see Chapter 2). Stimulation of immune cells can, with the use of proper controls, give an indication of cell function (e.g., the response to external activation). A common approach is to incubate mitogens with cells of interest over a specific period of time and the release of cytokines is assessed via ELISA, ELISpot, or flow cytometry. Table 3.8 summarises commonly used mitogens in exercise immunology studies. To give an example in exercise immunology, the capacity of immune cells to produce cytokines following lipopolysaccharide (LPS) stimulation has been examined before and after a marathon (Nielsen et al., 2016).

Soluble measurements (ex vivo–in vitro) and associated laboratory techniques

Exercise has a profound effect on soluble components of the immune system, such as acute phase proteins, cytokines, myokines, adipokines, hepatokines, and hormones. Many of these molecules can be assessed in plasma, serum, or in the supernatants of cells or tissue explants in culture. The general principle to measure any soluble molecule (also called an "analyte") is that is must be bound to a specifically designed capture antibody which is then used to visualise the molecule and provide a quantitative measurement. A vast number of assays for the detection of soluble components are available commercially, with the option to use pre-designed "off-the-shelf" kits or custom-made detection kits to measure a specific set of analytes. The choice of assay often depends on the equipment in the laboratory and the expertise of the researchers, but also considerations over sensitivity (i.e., ability to detect very low levels of a soluble molecule).

Table 3.8 Overview of mitogens and their function

Mitogen	Function	Primary target cell[b]
PHA-L[a]	Protein (lectin) from plants (e.g., red kidney beans). Binds and crosslinks white blood cells	T cells
PMA + ionomycin	Activation of protein kinase C and the intracellular Ca^{2+} increase leads to the activation of signalling pathways	T cells
LPS	LPS is part of the outer cell membrane of most Gram-negative bacteria. LPS binding includes CD14, toll-like receptor 4, and the B cell receptor	Monocytes & Macrophages; B cells
PWM	Protein (lectin) from plants (i.e., pokeweed). Mechanism largely unknown but results in B cell proliferation when T cells are present	T and B cells
Con-A	Protein (lectin) from plants. Binds irreversibly to T cell surface and activates proliferation	T cells

Legend: [a] PHA-P contains PHA-E and PHA-L where PHA-E induces red blood cell activation; [b]cross-stimulation of other cell types is possible; abbreviations: phytohemagglutinin (PHA), phorbol-myristic-acid-calcium ionophore (PMA), lipopolysaccharide (LPS), *Staphylococcus aureus* Cowan, pokeweed mitogen (PWM), concanavalin-A (Con-A).

Enzyme-linked immunosorbent assays

ELISAs are essential for measuring soluble immune system components. Prior to widespread application of the ELISA, radio-immunoassays were used, in which the analyte of interest competes with a radioactive isotope of the analyte for binding sites on specific antibodies. The more radioactivity in wells – the lower the concentration of the analyte in the sample. However, due to safety and waste disposal complications due to radioactivity, many laboratories switched to non-radioactive alternatives. ELISAs can be direct, indirect, sandwich, or competitive (Figure 3.3A). In exercise immunology, where cytokines and similar molecules are examined regularly, the sandwich ELISA is most common due to the direct application potential of complex samples such as plasma without prior purification. For most ELISAs, antibodies conjugated to enzymes are used to generate a signal by the enzymatic conversion of substrate into

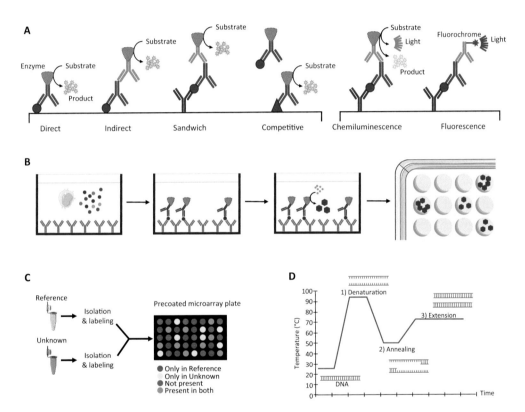

Figure 3.3 Detection of soluble components: ELISA, ELISpot, microarray, and PCR.

Legend: (A) Available ELISA formats: (i) For the direct ELISA, targets are immobilised on a surface and specific antibodies are used for detection. (ii) Indirect ELISAs utilise a secondary antibody to lessen background noise. (iii) Sandwich ELISAs improve specificity and sensitivity via specific capture antibodies. Therefore, protocols take longer and only molecules large enough for two specific binding sites can be used. (iv) The competitive ELISA is primarily used for small molecules not usable in sandwich approaches. Here, the analyte of interest (round shape) competes with analytes immobilised to the surface (triangle shape). The more analyte of interest is present the less enzyme will bind to immobilised analytes resulting in a lower enzymatic conversion and lower signal after washing. Alternatives to the colorimetric ELISA method, shown in (A), include signal generation by chemiluminescence and fluorescence. Graphical presentation of (B) ELISpot, (C) microarray, and (D) PCR techniques. This figure was partly generated using Servier Medical Art, provided by Servier, licensed under a Creative Commons Attribution 3.0 unported license. *Abbreviations: polymerase chain reaction (PCR), Enzyme-Linked Immunosorbent Assay (ELISA), Enzyme-linked immune absorbent spot (ELISpot), Deoxyribonucleic acid (DNA).*

a detectable dye (colourimetric assays). When compared to a standard curve of known concentrations and therefore known colour absorption, the concentration of the analyte of interest can be extrapolated. For example, the colour change in a sandwich ELISA is directly proportional to the concentration of the analyte (i.e., the higher the concentration, the more light is absorbed and visualised as being darker in colour).

The principles of an ELISA have been used to develop many other laboratory techniques. For example, a very simple example of an assay with a similar principle to an ELISA is a so-called lateral flow test, which can be used in almost any setting without specialist equipment or training. In lateral flow tests, fluid containing the analyte of interest, passively flows down a test strip and the analyte of interest is visualised in a similar fashion to a sandwich or competitive ELISA. The most prominent examples are pregnancy tests or the rapid antigen test for COVID-19. In more advanced versions of ELISAs, signal generation is not by a colour change, but is instead via fluorescence or (electro-) chemiluminescence, and this can increase the sensitivity due to signal amplification and background noise reduction (Figure 3.3A). Using fluorescence or (electro-) chemiluminescence allows for multiplexing (i.e., measuring more than one analyte of interest in the same sample) using either non-overlapping fluorescent signals or spatial separation of (electro-) chemiluminescent signals. Due to the increased complexity of multiplex assay, stringent validation and quality controls are necessary (Tighe et al., 2015). The ELISA format has also been adapted to measure immune cell function in ELISpot assays by which cytokine production of cells is assessed. Based on the sandwich ELISA method, cells are cultured in the presence of specific antibodies coated to the surface of a 96-well plate which can bind to analytes of interest. The enzymatic conversion of substrate results in the typical insoluble precipitate or spots seen in ELISpot assays (Figure 3.3B). In turn, ELISpot has been adapted to measure multiple parameters using the so-called FluoroSpot assay, which allows up to three fluorescently tagged antibodies (i.e., of different "colours") targeting different cytokines to be coated onto plates. Finally, multiplex assays can also come in different forms, and rather than multiplexing in a single well of a plate, microarrays have specific capture antibodies (or RNA/DNA sequences) coated in a spot-like pattern on a solid surface. Microarrays can measure hundreds of analytes and typically use the sandwich or competitive ELISA method. Antibody microarrays are often applied in biomarker discovery, protein profiling, or disease screening and have also been adapted for DNA and mRNA assessment to evaluate whether specific DNA sequences are present or whether genes are up- or downregulated (Figure 3.3C).

DNA content

Deoxyribonucleic acid (DNA) stores all genetic information of an organism and consists of a double-stranded (double helix) structure of nucleotides. DNA is the instruction template for ribonucleic acid (RNA) molecules and, subsequently, the amino acid sequence of proteins. DNA composition can change slowly over time by the accumulation of mutations. The process of DNA isolation from cells is divided into cell lysis, DNA precipitation and purification. DNA isolation can be performed manually or more conveniently using commercially available spin column kits in which the DNA is trapped and purified in a membrane complex designed to fit in a 1.5 mL tube that is centrifuged. DNA is relatively stable and can be isolated from whole blood, immune cells, and even from plasma. Selected applications for quantifying DNA are listed in Table 3.9 but also include microarrays, cell cycle assessment using flow cytometry, or the assessment of telomere length. As an example, regular exercise has been linked with longer telomeres, a marker for biological age (Friedenreich et al., 2018).

Table 3.9 Overview of common methods for DNA, RNA, and protein quantification

Method[a]	Target	Description
Gel electro-phoresis	DNA, RNA, protein	An electric field pushes charged molecules through a gel matrix. Molecules are separated based on pore size of the gel, molecule size, and charge
Blots	DNA, RNA, protein	After gel electrophoresis, separated molecules can be transferred to membranes using blotting techniques. In principle, capillary effects or electric current are used to push molecules out of the gel and onto a membrane. Analytes can be visualised using specific antibodies linked to luminescent or fluorescent signals similar to ELISA. Protocols depend on target size and type: southern blots are used for DNA, northern blot for RNA, and western blots for proteins
Polymerase chain reaction (PCR)	DNA, RNA	PCR is used to amplify specific regions of DNA (Figure 3.3D). Qantitative PCR (qPCR) allows for real-time assessment of the amount of DNA using DNA-binding dyes. Reverse transcriptase qPCR (RT-qPCR) allows for the quantification of RNA Consideration: RT-qPCR describes an advanced technique and researchers should adhere to The Minimum Information for Publication of Quantitative Real-Time PCR Experiments (MIQE) guidelines (Bustin et al., 2009), including their practical application for RT-qPCR (Nolan et al., 2006) for validity and reproducibility

Legend: [a]Many suppliers or manufacturers provided "off-the-shelf" kits with optimised protocols.

RNA content

DNA is transcribed into messenger RNA (mRNA) which consists of a single-stranded sequence of nucleotides and serves as a messenger between DNA and the sequence of amino acids which makes the final protein. RNA is unstable, partly because RNA degrading enzymes (RNases) are commonly found in all biological samples and on the skin. Therefore, storage of samples designated for RNA isolation and the RNA isolation itself are undertaken in the presence of RNases inhibitors (i.e., tubes containing RNases) and using work surfaces that are free from contaminating RNA and RNases. RNA is typically isolated after cell lysis using organic extraction or spin column technologies, similar to DNA. The mRNA content may change rapidly during exercise and can be used to quantify changes to cellular processes (e.g., the mRNA content of PGC-1α rapidly increases during exercise). PGC-1α is one of the master regulators for the energy metabolism of skeletal muscle and mediates exercise-induced changes (Pilegaard et al., 2003). A selection of methods used for examining RNA are listed in Table 3.9 with RT-qPCR being the most common method. Other methods include microarrays or untargeted transcriptomic approaches.

Protein content

Proteins are created after translating mRNA. Proteins contribute to the structure and function of all cells, tissues, and organs. The presence, absence, or levels of proteins represent biological processes better than the RNA content. As with DNA and RNA, proteins can be isolated and purified from cells or plasma using spin column techniques, but the methodology depends largely on protein size and origin. Alternatively, more complicated techniques include using chromatography-based methods such as high-performance liquid chromatography to separate, identify, and quantify proteins in targeted and untargeted ways. A very common technique in exercise physiology to identify proteins is gel electrophoresis and immuno-blotting (see Table 3.9). An application of this methodology includes analysis of specific proteins in muscle, such as citrate synthase or mitogen-activated protein kinase which can be used to examine the effect of exercise on muscle properties and signalling processes (Mijwel et al., 2018).

The untargeted approach – "Omics" techniques and sequencing

In contrast to traditional methods which require researchers to have decided upon specific targets, untargeted approaches use an unbiased method to generate large amounts of data. This approach is termed "Omics" and is used across a variety of fields such as genomics (study of DNA), transcriptomics (study of RNA), proteomics (study of proteins), metabolomics (study of metabolites), phosphoproteomics (study of post-translational modifications), epigenomics (study of epigenetic modifications), or lipidomics (study of lipids). For example, proteomics maps protein content within a sample which is then typically compared to a reference condition (e.g., pre-exercise, or to a different group of people) to identify differently regulated proteins. Due to the ever-increasing availability of omics-based techniques and sequencing approaches, studies in exercise immunology will more commonly use these methods, especially in screening and understanding novel pathways. The general principle of most omics-based techniques is the isolation and categorisation of analytes followed by bio-statistical quality control and comparison with a reference data set. Due to the amount of data generated, bioinformatics processes are a key component of all omics-based techniques. Due to the multitude of changes during just one bout of exercise, omics-based approaches are ideal to identify novel pathways and may help to identify a baseline blood-based biomarker signature which can predict aerobic fitness for example (Contrepois et al., 2020).

Nucleic acid sequencing

Genomics enables careful examination of the entire genome. Examples include genome-wide association studies (GWAS), which aim to identify genetic variants associated with a specific disease. GWAS often report single-nucleotide polymorphisms (SNPs) (i.e., a difference in the DNA sequence within genes between individuals and their association with disease) (Uffelmann et al., 2021). The most common approach to analyse the genome and transcriptome is next-generation sequencing. This methodology is often referred to as DNA/RNA-sequencing (or DNA/RNA-seq). Most methods are based on the Illumina technology (USA, San Diego) but others are available (Mardis, 2008). Typically, RNA-sequencing can be performed as bulk RNA-sequencing in which the generated data represents a bulk of different cells; however, more recently, single-cell RNA-sequencing (scRNA-seq) in which the generated data can be assigned on a per cell basis (Haque et al., 2017) has been developed. For example, scRNA-seq may be used to identify different NK or T cell subsets which are mobilised and redistributed during exercise.

Proteomics

Proteomics assess the identity, structure, and function of all proteins in an organism. Two main methods are currently in use, bottom-up proteomics and top-down proteomics (Catherman et al., 2014). In both cases, mass spectrometry is used to analyse the molecular weight of proteins based on the mass-to-charge ratio of proteins after ionisation. The data generated is compared to large proteome repositories to identify individual proteins. To date, the assessment of the human plasma proteome is challenging as it contains a very dynamic range of highly abundant proteins, such as albumin, compared with very low quantities of other proteins, such as cytokines.

Table 3.10 Assay controls

Type	Description	Consideration
Positive and negative control	Provides guaranteed positive and negative signals to judge whether the assay works properly; usually these reagents are included in the assay kit. A negative control can also be used to correct for background signal if the analyte of interest is left out (blank)	Crucial for assay and equipment validity and the comparability between different experiments (inter-assay)
Biological control	Biological specimen similar to the measured analyte with known or expected concentration	Recommended; used for inter-assay comparability
Replicas: duplicates or triplicates	Can be technical (identical sample) to assess measurement error of equipment or biological (same individual, multiple samples) to assess variation within the same experiment (intra-assay)	Recommended; trial experiments provide information about necessity depending on variation
Standard curve	Usually included in assay kit. Serial dilution of a reagent with known concentration. Should span higher and lower than the concentration of samples	Crucial for assay range validity (e.g., for ELISA)

Legend: Abbreviation: Enzyme-Linked Immunosorbent Assay (ELISA)

Metabolomics

Metabolites are small molecules, such as substrates or products of cell metabolism (e.g., lactate, pyruvate, creatine, or ketone bodies). Metabolomics describes the assessment of all metabolites in an organism to estimate biochemical activity (Alonso et al., 2015) (e.g., substrate availability after a bout of exercise). For example, metabolomic studies examining exercise bouts have shown an increase in neuroactive metabolites such as acetylcholine and kynurenic acid which are linked to mental health (Contrepois et al., 2020).

Assay controls

A crucial consideration with every method or laboratory technique is the design and inclusion of assay controls to assess validity and for general quality control. This is particularly important if the same measurement is repeated on multiple occasions and compared (e.g., before and after an intervention). Table 3.10 provides an overview of different types of controls.

Key Points

- A stress-induced immune response can be induced by both physical and psychological stressors.
- Stressors need to be standardised to compare immune function between participants or studies.
- Stressors and sampling techniques vary greatly between human and animal models whereas methods to assess aspects of immunity between human and animal models are similar.
- Exercise immunology studies rely on sampling of biological material from blood and tissues, including adipose and muscle.
- The effect of exercise on the immune system can be assessed by quantifying changes in cellular and soluble factors.
- Common cell-based methods include flow cytometry, cell culture, microscopy, and measurements of cell function (e.g., cytokine production, proliferation, cytotoxicity).

- Common methods to assess soluble components of the immune system include ELISA and other techniques that work with a similar principle.
- Other soluble components such as DNA, RNA, and proteins are commonly quantified using a variety of laboratory techniques, including blotting and polymerase chain reaction (PCR).
- Untargeted omics-based approaches including genomics (study of DNA), transcriptomics (study of RNA), proteomics (study of proteins), metabolomics (study of metabolites), phosphoproteomics (study of post-translational modifications), epigenomics (study of epigenetic modifications), or lipidomics (study of cellular lipids) will become more common in exercise immunology.

References

Albers, R., Bourdet-Sicard, R., Braun, D., Calder, P. C., Herz, U., Lambert, C., … Sack, U. (2013). Monitoring immune modulation by nutrition in the general population: Identifying and substantiating effects on human health. *British Journal of Nutrition, 110*, S1–S30.

Alderete, T. L., Sattler, F. R., Sheng, X., Tucci, J., Mittelman, S. D., Grant, E. G., & Goran, M. I. (2015). A novel biopsy method to increase yield of subcutaneous abdominal adipose tissue. *International Journal of Obesity, 39(1)*, 183–186.

Allen, A. P., Kennedy, P. J., Dockray, S., Cryan, J. F., Dinan, T. G., & Clarke, G. (2017). The trier social stress Test: Principles and practice. *Neurobiology of Stress, 6*, 113–126.

Allen, I. C. (2013). Delayed-type hypersensitivity models in mice. In I. C. Allen (Ed.), *Mouse Models of Innate Immunity* (pp. 101–109). Springer Protocols.

Alonso, A., Marsal, S., & Julià, A. (2015). Analytical methods in untargeted metabolomics: State of the art in 2015. *Frontiers in Bioengineering and Biotechnology, 3*, 1–20.

American College of Sports Medicine. (2016). *ACSM's Guidelines for Exercise Testing and Prescription* (10th ed.). Wolters Kluwer.

Bagchi, D. P., & Macdougald, O. A. (2019). Identification and dissection of diverse mouse adipose depots. *Journal of Visualized Experiments, 149*, doi: 10.3791/59499.

Balady, G. J., Arena, R., Sietsema, K., Myers, J., Coke, L., Fletcher, G. F., … Milani, R. V. (2010). Clinician's guide to cardiopulmonary exercise testing in adults: A scientific statement from the American Heart Association. *Circulation, 122(2)*, 191–225.

Baldelli, G., De Santi, M., Gervasi, M., Annibalini, G., Sisti, D., Højman, P., … Brandi, G. (2021). The effects of human sera conditioned by high-intensity exercise sessions and training on the tumorigenic potential of cancer cells. *Clinical and Translational Oncology, 23(1)*, 22–34.

BD Bioscience. (2002). *Introduction to Flow Cytometry: A Learning Guide*. Retrieved from: https://www.bu.edu/flow-cytometry/files/2010/10/BD-Flow-Cytom-Learning-Guide.pdf

Beltz, N. M., Gibson, A. L., Janot, J. M., Kravitz, L., Mermier, C. M., & Dalleck, L. C. (2016). Graded exercise testing protocols for the determination of VO2 max: Historical perspectives, progress, and future considerations. *Journal of Sports Medicine, 2016*, 1–12.

Bermon, S., Castell, L. M., Calder, P. C., Bishop, N. C., Blomstrand E., … Nagatomi, R. (2017). Consensus statement immunonutrition and exercise. *Exercise Immunology Review, 23*, 8–50.

Bustin, S. A., Benes, V., Garson, J. A., Hellemans, J., Huggett, J., Kubista, M., … Wittwer, C. T. (2009). The MIQE guidelines: Minimum information for publication of quantitative real-time PCR experiments. *Clinical Chemistry, 55(4)*, 611–622.

Campos, A. C., Fogaça, M. V., Aguiar, D. C., & Guimarães, F. S. (2013). Animal models of anxiety disorders and stress. *Brazilian Journal of Psychiatry, 35*(Suppl 2), S101–S111.

Catherman, A. D., Skinner, O. S., & Kelleher, N. L. (2014). Top down proteomics: Facts and perspectives. *Biochemical Biophysical Research Communications, 455(4)*, 683–693.

Contrepois, K., Wu, S., Moneghetti, K. J., Hornburg, D., Ahadi, S., Tsai, M. S., … Snyder, M. P. (2020). Molecular choreography of acute exercise. *Cell, 181(5)*, 1112–1130.e16.

Cossarizza, A., Chang, H. D., Radbruch, A., Acs, A., Adam, D., Adam-Klages, S., ... Zychlinsky, A. (2019). Guidelines for the use of flow cytometry and cell sorting in immunological studies (second edition). *European Journal of Immunology, 49*(10), 1457–1973.

D'Anna-Hernandez, K. L., Ross, R. G., Natvig, C. L., & Laudenslager, M. L. (2011). Hair cortisol levels as a retrospective marker of hypothalamic-pituitary axis activity throughout pregnancy: Comparison to salivary cortisol. *Physiology and Behavior, 104*(2), 348–353.

Dawson, J., & Vogelsanger, M. (2021). Cantharidin-Induced skin blister as an In Vivo model of inflammation. *Current Protocols, 1*(2), 1–14.

Day, R. M., Harbord, M., Forbes, A., & Segal, A. W. (2001). Cantharidin blisters: A technique for investigating leukocyte trafficking and cytokine production at sites of inflammation in humans. *Journal of Immunological Methods, 257*(1–2), 213–220.

Dubowitz, V., Sewry, C. A., & Oldfors, A. (2021). The procedure of muscle biopsy. In Dubowitz, V., Sewry, C. A., & Oldfors, A. (Eds.), *Muscle Biopsy: A Practical Approach* (5th ed., pp. 2–14). Elsevier.

Friedenreich, C. M., Wang, Q., Ting, N. S., Brenner, D. R., Conroy, S. M., McIntyre, J. B., ... Beattie, T. (2018). Effect of a 12-month exercise intervention on leukocyte telomere length: Results from the ALPHA Trial. *Cancer Epidemiology, 56*, 67–74.

Fuller, K. N. Z., & Thyfault, J. P. (2021). Barriers in translating preclinical rodent exercise metabolism findings to human health. *Journal of Applied Physiology, 130*(1), 182–192.

Gan, Z., Fu, Z., Stowe, J. C., Powell, F. L., & McCulloch, A. D. (2016). A protocol to collect specific mouse skeletal muscles for metabolomics studies. *Methods in Molecular Biology, 1375*, 169–179.

Granger, D. A., Johnson, S. B., Szanton, S. L., Out, D., & Schumann, L. L. (2012). Incorporating salivary biomarkers into nursing research: An overview and review of best practices. *Biological Research for Nursing, 14*(4), 347–356.

Haack, M., Kraus, T., Schuld, A., Dalal, M., Koethe, D., & Pollmächer, T. (2002). Diurnal variations of interleukin-6 plasma levels are confounded by blood drawing procedures. *Psychoneuroendocrinology, 27*(8), 921–931.

Haque, A., Engel, J., Teichmann, S. A., & Lönnberg, T. (2017). A practical guide to single-cell RNA-sequencing for biomedical research and clinical applications. *Genome Medicine, 9*(1), 1–12.

Hogue, S. R., Gomez, M. F., da Silva, W. V., & Pierce, C. M. (2019). A customized at-home stool collection protocol for use in microbiome studies conducted in cancer patient populations. *Microbial Ecology, 78*(4), 1030–1034.

House, R. V. (2010). Fundamentals of clinical immunotoxicology. In R. R. Dietert (Ed.), *Immunotoxicity Testing* (pp. 363–384). Springer Protocols.

Kaufmann, L., Syedbasha, M., Vogt, D., Hollenstein, Y., Hartmann, J., Linnik, J. E., ... Egli, A. (2017). An optimized hemagglutination inhibition (HI) assay to quantify influenza-specific antibody titers. *Journal of Visualized Experiments, 2017*(130), 1–10.

Kemi, O. J., Loennechen, J. P., Wisløff, U., & Ellingsen, Y. (2002). Intensity-controlled treadmill running in mice: Cardiac and skeletal muscle hypertrophy. *Journal of Applied Physiology, 93*(4), 1301–1309.

Kim, S. W., Roh, J., & Park, C. S. (2016). Immunohistochemistry for pathologists: Protocols, pitfalls, and tips. *Journal of Pathology and Translational Medicine, 50*(6), 411–418.

Knoblaugh, S. E., & Randolph-Habecker, J. (2018). Necropsy and histology. In P. M. Treuting, S. M. Dintzis & K. S. Montine, *Comparative Anatomy and Histology* (2nd ed., pp. 23–51). Elsevier Inc.

Krüger, K., Gessner, D. K., Seimetz, M., Banisch, J., Ringseis, R., Eder, K., ... Mooren, F. C. (2013). Functional and muscular adaptations in an experimental model for isometric strength training in mice. *PLoS One, 8*(11), e79069.

Kurz, E., Hirsch, C. A., Dalton, T., Shadaloey, S. A., Vucic, E., Winograd, R., ... Bar-Sagi, D. (2022). Exercise-induced engagement of the IL-15/IL15Rα axis promotes anti-tumor immunity. *Cancer Cell, 40*(7), 720–737.

Labuschagne, I., Grace, C., Rendell, P., Terrett, G., & Heinrichs, M. (2019). An introductory guide to conducting the trier social stress test. *Neuroscience and Biobehavioral Reviews, 107*, 686–695.

Lee, J. A., Spidlen, J., Boyce, K., Cai, J., Crosbie, N., Dalphin, M., ... Brinkman, R. R. (2008). MIFlowCyt: The minimum information about a flow cytometry experiment. *Cytometry Part A, 73*(10), 926–930.

Lowe, D. A., & Alway, S. E. (2002). Animal models for inducing muscle hypertrophy: Are they relevant for clinical applications in humans? *Journal of Orthopaedic and Sports Physical Therapy, 32*(2), 36–43.

Mardis, E. R. (2008). Next-generation DNA sequencing methods. *Annual Review of Genomics and Human Genetics, 9*, 387–402.

Midgley, A. W., Carroll, S., Marchant, D., McNaughton, L. R., ... Siegler, J. (2009). Evaluation of true maximal oxygen uptake based on a novel set of standardized criteria. *Applied Physiology Nutrition and Metabolism, 34*(2), 115–123.

Mijwel, S., Cardinale, D. A., Norrbom, J., Chapman, M., ... Rundqvist, H. (2018). Exercise training during chemotherapy preserves skeletal muscle fiber area, capillarization, and mitochondrial content in patients with breast cancer. *FASEB Journal, 32*(10), 5495–5505.

Neri, S., Mariani, E., Meneghetti, A., Cattini, L., ... Facchini, A. (2001). Calcein-acetyoxymethyl cytotoxicity assay: Standardization of a method allowing additional analyses on recovered effector cells and supernatants. *Clinical and Diagnostic Laboratory Immunology, 8*(6), 1131–1135.

Nielsen, H. G., Øktedalen, O., Opstad, P.-K., ... Lyberg, T. (2016). Plasma cytokine profiles in long-term strenuous exercise. *Journal of Sports Medicine, 2016*, 1–7.

Nolan, T., Hands, R. E., & Bustin, S. A. (2006). Quantification of mRNA using real-time RT-PCR. *Nature Protocols, 1*(3), 1559–1582.

Olivera, S., & Tomic-Canic, M. (2013). *Human Ex Vivo wound healing model*. In J. M. Walker (Ed.), *Wound Regeneration and Repair* (pp. 255–265). Springer Protocols.

Omary, M. B., Cohen, D. E., El-Omar, E. M., Jalan, R., ... Turner, J. R. (2016). Not all mice are the same: Standardization of animal research data presentation. *Gastroenterology, 150*(7), 1503–1504.

Parasuraman, S., Raveendran, R., & Kesavan, R. (2010). Blood sample collection in small laboratory animals. *Journal of Pharmacology and Pharmacotherapeutics, 1*(2), 87–93.

Pedersen, L., Idorn, M., Olofsson, G. H., Lauenborg, B., Nookaew, I., Hansen, R. H., ... Hojman, P. (2016). Voluntary running suppresses tumor growth through epinephrine- and IL-6-dependent NK cell mobilization and redistribution. *Cell Metabolism, 23*(3), 554–562.

Philippeos, C., Hughes, R. D., Dhawan, A., & Mitry, R. R. (2012). Introduction to cell culture. In R. R. Mitry & R. D. Hughes (Eds.), *Human Cell Culture Protocols* (3rd ed., pp. 1–14). Springer Protocols.

Pilegaard, H., Saltin, B., & Neufer, D. P. (2003). Exercise induces transient transcriptional activation of the PGC-1α gene in human skeletal muscle. *Journal of Physiology, 546*(3), 851–858.

Plouffe, B. D., Murthy, S. K., & Lewis, L. H. (2015). Fundamentals and application of magnetic particles in cell isolation and enrichment. *Reports on Progress in Physics, 78*(1), 016601.

Poole, D. C., Copp, S. W., Colburn, T. D., Craig, J. C., Allen, D. L., Sturek, M., ... Musch, T. I. (2020). Guidelines for animal exercise and training protocols for cardiovascular studies. *American Journal of Physiology - Heart and Circulatory Physiology, 318*(5), H1100–H1138.

Poole, D. C., & Jones, A. M. (2017). Measurement of the maximum oxygen uptake Vo2max: Vo2peak is no longer acceptable. *Journal of Applied Physiology, 122*(4), 997–1002.

Poole, D. C., Rossiter, H. B., Brooks, G. A., & Gladden, L. B. (2021). The anaerobic threshold: 50+ years of controversy. *Journal of Physiology, 599*(3), 737–767.

Rentka, A., Koroskenyi, K., Harsfalvi, J., Szekanecz, Z., Szucs, G., Szodoray, P., & Kemeny-Beke, A. (2017). Evaluation of commonly used tear sampling methods and their relevance in subsequent biochemical analysis. *Annals of Clinical Biochemistry, 54*(5), 521–529.

Reynolds, J. M., Gordon, T. J., & Roberts, R. A. (2006). Prediction of one repetition maximum strength from multiple repetition maximum testing and anthropometry. *Journal of Strength and Conditioning Research, 20*(3), 584–592.

Roestenberg, M., Hoogerwerf, M. A., Ferreira, D. M., Mordmüller, B., & Yazdanbakhsh, M. (2018). Experimental infection of human volunteers. *The Lancet Infectious Diseases, 18*(10), e312–e322.

Schauer, T., Djurhuus, S. S., Simonsen, C., Brasso, K., & Christensen, J. F. (2022). The effects of acute exercise and inflammation on immune function in early-stage prostate cancer. *Brain, Behavior, and Immunity - Health, 25*, November 2022, 100508.

Scheele, C., Henriksen, T. I., & Nielsen, S. (2022). Isolation and characterization of human brown adipocytes. In D. A. Guertin & C. Wolfrum (Eds.), *Brown Adipose Tissue: Methods and Protocols* (pp. 217–234). Springer US.

Schwabe, L., Haddad, L., & Schachinger, H. (2008). HPA axis activation by a socially evaluated cold-pressor test. *Psychoneuroendocrinology, 33*(6), 890–895.

Segerstrom, S. C., & Miller, G. E. (2004). Psychological stress and the human immune system: A meta-analytic study of 30 years of inquiry. *Psychological Bulletin, 130*(4), 601–630.

Shephard, R. J. (2017). Open-circuit respirometry: A brief historical review of the use of Douglas bags and chemical analyzers. *European Journal of Applied Physiology, 117*(3), 381–387.

Shields, G. S., & Slavich, G. M. (2017). Lifetime stress exposure and health: A review of contemporary assessment methods and biological mechanisms. *Social and Personality Psychology Compass, 11*(8), e12335.

Smeets, T., Cornelisse, S., Quaedflieg, C. W. E. M., Meyer, T., Jelicic, M., & Merckelbach, H. (2012). Introducing the Maastricht Acute Stress Test (MAST): A quick and non-invasive approach to elicit robust autonomic and glucocorticoid stress responses. *Psychoneuroendocrinology, 37*(12), 1998–2008.

Striedinger, K., Barruet, E., & Pomerantz, J. H. (2021). Purification and preservation of satellite cells from human skeletal muscle. *STAR Protocols, 2*(1), 100302.

Terra, R., Alves, P. J. F., Gonçalves Da Silva, S. A., Salerno, V. P., & Dutra, P. M. L. (2013). Exercise improves the Th1 response by modulating cytokine and NO production in BALB/c mice. *International Journal of Sports Medicine, 34*(7), 661–666.

Thomas, C. E., Sexton, W., Benson, K., Sutphen, R., & Koomen, J. (2010). Urine collection and processing for protein biomarker discovery and quantification. *Cancer Epidemiology Biomarkers and Prevention, 19*(4), 953–959.

Tighe, P. J., Ryder, R. R., Todd, I., & Fairclough, L. C. (2015). ELISA in the multiplex era: Potentials and pitfalls. *Proteomics - Clinical Applications, 9*(3–4), 406–422.

Uffelmann, E., Huang, Q. Q., Munung, N. S., de Vries, J., Okada, Y., Martin, A. R., ... Posthuma, D. (2021). Genome-wide association studies. *Nature Reviews Methods Primers, 1*(59), 1–21.

Wang, N., Strugnell, R. A., Wijburg, O. L., & Brodnicki, T. C. (2013). Systemic infection of mice with Listeria monocytogenes to characterize host immune responses. In I. C. Allen (Ed.), *Mouse Models of Innate Immunity* (pp. 125–144). Springer Protocols.

Westenberg, P. M., Bokhorst, C. L., Miers, A. C., Sumter, S. R., Kallen, V. L., van Pelt, J., & Blöte, A. W. (2009). A prepared speech in front of a pre-recorded audience: Subjective, physiological, and neuroendocrine responses to the Leiden Public Speaking Task. *Biological Psychology, 82*(2), 116–124.

World Health Organization. (2010). *WHO Guidelines on Drawing Blood: Best Practices in Phlebotomy.* Retrieved from: https://www.who.int/publications/i/item/9789241599221

Zaritskaya, L., Shurin, M. R., Sayers, T. J., & Malyguine, A. M. (2010). New flow cytometric assays for monitoring cell-mediated cytotoxicity. *Expert Review of Vaccines, 9*(6), 601–616.

Zuber, T. J. (2002). Punch biopsy of the skin. *American Family Physician, 65*(6), 1155–1158.

4 Cellular immunity and exercise

Karsten Krüger and Philipp Zimmer

Introduction

In 1902, Ralph Larrabee comprehensively described high leucocyte counts after the Boston Marathon in the United States (see Chapter 1). In the following decades, and mainly in the 1980s and 1990s, a growing body of literature started examining circulating leucocyte numbers and functions in response to individual bouts of exercise (sometimes called "acute" exercise or at least "acute effects" of exercise) and longer-term exercise training (sometimes called "chronic" exercise, or at least "chronic effects" of longer-term exposure to exercise). Resting blood leucocyte numbers, which usually range between 4 and 11 × 10⁹ per litre of blood can increase up to four-fold in response to individual bouts of exercise. Today, we know that these effects strongly depend on the mode, duration, and intensity of exercise, which influence both the short-term effects of individual exercise bouts and the long-term effects of exercise training. Many of these effects are also influenced by sex, age, infection history (see Chapter 8), health, fitness, body composition and chronic disease (see Chapters 10–12), environmental conditions (see Chapter 13), nutrition (see Chapter 14), and psychological stress (see Chapter 15).

This chapter will primarily focus on shifts in leucocyte numbers in blood. However, it must be considered that leucocytes in blood account for around 2% of all leucocytes in the body – with most cells residing in lymphoid tissue (Ganusov & De Boer, 2007). Thus, the observed systemic alterations in response to exercise may not reflect tissue-specific changes. However, blood provides a key route for redistributing leucocytes, facilitating their trafficking to tissues as part of immune-surveillance, and even small effects in blood may have clinical relevance. This chapter primarily focuses on human research; however, animal models as well as cell-culture experiments are summarised where relevant, because they provide detailed mechanistic insights, but the transferability of these findings to humans needs to be considered carefully. Resting blood leucocyte numbers can also serve as important markers for acute infections and provide diagnostic and prognostic indicators for several chronic diseases. In settings where access to advanced high-resolution flow cytometry is not available, integrative clinical markers – assessed using automated haematology analysers that can be used with almost no training – such as the neutrophil-to-lymphocyte ratio, can be derived from routinely measured complete blood counts (also called the leucocyte differential) and has prognostic value in several diseases (Walzik et al., 2021).

This chapter is divided into two sections. The first section describes the short-term effects of individual exercise bouts, and the second section describes the effects of longer-term exercise training. Within each section, the effects of exercise are summarised for innate immune cells followed by adaptive immune cells. For each cell type, most focus is on exercise-induced shifts in leucocyte numbers measured in blood, but where possible, alterations to the function of cells

DOI: 10.4324/9781003256991-4

are also covered. Due to the number of leucocyte populations summarised, the text provides an overview and summary of results and does always provide methodological details for each study. In places, the reader is directed towards specific experimental articles that have particularly robust methods and provide very clear insight into the exercise-induced time course, pattern, or magnitude of change in leucocytes.

Individual exercise bouts and leucocytes – general considerations

Individual exercise bouts induce temporary changes to the number of leucocytes in blood, which is characterised by an immediate mobilisation of cells. This process, termed leucocytosis (see Chapter 1, Box 1.1), is characterised by an increase in granulocytes, monocytes, and lymphocytes. The extent of mobilisation is determined primarily by the mode, intensity, and duration of exercise, but also by other factors such as age, nutritional status, and ambient temperature (Sand et al., 2013). Thus, the combination of both long (e.g., > 1 hour) and vigorous (e.g., >70% of maximum oxygen uptake) exercise invokes the strongest and longest leukocytosis. Events such as marathons which are particularly long (e.g., at least 2.5 hours) – but vary in intensity – also invoke strong responses. In general, endurance exercise provokes a stronger response than resistance exercise (Schlagheck et al., 2020).

Physiological reasons for exercise-induced leukocytosis have long been debated but are now thought to be an immunological correlate of a fight or flight response, which prepares the immune system to respond to a potential violation of tissue integrity or pathogen invasion. Correlation analyses, preclinical, and clinical work on the effect of exogenously administered agents have shown that stress hormones (e.g., catecholamines and cortisol) contribute significantly to the mobilisation of immune cells into blood. Moreover, mechanical forces, such as increased cardiac output and shear stress, further support this effect. In general, leucocytes can be mobilised from several sites, including the endothelium, the lymphatic system, including lymph nodes, the thymus, the bone marrow, the spleen, the liver, and the lungs. Indeed, evidence suggests that a variety of tissues have different responsibilities and contributions in deploying specific leucocytes. In this context, the strongest effects of exercise on cell mobilisation occur in cells of the innate immune system, such as neutrophils and NK cells.

It can take up to 48 hours or longer for some leucocyte sub-types to return to pre-exercise levels after very prolonged and vigorous endurance exercise. However, distinct differences can be observed between kinetic patterns of granulocytes, monocytes, and lymphocytes and among their sub-types. Generally, circulating granulocyte numbers further increase for up to 6 hours after cessation of exercise – driven by neutrophils. In contrast, lymphocyte numbers peak within the first few minutes after exercise cessation, before returning to pre-exercise levels, or falling below resting levels, in the following few hours. Without differentiating between lymphocyte sub-types, lymphocytopenia, can be observed for up to 6 hours after exercise. During this time, lymphocytes can decrease to 50% below resting levels. Lymphocytopenia is strongly driven by NK cells and effector-memory T cells, whereas other lymphocyte sub-types (e.g., B cells) are less responsive and even show changes in the opposite direction. Monocytes exhibit an immediate increase during exercise and usually return to baseline levels within the 3 hours of exercise cessation (Risøy et al., 2003).

While perturbations in circulating granulocyte numbers are suspected to be mainly driven by exercise duration, alterations of lymphocyte and monocyte numbers are thought to be mediated by exercise intensity. However, these assumptions are based on a few small studies and warrant further investigation. In general, studies have compared exercise modalities (e.g., running compared with cycling, or aerobic exercise compared with resistance exercise) and different intensities (e.g., moderate compared with vigorous), but a challenge is how to match the conditions for a true comparison (e.g., duration, intensity, or energy expenditure). Without examining leucocyte sub-types – and keeping in mind the differential response among the broader populations of granulocytes, monocytes, and lymphocytes and the effects of different modes, intensities, and durations of exercise – conclusions over exercise-induced changes to leucocytes can only be made on a cell sub-type and tissue-specific basis, especially when interpreting the clinical relevance of these effects. For example, exercise-driven changes in leucocyte numbers in adipose tissue and skeletal muscle (see Chapter 6) or tumour tissue (see Chapter 10) indicate specific patterns among sub-types which differ in a tissue-specific manner and when compared with changes in blood. More detailed information on exercise-induced alterations of circulating leucocyte sub-types is provided in the sections below, and for a summary, see Figure 4.1.

Individual exercise bouts and cells of the innate immune system

Innate immune cells contribute towards the primary defence against bacterial, viral, fungal, or parasitic infections. The innate immune response, consisting of many components, usually becomes active within minutes or hours upon pathogen encounter. Innate immune cells include neutrophils, mast cells, eosinophils, monocytes, macrophages, and NK cells. At the molecular level, innate immune cells respond to pathogens after recognising pathogen-associated molecular patterns via their pattern recognition receptors. Among innate immune cells, neutrophils are most common, comprising around 50–60% of leucocytes in blood. They have an important role in the non-specific defence of the body against microbial pathogens and perform repair and remodelling work when tissue integrity is lost. Likewise, NK cells are very important cells for recognising virus-infected cells or tumour cells. Partly for these reasons, neutrophils and NK cells have been studied extensively in exercise immunology, and both cell types are very sensitive to individual bouts of exercise and are mobilised mainly by adrenergic mechanisms (Malaguarnera et al., 2008).

Individual exercise bouts and neutrophils

Neutrophils are demarginated from endothelial cells and mobilised from other body sites immediately during exercise. In contrast to lymphocytes, neutrophils often continue to rise for 2 or 3 hours after exercise before returning to baseline levels in the following hours, depending on the duration of exercise. The changes in the post-exercise period have been described as a delayed second neutrophilia which is mainly triggered by an additional mobilisation of cells from the bone marrow (Alack et al., 2019). The degree of neutrophilia seems to be dependent on the duration of exercise and is mediated by adrenocorticotropic hormones as well as cortisol and catecholamines. Similar mobilisation effects are shown after infusion of adrenaline, providing further evidence that adrenergic mechanisms play a role in the trafficking of these cells (Simpson et al., 2021). Neutrophils can broadly be divided into non-segmented or "band" cells – which represent a younger population – and segmented cells, which represent more mature and functionally effective cells. One way to differentiate neutrophils is based on their expression of CD16, but a variety of identification strategies and stages of neutrophil development have been established (for a review, see: McKenna et al., 2021). The neutrophil sub-types that are mobilised are mature and differentiated. Studies have shown that the magnitude of neutrophil mobilisation during exercise is dependent on exercise intensity. For example, Robson et al. (1999) showed that neutrophil count was greater following cycling to exhaustion at 80% of maximum oxygen uptake (duration: 37 ± 19 minutes) compared with cycling at 55% of

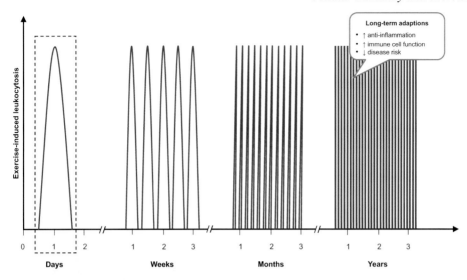

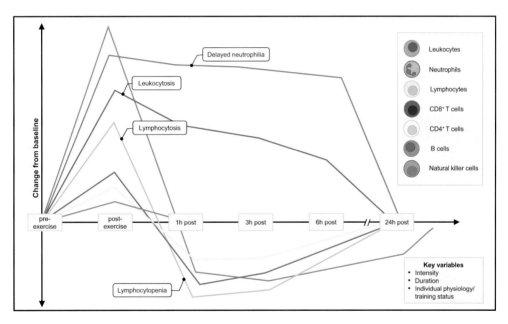

Figure 4.1 The time course of leucocyte and major leucocyte sub-type kinetics in response to an individual bout of vigorous endurance exercise. Individual bouts of vigorous (e.g., around 70% of maximum oxygen update) endurance exercise such as running or cycling (e.g., of around 1 hour in duration) cause substantial changes to the number of leucocytes circulating in blood. Total leucocytes increase 2- to 4-fold during exercise, but there is a differential response among leucocyte sub-types during and following exercise. During exercise, the strongest responses are shown by neutrophils, NK cells, and T cells. In the post-exercise period, neutrophils often remain above pre-exercise values, and NK cells and T cells fall below pre-exercise levels for several hours, whereas B cells typically return to pre-exercise values in the minutes after exercise cessation. The mobilisation of leucocytes is referred to as leukocytosis; the mobilisation of lymphocytes is referred to as lymphocytosis, and when lymphocytes fall below normal levels, this is referred to as lymphocytopenia.

maximum oxygen uptake (duration: 164 ± 23 minutes). After prolonged bouts of exercise, cells in the hours after exercise are typically CD16^{dim}/CD62L^{bright} (i.e., a low level of CD16 expression and a high level of CD62 expression) which represent young neutrophils that are non-segmented with a "band"-shaped nucleus and thought to come from the bone marrow. However, CD16^{bright}/CD62L^{dim} neutrophils (i.e., with a high level of CD16 expression, and with a low level of CD62 expression) – which make up a smaller proportion of neutrophils post-exercise – are thought to represent an immunosuppressive sub-type, because these cells can inhibit T cell proliferation (Kamp et al., 2012).

A substantial number of studies have examined whether exercise alters neutrophil function. The short-term exercise-induced mobilisation of neutrophils is often interpreted as a pro-inflammatory response, given that microtrauma and tissue damage can occur. Indeed, following vigorous exercise, neutrophils have been shown to produce IL-8 and TNF-α, which return to baseline levels after several hours (Borges et al., 2018). Exercise can also be a stimulus for neutrophils to migrate to tissues. After particularly long and vigorous bouts of exercise comprising eccentric muscle contractions, an accumulation of neutrophils can be detected in the skeletal muscle – showing they are attracted to sites of inflammation (Peake et al., 2017) (see Chapter 6). Indeed, neutrophils leave the blood in response to a chemotactic stimulus. Studies examining whether exercise influences neutrophil chemotaxis are not all consistent, but generally, it seems that individual bouts of moderate-intensity exercise promote chemotaxis (Peake, 2002). In addition, moderate-intensity exercise has been shown to increase the neutrophil oxidative burst, but long and vigorous bouts of exercise can temporarily impair oxidative burst and degranulation. Mediators of these processes appear to be cytokines, catecholamines, and cortisol (Peake, 2002). For example, it is assumed that the exercise-induced release of muscle-derived IL-6, which has a broad anti-inflammatory effect, may represent an inhibitory mechanism.

A crucial process for pathogen elimination is phagocytosis. Several older studies show that phagocytic activity of neutrophils is impaired by individual bouts of moderate-to-vigorous-intensity exercise (Ortega Rincón, 1994). However, the reason for these observations is due to lower numbers of phagocytically active cells in blood after exercise. More recent studies have shown that during and following exercise bouts, the number of phagocytically active cells increases, but the phagocytic activity per cell is reduced when accounting for a large increase in cell number (Peake, 2002; Malaguarnera et al., 2008). Indeed, one of the most important principles when interpreting any measurements of cell function, among any population of immune cells in the post-exercise period, is accounting for the shifts in numbers, proportions, and phenotypes of cells post-exercise (see Box 4.1). The most recent studies investigating neutrophil function following exercise have examined neutrophil extracellular traps (NETs) which capture microorganisms for killing via oxidative and non-oxidative methods (see Chapter 2). Studies have shown that exercise bouts promote a mild release of NETs, interpreted as a pro-inflammatory response to exercise, but it is thought to be not harmful, given that these structures are degraded by DNAses and proteases after exercise (Valeria Oliveira de Sousa et al., 2021).

Box 4.1   Individual exercise bouts and leucocyte function

In vitro cell-based assays can examine the viability, maturation, activation, and functional capacity of cells – including proliferation, migration, cytokine production, and killing. In exercise immunology, most focus has been on measurements reflecting the cytotoxicity of T cells and NK cells and phagocytic activity or production of reactive oxygen species (ROS) by neutrophils and monocytes. The principles of these methods are outlined in

Chapter 3. When interpreting functional assays, the profound cellular shifts observed in blood during and following exercise must be considered. Indeed, the phenotypic composition of leucocytes in blood changes, whereby cells with particular functional characteristics are more – or less – susceptible to exercise-induced leukocytosis and leukopenia. Thus, the relationships between cell frequency, cell phenotypes, and cell function must be taken into account at each time point. When conducting *in vitro* cell function assays, a stimulant is typically added to a set number of cells in blood collected before, immediately after exercise, and in the hours later. However, if the phenotypic composition of those cells is different at each time point (e.g., the proportion of cytotoxic cells in each sample), the results of the functional assay (e.g., a cytotoxicity assay) will be altered accordingly. This has been known for over three decades. For example, Shinkai and colleagues (1992) assessed changes in the number of different cell populations and their function in response to 60 minutes of cycling at 60% of maximum oxygen uptake. It was concluded that; *Functional capability correlated well with the percentage of each major responder subset in the assay, suggesting that the in vitro lymphocyte responsiveness and NK activity did not change significantly on a per-cell basis.*

This concept has not always been considered or accounted for in exercise immunology studies published in the following decades. To demonstrate the importance of shifting cell numbers during and following bouts of exercise, Lester et al. (2021) compared two 30-minute cycling bouts at moderate or vigorous intensity, measuring mitogen-stimulated ROS production in whole blood. Higher ROS production was shown immediately post-exercise compared with baseline (moderate intensity: +45% increase in ROS; vigorous intensity: +35% increase in ROS) and 60 minutes post-exercise compared with baseline (moderate intensity: +64% increase in ROS; vigorous intensity: +85% increase in ROS). These changes correlated positively with the change in number of cells known to be the primary producers of ROS – neutrophils and monocytes (and platelets). When ROS production was corrected for using a composite score (combined counts of neutrophils, monocytes, and platelets) or when data were expressed on a per-neutrophil basis (i.e., correcting for absolute counts of the most responsive cells), the exercise-associated increase in ROS production was unchanged from pre-exercise values. Thus, if studies conclude that bouts of exercise lead to increased or impaired immune cell function, readers should critically evaluate whether cell numbers and phenotypes have been accounted for. It is worth emphasising though, that it has indeed been shown that neutrophil function is impaired 3 hours after a 2-hour period of cycling at 55% of peak aerobic power when expressing results on a per-cell basis (Li & Cheng, 2007). It is thought this finding reflects the secondary mobilisation of immature neutrophils from bone marrow, which have lower content of granular digestive enzymes and lower ROS production. Thus, the study by Li and Cheng (2007) provides further rationale for needing to account for cell phenotypes in functional assays.

Individual exercise bouts and monocytes

Monocytes have an important role in combating pathogens and regulate adaptive immune responses. Their reactivity is triggered primarily by pattern recognition receptors such as Toll-like-receptor-4 (TLR-4) which recognises the bacterial membrane component lipopolysaccharide (LPS). Upon stimulation, this receptor associates with different adaptor proteins such as myeloid differentiation primary response 88 (MyD88) or TIR domain-containing adaptor-inducing interferon-β (TRIF), leading to the activation of signalling molecules such as mitogen-activated protein kinases or nuclear factor-kappa-B (NF-kB). As antigen-presenting cells, they also activate cells of the adaptive immune system and therefore monocytes link innate and adaptive immunity. Circulating monocytes can be subdivided by their variable expression of the LPS receptor CD14 and Fcγ-III receptor CD16 which mediates antibody-dependent cytotoxicity. Flow cytometry can be used to distinguish so-called "classical" monocytes, which are characterised as CD14^{bright}CD16– (i.e., a high expression of CD14 and no expression of CD16) and represent about 85% of monocytes in blood. Other subpopulations are so-called "intermediate" monocytes and defined as CD14^{bright}CD16+, and "non-classical" monocytes characterised as CD14^{dim}CD16+ (i.e., a lower expression of CD14). Individual bouts of exercise induce a transient monocytosis of about 1–3 hours, most likely due to a shift of monocytes from the marginalised pool. Exercise primarily mobilises CD14^{dim}CD16+ "non-classical" monocytes with a pro-inflammatory phenotype, whereas the proportion of "classical" CD14^{bright}CD16– monocytes declines. In addition, exercise has been shown to decrease the expression of TLR4 receptors among "classical" CD14^{bright}CD16– and "intermediate" CD14^{bright}CD16+, monocytes, but not in "nonclassical" CD14^{dim}CD16+ monocytes. It is thought that once monocytes have been mobilised by exercise, they migrate to tissues, differentiating into macrophages, to repair tissue damage and contributing to adaptation by promoting angiogenesis. These processes are initially stimulated by pro-inflammatory M1-phenotype macrophages, which stimulate the proliferation of satellite cells and subsequent differentiation into myotubes, before the macrophage adopts an anti-inflammatory M2-phenotype, contributing to tissue remodelling (Wynn & Vannella, 2016). M2-phenotype macrophages are involved in inflammation resolution, angiogenesis, extracellular matrix formation, myoblast fusion, and myotube growth (Tarban et al., 2022). The chronology and the central role of monocytes and macrophages in these processes have been shown predominantly in studies of experimentally induced muscle damage but are likely to be similar following other forms of exercise. For more details on tissue macrophages, see Chapter 6.

Individual exercise bouts and dendritic cells

Dendritic cells are professional antigen-presenting cells which initiate and regulate antigen-specific immunity via their interactions with T cells. In blood, two major sub-types of dendritic cells exist, referred to as myeloid and conventional dendritic cells. They differ primarily in the expression patterns of cytokines, toll-like receptors and migratory behaviour. Very few studies have examined how exercise affects dendritic cells, but these cells also mobilise during exercise. Like other immune cells, the magnitude of change appears to correlate with exercise intensity and catecholamine release. Mobilised dendritic cells exhibit increased expression of CCR5 and CD62L, suggesting that the cells originate from lymphoid tissues or the marginal pool (Suchánek et al., 2010). One of the most comprehensive analyses of dendritic cells was published by Brown and colleagues (2018). In this study, nine healthy men exercised on a cycle ergometer for 20 minutes at 80% of maximum oxygen uptake. Blood was collected before, in the last minute of exercise and 30 minutes later. Total dendritic cells increased by +150% during exercise, and plasmacytoid dendritic cells were mobilised to a greater extent than myeloid dendritic cells. Among myeloid dendritic cells, CD1c–CD141+ cells showed the greatest exercise-induced mobilisation, and CD205+ myeloid cells, which recognise apoptotic cells, responded most strongly. Interestingly, CD209+ myeloid cells, which recognise virus-infected cells, were very rare in blood and did not mobilise with exercise. All dendritic cells returned to their resting levels within 30 minutes of exercise. Given the importance of plasmacytoid dendritic cells in producing interferons against viral and bacterial pathogens, mobilisation of these effector cells is most likely reflective of an immune-surveillance response (Brown et al., 2018).

Two other studies have reported similar increases in the numbers of myeloid dendritic cells up to 72 hours after a marathon; however, this was accompanied by a decrease in the number of plasmacytoid dendritic cells, and lower expression levels of TLR7 – an important molecule for detecting viral infections (Lackermair et al., 2017; Nickel et al., 2012).

Individual exercise bouts and NK cells

Unlike other cells of the innate immune system, NK cells are a type of lymphocyte and comprise 5–20% of the lymphocyte pool. NK cells are characterised by the lack of CD3 and the co-expression of CD16 and CD56. NK cells are further divided into CD56^{bright} (i.e., a high expression of CD56) and CD56^{dim} (i.e., a low expression of CD56) sub-types, accounting for 10% and 90% of circulating NK cells, respectively. CD56^{bright} cells mainly produce cytokines (IFN-γ and TNF-α); they are weakly cytotoxic prior to activation and are thought to have immunoregulatory properties. On the other hand, CD56^{dim} cells are effector cells with immediate cytotoxic potential. NK cells have the ability to recognise and eliminate virus-infected cells and tumours without prior antigen exposure. NK cells are highly responsive to exercise and mainly contribute to the biphasic reaction within the lymphocyte compartment.

Circulating NK cell numbers quickly increase after the onset of exercise increasing by between 50% and 400% (Shephard & Shek, 1999). NK cells are a substantial contributor exercise-induced lymphocytosis, and studies have shown that this increase in NK cell numbers is driven by catecholamines (Graff et al., 2018; Pedersen et al., 2016). Within 15 minutes of exercise cessation, NK cells leave blood, migrating to tissues. Circulating NK cell numbers may be reduced by around 40% for 24–48 hours after prolonged and vigorous endurance exercise (Shek et al., 1995). A small number of studies even report reduced NK cell numbers after extreme exercise for up to seven days. The movement of NK cells from blood to the tissues is the most relevant contributor to exercise-induced lymphocytopenia. It was proposed in the 1990s that low blood NK cells represented an 'open window' to infection (see Chapters 1 and 9), but more recently, it has generally been agreed that this likely represents enhanced immune-surveillance (Campbell & Turner, 2018). Indeed, seminal animal studies have indicated that NK cells mobilised from the spleen may even migrate into tumour tissue after exercise (Pedersen et al., 2016). However, although NK cell recruitment to tumour tissue is one of the most frequently discussed anti-oncogenic effects of exercise, clinical data in humans is rare and small trials in humans have not confirmed these findings (Djurhuus et al., 2023; Schenk et al., 2022) (see Chapter 10). In addition, exercise-responsive mediators, such as serotonin, have also been shown to potentially alter NK cell migration suggesting that NK cell redeployment is a multifactorial process (Zimmer et al., 2017). Studies have also examined which cells are most exercise-responsive, and it is widely accepted that exercise predominantly affects the cytotoxic CD56^{dim} subset (Campbell et al., 2009). Recent work has advanced these findings to show that individual bouts of exercise mobilises highly differentiated NK cells (CD158b+/ NKG2A−, CD57+, NKG2C−) independent of exercise intensity. Moreover, other factors such as latent herpes virus infections, including *Cytomegalovirus* (CMV) infection (see Chapter 9) strongly impair exercise-driven NK cell mobilisation and should be considered when interpreting and discussing study results (Bigley et al., 2014).

Many studies have examined whether exercise bouts influence NK cell cytotoxicity. This extensive body of research has been summarised in a recent meta-analysis, which confirmed that individual bouts of exercise increase NK cell cytotoxicity for 1–2 hours after exercise cessation (Rumpf et al., 2021). In addition, responses are stronger with aerobic exercise compared with resistance exercise, with stronger cytotoxicity responses induced by vigorous-intensity exercise compared with moderate-intensity exercise. Importantly, these conclusions are not affected by absolute NK cell numbers, suggesting that the relative proportional increase in CD56^{dim} cells plays a role. Finally, it must be considered that such analyses refer to the cytotoxicity of NK cells in blood, and potential tissue-specific changes have not been adequately examined.

Individual exercise bouts and other innate ILCs

Innate lymphoid cells (ILCs) are types of lymphocytes that do not express diversified antigen receptors. Thus, they are distinct from classical lymphocytes, such as T and B cells. ILCs are largely tissue-resident, maintaining homeostasis and producing regulatory cytokines that control immune responses at mucosal barriers, contributing to adaptive immunity and regulating tissue inflammation (Vivier et al., 2018). Type-1 ILCs include NK cells, which – as outlined in the prior section – are the most studied ILC and, unlike other ILCs, are present in large numbers in the bloodstream. Type-2 ILCs play a critical role in curtailing infections by helminths while also aggravating allergic asthma and respiratory disorders as they tend to reside in the lungs and nasal mucosa. Type-3 ILCs secrete IL-22, an important cytokine that regulates epithelial cell repair and secretion of antimicrobial peptides, such as β-defensin and lipcalin-2, by mucosal epithelial cells. A sub-population of Type-3 ILCs can secrete IL-17, a key cytokine for neutrophil recruitment to sites of infection. Loss of tissue integrity and activation of stress-signalling pathways, such as through the release of DAMPs, activate PRRs on ILCs. Historically, only NK cells were thought to be impacted by individual bouts of exercise, while the remaining ILCs sub-types were believed to remain in tissues. However, a recent study assessing the impact of 30 minutes of vigorous-intensity cycling (~70% maximum oxygen uptake) on the number of circulating ILCs highlighted that the number of tissue-resident progenitor ILCs, along with Type-1 helper ILCs and Type-3 ILCs increased in circulation in response to exercise. Conversely, the same bout of exercise led to significant reductions in the number of circulating Type-2 ILCs (Cho et al., 2021). ILCs typically leave blood within the first minutes of exercise cessation. Further research is required to fully understand how individual bouts of exercise affect the less-studied ILC populations.

Individual exercise bouts and haematopoietic stem and progenitor cells

Haematopoietic stem and progenitor cells (HSPC) are commonly found in bone marrow and are responsible for the formation of all blood cells. It is assumed that these cells represent a "reserve" of leucocytes in an immature form and contribute to repair and adaptation processes following exercise. Individual bouts of exercise mobilise most sub-types of HSPCs, and exercise intensity and duration appear not to influence this response (Niemiro et al., 2017). Several human studies have characterised the time course and magnitude of HSPC mobilisation in response to individual bouts of exercise, given the potential relevance and clinical utility of exercise for improving cell yields during apheresis in the context of stem-cell transplant for treating haematological malignancies – for a review, see Emmons et al. (2016). Generally, short bouts of moderate-to-vigorous-intensity exercise lead to a transient two- or four-fold increase in HSPCs, which return to pre-exercise levels shortly after exercise cessation (Emmons et al., 2016). The function of HSPCs has also been examined following exercise bouts, and some studies have demonstrated improved proliferation or clonogenic capacity, expressed as colony-forming units (CFUs) which is a measure of the number of viable colonogenic cells. Colonogenic cells provide an indication of the number of viable cells which proliferate to form small colonies – a hallmark of HSPCs. In contrast, the number of progenitor cells in the bone marrow does not appear to change after an exercise bout. Possible signalling molecules responsible for the effects of exercise on HSPCs include granulocyte-colony stimulating factor (G-CSF), stem-cell factor (SCF), interleukin-3, and thrombopoeitin, which are released from bone marrow (Emmons et al., 2016) and also see Chapter 12, Box 12.2.

Individual exercise bouts and cells of the adaptive immune system

The cells of the adaptive immune system are B cells and T cells. B cells represent 5–15% of lymphocytes and express the cell surface markers CD19 and CD20. B cells can be differentiated into naive and memory sub-types, and upon activation and antigen recognition, naive B cells differentiate to memory cells and antibody-producing plasma cells, which can be further examined for the type of antibody produced (i.e., IgA, IgD, IgE, IgG, IgM). Following an immune response, long-lived memory B cells provide a faster and more vigorous response upon encountering the same pathogen. T cells account for between 60% and 80% of lymphocytes and respond to viral or bacterial infection, or tumours, by either producing cytokines as part of orchestrating the immune response or directly killing affected cells. Major T cell sub-types include CD4+ helper T cells which orchestrate immune responses by producing cytokines, and CD8+ cytotoxic T cells which are highly functional effector cells. Among CD4+ helper T cells, there are several sub-types, differentiated by their production of cytokines. Exercise immunology studies examining CD4+ T-helper cells have focused mostly on so-called Th1 cells, which produce cytokines upon activation that support innate cells during pathogen clearance, and so-called Th2 cells, which are involved in T-cell-dependent B cell activation and antibody production. However, many other sub-types exist, including Th9, Th17, Th22, T-follicular helper cells, and regulatory T cells, but these cell types have received little if any coverage in exercise immunology – except for some attention on regulatory T cells and Th17 cells. Readers are directed to comprehensive reviews covering these sub-types (Jiang & Dong, 2013; Cenerenti et al., 2022). The studies that have examined regulatory T cells typically identify these cells as CD25+ and FOXP3+ with a downregulation of CD127 (i.e., CD127^{dim}). Their anti-inflammatory properties include producing IL-10 and TGF-ß. A final sub-type of so-called "unconventional" T cells includes gamma-delta (γδ) T cells. These cells are found in small numbers in blood and are characterised by the expression of heterodimeric T cell receptors (TCRs) consisting of γ- and δ-chains, distinguishing them from the classical CD4+ helper T cells and CD8+ cytotoxic T cells, which express αβ-TCRs (Kumar et al., 2018). Overall, both B and T cells have been examined considerably in exercise immunology, with most focus on T cells.

Individual exercise bouts and T cells

Exercise-induced lymphocytosis is most prominent among effector cells, such as CD8+ cytotoxic T cells and γδ T cells, whereas CD4+ helper T cells and regulatory T cells exhibit small changes. Indeed, the magnitude of exercise-induced mobilisation is largely an effect of memory status (Natale et al., 2003). Indeed, antigen-specific T cells with a differentiated effector-memory phenotype exhibit the largest responses, and these cells re-express the naive T cell marker CD45RA, often called effector-memory "RA" expressing cells (EMRA). These cells are redistributed to a greater magnitude than effector-memory cells (EM) followed by Central Memory cells (CM) and finally naive cells (NA) (Campbell et al., 2009). The mobilisation profile is also dependent on CMV status – individuals infected with this virus exhibit larger responses than those who are not infected with CMV (Turner et al., 2010; Simpson et al., 2016). Lymphocytosis transitions to lymphocytopenia around 30–60 minutes after exercise cessation. Accordingly, lymphocytes seem to "disappear" from blood, which is now known to represent trafficking of cells to the tissues, first by CD8+ cytotoxic T cells, then γδ T cells, followed by CD4+ helper T cells. Lymphocytopenia is linked with two processes. First, there is a redistribution in lymphatic and non-lymphatic organs, such as lungs and gut. A small percentage of the mobilised cells become apoptotic, especially during vigorous exercise, which mainly affects highly differentiated cells (Krüger et al., 2008; Krüger et al., 2016). One theory of exercise-induced apoptosis is that senescent cell types may be removed from blood, making space in the immunological repertoire for more naive cells (Simpson, 2011) (see Chapter 9). In the 1980s and 1990s, it was first thought that lymphocytopenia represented a period of broad immunosuppression, also referred to as the "open window hypothesis" – a concept explaining the apparent increased susceptibility to infections following periods of vigorous and prolonged exercise. This concept remains controversial to date and is reviewed thoroughly in Chapters 1 and 9. Indeed, the contribution of lymphocytopenia to infection susceptibility has been thoroughly debated (Simpson et al., 2020). If exercise can cause temporary immunosuppression, it is likely that the mechanism is not as simple as lower cell counts in blood and minor fluctuations in cell function. For example, after vigorous exercise, such as running a marathon, a temporary imbalance of the Th1/Th2 ratio provides a potential explanation for altered aspects of immune function (Xiang et al., 2014).

Studies have also examined the effects of individual bouts of exercise on aspects of cell function. A common finding is that mitogen-stimulated T cell proliferation and production of cytokines such as IL-2 and IFN-γ are impaired immediately after individual bouts of vigorous exercise (Gleeson & Bishop, 2005). However, as outlined in Chapter 1 and Box 4.1 in this chapter, an important consideration when interpreting the results of these experiments, is whether authors have adequately accounted for the shifts in cell number and phenotypes during and following exercise. Thus, in many cases, when a decrease in function is reported post-exercise, this is usually due to a lower frequency of responding cells in samples (see Box 4.1). Indeed, cytokine production is strongly influenced by the cells present in blood at the time of sampling, and studies have shown that prolonged vigorous exercise decreases the proportion of Th1 cells but has less effect on Th2 cells (Steensberg et al., 2001). In turn, studies have shown that the release of cytokines when stimulated *in vitro* follows a similar pattern. For example, 1 hour of cycling at 75% maximum oxygen uptake decreased IL-2 production from stimulated T cells (Tvede et al., 1993), yet IL-4 production by stimulated lymphocytes was unaffected by 18 minutes of incremental exercise ranging from 55% to 85% maximum oxygen uptake (Moyna et al., 1996). Likewise, while studies have reported significant decreases in mitogen-stimulated T cell proliferation following prolonged vigorous exercise (Henson et al., 1999), these findings need to be interpreted with caution considering the points made in Box 4.1. However, if individual bouts of exercise do influence aspects of cell function on a per-cell basis, it is possible the mechanistic explanation is related to changes in cellular metabolism (see Chapter 14). For example, there is evidence that intracellular calcium homeostasis is affected by exercise, probably by increased transmembrane $Ca2+$ influx into cells and by decreased expression of plasma membrane calcium ATPases and sarco(endo) reticulum calcium ATPases (SERCA). However, a direct correlation between altered $Ca2+$ signalling and proliferation was not found (Liu et al., 2017), but other aspects of cellular immunometabolism may influence cell function (Rosa-Neto et al., 2022) (see Chapter 14).

Rather than examining whether T cell function is impaired in blood following exercise bouts, more recent studies have focused on whether exercise bouts can facilitate the process of manufacturing virus-specific T cells for use in clinical settings, such as improving the capacity of T cells to differentiate *ex vivo*. For example, vigorous exercise improves the *ex vivo* manufacture of T cells specific for adenovirus, CMV, and Epstein–Barr Virus (EBV) (Spielmann et al., 2016;

Kunz et al., 2018). This has clinical relevance because these cells can be infused into patients to manage severe viral infections, especially in the context of organ transplantation or cancer treatment. Similar work has shown that an individual bout of exercise improves the cytotoxic activity of T cells towards tumour antigens, shown for the melanoma antigen gene (MAGE) and preferentially expressed antigen in melanoma (PRAME). Both MAGE and PRAME are aberrantly expressed in a variety of cancers (see Chapter 10). Indeed, an individual bout of exercise increased the number of T cells specific for MAGE-A4 and PRAME that could be harvested from blood, which could be re-infused into patients as part of immunotherapy (LaVoy et al., 2015).

During the past decade, exercise immunology studies have broadened to examine many other sub-types of T cells. Some focus has been on regulatory T cells, given their likely role in regulating anti-inflammatory responses to exercise, and because their numbers and functions are differentially disturbed in chronic diseases, such as cancer (see Chapter 10), central nervous system pathologies and autoimmunity (see Chapter 11), and cardiovascular disease (see Chapter 12). For example, Wilson et al. found a significant increase in CD4+CD25+FOXP3 regulatory T cells following a single bout of exercise in elite adolescent swimmers (Wilson et al., 2009). This was later confirmed by a study from Minuzzi and colleagues (2017), in active older adults where cycling to exhaustion led to increases in numbers and proportions of CD4+CD25+CD127^{dim} regulatory T cells (Minuzzi et al., 2017). However, most studies make measurements in blood, which may not reflect tissue-specific changes that are more relevant to the function of regulatory T cells. For a more comprehensive review of the effects of exercise on regulatory T cells, see Proschinger et al. (2021). Other recent studies have examined Th17 cells given their pro-inflammatory role in auto-immune diseases (see Chapter 11). Although individual bouts of vigorous exercise provoke a short-term pro-inflammatory response, studies examining the mobilisation of Th17 cells have provided conflicting results. A marathon run has been shown to increase the number of Th17 cells in blood which return to pre-exercise levels within 10 days (Perry et al., 2013). However, the marathon induced a 10-day fall in the number of regulatory T cells in blood, which culminated in an elevated Th17 to regulatory T cell ratio. The authors suggested that this imbalance might increase the risk of auto-immune reactions (Perry et al., 2013). However, another study observed a decrease in circulating Th17 cells after endurance exercise, although an increase in Th17 cells was found after resistance exercise (Graff et al., 2022).

Individual exercise bouts and B cells

B cell mobilisation and redistribution in response to exercise bouts has received less focus than T cells. Overall, studies show an increase in B cell numbers in a dose-dependent response to aerobic exercise, albeit the magnitude of increase is considerably smaller than T cells and NK cells. This differential response was shown in a cross-over study (Campbell et al., 2009), whereby 13 healthy and physically active young males attended three sessions: a control session (no exercise) for 20 minutes, cycling at 35% Watt(max) for 20 minutes, and 85% Watt(max) for 20 minutes. B cells (CD3−CD19+) were unchanged throughout the control condition, and no significant changes were observed in response to the low-intensity cycling condition; however, the higher-intensity cycling condition caused an approximate doubling of B cells in blood. For comparison, the same high-intensity condition evoked a near +1000% increase in NK cells. This finding regarding B cells also aligns with earlier findings, indicating that exercise induces a lesser mobilisation effect on B cells than on other immune cell types (Shinkai et al., 1992; Shek et al., 1995). Generally, it is also consistently observed that exercise at lower intensities, and resistance exercise, does not increase B cell counts in blood. Thus, given their relatively small numbers in blood, and their relatively small magnitude change in response to exercise, the contribution of B cells to exercise-induced lymphocytosis is small.

Due to their relatively modest increase in response to exercise – at least in comparison with other types of immune cells, like NK cells – B cell subset mobilisation responses to exercise have been less well-studied. A recent study has advanced phenotyping considerably and characterised the exercise response among numerous B cell sub-types. The authors of this study reported that an individual bout of exercise primarily mobilised immature CD27−IgD−CD10+ B cells, whereas more mature B cells (i.e., CD27+) were mobilised to a lesser extent (Turner et al., 2016). Interestingly, the same study found that B1 cells – a type of B cell that is associated with the secretion of natural antibodies – increased during exercise. The same study also showed that plasmablasts – the precursors to plasma cells – were largely undetectable before and after exercise. Following vigorous aerobic exercise lasting 20 minutes, it was shown that B cell numbers returned to baseline within 15 minutes (Campbell et al., 2009). However, when expressed as a percentage of lymphocytes at the time of sampling, the contribution of B cells to total lymphocytes in the post-exercise period (e.g., 1 hour after exercise) increases, due to T cells and NK cells leaving the circulation. Nevertheless, it has been shown that B cells also leave blood 60 minutes after vigorous exercise, albeit to a lesser extent than some other immune cell types, yet the contribution of different B cell subsets to this lymphopenia remains unclear (Turner et al., 2016). Indeed, a recent systematic review has highlighted the relative dearth of research in exercise immunology on B cell sub-types (Walzik et al., 2023). Thus, because B cells have generally been less well-studied than other immune sub-types, further research is needed to provide more detailed information about exercise-induced shifts in the numbers of B cells and their sub-types as well as potential alterations to B cell function (e.g., after stimulation by antigens for example, or following vaccination) (Drummond et al., 2022).

Exercise training and leucocytes – general considerations

Compared with studies examining individual bouts of exercise, there are fewer studies in the context of longer-term exercise training, and not all leucocyte sub-types have been examined. Generally, exercise training is associated with fluctuations in the numbers, proportions and functions of leucocytes – but consistent patterns have not been established. While some cells – including neutrophils, monocytes, NK cells, T cells, and B cells – have been investigated, other cells, such as eosinophils, basophils (or mast cells), dendritic cells, and ILCs, have received almost no attention. Briefly, it is thought that because individual bouts of exercise may cause changes to the number and functions of cells, if exercise is undertaken on a regular basis, it is assumed that the sum of these short-term effects can lead to benefits (see Figure 4.1). For example, individual bouts of exercise may induce molecular signalling – for example, via secretion of 'exerkines' (Chow et al., 2022) – which results in long-term adaptions within the immune system, and long-term changes to resting leucocyte characteristics. It is well established that individual bouts of exercise induce a systemic short-term 'pro-inflammatory' environment, which is characterised by the mobilisation of effector cells (e.g., neutrophils, NK cells, and CD8+ cytotoxic T cells) as well as causing production of pro-inflammatory cytokines. However, in the long-term, these repeated pro-inflammatory stimuli might invoke a compensatory anti-inflammatory response. Although mechanisms underlying long-term adaptations have been proposed, linking the response to each individual bout of exercise with these adaptations remains a challenge. In addition, adaptations within the immune system strongly depend on the people taking part in the studies, including sex, age, lifestyle factors, and health status. Nevertheless, a recent meta-analysis in humans and rodents has linked lower resting lymphocyte counts to exercise training (Baker et al., 2022). The next sections will summarise the effects of exercise training on leucocyte sub-types that have received attention with regard to exercise immunology, including innate immune cells – such as neutrophils, monocytes, dendritic cells, and NK cells – and cells of the adaptive immune system, including T cells and B cells.

Exercise training and innate immune cells

Exercise training and neutrophils

Long-term exercise training induced adaptation among neutrophils is not established and the literature is not consistent. Some studies have shown a change in neutrophil number among athletes and in response to hypoxic training, but these findings have not been confirmed by longitudinal training interventions (Chen et al., 2018). Cross-sectional studies and exercise-training studies have examined neutrophil function, including chemotaxis and phagocytosis, which is particularly important, considering that ageing is characterised by impaired neutrophil function (Shaw et al., 2010). Indeed, a cross-sectional study showed that phagocytosis was maintained in physically active older adults compared with inactive counterparts (Yan et al., 2001). Intervention studies have shown that neutrophil chemotaxis and phagocytosis improved in young people after two months of moderate-intensity endurance training (Syu et al., 2012), and similar intervention studies have shown that exercise improves other neutrophil functions in healthy older adults (Bartlett et al., 2016). Studies have also compared different modes of exercise. For example, one study compared a 10-week intervention of moderate-intensity continuous cycling (30–45 minutes per session, five times a week) or high-intensity interval training (18–25 minutes per session, 15–60 second sprints, 45–120 seconds of active rest, three times a week). Both interventions improved phagocytic activity of neutrophils and ROS production in previously sedentary adults (Bartlett et al., 2017) with similar effects in people with type 2 diabetes (Bartlett et al., 2020). While formation of NETs is an important function of neutrophils for defence against pathogens, some chronic diseases are characterised by greater neutrophil activation and NET formation. However, endurance training has been shown to inhibit NET formation, and generally seems to have a regulating function on activated neutrophils (Bartlett et al., 2017).

Exercise training and monocytes

The number of circulating monocytes, their sub-types, and functions have been examined in exercise-training studies comprehensively – for a review, see Yarbro and Pence (2018). The total number of circulating monocytes does not seem to change with exercise training, but the composition of the monocyte pool does when monocyte sub-types are examined. For example, early work from Woods et al. (1999) demonstrated that a six-month exercise intervention in sedentary older adults did not alter total circulating monocyte numbers. However, Timmerman et al. (2008) later demonstrated that exercise training in older adults decreased the proportion of what were referred to at the time as "inflammatory" monocytes which were CD14+CD16+; however, this was before the establishment of the CD14^{dim}CD16+ "non-classical" monocyte population. More recently, it has been demonstrated that exercise training reduced CD14^{bright}CD16+ "intermediate" but not CD14^{dim}CD16+ "non-classical monocyte" proportions in haemodialysis patients (Dungey et al., 2017). Exercise training also modulates other receptors on monocytes that are important for inflammatory activation. For example, in previously active older adults, TLR4 expression on CD14^{bright}CD16− "classical" monocytes was reduced compared with sedentary controls (Timmerman et al., 2008), and both high-intensity interval training and moderate continuous exercise training have been shown to reduce TLR4 but not TLR2 expression on total monocytes (Robinson et al., 2015). Other functions have also been examined with mixed results, for example, a 12-week walking intervention in middle-aged obese women did not change monocyte phagocytosis or ROS production, whereas 12 weeks of either moderate-intensity exercise or high-intensity interval training among healthy sedentary adults improved the same parameters (Yarbro & Pence, 2018). A cross-sectional study has examined the relationship between accelerometer-measured habitual physical activity and *ex vivo* monocyte tethering and migration comparing 12 obese and 12 non-obese adults (Wadley et al., 2021). Tethering and migration for most monocyte sub-types was greater among obese adults compared to non-obese adults, but this was not affected by physical activity level. However, among obese adults, higher physical activity was associated with lower migration and tethering of CD16+ monocytes. The function of monocytes is strongly influenced by immunometabolism (see Chapter 14); therefore, interaction between exercise and immune cell metabolic pathways is a possible mechanism underlying the beneficial effects of exercise (Rosa-Neto et al., 2022; Yarbro & Pence, 2018).

Exercise training and dendritic cells

Exercise-training studies examining dendritic cells have been studied more extensively in animal models, despite their importance in regulating immunity in humans. In one rat model, five weeks of endurance exercise training modulated the development of dendritic cells and caused a shift towards a more mature form, characterised by greater MHC-class II expression and IL-12 production (Chiang et al., 2007). Another rat study, examining a similar intervention, showed that dendritic cell numbers increased following exercise training, but this was not accompanied by a change to co-stimulatory molecule expression (i.e., CD80 or CD86) (Liao et al., 2006). Finally, a mouse model has shown that four weeks of aerobic exercise training on a treadmill decreases the activation and maturation of dendritic cells (Mackenzie et al., 2016). Human studies have also been conducted, albeit at a less detailed level than in animal studies. For example, in one study of elite swimmers monitored at multiple timepoints throughout a competitive season, it was found that athletes – but not non-athlete controls – showed a decrease in dendritic cell numbers over time (Morgado et al., 2012). Correspondingly, there was also a decrease in cytokines produced by immune cells – including dendritic cells – over the course of the season, such as IL-1β, IL-6, IL-12, TNF-α, and MIP-1β. These collective changes were associated with cortisol levels in blood, which can be heightened in chronic psychological stress and is in turn associated with decreased immune function (see Chapter 15). These findings regarding the lower dendritic cell output of cytokines have been reproduced in a separate small study more recently (Evstratova et al., 2016). In addition, one human study has shown that the regular practice of Tai Chi is associated with a higher count of myeloid dendritic cells in blood compared with controls, but there were no differences in plasmacytoid dendritic cells (Chiang et al., 2010). Nevertheless, very few studies have explored the relationship between longer-term exercise training and dendritic cell sub-type frequency and function, and more research is needed to understand how exercise affects this important group of immune cells.

Exercise training and NK cells

NK cell counts and function have been examined before and after periods of exercise training. For example, eight female college-level volleyball players undertook one month of heavy training (five hours per day, six days per week). The circulating count of CD56^{dim} NK cells did not change but counts for CD56^{bright} increased (Suzui et al., 2004). However, despite studies showing changes to NK cell counts after exercise training, it has been concluded in a recent meta-analysis that periods of exercise training do not alter resting NK cell counts (Baker et al., 2022). In terms of NK cell function, two of the largest exercise-training studies report conflicting results. For example, one study found that 15 weeks of moderate exercise training increased NK cytolytic activity compared with sedentary controls (Fairey et al., 2005), while another 12-month trial found no change in NK cytolytic activity in 115 post-menopausal women (Campbell et al., 2008). Indeed, the literature is far from consistent, and some exercise-training studies have even reported a reduction in NK cell cytolytic activity and no correlation with aerobic capacity (Bigley & Simpson, 2015). Evidence regarding the longer-term effects of exercise on NK cell function can also be derived from cross-sectional studies. Indeed, NK cytotoxicity has been shown to be higher in older adults who are physically active. For example, Nieman et al. (1993) showed 54% higher resting NK cytotoxicity in older athletes when compared with sedentary controls, without a change in NK cell number. Therefore, it is plausible that long-term exposure to exercise improves NK cell function, but further research is needed.

Exercise training and cells of the adaptive immune system

Exercise training and T cells

In the context of exercise training, T cells have received considerable focus in exercise immunology. However, findings from these studies indicate there appears to be little or no effect on the total number of T cells in blood – or at least, exercise-training studies have reported only very small fluctuations. A more consistent finding is that exercise training influences the proportions of T cell sub-types – and specifically those associated with an ageing immune system (see Chapter 8). For example, a cross-sectional study has shown a negative association between cardiorespiratory fitness and so-called "senescent" T cells (e.g., defined as CD4+/CD8+ KLRG1+CD28−/CD57+) (Spielmann et al., 2011). In addition, another cross-sectional study has shown that older adults who have cycled competitively for most of their life have higher proportions of naive T cells than less active age-matched controls, perhaps mediated by lower levels of IL-6 and higher levels of IL-7 and IL-15 which preserve thymic output and T cell survival (Duggal et al., 2018). Exercise-training studies have also been conducted to examine whether cell profiles associated with immunosenescence change. For example, it has been shown among 32 healthy women aged 18–29 years that 14 weeks of aerobic exercise for 90 minutes each week, decreased the percentage of CD4+ CM T cells and CD8+ EMRA T cells in blood and increased the percentage of CD8+ NA T cells (Balogh et al. 2022). In addition, another study randomised 40 inactive people aged 60–75 years into either a six-week low-dose combined resistance and endurance training exercise group or a control group (Despeghel et al., 2021). Exercise training returned the CD4+/CD8+ ratio to normal values of between 1.5 and 2.5, whereas the control group exhibited increases in "senescent-prone" CD8+ T cells defined using CD57 and CD28 (see Chapter 8 for more details).

Aligned with the focus of exercise training inducing anti-inflammatory effects, studies on rodents and humans have shown that regulatory T cells increase in numbers and proportions with exercise training, and cross-sectional studies have shown a positive association with cardiorespiratory fitness (Proschinger et al., 2021). A model has proposed that regulatory T cells differentiate from naive T cells due to an exercise-induced activation of aryl hydrocarbon receptor (AhR) signalling because exercise strongly increases systemic concentrations of AhR ligands, such as the tryptophan metabolites kynurenic acid and xanthurenic acid (Koliamitra et al., 2019). However, it should be considered that regulatory T cells are most relevant in the tissues rather than blood, and possible exercise-induced changes in the tissues might have most relevance to the tumour microenvironment or in neuro-inflammatory pathologies.

Many studies in the 1980s and 1990s examined aspects of T cell function following exercise training, generally focusing on T cell proliferation or cytokine production. However, most of these studies were conducted with elite athletes, and in the context of intensified periods of training (i.e., examining whether very demanding exercise training impairs immunity). For example, in a study recruiting competitive cyclists, T cell numbers, mitogen-induced lymphocyte proliferation, and IL-2 production, were measured at rest at the beginning of a training season and after six months of intensive training where participants cycled approximately 500 km a week (Baj et al., 1994). Baseline values were within the range observed in non-trained healthy controls, but at the end of the training season, significant decreases in absolute numbers of CD3+ and CD4+ T-helper cells were reported alongside diminished IL-2 production, but a marked increase in lymphocyte proliferation was shown. Other studies have reported that intensive training impairs T cell proliferation (Bury et al., 1998) and decreases the CD4+ T cell to CD8+ T cell ratio (Rebelo et al., 1998). However, studies in non-elite settings show that T cell proliferation is positively correlated with cardiorespiratory fitness (Dorneles et al., 2021). Indeed, cross-sectional studies show that physically active older adults have greater T cell proliferative responses to mitogens compared to their sedentary counterparts. For example, Nieman et al. (1993) showed that mitogen-induced T cell proliferation was greater in aerobically conditioned compared to sedentary older women aged 65–85 years, while Shinkai et al. (1995) also reported that T cell proliferation was 44% higher in recreational runners compared to non-runners aged 60 years and over. However, longitudinal exercise-training studies have not shown consistent positive effects on T cell proliferation in non-athletes (Simpson & Guy, 2010).

In the most recent exercise immunology studies, rather than examining individual cell functions *in vitro*, there has been increasing focus on broader and clinically meaningful processes, using a combination of human trials and animal models. For example, in the context of cancer, there has been great interest in examining whether repeated bouts of exercise facilitate the infiltration of effector cells, such as CD8+ cytotoxic T cells and NK cells, into tumours (see Chapter 10). Indeed, exercise training increases the anti-tumour response of CD8+ cytotoxic T cells by releasing metabolites or molecules that influence energy metabolism (Rundqvist et al., 2020). In addition, exercise training in mouse models of pancreatic cancer and patients has shown that exercise bouts mobilise CD8+ cytotoxic T cells into blood due to adrenaline and that CD8+ cytotoxic T cells expressing the IL15-receptor infiltrate tumour tissue, mediated by the myokine IL-15 (Kurz et al., 2022) (see Chapter 10).

Exercise training and B cells

Few studies have robustly examined how exercise training influences the numbers, proportions, and functions of B cells. One of the first randomised-controlled trials examining the influence of exercise training on B cell counts was conducted in sedentary, mildly obese, young women, and the exercise intervention involved walking exercise on five days of the week for 15 weeks (Nehlsen-Cannarella et al., 1991). It was found that B cell counts in the control group gradually increased, whereas the exercise group generally showed maintenance of B cell counts. Other studies of similar design have not corroborated these results. For example, it was shown that aerobic exercise training on five days of the week for 12 months did not result in changes to B cell counts compared to a control group (Campbell et al., 2008). A recent systematic review has highlighted the dearth of well-designed studies examining the impact of exercise training on B cell counts (Walzik et al., 2023). Well-designed studies that met the criteria for inclusion in that systematic review, report inconclusive findings, as studies show that exercise training increases, decreases, or does not change B cell counts in blood. Therefore, future research is needed to explore the impact of exercise training on B cell phenotypes in blood. Beyond exercise-training studies, insight can also be obtained from cross-sectional studies exploring the relationship between a physically active lifestyle and B cell phenotypes in blood. One seminal study investigated a diverse array of B cell phenotypes among healthy young adults, healthy older adults, and a cohort of older adults who had engaged in lifelong physical activity (Duggal et al., 2018). It was found that naive B cell output appeared to be lower in healthy older adults compared to younger adults, yet the physically active group of older adults had a greater proportion of naive B cells than their less active age-matched counterparts. Interestingly, memory B cells were also different between groups, as it was shown that younger adults had lower counts than older adults, yet physically active older adults had fewer memory B cells than their less active older counterparts. The meaning of these findings is unclear but they may indicate that lifelong exercise ameliorates features of immunosenescence beyond T cells, by preserving naive B cell output and limiting the accumulation of oligoclonal memory B cells (see Chapter 8).

Future directions

Not all leucocyte sub-types have been examined thoroughly in exercise immunology. Although there has been significant focus on the most predominant granulocytes – neutrophils – eosinophils and basophils have received very little attention. Eosinophils primarily combat parasites but also have a role in allergy and asthma. In the context of exercise immunology, eosinophils are occasionally reported in studies alongside other parameters when describing changes to leucocytes in general, probably because a value for these cells is provided in a five-part leucocyte differential when assessed by automated haematology analysers. Although eosinophils and their function have not been the primary focus of any investigations, in response to an individual bout of exercise, eosinophils have been reported to increase acutely after downhill running (Alizadeh & Alizadeh, 2022) and decrease acutely after intermittent team sport activity (Avloniti et al., 2007). In addition, six months of moderate-intensity exercise training in older adults did not change eosinophil counts (Woods et al., 1999), but five to seven days of military training was associated with a 30% fall in eosinophils (Bøyum et al., 1996). Basophils have also received very little attention in exercise immunology. These rare cells in blood migrate to tissues to form mast cells, and are primarily implicated in allergic responses, but also have roles in wound healing. Individual bouts of exercise are thought to activate mast cells in skeletal muscle, which release histamine, and contribute to exercise-induced adaptations, including metabolic effects and angiogenesis (Luttrell & Halliwill, 2017). However, this area has not been explored thoroughly, and the very few studies that have examined basophils are in the context of allergy. Indeed, a rationale for studying basophils further are the links between these cells and exercise-associated asthma, urticarial, and anaphylaxis (Cooper et al., 2007). For example, there is preliminary evidence that exercise-induced alterations to plasma osmolarity activate basophils, contributing to food-dependent exercise-induced anaphylaxis (Barg et al., 2008). Similarly, highly trained athletes have been shown to exhibit exacerbated histamine release during exercise compared with untrained controls (Mucci et al., 2000) which may be related to basopenia during exercise and basophil degranulation (Mucci et al., 2001). In support, high systemic tryptase levels – a biomarker of mast-cell activity and degranulation – have been reported among marathon runners, despite no differences in histamine (Spooner et al., 2003). However, another study reported that short duration exercise had no effect on tryptase or histamine levels (Kulinski et al., 2019). Importantly, there have been no exercise-training studies in humans examining these cells.

One animal model has shown that regular exercise does not change the number of cardiac mast cells, but degranulation was reduced, preventing hyperactivation (Jitmana et al., 2019).

Although T cells have received a lot of attention in exercise immunology, most focus has been on conventional T cells; however, other T cell sub-types, including natural killer T cells (NKT cells) and the functional characteristics of CD4+ T-helper sub-types need more attention. NKT cells are a distinct population of T cells that express a TCR and a set of cell surface molecules that they share with NK cells. Two categories of NKT cells exist: type I invariant natural killer T cells (iNKT cells) and type II nonclassical NKT cells. Together, NKT cells make up about 0.01–1% of lymphocytes in human blood but are important immunoregulatory cells that rapidly produce large amounts of cytokines that can influence other immune cells. A recent review article has highlighted that no studies have investigated the effects of exercise on NKT cells in humans, using the most appropriate phenotypic markers – therefore, the effects of exercise on these cells are not well understood (Hanson et al., 2021). In addition, more studies are needed to examine the cytokine secretion profiles of CD4+ T-helper cells, particularly in the context of exercise training. One study has investigated the effect of 12 weeks of Tai Chi exercise on CD4+ T-helper cells in patients with type 2 diabetes, showing that T-box transcription factor (T-bet) declined, which is important for Th1 cells, whereas there was no change in GATA3, which is important for Th2 cells (Yeh et al., 2009). In another study of 32 lung cancer survivors who practiced Tai Chi during a 16-week period following surgery, the percentages of IFN-γ producing CD3+ T cells (assumed to be Th1 cells) and IL-4 producing CD3+ T cells (assumed to be Th2 cells) were examined. Whereas the ratio of Th1 cells to Th2 cells decreased in the control group, no changes were observed in the Tai Chi group (Wang et al., 2013). Thus, further studies are required because exercise training could be particularly important for diseases characterised by Th2 immune profiles, such as asthma or allergies (Zarneshan & Gholamnejad, 2019). Likewise, studies examining whether exercise training affects Th17 cells are mostly limited to animal models, showing that endurance training decreases Th17 cells within the central nervous system in mice with neuro-inflammatory disease (Xie et al., 2019) and among animals with ischaemic cardiomyopathy (Chen et al., 2018). However, one human study has examined the effects of normoxic or hypoxic treadmill walking (1 hour, three times a week, for four weeks) among 34 relapsing-remitting patients with multiple sclerosis, showing that normoxic training, but not hypoxic training, decreased IL-17A-producing CD4+ T-helper cells (Mähler et al., 2018).

Finally, although HSPCs have been thoroughly examined following individual bouts of exercise, research examining the effects of exercise training is limited. Cross-sectional studies have shown that endurance athletes have higher levels of circulating HSPCs than sedentary controls as reviewed by Emmons et al. (2016). However, in this review, four out of the five human studies examining whether exercise training influences HSPC counts, showed no change, with one study showing a small increase, but most of these studies most likely did not provide a large enough exercise stimulus to show effects (Emmons et al., 2016). One animal study has shown that exercise training over eight weeks increases the quantity of HSPCs in the vascular niche but has no effect on their functional properties such as homing or long-term engraftment (De Lisio & Parise, 2012; DiLisio & Parise, 2013).

Summary

Individual bouts of exercise and longer-term exercise training affect the numbers and functions of leucocytes in blood. A prolonged, intensive bout of exercise evokes a profound increase in leucocytes in blood. Innate immune cells, primarily neutrophils and NK cells, exhibit the largest responses to exercise. Neutrophil numbers remain elevated for several hours after exercise cessation, whereas T cells and NK cells – which also increase during exercise – fall below pre-exercise values around 1–2 hours after exercise cessation. The NK cells and T cells which leave the blood post-exercise migrate to tissues for hours or days as part of immune-surveillance, which corresponds to a decline in measurements of cell function in blood during that time period. Although this was once thought to represent immunosuppression, it is now generally agreed that these findings are largely caused by a lower proportion of effector cells in blood. In contrast to the short-term effects of individual exercise bouts, knowledge on long-term adaptions to exercise training is not well understood. Studies show that regular exercise causes changes to the numbers and proportions of some cell types, and some cell functions are changed (e.g., cytokine production, proliferation) but further research is required to confirm these effects.

Key points

- Individual bouts of exercise change the numbers and proportions of leucocytes in blood and these effects are influenced by the mode, duration, and intensity of exercise
- Vigorous and prolonged endurance exercise stimulates the largest changes to leucocytes in blood
- Innate immune cells – neutrophils and NK cells – exhibit the largest response to exercise, followed by the adaptive immune cells, T cells, and B cells
- The number of neutrophils, NK cells, T cells, and B cells increases almost immediately in blood with exercise
- Neutrophils continue to increase after exercise cessation for several hours, whereas NK cells and T cells quickly return to pre-exercise levels, but can fall lower than baseline for several hours or even days
- Fewer NK cells and T cells in blood in the hours after individual bouts of exercise corresponds to lower measurements of cell function (e.g., proliferation, cytokine production, cytotoxicity) due to a lower proportion of effector cells in samples
- The migration of NK cells and T cells into the tissues in the hours following individual bouts of exercise is thought to be an immune-surveillance response
- Exercise training is associated with small fluctuations in the numbers, proportions, and functions of most leucocytes – but consistent patterns have not been established
- Exercise training induces anti-inflammatory effects and might limit or reverse changes associated with immunosenescence.

References

Alack, K., Pilat, C., & Krüger, K. (2019). Current knowledge and new challenges in exercise immunology. *Deutsche Zeitschrift für Sportmedizin, 70*(10), 250–260.

Alizadeh, A., Alizadeh, H. (2022). Downhill running exercise increases circulating level of myokine meteorin-like hormone in humans. *Journal of Sports Medicine and Physical Fitness, 62*(5), 700–704.

Avloniti, A. A., Douda, H. T., Tokmakidis, S. P., Kortsaris, A. H., Papadopoulou, E. G., Spanoudakis, E. G. (2007). Acute effects of soccer training on white blood cell count in elite female players. *International Journal of Sports Physiology and Performance, 2*(3), 239–249.

Baj, Z., Kantorski, J., Majewska, E., Zeman, K., Pokoca, L., Fornalczyk, E.,... Lewicki, R. (1994). Immunological status of competitive cyclists before and after the training season. *International Journal of Sports Medicine, 15*, 319–324.

Baker, C., Hunt, J., Piasecki, J., & Hough, J. (2022). Lymphocyte and dendritic cell response to a period of intensified training in young healthy humans and rodents: A systematic review and meta-analysis. *Frontiers in Physiology, 13*, 998925.

Balogh, L., Szabó, K., Pucsok, J. M., Jámbor, I., Gyetvai, Á., Mile, M.,... et al. (2022). The effect of aerobic exercise and low-impact pilates workout on the adaptive immune system. *Journal of Clinical Medicine, 11*(22), 6814.

Barg, W., Wolanczyk-Medrala, A., Obojski, A., Wytrychowski, K., Panaszek, B., & Medrala, W. (2008). Food-dependent exercise-induced anaphylaxis: possible impact of increased basophil histamine releasability in hyperosmolar conditions. *Journal of Investigative Allergology and Clinical Immunology, 18*(4), 312–315.

Bartlett, D. B., Fox, O., McNulty, C. L., Greenwood, H. L., Murphy, L., Sapey, E., et al. (2016). Habitual physical activity is associated with the maintenance of neutrophil migratory dynamics in healthy older adults. *Brain, Behavior, and Immunity, 56*, 12–20.

Bartlett, D. B., Shepherd, S. O., Wilson, O. J., Adlan, A. M., Wagenmakers, A. J. M., & Shaw, C. S. (2017). Neutrophil and monocyte bactericidal responses to 10 weeks of low-volume high-intensity interval or moderate-intensity continuous training in sedentary adults. *Oxidative Medicine and Cellular Longevity, 2017*(8148742).

Bartlett, D. B., Slentz, C. A., Willis, L. H., Hoselton, A., Huebner, J. L., Kraus, V. B.,... Kraus, W. E. (2020). Rejuvenation of neutrophil functions in association with reduced diabetes risk following ten weeks of low-volume high intensity interval walking in older adults with prediabetes – A pilot study. *Frontiers in Immunology, 11*.

Bigley, A. B., Rezvani, K., Chew, C., Sekine, T., Pistillo, M., Crucian, B., … Simpson, R. J. (2014). Acute exercise preferentially redeploys NK-cells with a highly-differentiated phenotype and augments cytotoxicity against lymphoma and multiple myeloma target cells. *Brain, Behavior, and Immunity, 39*, 160–171.

Bigley, A. B., & Simpson, R. J. (2015). NK cells and exercise: Implications for cancer immunotherapy and survivorship. *Discovery Medicine, 19*(107), 433–445.

Borges, L., Dermargos, A., Gray, S., Barros Silva, M. B., Santos, V., Pithon-Curi, T. C.,... Hatanaka, E. (2018). Neutrophil migration and adhesion molecule expression after acute high-intensity street dance exercise. *Journal of Immunology Research, 2018*, 1684013.

Bøyum, A., Wiik, P., Gustavsson, E., Veiby, O. P., Reseland, J., Haugen, A. H., & Opstad, P. K. (1996). The effect of strenuous exercise, calorie deficiency and sleep deprivation on white blood cells, plasma immunoglobulins and cytokines. *Scandinavian Journal of Immunology, 43*(2), 228–235.

Brown, F. F., Campbell, J. P., Wadley, A. J., Fisher, J. P., Aldred, S., & Turner, J. E. (2018). Acute aerobic exercise induces a preferential mobilization of plasmacytoid dendritic cells into the peripheral blood in man. *Physiology and Behaviour, 194*, 191–198.

Bury, T., Marechal, R., Mahieu, P., & Pirnay, F. (1998). Immunological status of competitive football players during the training season. *International Journal of Sports Medicine, 19*, 364–368.

Campbell, P. T., Wener, M. H., Sorensen, B., Wood, B., Chen-Levy, Z., Potter, J. D.,... Ulrich, C. M. (2008). Effect of exercise on in vitro immune function: A 12-month randomized, controlled trial among postmenopausal women. *Journal of Applied Physiology, 104*, 1648–1655.

Campbell, J. P., Riddell, N. E., Burns, V. E., Turner, M., van Zanten, J. J., Drayson, M. T.,... Bosch, J. A. (2009). Acute exercise mobilises CD8+ T lymphocytes exhibiting an effector-memory phenotype. *Brain, Behavior, and Immunity, 23*(6), 767–775.

Campbell, J. P., & Turner, J. E. (2018). Debunking the myth of exercise-induced immune suppression: Redefining the impact of exercise on immunological health across the lifespan. *Frontiers in Immunology, 9*, 648.

Cenerenti, M., Saillard, M., Romero, P., & Jandus, C. (2022). The era of cytotoxic CD4 T cells. *Frontiers in Immunology, 13*, 867189.

Chen, Z., Yan, W., Mao, Y., Ni, Y., Zhou, L., Song, H., … Shen, Y. (2018). Effect of aerobic exercise on treg and Th17 of rats with ischemic cardiomyopathy. *Journal of Cardiovascular Translational Research, 11*(3), 230–235.

Chiang, L.-M., Chen, Y.-J., Chiang, J., Lai, L.-Y., Chen, Y.-Y., & Liao, H.-F. (2007). Modulation of dendritic cells by endurance training. *International Journal of Sports Medicine, 28*(9), 798–803.

Cho, E., Theall, B., Stampley, J., Granger, J., Johannsen, N. M., Irving, B. A.,... Spielmann, G. (2021). Cytomegalovirus infection impairs the mobilization of tissue-resident innate lymphoid cells into the peripheral blood compartment in response to acute exercise. *Viruses, 13*(8), 1535.

Cooper, D. M., Radom-Aizik, S., Schwindt, C., & Zaldivar, F., Jr. (2007). Dangerous exercise: Lessons learned from dysregulated inflammatory responses to physical activity. *Journal of Applied Physiology (1985), 103*(2), 700–709.

Chiang, J., Chen, Y. Y., Akiko, T., Huang, Y. C., Hsu, M. L., Jang, T. R.,... Chen, Y. J. (2010). Tai Chi Chuan increases circulating myeloid dendritic cells. *Immunological Investigations, 39*(8), 863–873.

Chow, L. S., Gerszten, R. E., Taylor, J. M., Pedersen, B. K., van Praag, H., Trappe, S.,... Snyder, M. P. (2022). Exerkines in health, resilience and disease. *Nature Reviews. Endocrinology, 18*(5), 273–289.

De Lisio, M., & Parise, G. (2012). Characterization of the effects of exercise training on hematopoietic stem cell quantity and function. *Journal of Applied Physiology (Bethesda, Md.: 1985), 113*(10), 1576–1584.

Despeghel, M., Reichel, T., Zander, J., Krüger, K., & Weyh, C. (2021). Effects of a 6-week low-dose combined resistance and endurance training on T cells and systemic inflammation in the elderly. *Cells, 10*(4), 843.

Djurhuus, S. S., Simonsen, C., Toft, B. G., Thomsen, S. N., Wielsøe, S., Røder, M. A., … Christensen, J. F. (2023). Exercise training to increase tumour natural killer-cell infiltration in men with localised prostate cancer: A randomised controlled trial. *BJU International, 131*(1), 116–124.

Dorneles, G. P., Lira, F. S., Romão, P. R. T., Krüger, K., Rosa-Neto, J. C., Peres, A.,... Mooren, F. C. (2021). Levels of cardiorespiratory fitness in men exert a strong impact on lymphocyte function after mitogen stimulation. *Journal of Applied Physiology (Bethesda, Md.: 1985), 130*(4), 1133–1142.

Drummond, L. R., Campos, H. O., Drummond, F. R., de Oliveira, G. M., Fernandes, J. G. R. P., Amorim, R. P., … Coimbra, C. C. (2022). Acute and chronic effects of physical exercise on IgA and IgG levels and susceptibility to upper respiratory tract infections: A systematic review and meta-analysis. *Pflugers Archiv, 474*(12), 1221–1248.

Duggal, N. A., Pollock, R. D., Lazarus, N. R., Harridge, S., & Lord, J. M. (2018). Major features of immune senescence, including reduced thymic output, are ameliorated by high levels of physical activity in adulthood. *Aging Cell, 17*(2), e12750.

Dungey, M., Young, H. M. L., Churchward, D. R., Burton, J. O., Smith, A. C., & Bishop, N. C. (2017). Regular exercise during hemodialysis promotes an anti-inflammatory leukocyte profile. *Clinical Kidney Journal, 10*(6), 813–821.

Emmons, R., Niemiro, G. M., & De Lisio, M. (2016). Exercise as an adjuvant therapy for hematopoietic stem cell mobilization. *Stem Cells International, 2016*, 7131359.

Evstratova, V. S., Nikityuk, D. B., Riger, N. A., Fedyanina, N. V., & Khanferyan, R. A. (2016). Evaluation in vitro of immunoregulatory cytokines secretion by dendritic cells in mountain skiers. *Bulletin of Experimental Biology and Medicine, 162*(1), 60–62.

Fairey, A. S., Courneya, K. S., Field, C. J., Bell, G. J., Jones, L. W., & Mackey, J. R. (2005). Randomized controlled trial of exercise and blood immune function in postmenopausal breast cancer survivors. *Journal of Applied Physiology, 98*(4), 1534–1540.

Ganusov, V. V., & De Boer, R. J. (2007). Do most lymphocytes in humans really reside in the gut? *Trends in Immunology, 28*(12), 514–518.

Gleeson, M., & Bishop, N. C. (2005). The T cell and NK cell immune response to exercise. *Annals of Transplantation, 10*(4), 43–48.

Graff, R. M., Jennings, K., LaVoy, E. C. P., Warren, V. E., Macdonald, B. W., Park, Y., & Markofski, M. M. (2022). T-cell counts in response to acute cardiorespiratory or resistance exercise in physically active or physically inactive older adults: A randomized crossover study. *Journal of Applied Physiology (Bethesda, Md.: 1985), 133*(1), 119–129.

Graff, R. M., Kunz, H. E., Agha, N. H., Baker, F. L., Laughlin, M., Bigley, A. B., … Simpson, R. J. (2018). B2-Adrenergic receptor signalling mediates the preferential mobilization of differentiated subsets of CD8+ T-cells, NK-cells and non-classical monocytes in response to acute exercise in humans. *Brain, Behavior, and Immunity, 74*, 143–153.

Hanson, E. D., Bates, L. C., Bartlett, D. B., & Campbell, J. P. (2021). Does exercise attenuate age- and disease-associated dysfunction in unconventional T cells? Shining a light on overlooked cells in exercise immunology. *European Journal of Applied Physiology, 121*(7), 1815–1834.

Henson, D. A., Nieman, D. C., Blodgett, A. D., Butterworth, D. E., Utter, A. C., Davis, J. M., Sonnefeld, G., Morton, S. D., Fagoaga, O. R., & Nehlsen-Cannarella, S. L. (1999). Influence of exercise mode and carbohydrate on the immune response to prolonged exercise. *International Journal of Sports Nutrition, 9*, 213–228.

Jiang, S., & Dong, C. (2013). A complex issue on CD4(+) T-cell subsets. *Immunological Reviews, 252*(1), 5–11.

Jitmana, R., Raksapharm, S., Kijtawornrat, A., Saengsirisuwan, V., & Bupha-Intr, T. (2019). Role of cardiac mast cells in exercise training-mediated cardiac remodeling in angiotensin II-infused ovariectomized rats. *Life Sciences, 219*, 209–218.

Kamp, V. M., Pillay, J., Lammers, J.-W. J., Pickkers, P., Ulfman, L. H., & Koenderman, L. (2012). Human suppressive neutrophils CD16bright/CD62Ldim exhibit decreased adhesion. *Journal of Leukocyte Biology, 92*(5), 1011–1020.

Koliamitra, C., Javelle, F., Joisten, N., Shimabukuro-Vornhagen, A., Bloch, W., Schenk, A., & Zimmer, P. (2019). Do acute exercise-induced activations of the kynurenine pathway induce regulatory T-Cells on the long-term? - A theoretical framework supported by pilot data. *Journal of Sports Science & Medicine, 18*(4), 669–673.

Krüger, K., Alack, K., Ringseis, R., Mink, L., Pfeifer, E., Schinle, M.,... Mooren, F. C. (2016). Apoptosis of T-cell subsets after acute high-intensity interval exercise. *Medicine and Science in Sports and Exercise, 48*(10), 2021–2029.

Krüger, K., Lechtermann, A., Fobker, M., Völker, K., & Mooren, F. C. (2008). Exercise-induced redistribution of T lymphocytes is regulated by adrenergic mechanisms. *Brain, Behavior, and Immunity, 22*(3), 324–338.

Kulinski, J. M., Metcalfe, D. D., Young, M. L., Bai, Y., Yin, Y., Eisch, R., … Komarow, H. D. (2019). Elevation in histamine and tryptase following exercise in patients with mastocytosis. *The Journal of Allergy and Clinical Immunology. In Practice, 7*(4), 1310–1313.e2.

Kumar, B. V., Connors, T. J., & Farber, D. L. (2018). Human T cell development, localization, and function throughout life. *Immunity, 48*(2), 202–213.

Kunz, H. E., Spielmann, G., Agha, N. H., O'Connor, D. P., Bollard, C. M., & Simpson, R. J. (2018). A single exercise bout augments adenovirus-specific T-cell mobilization and function. *Physiology and Behaviour, 194*, 56–65.

Kurz, E., Hirsch, C. A., Dalton, T., Shadaloey, S. A., Khodadadi-Jamayran, A., Miller, G., Pareek, S., Rajaei, H., Mohindroo, C., Baydogan, S., Ngo-Huang, A., Parker, N., Katz, M. H. G., Petzel, M., Vucic, E., McAllister, F., Schadler, K., Winograd, R., & Bar-Sagi, D. (2022). Exercise-induced engagement of the IL-15/IL-15Rα axis promotes anti-tumor immunity in pancreatic cancer. *Cancer Cell, 40*(7), 720–737.e5.

Lackermair, K., Scherr, J., Waidhauser, G., Methe, H., Hoster, E., Nieman, D. C., Hanley, A., Clauss, S., Halle, M., & Nickel, T. (2017). Influence of polyphenol-rich diet on exercise-induced immunomodulation in male endurance athletes. *Applied Physiology, Nutrition, and Metabolism, 42*(10), 1023–1030.

LaVoy, E. C. P., Bollard, C. M., Hanley, P. J., Blaney, J. W., O'Connor, D. P., Bosch, J. A.,... Simpson, R. J. (2015). A single bout of dynamic exercise enhances the expansion of MAGE-A4 and PRAME-specific cytotoxic T-cells from healthy adults. *Exercise Immunology Review, 21*, 144–153.

Lester, A., Vickers, G. L., Macro, L., Gudgeon, A., Bonham-Carter, A., Campbell, J. P., & Turner, J. E. (2021). Exercise-induced amplification of mitogen-stimulated oxidative burst in whole blood is strongly influenced by neutrophil counts during and following exercise. *Physiological Reports, 9*(17), e15010.

Liao, H. F., Chiang, L. M., Yen, C. C., Chen, Y. Y., Zhuang, R. R., Lai, L. Y., Chiang, J., & Chen, Y. J. (2006). Effect of a periodized exercise training and active recovery program on antitumor activity and development of dendritic cells. *Journal of Sports Medicine and Physical Fitness, 46*, 307–314.

Liu, R., Fan, W., Krüger, K., Xiao, Y. U., Pilat, C., Seimetz, M.,... Mooren, F. C. (2017). Exercise affects T-Cell function by modifying intracellular calcium homeostasis. *Medicine and Science in Sports and Exercise, 49*(1), 29–39.Li, T.-L., & Cheng, P.-Y. (2007). Alterations of immunoendocrine responses during the recovery period after acute prolonged cycling. *European Journal of Applied Physiology, 101*(5), 539–546.

Luttrell, M. J., & Halliwill, J. R. (2017). The intriguing role of histamine in exercise responses. *Exercise and Sport Sciences Reviews, 45*(1), 16–23.

Mackenzie, B., Andrade-Sousa, A. S., Oliveira-Junior, M. C., Assumpção-Neto, E., Brandão-Rangel, M. A. R., Silva-Renno, A.,... Simpson, R. J. (2016). Dendritic cells are involved in the effects of exercise in a model of asthma. *Medicine and Science in Sports and Exercise, 48*(8), 1459–1467.

Mähler, A., Balogh, A., Csizmadia, I., Klug, L., Kleinewietfeld, M., Steiniger, J., Šušnjar, U., Müller, D. N., Boschmann, M., & Paul, F. (2018). Metabolic, mental and immunological effects of normoxic and hypoxic training in multiple sclerosis patients: A pilot study. *Frontiers in Immunology, 9*, 2819.

Malaguarnera, L., Cristaldi, E., Vinci, M., & Malaguarnera, M. (2008). The role of exercise on the innate immunity of the elderly. *European Review of Aging and Physical Activity, 5*(1), 43–49.

McKenna, E., Mhaonaigh, A. U., Wubben, R., Dwivedi, A., Hurley, T., Kelly, L. A., … Molloy, E. J. (2021). Neutrophils: Need for standardized nomenclature. *Frontiers in Immunology, 12*, 602963.

Minuzzi, L. G., Rama, L., Bishop, N. C., Rosado, F., Martinho, A., Paiva, A., & Teixeira, A. M. (2017). Lifelong training improves anti-inflammatory environment and maintains the number of regulatory T cells in masters athletes. *European Journal of Applied Physiology, 117*(6), 1131–1140.

Morgado, J. M., Rama, L., Silva, I., de Jesus Inácio, M., Henriques, A., Laranjeira, P., … Teixeira, A. M. (2012). Cytokine production by monocytes, neutrophils, and dendritic cells is hampered by long-term intensive training in elite swimmers. *European Journal of Applied Physiology, 112*(2), 471–482.

Moyna, N. M., Acker, G. R., Fulton, J. R., Weber, K., Goss, F. L., Robertson, R. J., Tollerud, D. J.,... Rabin, B. S. (1996). Lymphocyte function and cytokine production during incremental exercise in active and sedentary males and females. *International Journal of Sports Medicine, 17*, 585–591.

Mucci, P., Durand, F., Lebel, B., Bousquet, J., & Préfaut, C. (2000). Interleukins 1-beta, -8, and histamine increases in highly trained, exercising athletes. *Medicine and Science in Sports and Exercise, 32*(6), 1094–1100.

Mucci, P., Durand, F., Lebel, B., Bousquet, J., & Préfaut, C. (2001). Basophils and exercise-induced hypoxemia in extreme athletes. *Journal of Applied Physiology (1985), 90*(3), 989–996.

Natale, V. M., Brenner, I. K., Moldoveanu, A. I., Vasiliou, P., Shek, P., & Shephard, R. J. (2003). Effects of three different types of exercise on blood leukocyte count during and following exercise. *Sao Paulo Medical Journal = Revista Paulista De Medicina, 121*(1), 9–14.

Nehlsen-Cannarella, S. L., Nieman, D. C., Balk-Lamberton, A. J., Markoff, P. A., Chritton, D. B., Gusewitch, G., & Lee, J. W. (1991). The effects of moderate exercise training on immune response. *Medicine and Science in Sports and Exercise, 23*(1), 64–70.

Nickel, T., Emslander, I., Sisic, Z., David, R., Schmaderer, C., Marx, N., Schmidt-Trucksäss, A., Hoster, E., Halle, M., Weis, M., & Hanssen, H. (2012). Modulation of dendritic cells and toll-like receptors by marathon running. *European Journal of Applied Physiology, 112*(5), 1699–1708.

Nieman, D. C., Henson, D. A., & Gusewitch, G. (1993). Physical activity and immune function in elderly women. *Medicine and Science in Sports and Exercise, 25*, 823–831.

Niemiro, G. M., Parel, J., Beals, J., van Vliet, S., Paluska, S. A., Moore, D. R.,... De Lisio, M. (2017). Kinetics of circulating progenitor cell mobilization during submaximal exercise. *Journal of Applied Physiology (1985), 122*(3), 675–682.

Ortega Rincón, E. (1994). Physiology and biochemistry: Influence of exercise on phagocytosis. *International Journal of Sports Medicine, 15*(Suppl 3), S172–178.

Peake, J. M. (2002). Exercise-induced alterations in neutrophil degranulation and respiratory burst activity: Possible mechanisms of action. *Exercise Immunology Review, 8*, 49–100.

Peake, J. M., Neubauer, O., Walsh, N. P., & Simpson, R. J. (2017). Recovery of the immune system after exercise. *Journal of Applied Physiology (Bethesda, Md.: 1985), 122*(5), 1077–1087.

Pedersen, L., Idorn, M., Olofsson, G. H., Lauenborg, B., Nookaew, I., Hansen, R. H., … Hojman, P. (2016). Voluntary running suppresses tumor growth through Epinephrine- and IL-6-Dependent NK cell mobilization and redistribution. *Cell Metabolism, 23*(3), 554–562.

Perry, C., Pick, M., Bdolach, N., Hazan-Halevi, I., Kay, S., Berr, I., … Grisaru, D. (2013). Endurance exercise diverts the balance between Th17 cells and regulatory T cells. *PloS One, 8*(10), e74722.

Proschinger, S., Winker, M., Joisten, N., Bloch, W., Palmowski, J., & Zimmer, P. (2021). The effect of exercise on regulatory T cells: A systematic review of human and animal studies with future perspectives and methodological recommendations. *Exercise Immunology Review, 27*, 142–166.

Rebelo, A. N., Candeias, J. R., Fraga, M. M., Duarte, J. A., Soares, J. M., Magalhaes, C., & Torrinha, J. A. (1998). The impact of soccer training on the immune system. *Journal of Sports Medicine and Physical Fitness, 38*, 258–261.

Risøy, B. A., Raastad, T., Hallén, J., Lappegård, K. T., Bæverfjord, K., Kravdal, A., … Benestad, H. B. (2003). Delayed leukocytosis after hard strength and endurance exercise: Aspects of regulatory mechanisms. *BMC Physiology*, *3*, 14.

Robson, P. J., Blannin, A. K., Walsh, N. P., Castell, L. M., & Gleeson, M. (1999). Effects of exercise intensity, duration and recovery on in vitro neutrophil function in male athletes. *International Journal of Sports Medicine, 20*(2), 128–135.

Robinson, E., Durrer, C., Simtchouk, S., Jung, M. E., Bourne, J. E., Voth, E., & Little, J. P. (2015). Short-term high-intensity interval and moderate-intensity continuous training reduce leukocyte TLR4 in inactive adults at elevated risk of type 2 diabetes. *Journal of Applied Physiology (1985), 119*(5), 508–516.

Rosa-Neto, J. C., Lira, F. S., Little, J. P., Landells, G., Islam, H., Chazaud, B.,… Krüger, K. (2022). Immunometabolism-fit: How exercise and training can modify T cell and macrophage metabolism in health and disease. *Exercise Immunology Review, 28*, 29–46.

Rumpf, C., Proschinger, S., Schenk, A., Bloch, W., Lampit, A., Javelle, F., & Zimmer, P. (2021). The effect of acute physical exercise on NK-cell cytolytic activity: A systematic review and meta-analysis. *Sports Medicine (Auckland, N.Z.), 51*(3), 519–530.

Rundqvist, H., Veliça, P., Barbieri, L., Gameiro, P. A., Bargiela, D., Gojkovic, M.,… Johnson, R. S. (2020). Cytotoxic T-cells mediate exercise-induced reductions in tumor growth. *eLife, 9*, e59996.

Sand, K. L., Flatebo, T., Andersen, M. B., & Maghazachi, A. A. (2013). Effects of exercise on leukocytosis and blood hemostasis in 800 healthy young females and males. *World Journal of Experimental Medicine, 3*(1), 11–20.

Schlagheck, M. L., Walzik, D., Joisten, N., Koliamitra, C., Hardt, L., Metcalfe, A. J.,… Zimmer, P. (2020). Cellular immune response to acute exercise: Comparison of endurance and resistance exercise. *European Journal of Haematology, 105*(1), 75–84.

Schenk, A., Esser, T., Belen, S., Gunasekara, N., Joisten, N., Winker, M. T., … Zimmer, P. (2022). Distinct distribution patterns of exercise-induced natural killer cell mobilization into the circulation and tumor tissue of patients with prostate cancer. *American Journal of Physiology. Cell Physiology, 323*(3), C879–C884.

Shaw, A. C., Joshi, S., Greenwood, H., Panda, A., & Lord, J. M. (2010). Aging of the innate immune system. *Current Opinion in Immunology, 22*(4), 507–513.

Shek, P. N., Sabiston, B. H., Buguet, A., & Radomski, M. W. (1995). Strenuous exercise and immunological changes: A multiple-time-point analysis of leukocyte subsets, CD4/CD8 ratio, immunoglobulin production and NK cell response. *International Journal of Sports Medicine, 16*(7), 466–474.

Shephard, R. J., & Shek, P. N. (1999). Effects of exercise and training on natural killer cell counts and cytolytic activity: A meta-analysis. *Sports Medicine (Auckland, N.Z.), 28*(3), 177–195.

Shinkai, S., Kohno, H., Kimura, K., Komura, T., Asai, H., Inai, R., Oka, K., Kurokawa, Y., & Shephard, R. (1995). Physical activity and immune senescence in men. *Medicine and Science in Sports and Exercise, 27*, 1516–1526.

Shinkai, S., Shore, S., Shek, P. N., & Shephard, R. J. (1992). Acute exercise and immune function. Relationship between lymphocyte activity and changes in subset counts. *International Journal of Sports Medicine, 13*(6), 452–461.

Simpson, R. J. (2011). Aging, persistent viral infections, and immunosenescence: Can exercise "make space"? *Exercise and Sport Sciences Reviews, 39*(1), 23–33.

Simpson, R. J., Bigley, A. B., Spielmann, G., LaVoy, E. C. P., Kunz, H., & Bollard, C. M. (2016). Human cytomegalovirus infection and the immune response to exercise. *Exercise Immunology Review, 22*, 8–27.

Simpson, R. J., Boßlau, T. K., Weyh, C., Niemiro, G. M., Batatinha, H., Smith, K. A., & Krüger, K. (2021). Exercise and adrenergic regulation of immunity. *Brain, Behavior, and Immunity, 97*, 303–318.

Simpson, R. J., Campbell, J. P., Gleeson, M., Krüger, K., Nieman, D. C., Pyne, D. B.,... Walsh, N. P. (2020). Can exercise affect immune function to increase susceptibility to infection? *Exercise Immunology Review, 26*, 8–22.

Simpson, R. J., & Guy, K. (2010). Coupling aging immunity with a sedentary lifestyle: Has the damage already been done? A mini-review. *Gerontology, 56*(5), 449–458.

Spielmann, G., Bollard, C. M., Kunz, H., Hanley, P. J., & Simpson, R. J. (2016). A single exercise bout enhances the manufacture of viral-specific T-cells from healthy donors: implications for allogeneic adoptive transfer immunotherapy. *Scientific Reports, 6*, 25852.

Spielmann, G., McFarlin, B. K., O'Connor, D. P., Smith, P. J. W., Pircher, H., & Simpson, R. J. (2011). Aerobic fitness is associated with lower proportions of senescent blood T-cells in man. *Brain, Behavior, and Immunity, 25*(8), 1521–1529.

Spooner, C. H., Spooner, G. R., & Rowe, B. H. (2003). Mast-cell stabilising agents to prevent exercise-induced bronchoconstriction. *The Cochrane Database of Systematic Reviews, 2003*(4), CD002307.

Steensberg, A., Toft, A. D., Bruunsgaard, H., Sandmand, M., Halkjaer-Kristensen, J., & Pedersen, B. K. (2001). Strenuous exercise decreases the percentage of type 1 T cells in the circulation. *Journal of Applied Physiology, 91*, 1708–1712.

Suchánek, O., Podrazil, M., Fischerová, B., Bočínská, H., Budínský, V., Stejskal, D.,... Kolář, P. (2010). Intensive physical activity increases peripheral blood dendritic cells. *Cellular Immunology, 266*(1), 40–45.

Suzui, M., Kawai, T., Kimura, H., Takeda, K., Yagita, H., Okumura, K., Shek, P. N., & Shephard, R. J. (2004). Natural killer cell lytic activity and CD56(dim) and CD56(bright) cell distributions during and after intensive training. *Journal of Applied Physiology (1985), 96*(6), 2167–2173.

Syu, G.-D., Chen, H.-I., & Jen, C. J. (2012). Differential effects of acute and chronic exercise on human neutrophil functions. *Medicine and Science in Sports and Exercise, 44*(6), 1021–1027.

Tarban, N., Halász, H., Gogolák, P., Garabuczi, É., Moise, A. R., Palczewski, K.,... Szondy, Z. (2022). Regenerating skeletal muscle compensates for the impaired macrophage functions leading to normal muscle repair in retinol saturase null mice. *Cells, 11*(8), 1333.

Timmerman, K. L., Flynn, M. G., Coen, P. M., Markofski, M. M., & Pence, B. D. (2008). Exercise training-induced lowering of inflammatory (CD14+CD16+) monocytes: A role in the anti-inflammatory influence of exercise? *Journal of Leukocyte Biology, 84*(5), 1271–1278.

Turner, J. E., Aldred, S., Witard, O. C., Drayson, M. T., Moss, P. M., & Bosch, J. A. (2010). Latent cytomegalovirus infection amplifies CD8 T-lymphocyte mobilisation and egress in response to exercise. *Brain, Behavior, and Immunity, 24*(8), 1362–1370.

Turner, J. E., Spielmann, G., Wadley, A. J., Aldred, S., Simpson, R. J., & Campbell, J. P. (2016). Exercise-induced B cell mobilisation: Preliminary evidence for an influx of immature cells into the bloodstream. *Physiology & Behavior, 164*(Pt A), 376–382.

Tvede, N., Kappel, M., Halkjaer-Kristensen, J., Galbo, H., & Pedersen, B. K. (1993). The effect of light, moderate and severe bicycle exercise on lymphocyte subsets, natural and lymphokine activated killer cells, lymphocyte proliferative response and interleukin 2 production. *International Journal of Sports Medicine, 14*, 275–282.

Valeria Oliveira de Sousa, B., de Freitas, D. F., Monteiro-Junior, R. S., Mendes, I. H. R., Sousa, J. N., Guimarães, V. H. D., & Santos, S. H. S. (2021). Physical exercise, obesity, inflammation and neutrophil extracellular traps (NETs): A review with bioinformatics analysis. *Molecular Biology Reports, 48*(5), 4625–4635.

Vivier, E., Artis, D., Colonna, M., Diefenbach, A., Di Santo, J. P., Eberl, G.,... Powrie, F., Spits, H. (2018). Innate lymphoid cells: 10 Years On. *Cell, 174*(5), 1054–1066.

Wadley, A. J., Roberts, M. J., Creighton, J., Thackray, A. E., Stensel, D. J., & Bishop, N. C. (2021). Higher levels of physical activity are associated with reduced tethering and migration of pro-inflammatory monocytes in males with central obesity. *Exercise Immunology Review, 27*, 54–66.

Walzik, D., Belen, S., Wilisch, K., Kupjetz, M., Kirschke, S., Esser, T., Joisten, N., Schenk, A., Proschinger, S., & Zimmer, P. (2023). Impact of exercise on markers of B cell-related immunity: A systematic review. *Journal of Sport and Health Science*. Advance online publication.

Walzik, D., Joisten, N., Zacher, J., & Zimmer, P. (2021). Transferring clinically established immune in-flammation markers into exercise physiology: focus on neutrophil-to-lymphocyte ratio, platelet-to-lymphocyte ratio, and systemic immune-inflammation index. *European Journal of Applied Physiology, 121*(7), 1803–1814. Epub 2021 March 31.

Wang, R., Liu, J., Chen, P., & Yu, D. (2013). Regular tai chi exercise decreases the percentage of type 2 cytokine-producing cells in postsurgical non-small cell lung cancer survivors. *Cancer Nursing, 36*(4), E27–E34.

Wilson, L. D., Zaldivar, F. P., Schwindt, C. D., Wang-Rodriguez, J., & Cooper, D. M. (2009). Circulating T-regulatory cells, exercise and the elite adolescent swimmer. *Pediatric Exercise Science, 21*(3), 305–317.

Woods, J. A., Ceddia, M. A., Wolters, B. W., Evans, J. K., Lu, Q., & McAuley, E. (1999). Effects of 6 months of moderate aerobic exercise training on immune function in the elderly. *Mechanisms of Ageing and Development, 109*(1), 1–19.

Wynn, T. A., & Vannella, K. M. (2016). Macrophages in tissue repair, regeneration, and fibrosis. *Immunity, 44*(3), 450–462.

Xiang, L., Rehm, K. E., & Marshall, G. D. (2014). Effects of strenuous exercise on Th1/Th2 gene expression from human peripheral blood mononuclear cells of marathon participants. *Molecular Immunology, 60*(2), 129–134.

Xie, Y., Li, Z., Wang, Y., Xue, X., Ma, W., Zhang, Y., & Wang, J. (2019). Effects of moderate- versus high-intensity swimming training on inflammatory and CD4+ T cell subset profiles in experimental autoimmune encephalomyelitis mice. *Journal of Neuroimmunology, 328*, 60–67.

Yan, H., Kuroiwa, A., Tanaka, H., Shindo, M., Kiyonaga, A., & Nagayama, A. (2001). Effect of moderate exercise on immune senescence in men. *European Journal of Applied Physiology, 86*(2), 105–111.

Yarbro, J. R., & Pence, B. D. (2018). Monocytes in aging and exercise. *Exercise Medicine, 2*.

Yeh, S. H., Chuang, H., Lin, L. W., Hsiao, C. Y., Wang, P. W., Liu, R. T., & Yang, K. D. (2009). Regular Tai Chi Chuan exercise improves T cell helper function of patients with type 2 diabetes mellitus with an increase in T-bet transcription factor and IL-12 production. *British Journal of Sports Medicine, 43*(11), 845–850.

Zarneshan, A., & Gholamnejad, M. (2019). Moderate aerobic exercise enhances the Th1/Th2 ratio in women with asthma. *Tanaffos, 18*(3), 230–237.

Zimmer, P., Bloch, W., Kieven, M., Lövenich, L., Lehmann, J., Holthaus, M … Schenk, A. (2017). Serotonin shapes the migratory potential of NK Cells—An in vitro approach. *International Journal of Sports Medicine, 38*(11), 857–863.

5 Soluble immunity and exercise

Arwel W. Jones and Glen Davison

Introduction

Both individual bouts of exercise and regular exercise training can exert profound effects on a diverse array of soluble immune factors, including immunoglobulins, complement, cytokines, antimicrobial peptides, and acute-phase proteins. These effects have been measured in a range of biofluids including those of the mucosal immune system (e.g., saliva, tear fluid, nasal lavage, sputum) and in blood. This chapter will outline how individual bouts of exercise and regular exercise training affect soluble factors in those different biofluids, and the clinical implications of these effects will be summarised.

Soluble immunity at the mucosae

Common mucosal immune system

The human mucosae are the first line of defence where most pathogens enter the body (Macpherson et al., 2012). The gut-associated lymphoid tissue, lacrimal glands, and respiratory tracts which include the bronchus-associated lymphoid tissue, salivary glands, and nasal-associated lymphoid tissue are all mucosal surfaces which fall under the network of immune structures known as the common mucosal immune system (Gleeson and Pyne, 2000). The immunological protection provided by this network may be via organised tissue with well-formed follicles (mucosa-associated lymphoid tissue) or as a diffuse accumulation of leucocytes (for example, plasma cells and phagocytes) that produce soluble factors.

The local production of soluble factors including immunoglobulins represents key immunological barriers at mucosal surfaces (Bishop and Gleeson, 2009; Brandtzaeg et al., 1999). IgA is the predominant immunoglobulin in mucosal secretions (McClelland et al., 1976) and unlike its usual monomeric peptide structure in the bloodstream, it is found as a dimeric protein. IgA can be further divided into subclasses, where IgA2 is the most abundant in the distal gastrointestinal tract (60%), and IgA1 predominates in the salivary glands (60–80%) and nasal lymphoid tissue (NALT) (> 90%) (Gleeson, 2000). IgA in mucosal secretions is covalently linked by a J chain and contains another peptide termed the secretory component (Gleeson and Pyne, 2000) (Figure 5.1). The secretory component wrapped around the J chain-linked dimeric IgA forms secretory IgA (SIgA) which is resistant to proteases secreted at mucosal sites (Johansen et al., 2001; Lindh, 1975; Strugnell and Wijburg, 2010; Underdown and Dorrington, 1974). It is this local production of SIgA (by plasma cells) that forms the major effector function of mucosal immunity (Bishop and Gleeson, 2009) (Figure 5.2).

DOI: 10.4324/9781003256991-5

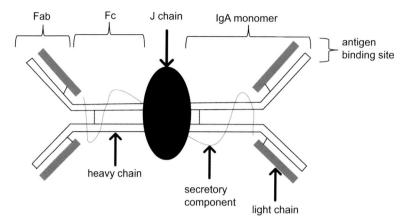

Figure 5.1 Structure of secretory IgA (SIgA). Fab = fragment antigen unit, Fc = fragment crystallisable unit. Adapted from Bishop and Gleeson (2009).

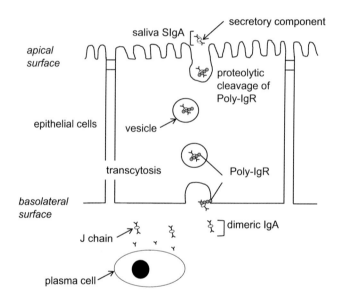

Figure 5.2 Production and release of SIgA in saliva. IgA monomers are secreted by plasma cells and joined together by the J-chain to form dimeric IgA. Polymeric immunoglobulin receptor (pIgR) binds to dimeric IgA and transports it through epithelial cells on to the mucosal surface. pIgR is proteolytically cleaved in the process, leaving only the secretory component bound to the secreted IgA. With this secretory component the SIgA is more resistant to protease degradation in mucosal secretions (e.g. saliva). Adapted from Bishop and Gleeson (2009).

SIgA provides this first line of defence at mucosal surfaces with the support of antimicrobial peptides (AMPs) that represent small cationic peptides (< 100 amino acids) and inducible, constituent factors of mucosal secretions (West et al., 2006). There is a wide variety of other AMPs, most of which have been grouped into three main families; cathelicidins (e.g. LL37), defensins and histatins (Bals, 2000). These are primarily released by epithelial cells, submucosal glands, and/or phagocytes (De Smet and Contreras, 2005; Fábián et al., 2012). The AMPs work synergistically in low concentrations to provide a broad spectrum of activity against Gram-positive and Gram-negative bacteria (Bals, 2000). The most abundant AMPs in the secretions of the upper respiratory tract are lysozyme and lactoferrin (Fullard and Snyder, 1990; Singh et al., 2000). Lysozyme is released by neutrophils, macrophages, and the submucosal glands and possesses bactericidal capacity through hydrolysing the polysaccharide of bacterial cell walls (Bosch et al., 2002; Fábián et al., 2012; Jollès and Jollès, 1984; Travis et al., 2001; West et al., 2006). Lactoferrin, produced by neutrophils and submucosal glands binds free iron, which is a nutrient essential for the growth and multiplication of bacteria (Bowdish et al., 2005; Legrand et al., 2004; Ward et al., 2005). Lactoferrin also has anti-viral properties, which are thought to act via the prevention of virus entry into the host cell (van der Strate et al., 2001).

Given that mucosal surfaces of the upper and respiratory tract account for ~ 50–60% of total immune protection by the body (Kudsk, 2002), a range of biofluids (Table 5.1) have been collected to investigate soluble factors as measures of immune competency and inflammation at these mucosae.

Saliva

Saliva has tended to be the biofluid of choice for measuring soluble immune factors in the upper airway due to the ease of collection (Bishop and Gleeson, 2009; Gleeson, 2000; Korsrud and Brandtzaeg, 1980; Sari-Sarraf et al., 2006). Notably, saliva is produced in vast quantities with normal daily production varying between 0.5 and 1.5 litres (Iorgulescu, 2009). Most of the salivary fluid (99% water and 1% protein and salts) is formed by three pairs of major salivary glands (parotid, submandibular, sublingual), but production is supplemented by a vast amount of small submucosal glands that lie on and around the tissue (e.g. palate, tongue) within the oral cavity (Proctor and Carpenter, 2007). Although saliva drains from the acini (secretory cells) of each of these glands into the mouth via striated and excretory ducts, the nature of the secretion differs whereby serous (watery), mucous, and sero-mucous fluids are produced by the parotid gland, submandibular gland, and sublingual glands respectively (Aps and Martens, 2005). In unstimulated saliva secretion, the proportion of the fluid provided by parotid, submandibular, sublingual, and the remaining submucosal glands are on average suggested to be 25%, 60%, 8%, and 8% respectively (Dawes, 2008).

The whole unstimulated saliva flow rate can have large between-person variability, but it is approximately around 0.3–0.4 ml per minute, which decreases to 0.1 ml per minute during sleep and increases during eating, chewing, and other stimulating activities (Iorgulescu, 2009). The most reproducible collection method of saliva appears to be unstimulated whole saliva collection (also termed passive drool), due to the potential for stimulated saliva flow (e.g. chewing) with other collection methods (e.g. swabs) and hence preferentially inducing secretion from certain glands and/or influence saliva composition once secreted (Allgrove et al., 2014; Beltzer et al., 2010; Bishop and Gleeson, 2009; Granger et al., 2012; Harmon et al., 2007; Navazesh and Christensen, 1982; Navazesh, 1993; Proctor and Carpenter, 2007).

The variability reported in the literature has not all been due to the method of collection (i.e. unstimulated vs stimulated) but also partly related to how soluble immune factors are expressed relative to the other exercise-induced changes (Walsh et al., 2002). Physiological changes (e.g. nervous stimulation, dehydration) during exercise can all influence the secretion of saliva and its protein components (Bishop and Gleeson, 2009; Walsh et al., 2004). To account for such changes, concentrations of parameters have been expressed relative to total salivary protein/albumin (Gleeson, 2000), but this has made the comparison between studies difficult (Bishop and Gleeson, 2009). As saliva osmolality reflects the inorganic electrolyte concentration (rather than

Table 5.1 Mucosal or respiratory tract fluids collected to measure immune competency and inflammation in response to exercise

	Immune competency		Inflammation	
	Individual bouts of exercise	*Regular exercise*	*Individual bouts of exercise*	*Regular exercise*
Tear fluid (eyes)	Prolonged strenuous: − SIgA (concentration & secretion rate) − lactoferrin & lysozyme (secretion rate)	n/a	n/a	n/a
Nasal lavage (nose)			Prolonged strenuous: + nitric oxide + IL-6	Competition / Intensified training: + IL-1ra Intensified training: + free light chains
Saliva (mouth)	Moderate: ↔ or + SIgA (concentration) ↔ or + lactoferrin & lysozyme (concentration & secretion rate) Prolonged strenuous: ↔ or − SIgA (concentration) − lactoferrin & lysozyme (concentration & secretion rate)	Moderate: + SIgA (concentration) Competition/intensified training: − SIgA (concentration & secretion rate) − lactoferrin (concentration & secretion rate) & lysozyme (concentration)	n/a	
Sputum (a mixture of saliva and mucous coughed up from the respiratory tract)	n/a	n/a	Exercise challenge/ competitive event: + IL-6, IL-8 + leukotrienes Moderate:[b] + IL-6, IL-8	Competition: + IL-8
Bronchoalveolar lavage (fluid collected from the lungs during a bronchoscopy)	n/a	n/a	n/a	Moderate:[c] + IL-6, IL-8, TNFa + IL-10, IL-1ra + IL-4, IL-5, IL-13 + leukotrienes

Table reports a summary of key changes in immune measures in these fluids. − decrease; + increase; ↔ no change; [a]studies of exercise-induced bronchoconstriction; [b]populations with chronic lung disease (chronic obstructive pulmonary disease); [c]animal models of asthma and COPD. n/a = not applicable or not investigated.

protein content) and hence falls in proportion with decreases in flow rate, expression according to osmolality has been an alternative method for measuring changes in salivary immune factors (Bishop and Gleeson, 2009; Blannin et al., 1998). The secretion rate or expression relative to saliva osmolality is preferred over measures as ratio to total protein (Blannin et al., 1998). One reason for this is that other salivary proteins (e.g. amylase) are known to increase with exercise without a change in the studied immune factor (Walsh et al., 1999). Furthermore, expression as a secretion rate may better reflect the amount of available defence on the mucosal surface (MacKinnon et al., 1993). Some may consider that the secretion rate only provides an explanation to how salivary flow rate has changed, and it is the absolute concentration that is of greater biological significance (Bishop and Gleeson, 2009). However, as the majority of saliva is water, expression relative to secretion rate also accounts for the concentrating effect of other salivary components following any dehydration (Bishop et al., 2000; Oliver et al., 2007). Over time, absolute concentration and secretion rate have become the most common methods to report changes in immune factors in saliva.

The most popular soluble factor of mucosal surfaces that has been investigated in an exercise setting has been salivary SIgA. One of the earliest studies was done by Tomasi et al. (1982) where salivary SIgA was shown to decrease 20% following 2–3 hours of competition in elite cross-country skiers. Although there was recognised inconsistency in study design to draw a definitive conclusion, Walsh et al., (2011) suggested that SIgA concentration in saliva generally decreases (e.g. Nieman et al., 2002, 2003; Palmer et al., 2003; Tomasi et al., 1982) or remains unchanged (e.g. MacKinnon and Hooper, 1994; Sari-Sarraf et al., 2006) following prolonged exercise ($\geq$ 1.5 hours at 50–75% maximum oxygen uptake, VO_2 max). It seems that the combination of high-intensity and long duration is what results in the most significant decline in salivary SIgA concentration with the former having the most impact (Mackinnon and Hooper, 1994; Nieman et al., 2002). Lower concentrations and secretion rate of saliva SIgA have also been observed in response to some intensified training periods (Moraes et al., 2017; Morgans et al., 2014; Tiollier et al., 2005a) but not all (Halson et al., 2003). Meanwhile, salivary SIgA response remains unchanged or increases following individual bouts of moderate or intermittent aerobic exercise or resistance exercise (Carlson et al., 2013; Leicht et al., 2018; Nunes et al., 2011; Sari-Sarraf et al., 2007). The effects of moderate exercise on salivary IgA are more pronounced when performed on a regular basis whereby Chastin et al. (2021) in their meta-analysis of data from seven studies reported a large and statistically significant effect of exercise interventions on salivary IgA (median characteristics: three times per week, moderate-to-vigorous intensity, 30 minutes in length for 15 weeks), albeit there was substantial heterogeneity in intervention effect across the studies.

Despite these possible exercise-induced changes, the underlying mechanisms of effects on salivary SIgA remain unclear (Box 5.1 introduces the key purported mechanisms). The salivary glands are innervated by both the parasympathetic and sympathetic nervous system, with changes in the stimulation of either of these having an influence on the volume, viscosity, protein, and mucin concentration (Aps and Martens, 2005). A high volume of watery saliva that is low in protein concentration has been associated with parasympathetic nervous stimulation and vasodilation of the salivary glands (Bishop and Gleeson, 2009). In contrast, sympathetic stimulation has been associated with salivary secretions that are low in volume but high in protein due to enhanced active transport of proteins from salivary cells (Proctor and Carpenter, 2007). The flow rate of saliva can also be a major source of variation in mucosal parameters (Gleeson and Pyne, 2000). Typically, saliva flow rate decreases in response to prolonged exercise (Bishop et al., 2000; Walsh et al., 1999), as a result of the withdrawal of parasympathetic stimulation rather than sympathetic-induced vasoconstriction of salivary glands (Bishop and Gleeson, 2009).

Box 5.1   Parasympathetic and sympathetic activation

Saliva components including SIgA are susceptible to the sympathetic and parasympathetic responses to exercise (Walsh et al., 2011b). Animal studies have shed light on why such mechanisms may be responsible for the effects of exercise on SIgA. SIgA contains the secretory component, which is the cleaved segment of the polymeric immunoglobulin (Ig) receptor (pIgR) and what makes SIgA more resistant to protease degradation in mucosal secretions (Figure 5.1, 5.2). pIgR is a transport protein produced by the mucosal epithelial and glandular cells and expressed on the basolateral membrane (Bishop and Gleeson, 2009; Teeuw et al., 2004). pIgR binds to dimeric IgA (produced by local plasma cells) to transport it through epithelial cells on to the mucosal surface and is proteolytically cleaved in the process (Strugnell and Wijburg, 2010). Hence, a crucial determinant of SIgA release into saliva is the presence of pIgR to permit transport across the mucosal epithelium (Bosch et al., 2002). Increased mobilisation of pIgR may only occur above a certain threshold of increased sympathetic stimulation (Proctor et al., 2003).

Kurimoto et al. (2016) placed six-week-old male Wistar rats in individual cages with or without access to exercise wheels for three weeks. Rats who engaged in voluntary exercise demonstrated significant increases in IgA concentration in saliva and submandibular gland tissue, as well as elevated pIgR mRNA expression in the submandibular gland tissue. This, however, would not explain the decrease found with prolonged exercise bouts or intensified training (Bishop and Gleeson, 2009). It has been speculated that stimulation over a longer period (i.e. prolonged exercise) may deplete the available IgA (Allgrove et al., 2008; Proctor et al., 2003) or there may be a further threshold or interaction with duration where pIgR mobilisation is downregulated (Walsh et al., 2011b). Kimura et al. (2008) undertook a study that involved eight-week-old male Wistar rats being randomised to control or an exercise group (ran on a treadmill at a speed of 15 m min^{-1} for the first 30 minutes prior to treadmill speed being gradually increased until they reached exhaustion). The exercise group had significant decreases in saliva IgA concentration, whereas IgA in the control group remained unchanged. These changes occurred in the absence of any changes in the saliva flow rate but a reduced expression of pIgR post-exercise.

Stimulation of HPA axis

Stimulation of the HPA axis and modulation of cortisol levels have also been postulated to be involved in changes in salivary SIgA as well as salivary AMPs. In a season-long study of a national basketball team, He et al. (2010) observed significant decreases in secretion rates and absolute concentrations of sIgA and lactoferrin during periods of intense training and competition that coincided with significant increases in salivary cortisol. An 11-month longitudinal study of elite rugby union players by Cunniffe et al. (2011) observed that periods of increasing training loads resulted in decreases in salivary SIgA and lysozyme concentrations, which were also associated with a corresponding increase in salivary cortisol. These studies found significant inverse correlation between sIgA secretion rate and cortisol, but they are not universal findings in training studies (McDowell et al., 1992; Tiollier et al., 2005b). In a study assessing the awakening cortisol response in 30 healthy young adults, Hucklebridge et al. (1998) reported a significant correlation between a rise in cortisol and fall in SIgA. None of these studies establish cause and effect.

Limited supporting evidence, however, may come from the study of Wira and Rossoll (1991) who demonstrated the administration of the glucocorticoid, dexamethasone, to rats resulted in a reduction in salivary IgA. More recently, Leicht et al. (2018) have suggested that cortisol is not an appropriate surrogate marker for sIgA concentration and vice versa. They instead suggested it is saliva flow rate that is associated with cortisol levels, whereby plasma cortisol concentration was able to explain 25% of the inter-individual variance found in saliva flow across all sampling time points.

Limited attempts have been made to identify changes in other salivary immunoglobulins with exercise. IgG concentration appears unaffected while IgM may parallel decreases in salivary SIgA (Bishop and Gleeson, 2009; Gleeson and Pyne, 2000). Other aspects of soluble immunity that have received little attention to date when compared with the investigations of SIgA are AMPs (Walsh et al., 2011b). To date, the few investigations that have been conducted suggest responses of AMPs may reflect other immunological measures of being dependent on the intensity and duration of the exercise bout. A 2-hour cycling bout at ~ 65% VO_2 max resulted in a significant decrease in salivary lysozyme (sLys) concentration, sLys secretion rate, and sLys:osmolality which improved after 1 hour of recovery (Davison and Diment, 2010). In the recovery period (0–1.5 hours post) following a 50-km mountain trail race (mean running time ~ 8 h) there was a significant decrease in salivary lactoferrin (sLac) concentration and non-significant decreases in sLac and sLys secretion rate (Gillum et al., 2013). Gill et al. (2014) reported that a 24-hour continuous overnight ultramarathon competition (distance range: 122–208 km) decreased saliva lysozyme secretion rate. In contrast, other AMPs (LL37 and defensins: human neutrophil peptide 1–3) in response to similar stressors (2.5 hours at ~ 60% VO_2 max) have been shown to significantly increase immediately post-exercise (Davison et al., 2009). The work of Allgrove et al. (2008) suggested exercise intensity may be an influential factor as saliva lysozyme secretion rate was found to significantly increase with cycling at 75% VO_2 max for ~ 22 minutes or until exhaustion but remain unchanged with ~ 22 minutes at 50% VO_2 max. Kunz et al. (2015) proposed that fitness level may also significantly impact saliva AMP responses as secretion rate and/or concentration of HNP1–3, LL-37, and lactoferrin in their study increased with 30 minutes of exercise at a greater magnitude in the highly fit compared to the less fit cyclists. The authors of these studies suggested that greater sympathetic activation may explain the findings (Allgrove et al., 2008: higher-intensity exercise; Kunz et al., 2015: higher catecholamine responses in trained individuals) whereby increases of saliva AMPs are observed when a specific threshold of exercise intensity or sympathetic nervous activity is met, but AMPs remain unchanged below that threshold. The similar pattern and magnitude of responses in AMPs between short (but intense) exercise stress and other stressors (e.g. rookie vs veteran crew members on spaceflight) may lend support to this (Agha et al., 2020).

Other salivary factors that have been proposed to have immune properties, or rather represent a marker of inflammation in the oral cavity, are a-amylase (Scannapieco et al., 1994) and free light chains. Many studies have shown increases in salivary a-amylase concentrations in response to exercise particularly at an intensity of >70% VO_2 max, whereby it is considered a marker of physical stress (Koibuchi and Suzuki, 2014). Immunoglobulin-free light chains have more recently been quantified for the first time in human saliva and emerged as biomarkers of oral inflammation (Heaney et al., 2018). Immunoglobulins are produced by plasma cells and comprise two identical heavy chains and two identical light chains, which can be either kappa or lambda isotypes (Rapson et al., 2020). Light chains are produced by plasma cells in excess of heavy chains, meaning that for every intact immunoglobulin, there is surplus light chains released into the circulation (Suki and Massry, 1998). Of the limited evidence available to date, free light chains appear sensitive to exercise stress whereby salivary concentrations were increased in highly trained male cyclists who underwent a nine-day period of intensified training (Heaney et al., 2018).

The investigations of biomarkers of oral inflammation have led to questions whether the lack of assessment of oral hygiene has been overlooked as a methodological limitation in studies of salivary immunity and exercise (Campbell and Turner 2018). Changes in oral hygiene practices or oral health status have been reported to influence mucosal homeostasis (da Silveira et al., 2023; Hall et al., 2023). A cross-sectional study held at the London 2012 Olympic Games demonstrated high prevalence of poor oral health in athletes attending a dental clinic, including dental caries (55%), dental erosion (45%), and periodontal disease (gingivitis: 76%, periodontitis: 15%) (Needleman et al., 2013). Almost half of these athletes had not undergone a dental examination or hygiene care in the previous year. Hence, unless there is adequate screening on entry to exercise studies, it is possible that oral hygiene may be an important source of between-person variation in salivary immune factors. An additional consideration relevant but not limited to oral hygiene is the potential for blood contamination of saliva. There are various assays of blood factors that have been used to measure contamination of saliva (e.g. transferrin, haemoglobin, albumin) (Kang and Kho, 2019), but this has not been routinely done across all exercise studies. Blood contamination in salivary samples can inevitably lead to errors in analytical results when the immune factor is present in both blood and saliva. To help overcome such issues, it is now often the case that studies utilise assays that only measure SIgA (i.e., IgA containing the secretory component) as this represents IgA produced by local plasma cells, and not IgA from the bloodstream transported via crevicular fluid (Campbell and Turner, 2018). This has not been possible for factors such as the AMPs, lysozyme, and lactoferrin, where the target analyte is similar across biofluids and thereby studies have reported loss of samples because of unforeseen blood contamination (e.g. Jones et al., 2023). However, such detail on laboratory assays (specific for SIgA) or screening for blood contamination has not always been reported in the literature. These sources of variation in salivary immune factors are partly why research is being undertaken to explore mucosal immunity in other biofluids.

Tear fluid

In the search for non-invasive measurements of soluble factors at clinically relevant surfaces, tear fluid has also attracted the interest of exercise immunologists. The transmission of viral upper respiratory tract infection (URTI) can occur at the ocular surface whereby self-inoculation at the eyes or nose has been suggested to occur more readily than oral transmission (Belser et al., 2009; Bischoff et al., 2011; Hendley and Gwaltney, 1998; Hendley, 1998). It is worthy to note that the measurement of soluble factors in tear fluid is not a new concept (Covey et al., 1971), but interest has grown in its utility in exercise settings as some of its constituents represent those that have been heavily studied in saliva. Unsurprisingly basal (unstimulated) tear flow rate (approximately $1–3\ \mu L \cdot min^{-1}$) is less than that of saliva, but like saliva flow rate, is under regulation of sympathetic and parasympathetic innervation (Dartt and Wilcox 2013). The tear film, comprising an inner mucin layer, middle aqueous layer, and outer lipid layer, covers the ocular surface. The main lacrimal gland and accessory lacrimal glands (along with a minor contribution from the conjunctival and corneal epithelium) produce the aqueous layer, whilst the conjunctival goblet cells and the meibomian glands secrete the mucous layer and lipid layer, respectively (Rocha et al., 2008). The aqueous layer is the largest component of the tear film and comprises a variety of proteins, electrolytes, and water (Dartt and Willcox, 2013; Stahl et al., 2012). The lacrimal gland represents the main supply of SIgA and AMPs to this aqueous layer, and hence the key contributor to the mucosal defence at the ocular defence (O'Sullivan and Montgomery, 2015). It is the involvement of fewer glands in the secretion of this aqueous component compared with the multiple glands that contribute to saliva composition is partly why some researchers have suggested that exercise-induced immune changes in tear fluid might be less variable than saliva (Hanstock et al., 2016).

The investigation of responses of tear fluid to exercise thus far has been limited to the effects of prolonged exercise. The available evidence, however, is consistent with the theory that exercise can suppress the immune system (the 'open window' hypothesis – see Chapter 9) that has been previously proposed with salivary immune parameters. In a study of 13 healthy, recreationally active adult males, Hanstock et al. (2016) observed that 2 hours of moderate intensity running (65% VO_2 peak) resulted in large and small changes in tear SIgA concentration (57% decrease, ES = 1.31) and secretion rate (44% decrease; ES = 0.47), respectively. The perturbation in tear SIgA continued for 30 minutes post-exercise with moderate changes in the secretion rate (55% decrease; ES, 0.67). In the same study, reported in a follow-up publication (Hanstock et al., 2019), the prolonged bout of exercise was also shown to induce large decreases in the secretion rate of tear lactoferrin (30 minutes post: 43% decrease, ES = 0.91; 1 hours post: 43% decrease, ES = 0.93) and lysozyme (30 minutes post: 49% decrease, ES = 0.80; 1 hour post: 48% decrease, ES = 0.81) albeit no changes in the concentration of either parameter. The strengths of these findings lie on the repeated-measures crossover design of the study where the comparator group comprised a period of seated rest and hence controlled for potential confounding effects on exercise-induced responses. This study was additionally able to provide preliminary insight into the utility of tear fluid by concurrent salivary measures of SIgA and assessment of hydration status. Tear AMPs and IgA were robust to alterations in hydration status unlike salivary SIgA concentration which was affected by dehydration. Collection of tear fluid is not without limitations, however, with large between-person variability in tear flow rates (Hanstock et al., 2016, 2019). As it is impractical to drain the eye of tears before starting sample collection, secretion rate measurements likely include basal tear fluid already residing on the lower lid of the eye. This contrasts with unstimulated whole saliva collection where there are more practical attempts available to clear saliva prior to measurement of the secretion rate (i.e. participants are encouraged to swallow to empty the mouth).

Nasal lavage

Although the transmission of pathogens may occur more readily at the nose than the oral cavity (Belser et al., 2009), there have been limited investigations of changes in immune competence following exercise in samples derived from the nasal cavity (e.g. lavage). One of the earliest studies of nasal lavage reported no change in lysozyme concentration in a cohort of recreational athletes immediately after and up to three days following a marathon race compared with a sedentary control group (Müns et al., 1996). A more recent study that investigated a moderate-to-vigorous bout of exercise (20 minutes at 70% heart rate reserve) reported significant increases in lysozyme concentration irrespective of whether participants were physically active or inactive (Kiwata et al., 2014).

Most of the investigations of nasal lavage have been conducted in the context of non-infectious causes of upper respiratory symptoms, including asthma, allergic rhinitis, and air pollution. The interest in such investigations arises from the high prevalence of such conditions in athletes and environments (e.g. aquatic, cold air) that have irritant effects on the nasal mucosa. For example, a review of studies of allergic rhinitis in athletes reported prevalence estimates between 21% and 56% where swimmers seem to be the most affected, followed by cross-country skiers (Surda et al., 2017). Allergic inflammation is the most common cause of chronic rhinitis and responsible for symptoms of nasal congestion, runny nose, nasal itch, and sneezing (Price et al., 2022b). A study of non-asthmatic middle-aged recreational runners reported that a marathon resulted in increased nasal nitric oxide (a non-invasive marker of airway inflammation) (Bonsignore et al., 2001). Vaisberg et al. (2013) performed measurements of cytokines in the protein extract of nasal mucosal cells and monitored upper respiratory symptoms following a marathon in recreational runners. Athletes who were symptomatic (mainly nasal congestion and an itchy and runny nose) following the marathon had greater concentrations of IL-6 but lower levels of IL-10 in the 72 hours following the marathon than those who were asymptomatic. Studies investigating cytokines (IL-6, IL-8, IL-10, and TNFα) within nasal fluid in response to 30–45-minute bouts of moderate-to-vigorous exercise have reported no changes (Cavalcante de Sá et al., 2016; Gomes et al., 2011; Min et al., 2020) or concentrations have fallen below detection limits (Kurowski et al., 2018; Gomez et al., 2011).

Kurowski et al. (2018) reported that competitive athletes (speed skaters and swimmers) during training and off-training periods, like those with asthma, have elevated levels of nasal IL-1ra compared with healthy controls. Only one of the athletes had been previously diagnosed with asthma, which led the authors to suggest that increased concentrations in nasal lavage may reflect the presence of airway inflammation in athletes. However, they were not able to detect any differences between groups in the changes in nasal cytokine response to an exercise challenge protocol often employed as a measure of bronchial hyperresponsiveness. Boyd et al. (2012) investigated whether cytokines in nasal lavage may be modifiable with moderate exercise training (12-week walking programme exercising at 60–75% of maximum heart rate) in adults with mild-to-moderate asthma but observed no changes in IL-1β, TNFα, IL-4, IL-5, IL-6, and IL-13. However, it is worthy to note that there was also no improvement in asthma control with the intervention compared with a non-exercising control group in this study.

Sputum or bronchoalveolar lavage

Much of the interest in the investigation of soluble factors in lower respiratory fluids has been to determine the role of inflammation in exercise-induced bronchoconstriction (EIB) or the use of exercise as an anti-inflammatory intervention in populations with chronic lung disease (e.g. asthma, COPD) where airway hyperresponsiveness and lower respiratory symptoms (e.g. sputum production, cough) are cardinal signs or symptoms of disease.

Lower airway dysfunction such as asthma and EIB is common in athletes where the highest prevalence is observed in elite endurance, winter, and aquatic disciplines (Price et al., 2022a). Asthma is typically characterised by recurrent respiratory symptoms and the presence of airway hyperresponsiveness (AHR) whereas EIB is defined as a transient narrowing of the lower airway during or following exercise in the presence or absence of clinically confirmed asthma (Boulet et al., 2017; Weiler et al., 2016). The exact pathophysiology of EIB remains to be fully established but is thought to involve exercise hyperpnoea and airway drying, causing increased osmolarity of airway-lining fluid and inflammatory mediator release from immune cells (e.g., histamine, cysteinyl leukotrienes, and prostaglandins) that results in airway smooth muscle contraction and bronchoconstriction (Hostrup et al., 2023).

Following exercise challenge, Hallstrand et al. (2005) reported that those with moderate-to-severe EIB had greater concentration of cysteinyl leukotrienes and ratio of cysteinyl leukotrienes to prostaglandin E$_2$ in induced sputum than asthmatics without a history of bronchoconstriction. This association of post-exercise challenge values of eicosanoids and propensity to develop EIB was replicated by Murphy et al. (2021). Greater concentrations of inflammatory mediators in induced sputum in EIB-positive versus EIB-negative athletes have also been observed in those without any history of asthma (Parsons et al., 2008). It is currently unclear if the associations between inflammatory mediators and EIB are dependent on the type of athletes. In a cross-sectional study of elite performance athletes self-reporting symptoms of EIB, there were no differences between pool-based and non-pool-based athletes in sputum concentrations of histamine or cysteinyl leukotrienes but pool-based athletes did have greater sputum concentrations of prostaglandin E2 (Martin et al., 2012). There are studies that have reported differences in the concentrations of other inflammatory mediators in respiratory fluids between different athletes. Seys et al. (2015) reported significant elevations in IL-8 at rest in sputum of elite swimmers (23% of which had EIB) compared with indoor athletes and controls, with a 90-minute intensive swimming training session leading to significant elevations in sputum IL-1β, TNFα, and IL-6. The authors suggested that their data demonstrated that intensive training, especially when combined with exposure to by-products of chlorination, induces airway epithelial damage and airway inflammation. In contrast, Škrgat et al. (2018) did not see any changes in sputum cytokines (IL-5, IL-6, IL-8, and TNFα) in healthy non-asthmatic elite swimmers during the high-intensity training period prior to a national championship. Belda et al. (2008) determined that elite athletes in water-based disciplines develop airway inflammation, with and without asthma, but its extent is related to the degree of bronchial hyperreactivity and the duration of training in pool water. Pedersen et al. (2008) determined that adolescent swimmers with two to three years of intense training and competition training showed no evidence of raised baseline airway inflammation. Such evidence has led to suggestions that a minimum of four years of competitive swimming, including several training hours per week (i.e. >10 hours week^{-1}) is required to detect airway dysfunction in otherwise healthy elite swimmers (Lomax, 2016). Indeed, there are multiple factors at play when it comes to detecting changes in baseline airway inflammation with athletes. Indeed, there also appears to be individual variation in the susceptibility to environmental stimuli, as some individuals do not appear to show airway changes at rest despite exposure to extreme temperatures or high concentrations of chemicals or pollutants (Bougault et al., 2022).

Transient changes in inflammatory mediators in response to acute exertion, may, however, occur with a wider range of exercise modalities. For example, Chimenti et al. (2010) observed increases in IL-8 in induced sputum from non-asthmatic runners following completing a half-marathon. Denguezli-Bouzgarrou et al. (2007) also reported that non-asthmatic runners had significant increases in leukotrienes (E4) and IL-8 in induced sputum samples collected at rest and 2 hours after an exercise session (~60 minutes at 80% of maximal aerobic speed) during competition periods compared with pre-competition and basic training. In contrast, Chimenti et al. (2009) did not see any significant changes in sputum IL-8 nor TNFα in recreational runners following races held in the fall (21 km), winter (12 km), and summer (10 km) compared with rest (baseline), but this may have been due to sample timing (20 hours after the races and hence may have not captured the transient inflammatory response).

Such increases in inflammatory mediators in the airways do question whether exercise may exaggerate pro-inflammatory responses in diseases such as COPD and severe asthma characterised by chronic airway inflammation, but human studies remain limited. Davidson et al. (2012) reported an incremental exercise test to symptom limitation resulted in reduced sputum IL-6 and IL-8 in people with COPD. There is a convincing body of evidence from animal models of COPD (lipopolysaccharide- or elastase-induced emphysema) and asthma (ovalbumin) that low-to-moderate intensity aerobic exercise training results in reduced bronchoalveolar lavage fluid concentrations of cysteinyl leukotrienes, leukotriene B4, TNFα, IL-6, and Th2 cytokines (IL-4, IL-5, and IL-13) as well as an increased expression of anti-inflammatory cytokines (IL-10) and cytokine inhibitors (IL-1ra) (Alberca-Custódio et al., 2016; Lee et al., 2019; Silva et al., 2010; Silva et al., 2016; Vieira et al., 2007; Vieira et al., 2011; Vieira et al., 2014; Wang et al., 2021).

Soluble immunity in the circulation

Humoral immunity

The major soluble components responsible for humoral immunity in the circulation are immunoglobulins (see Chapter 2). The circulating immunoglobulins fall into five major classes: IgM, IgG, IgA, IgD, and IgE and these can be divided further into subclasses. Following a lag period for accumulation of immunoglobulins within the blood, the IgM class predominates during any primary response to an antigen whereas IgG otherwise predominates including during rapid immuoglobulin responses to secondary antigen exposure (McKune et al., 2005; Stavnezer, 1996). Unlike immunoglobulins in saliva (discussed earlier in this chapter), the findings on the predominant circulating (plasma or serum) immunoglobin responses to exercise have been largely equivocal. Reports of IgA, IgG, and IgM concentrations include no change, increases, or decreases following strenuous prolonged exercise (Gunga et al., 2002; McKune et al., 2005; Petibois et al., 2003). Many of the early studies of brief moderate exercise that reported the most significant acute changes in immunoglobulins were largely explained by lack of adjustment for exercise-induced plasma volume changes. Modest effects have been observed when such plasma volume changes have been accounted for. Nehlsen-Cannarella et al. (1991b) reported that a walk of 45 minutes at 60% VO_2 max in a laboratory setting resulted in a 7% increase in IgG that returned to baseline 1.5 hours later. Nehlsen-Cannarella et al. (1991a) reported that 15 weeks of moderate exercise training led to increases in serum immunoglobulins (IgM, IgA, and IgG) compared with control, but the intervention group commenced the study with lower baseline concentrations. The circulating concentrations of IgD and IgE are small in comparison with others and a limited number of studies have assessed their response to exercise. The studies that have attempted to investigate them report conflicting findings or large amount of individual variability (McKune et al., 2005; Petibois et al., 2003). Over time there has been an increasing view that measurement of total immunoglobulin in the circulation may not be the most meaningful way to measure the effects of exercise on immune competency, in part because it lacks specificity. The shift has focused on immunoglobulin responses to vaccination and other immunological challenge models, which is presented in Chapter 7.

Cytokines and the acute-phase response

Unaccustomed, prolonged, and/or intense exertion is well established to trigger immune modulation that shares resemblance with an acute-phase response. This is a systemic response closely linked to inflammation and is an integral component of the innate immune response to protect the body in response to chemical/physical trauma or microbial invasion. Such disturbance to homeostasis results in changes in circulating concentrations of acute-phase proteins including C-reactive protein (CRP), fibrinogen, and complement proteins. The production and release of acute-phase proteins from the liver are induced by cytokines. The first account of exercise inducing a systemic release of cytokines is reported to be that by Cannon and Kluger (1983). In their study, rats injected intra-peritoneally with plasma collected from human participants after 1 hour of cycling at 60% VO_2 max resulted in elevated rectal temperature whereas plasma from the resting state (obtained pre-exercise) did not induce this response. The same effects on rats were observed when they were injected with supernatant from cultures of post-exercise peripheral mononuclear blood cells. We now know these endogenous pyrogens released following exercise as 'cytokines', a large family of polypeptides or proteins that can act in an autocrine, paracrine, or endocrine manner. As soluble factors go, they are considered key in regulating immune competency and inflammatory responses to exercise.

Northoff and Berg (1991) were the first to suggest that the cytokine IL-6 is a key systemic mediator of an acute-phase response following exercise due to their data of markedly increased levels of serum IL-6 concentrations after marathon running. Since then, increased circulating IL-6 concentrations following prolonged exercise is one of the most consistently reported findings in the literature (Box 5.2). Whilst we also now know that multiple other cytokines are also increased with prolonged exercise, the elevation of IL-6 is among the most marked and its release precedes other cytokines. It is the magnitude of and temporal pattern of the cytokine response to prolonged exercise that makes it distinct to other physical stressors such as infection, sepsis, burns, or trauma (Shephard, 2001, 2002). In response to prolonged exercise, the exponential rise of IL-6 is followed predominantly by anti-inflammatory responses which down-regulate inflammatory responses (Figure 5.3, panel A). In contrast, some of the clinical stressors listed above (e.g. sepsis) trigger excessive and overwhelming elevations in systemic pro-inflammatory responses (Figure 5.3, panel B) (de Jong et al., 2010; Netea et al., 2003; Shephard, 2002). This cytokine response was elegantly investigated in a model of low-grade inflammation by Starkie et al. (2003). Healthy adults were randomised to 3 hours of cycling or rest prior to intravenous administration of low-dose E-coli lipopolysaccharide endotoxin. A two-to-three-fold rise in circulating TNFα was observed in rested participants after endotoxin administration whilst this TNF response was totally blunted by those who undertook the prolonged bout of exercise prior to endotoxin administration.

Studies where investigators have measured IL-6 during exercise have demonstrated that IL-6 concentrations peak at end of a prolonged exercise bout or briefly thereafter, followed by a rapid decline towards pre-exercise concentrations. The increase in IL-6 concentrations with prolonged running bouts of 1 hour–3.5 hours have ranged from 4-fold to more than 100-fold, whilst more extreme ultramarathon events have been associated with an 8000-fold increase in systemic concentrations. Indeed, exercise duration is the single most influential factor of increases in IL-6 where it is suggested to account for more than 50% of the variation in exercise-induced changes (Figure 5.4) (Fischer 2006). The mode and intensity of exercise also contribute to this variability.

Early work suggested the changes in IL-6 with exercise were related to muscle damage (Bruunsgaard et al., 1997). We know now that the magnitude of increase in IL-6 is not reliant on exercise-induced muscle damage (Box 5.2). For example, eccentric exercise is not associated with more elevations than concentric exercise. Eccentric exercise may, however, lead to a delayed peak in IL-6 and slower return to pre-exercise levels during recovery. Similarly, although a rapid decline in IL-6 concentration is observed upon cessation of exercise, low sustained elevated IL-6 concentrations following prolonged running appear to be more related to muscle damage. Instead, the magnitude of change observed during exercise is dependent on the amount of skeletal muscle contraction involved. This is why studies of exercise involving limited muscle mass (e.g. upper extremity exercise only) often report lack of exercise-induced changes in plasma IL-6. It is also why you can expect higher-intensity exercise bouts to result in increased IL-6 concentrations. Cullen et al. (2016) demonstrated that in duration-matched bouts (35 minutes), increasing cycling intensity was associated with greater increases in plasma IL-6.

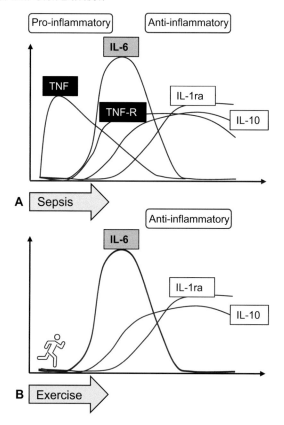

Figure 5.3 Cytokine responses to prolonged exercise (A) and sepsis (B). During exercise the marked increase in IL-6 is not preceded by elevated TNF and IL-1, with elevations in IL-6 driving increases in cytokine inhibitors and anti-inflammatory cytokines. During sepsis, a marked and rapid increase in circulating TNF and IL-1, precedes an increase in IL-6. Adapted from Pedersen (2017).

Box 5.2   Source of systemic changes in IL-6

A more extensive discussion of the local effects of acute and chronic exercise on skeletal muscle or adipose tissue and immune crosstalk is presented in Chapter 6. This section will rather focus on the studies that have led us to realise the key sources of the exercise-induced changes in systemic cytokine concentrations. Myokines has been adopted as the term to classify cytokines that exert paracrine or endocrine effects after production and release from skeletal muscle fibres. IL-6, was the first myokine to be identified and as we learnt earlier in this chapter, has been the most widely studied. Activated monocytes/macrophages are sources of IL-6 *in vivo* and early studies investigated whether they were the source of the exercise-induced changes in circulating IL-6.

Ullum et al. (1994) provided evidence to refute this hypothesis by showing 1 hour of cycling resulted in almost a two-fold increase in plasma IL-6 but no change in IL-6 mRNA within PBMC. These findings were replicated by Starkie et al. (2000) with a 2 hour bout of cycling. Focus shifted to the skeletal muscle whereby findings of greater plasma IL-6 following eccentric exercise than concentric exercise (Bruunsgaard et al., 1997) and mRNA for IL-6 being detectable in muscle biopsies after prolonged running (Ostrowski et al., 1998) led to the hypothesis that increases in plasma IL-6 were related to exercise-induced muscle damage. Croisier et al. (1999) soon delivered compelling findings that muscle damage was not a pre-requisite for elevated plasma IL-6 concentrations. Five moderately active males underwent two maximal eccentric exercise tests, separated by a period of three weeks of training sessions. As we may expect, the first bout of maximal eccentric contractions resulted in significant muscle damage (characterised by elevated serum myoglobin) and delayed-onset muscle soreness; DOMS), alongside increases in plasma IL-6. Our knowledge of classical adaptation to exercise tells us that if we allow sufficient recovery between that first exercise bout and then complete a second bout identical to the first, we will observe significantly lower muscle damage. Indeed, serum myoglobin and delayed-onset muscle soreness were significantly blunted following the second bout. There was, however, no significant difference between plasma IL-6 levels in response to the first and second exercise bout.

Steensberg et al. (2000) published a rapid report concluding that contracting skeletal muscle releases IL-6, and this release accounts for almost all of the exercise-induced increase in plasma IL-6. They recruited six healthy males to perform one-legged dynamic knee-extensor exercise for 5 hours at 25 W, which represented 40% of peak power output (Wmax). Hence, it was an exercise bout of relatively low intensity but importantly allowed the investigators to exclude the possible influence of eccentric exercise-induced muscle damage as it was purely a concentric exercise model. Catheters were placed into the femoral artery of one leg and the femoral vein of both legs. Blood for IL-6 measurement was obtained from catheters placed in both femoral veins and one artery, which provided the arterial–femoral venous (a–fv) difference. Despite the same supply of potential mediators of systemic IL-6 production from the femoral arteries, no release of IL-6 was detected in the resting leg. The plasma concentrations for IL-6 increased 19-fold with exercise. The measurements throughout the exercise bout were able to show that there was a rapid increase in plasma IL-6 concentration and release of IL-6 from the exercise leg in the final 2 hours of exercise. The authors concluded that exercise-induced changes in IL-6 are closely related to the duration of exercise and suggested increased production of IL-6 by contracting skeletal muscles during prolonged exercise is to maintain blood glucose, including signalling to the liver to increase glucose output. Keller et al. (2001) confirmed that 180 minutes of two-legged dynamic knee-extensor exercise results in a quicker and larger overall increase in IL-6 when commencing exercise in low-muscle glycogen levels compared with normal pre-exercise levels. It is worthy to note that these studies do not provide *prima facie* evidence that skeletal muscle cells are the main cell type involved in releasing IL-6, but this would come in a later study by Hiscock et al., (2004). It is worth noting that Starkie et al. (2001) determined that carbohydrate ingestion blunted the increase in plasma IL-6 concentration during both cycling and running exercise but had no effect on IL-6 mRNA expression in the muscle suggesting that carbohydrate ingestion may also affect IL-6 production by tissues other than the skeletal muscle.

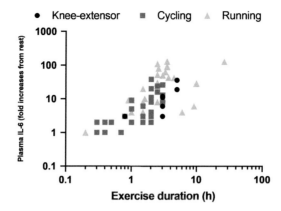

Figure 5.4 Changes in post-exercise plasma IL-6 with different modes of exercise (knee-extensor, cycling, running) and increasing duration of exercise. Produced from data reported of 67 studies in Fischer, 2006.

Given that high-intensity exercise will inherently be of shorter duration, these elevations do not compare to that seen by prolonged exercise.

Typically, the magnitude of change with IL-6 determines the appearance of the cytokine inhibitor, IL-1ra, and anti-inflammatory cytokine, IL-10, in the circulation (Steensberg et al., 2003). A systematic review and meta-analysis of long-distance running studies by Alves et al. (2022) reported that the majority of half-marathon, marathon, and ultramarathon events have resulted in significant increases in IL-6, IL-1ra, and IL-10. In a further systematic review of individual exercise bouts and IL-10 responses only, Cabral-Santos et al. (2019) reported that like IL-6, exercise duration is the single most important factor, accounting for 48% of the variation in exercise-induced changes in IL-10. Although the magnitude of change may not match that of IL-6, the increase in IL-1ra and IL-10 is largely regulated by the change in IL-6 concentrations. IL-10 concentrations appear to peak at the end of exercise or shortly thereafter whilst the peak for IL-1ra concentrations appear to peak during the early hours of recovery following exercise (Ostrowski et al., 1999; Nieman et al., 2001).

IL-8, MIP-1b, and MCP are chemokines that are commonly observed to increase in the circulation following prolonged exercise (Alves et al., 2022). In contrast, systemic concentrations of pro-inflammatory IL1-β, TNFα, and IFNγ have more modest increases following prolonged exercise or remain unchanged. Ostrowski et al. (1998) in their study of the Copenhagen marathon observed small but significant increases in IL-1β and TNFα immediately post-exercise, but these were matched by similar or greater increases of cytokine inhibitors (IL-1ra, sTNFr1, sTNFr2) during the hours following the race. Taken together, the available evidence tells us that the release of cytokine inhibitors and anti-inflammatory cytokines limit the magnitude and duration of pro-inflammatory responses following individual bouts of prolonged exercise.

Such promotion of an anti-inflammatory systemic compartment with each bout of exercise triggers interest in what occurs with regular exercise training. For the most part, it seems basal plasma IL-6 levels appear to be downregulated by exercise training and may represent a normal adaptation response to training. There generally have been fewer investigations of regular exercise on cytokine responses, which has sometimes been put down to methodological constraints, including but not limited to difficulty in establishing 'true' resting cytokine concentrations (Docherty et al., 2022). Most of the attention in this area has been placed on populations characterised by systemic inflammation where there may be dysregulated cytokine responses at rest. Obesity, type 2 diabetes, cardiovascular diseases, and chronic lung diseases are just some examples of disease groups associated with low-level sustained elevations of plasma IL-6. Given that exercise intolerance and physical inactivity are also common in these populations, studies have explored the relationship with cytokines and habitual physical activity or outcomes following exercise training interventions. For example, in people with COPD, Moy et al. (2014) reported that greater functional capacity (measured by 6-minute walk distance) and physical activity (measured in daily steps) were associated with lower plasma IL-6. Aerobic and/or resistance exercise training interventions have been shown to reduce IL-6, IL-8, MCP, and/or TNFα concentrations in COPD (Abd El-Kader et al., 2016; do Nascimento et al., 2015; Wang et al., 2014), severe asthma (França-Pinto et al., 2015), obesity (Freitas et al., 2017), type 2 diabetes (Miller et al., 2017), and heart failure (Adamopoulos et al., 2002). These findings have been reported alongside increases in IL-10. These effects are not consistently reported and the amount of evidence in different populations does vary. This is likely due to the outcomes not being routine parameters in clinical settings. IL-6, however, has often been measured alongside acute-phase proteins.

CRP is the most studied acute-phase protein in exercise settings and is often measured alongside IL-6 as CRP production in the liver (and hence release into the bloodstream) is stimulated by IL-6 (Del Giudice and Gangestad 2018). CRP has consistently been reported to be elevated for 24–48 hours following one-off endurance events such as a marathon (Weight et al., 1991), or is increased during the early stages of multi-day events (ultramarathons, cycling tours) and remains elevated throughout the competition (Gill et al., 2015; Robson-Ansley et al., 2009; Semple et al., 2006). Shorter, intense resistance exercise that is capable of inducing muscle damage has also been shown to induce elevations in CRP that may last for upto 24 hours or longer (Chatzinikolaou et al., 2010; Margaritelis et al., 2021). It appears that it is not the modality of resistance exercise (e.g. eccentric, plyometric) per se that affects changes in CRP but the muscle unaccustomedness to the type of exercise. In contrast to individual bouts of exercise, regular exercise reduces resting concentrations of CRP. In a systematic review of randomised and non-randomised controlled trials conducted between 1993 and 2015, Fedewa et al. (2017) reported that exercise training produces a small (mean ES of 0.26) but significant decrease in CRP levels. These effects were consistent for all individuals regardless of age and sex but exercise training led to a greater reduction in CRP when accompanied by a decrease in body mass index (ES=0.38).

Acute-phase proteins like fibrinogen have received less attention than CRP but when assessed look to follow a similar pattern. Weight et al. (1991) observed that plasma fibrinogen and CRP in 90 competitive distance runners, remained markedly increased up to 48 hours following a marathon. Plasma or serum fibrinogen is a biomarker linked with future risk of severe outcomes (hospitalisations and death) in chronic lung disease (e.g. exacerbations of respiratory symptoms caused by viral infections). There is preliminary evidence to suggest that short-term exercise training interventions can reduce fibrinogen concentrations in the circulation of people with chronic lung disease (Jenkins et al., 2020; Araújo et al., 2022).

Other components of the acute-phase response of the innate immune system such as complement proteins are also modulated by individual bouts of exercise and regular exercise training, where studies – as noted in a review by Rothschild-Rodriguez et al. (2022) – have mostly focused on the C3 and C4 family proteins. Similar to CRP, complement proteins are transiently activated immediately following a bout of exercise. These elevations are greatest in the 72 hours post-exercise when the exercise is prolonged in nature (e.g. ultra-endurance running) or with a bout designed to induce significant muscle damage (e.g. resistance training). In contrast, complement proteins are unchanged with bouts of cycling or running of 60 minutes or less (Castell et al., 1997; Hanson and Flaherty, 1981; Karacabey et al., 2005a, 2005b). Early studies in this area suggested that resting levels of C3 and/or C4 were lower in athletes (marathon runners) compared with sedentary controls (Nieman et al., 1989) or were downregulated in those involved in regular exercise (Smith et al., 1990). Such findings have been replicated elsewhere whereby higher levels of habitual physical activity and cardiorespiratory fitness are associated with reduced C3 in blood (Artero et al., 2014; Martinez-Gomez et al., 2010; Martinez-Gomez et al., 2012; Ruiz et al., 2007, 2008), effects that are consistent with anti-inflammatory effects of exercise discussed earlier. This was nicely illustrated in a population-based cohort study by Phillips et al. (2017) who undertook isotemporal substitution modelling to investigate replacing the amount of time spent in different physical activity intensities whilst total time remains constant. The analyses showed that re-allocation of 30 minutes of sedentary behaviour to moderate-vigorous physical activity per day was associated with lower C3 concentrations.

Clinical relevance of changes in soluble factors

In conjunction with determining what changes in soluble factors occur with individual bouts of exercise and regular exercise training, a keen interest for many in the field of exercise immunology has been to determine the clinical relevance of any changes to such soluble factors. The investigation of the clinical implications of changes in soluble factors can be broadly grouped into two broad areas, one of risk of URTI and the other of prevention or treatment of chronic diseases. Here we provide a synopsis of how exercise-induced changes to soluble factors – as mentioned previously in this chapter – might influence disease outcomes. For more detail regarding how exercise affects URTIs, see Chapter 9, and more detailed information about exercise and chronic diseases is available in Chapters 10–12.

Risk of URTI

Ever since the publication of studies reporting greater incidence of URTI following marathon or ultramarathon races, there have been investigations to identify an immune parameter that may help explain such susceptibility or has the most value in predicting those at greatest risk. Earlier in this chapter SIgA was introduced as the parameter of choice in many studies. In a cohort study of elite swimmers over a seven-month period, Gleeson et al. (1999) reported that the incidence of (physician verified) infections was best predicted by preseason salivary IgA levels, the mean pretraining IgA levels, and the rate of decrease in pretraining IgA levels over the seven-month training season. It was hypothesised that salivary IgA concentration below 40 mg L^{-1} may warrant close attention.

Others later suggested that SIgA secretion rate may be a better predictor of URTI (Gleeson et al., 2012). Nieman et al. (2003) found that a decline in SIgA secretion rate following a 160 km ultramarathon was 53% greater in runners who reported a URTI in the two weeks following the race compared with those not reporting a URTI. Fahlman and Engels (2005) also found that SIgA secretion rate to be a unique predictor of URTI throughout a competitive season in American College Football Players. The authors suggested that a secretion rate below 40 mg L^{-1} places athletes at most risk. In contrast, following a 50-week training study in America's Cup yacht racing athletes, Neville et al. (2008) suggested that SIgA concentrations that have taken into account each athlete's healthy baseline is what matters. A 28% reduction in individual's relative SIgA concentration occurred in the three weeks before URTI. On a group level, SIgA lower than 40% of their mean healthy SIgA concentration indicated a one in two chance of an athlete contracting a URI within three weeks. Although a smaller sample (15 male players from a professional English football League over a 16-week period), Perkins and Davison (2022) also found that the incidence of two URTI episodes in their study was preceded by individuals falling 40% below their healthy baseline in SIgA secretion rate (but not concentration).

It must be said that not all studies have been able to demonstrate a relationship between changes in SIgA and URTI (e.g. Cunniffe et al., 2011), leading some to conclude that SIgA is not a suitable biomarker for infection in athletes (Turner et al., 2021). All studies do, however, agree that SIgA measures have significant between-person (60%) and within-person (50%) variability (Rapson et al., 2020). This may partly explain why studies that have measured SIgA more frequently (e.g. weekly vs monthly) or in larger sample sizes have had better success. Preliminary evidence of lower incidence of URTI symptoms during 12 weeks of moderate activity being associated with an increase in SIgA (Klentrou et al., 2002), acts as further support to the clinical relevance of SIgA. Taken together, if frequent monitoring is feasible from a cost perspective and there is adequate time to establish a healthy baseline, SIgA can be a valuable part of athlete monitoring, particularly in those with a history of frequent infections. Point-of-care testing provides the potential to offer longitudinal tracking of such parameters in the sports setting (Turner et al., 2021), but it is important to confirm that any assays are specific for SgA or there is the potential for analytical errors as a result of blood contamination. It is unlikely, however, that SIgA alone can predict all URTI; recent work (Gleeson et al., 2017) on the development of multi-component immune models provides some insight into where evaluating the risk of respiratory illness in athletes will go in the future.

Prevention or treatment of inflammatory diseases

If we deem that establishing the evidence to conclusively support a link between exercise-induced changes in immune competence and risk of URTI has been a challenging endeavour, then we have barely scratched the surface with delineating the clinical significance of changes in soluble factors and the anti-inflammatory effects of exercise in the wider population. A significant part of this chapter has discussed how the systemic changes in cytokines following prolonged exercise, driven by IL-6 release from the contracting skeletal muscle, largely confers an anti-inflammatory response. Plasma concentrations of IL-6 in a rested state typically reflect release from sources (e.g. leucocytes, adipocytes) other than the skeletal muscle (Nash et al., 2023). In healthy individuals, plasma IL-6 concentration is usually less than 1 pg L^{-1} and diseases characterised by chronic low-grade systemic inflammation may see plasma IL-6 concentrations up to 10 pg L^{-1} or higher (Nash et al., 2023). The transient increase in IL-6 in response to strenuous exercise has seen plasma concentrations as high as 1,000 pg L^{-1} (Fischer, 2006). So, context is critical, and caution is advised when interpreting or extrapolating the clinical relevance of changes in cytokines across settings and populations.

The hypothesis that the health benefits of exercise are mediated by cytokines (or myokines) is based on the anti-inflammatory response elicited by bouts of exercise. In other words, regular exercise (or the adaptations it promotes) is anti-inflammatory in nature. Whilst this is appealing from a mechanistic perspective, the link between the responses to each bout of exercise translating to long-term benefit is only in the early stages of being substantiated (Peake et al., 2015). We must also acknowledge that the anti-inflammatory environment (increased plasma IL-1ra and IL-10) induced by IL-6 is heavily dependent on exercise duration where changes are most pronounced with strenuous prolonged exertion and modest, if any, changes occur in response to brief moderate bouts of exercise. Meanwhile, it is moderate bouts of physical activity that are most commonly recommended for the prevention and treatment of diseases associated with chronic low-grade systemic inflammation. Inflammation is known to have a pathophysiological role in the development of many chronic diseases and circulating factors like IL-6 and

CRP have been shown to be predictors of disease activity or severe outcomes in such populations (Pedersen, 2017). Physical inactivity is a well-accepted risk factor for the development of these diseases and exercise interventions or regular physical activity do improve clinical outcomes (symptoms of disease) (Pedersen, 2009). This chapter previously introduced how exercise training can result in a lowering of cytokine concentrations and acute-phase proteins, but these studies have largely involved older but otherwise healthy populations. What is missing are randomised controlled trials that concurrently measure clinical endpoints and circulating factors. There is some preliminary evidence, but the level of understanding on the mechanisms of exercise in a clinical context has not been studied in sufficient depth, or at least not to the extent that we have come to expect in the development of anti-inflammatory drug interventions. For more information about exercise and chronic disease, please refer to Chapters 10–12.

Key points

- Blood (plasma or serum) and mucosal or respiratory tract fluids (saliva, tear fluid sputum, or fluid obtained from nasal or bronchoalveolar lavage) have all been used to assess changes of soluble factors in response to individual bouts of exercise or regular exercise training.
- Saliva has often been the biofluid of choice for measuring soluble immune factors in the airway due to the ease of collection.
- Secretory IgA (SIgA) provides this first line of defence at mucosal surfaces with the support of AMPs.
- Self-inoculation at the eyes or nose is a common route of entry for pathogens and there is preliminary evidence that tear fluid may act as an alternative to saliva for measuring SIgA and AMPs' response to exercise.
- Monitoring salivary SIgA during periods of training and competition has achieved some success in predicting URTI in athletes.
- The between and within-person variability in SIgA means it is likely that multi-component immune models are necessary to improve on the evaluation of URTI risk.
- Prolonged exercise induces a pattern of cytokine response that is distinct to other stressors (e.g. infection) including an exponential increase in IL-6 that inhibits TNF and causes increased circulating concentrations of cytokine inhibitors (IL-1ra), anti-inflammatory cytokines (e.g. IL-10), acute-phase proteins (e.g. CRP) and complement.
- Contracting skeletal muscle is the main source of increased plasma IL-6 during exercise, with the magnitude mostly explained by the duration of exercise.
- Exercise training and regular physical activity can counteract the low-grade systemic inflammation involved in the development and progression of many chronic diseases.
- The mechanisms underpinning these anti-inflammatory effects remain understudied but such potential can be harnessed in an era of precision medicine.

References

Abd El-Kader, S. M., Al-Jiffri, O. H., & Al-Shreef, F. M. (2016). Plasma inflammatory biomarkers response to aerobic versus resisted exercise training for chronic obstructive pulmonary disease patients. *Afr Health Sci, 16*(2), 507–515.

Adamopoulos, S., Parissis, J., Karatzas, D., Kroupis, C., Georgiadis, M., Karavolias, G.,... Kremastinos, D. T. (2002). Physical training modulates proinflammatory cytokines and the soluble Fas/soluble Fas ligand system in patients with chronic heart failure. *J Am Coll Cardiol, 39*(4), 653–663.

Agha, N. H., Baker, F. L., Kunz, H. E., Spielmann, G., Mylabathula, P. L., Rooney, B. V.,... Simpson, R. J. (2020). Salivary antimicrobial proteins and stress biomarkers are elevated during a 6-month mission to the international space station. *J Appl Physiol, 128*(2), 264–275.

Alberca-Custódio, R. W., Greiffo, F. R., MacKenzie, B., Oliveira-Junior, M. C., Andrade-Sousa, A. S., Graudenz, G. S.,... Vieira, R. P. (2016). Aerobic exercise reduces asthma phenotype by modulation of the leukotriene pathway. *Front Immunol, 7*, 237.

Allgrove, J. E., Gomes, E., Hough, J., & Gleeson, M. (2008). Effects of exercise intensity on salivary antimicrobial proteins and markers of stress in active men. *J Sports Sci, 26*(6), 653–661.

Allgrove, J. E., Oliveira, M., & Gleeson, M. (2014). Stimulating whole saliva affects the response of anti-microbial proteins to exercise. *Scand J Med Sci Sports, 24*(4), 649–655.

Alves, M. D. J., Silva, D. D. S., Pereira, E. V. M., Pereira, D. D., de Sousa Fernandes, M. S., Santos, D. F. C.,... de Souza, R. F. (2022). Changes in cytokines concentration following long-distance running: A systematic review and meta-analysis. *Front Physiol, 13*, 838069.

Aps, J. K., & Martens, L. C. (2005). Review: The physiology of saliva and transfer of drugs into saliva. *Forensic Sci Int, 150*(2–3), 119–131.

Araújo, A. S., Figueiredo, M. R., Lomonaco, I., Lundgren, F., Mesquita, R., & Pereira, E. D. B. (2022). Effects of pulmonary rehabilitation on systemic inflammation and exercise capacity in bronchiectasis: A randomized controlled trial. *Lung, 200*(3), 409–417.

Artero, E. G., España-Romero, V., Jiménez-Pavón, D., Martinez-Gómez, D., Warnberg, J., Gómez-Martínez, S.,... Castillo, M. J. (2014). Muscular fitness, fatness and inflammatory biomarkers in adolescents. *Pediatr Obes, 9*(5), 391–400.

Bals, R. (2000). Epithelial antimicrobial peptides in host defense against infection. *Respir Res, 1*(3), 141–150.

Belda, J., Ricart, S., Casan, P., Giner, J., Bellido-Casado, J., Torrejon, M.,... Drobnic, F. (2008). Airway inflammation in the elite athlete and type of sport. *Br J Sports Med, 42*(4), 244–248; discussion 248–249.

Belser, J. A., Wadford, D. A., Xu, J., Katz, J. M., & Tumpey, T. M. (2009). Ocular infection of mice with Influenza A (H7) viruses: A site of primary replication and spread to the respiratory tract. *J Virol, 83*(14), 7075–7084.

Beltzer, E. K., Fortunato, C. K., Guaderrama, M. M., Peckins, M. K., Garramone, B. M., & Granger, D. A. (2010). Salivary flow and alpha-amylase: collection technique, duration, and oral fluid type. *Physiol Behav, 101*(2), 289–296.

Bischoff, W. E., Reid, T., Russell, G. B., & Peters, T. R. (2011). Transocular entry of seasonal influenza-attenuated virus aerosols and the efficacy of N95 respirators, surgical masks, and eye protection in humans. *J Infect Dis, 204*(2), 193–199.

Bishop, N. C., Blannin, A. K., Armstrong, E., Rickman, M., & Gleeson, M. (2000). Carbohydrate and fluid intake affect the saliva flow rate and IgA response to cycling. *Med Sci Sports Exerc, 32*(12), 2046–2051.

Bishop, N. C., & Gleeson, M. (2009). Acute and chronic effects of exercise on markers of mucosal immunity. *Front Biosci (Landmark Ed), 14*(12), 4444–4456.

Blannin, A. K., Robson, P. J., Walsh, N. P., Clark, A. M., Glennon, L., & Gleeson, M. (1998). The effect of exercising to exhaustion at different intensities on saliva Immunoglobulin A, protein, and electrolyte secretion. *Int J Sports Med, 19*(8), 547–552.

Bonsignore, M. R., Morici, G., Riccobono, L., Insalaco, G., Bonanno, A., Profita, M.,... Vignola, A. M. (2001). Airway inflammation in nonasthmatic amateur runners. *Am J Physiol Lung Cell Mol Physiol, 281*(3), L668–676.

Bosch, J. A., Ring, C., de Geus, E. J., Veerman, E. C., & Amerongen, A. V. (2002). Stress and secretory immunity. *Int Rev Neurobiol, 52*, 213–253.

Bougault, V., Adami, P. E., Sewry, N., Fitch, K., Carlsten, C., Villiger, B.,... Schobersberger, W. (2022). Environmental factors associated with non-infective acute respiratory illness in athletes: A systematic review by a subgroup of the IOC consensus group on "acute respiratory illness in the athlete." *J Sci Med Sport, 25*(6), 466–473.

Boulet, L. P., Turmel, J., & Côté, A. (2017). Asthma and exercise-induced respiratory symptoms in the athlete: New insights. *Curr Opin Pulm Med, 23*(1), 71–77.

Bowdish, D. M., Davidson, D. J., & Hancock, R. E. (2005). A re-evaluation of the role of host defense peptides in mammalian immunity. *Curr Protein Pept Sci, 6*(1), 35–51.

Boyd, A., Yang, C. T., Estell, K., Ms, C. T., Gerald, L. B., Dransfield, M.,... Schwiebert, L. M. (2012). Feasibility of exercising adults with asthma: a randomized pilot study. *Allergy Asthma Clin Immunol, 8*(1), 13.

Brandtzaeg, P., Farstad, I. N., & Haraldsen, G. (1999). Regional specialization in the mucosal immune system: Primed cells do not always home along the same track. *Immunol Today, 20*(6), 267–277.

Bruunsgaard, H., Galbo, H., Halkjaer-Kristensen, J., Johansen, T. L., MacLean, D. A., & Pedersen, B. K. (1997). Exercise-induced increase in serum interleukin-6 in humans is related to muscle damage. **J Physiol, 499*(Pt 3), 833–841.

Cabral-Santos, C., de Lima Junior, E. A., Fernandes, I. M. D. C., Pinto, R. Z., Rosa-Neto, J. C., Bishop, N. C., & Lira, F. S. (2019). Interleukin-10 responses from acute exercise in healthy subjects: A systematic review. *J Cell Physiol, 234*(7), 9956–9965.

Campbell, J. P., & Turner, J. E. (2018). Debunking the myth of exercise-induced immune suppression: Redefining the impact of exercise on immunological health across the lifespan. *Front Immunol, 9*, 648.

Cannon, J. G., & Kluger, M. J. (1983). Endogenous pyrogen activity in human plasma after exercise. *Science, 220*(4597), 617–619.

Carlson, L. A., Kenefick, R. W., & Koch, A. J. (2013). Influence of carbohydrate ingestion on salivary immunoglobulin A following resistance exercise. *J Int Soc Sports Nutr, 10*(1), 14.

Castell, L. M., Poortmans, J. R., Leclercq, R., Brasseur, M., Duchateau, J., & Newsholme, E. A. (1997). Some aspects of the acute phase response after a marathon race, and the effects of glutamine supplementation. *Eur J Appl Physiol Occup Physiol, 75*(1), 47–53.

Cavalcante de Sá, M., Nakagawa, N. K., Saldiva de André, C. D., Carvalho-Oliveira, R., de Santana Carvalho, T., Nicola, M. L.,... Vaisberg, M. (2016). Aerobic exercise in polluted urban environments: Effects on airway defense mechanisms in young healthy amateur runners. *J Breath Res, 10*(4), 046018.

Chastin, S. F. M., Abaraogu, U., Bourgois, J. G., Dall, P. M., Darnborough, J., Duncan, E.,... Hamer, M. (2021). Effects of regular physical activity on the immune system, vaccination and risk of community-acquired infectious disease in the general population: Systematic review and meta-analysis. *Sports Med, 51*(8), 1673–1686.

Chatzinikolaou, A., Fatouros, I. G., Gourgoulis, V., Avloniti, A., Jamurtas, A. Z., Nikolaidis, M. G.,... Taxildaris, K. (2010). Time course of changes in performance and inflammatory responses after acute plyometric exercise. *J Strength Cond Res, 24*(5), 1389–1398.

Chimenti, L., Morici, G., Paterno, A., Bonanno, A., Vultaggio, M., Bellia, V., & Bonsignore, M. R. (2009). Environmental conditions, air pollutants, and airway cells in runners: A longitudinal field study. *J Sports Sci, 27*(9), 925–935.

Chimenti, L., Morici, G., Paternò, A., Santagata, R., Bonanno, A., Profita, M.,... Bonsignore, M. R. (2010). Bronchial epithelial damage after a half-marathon in nonasthmatic amateur runners. *Am J Physiol Lung Cell Mol Physiol, 298*(6), L857–862.

Covey, W., Perillie, P., & Finch, S. C. (1971). The origin of tear lysozyme. *Proc Soc Exp Biol Med, 137*(4), 1362–1363.

Croisier, J. L., Camus, G., Venneman, I., Deby-Dupont, G., Juchmès-Ferir, A., Lamy, M.,... Duchateau, J. (1999). Effects of training on exercise-induced muscle damage and interleukin 6 production. *Muscle Nerve, 22*(2), 208–212.

Cullen, T., Thomas, A. W., Webb, R., & Hughes, M. G. (2016). Interleukin-6 and associated cytokine responses to an acute bout of high-intensity interval exercise: The effect of exercise intensity and volume. *Appl Physiol Nutr Metab, 41*(8), 803–808.

Cunniffe, B., Griffiths, H., Proctor, W., Davies, B., Baker, J. S., & Jones, K. P. (2011). Mucosal immunity and illness incidence in elite rugby union players across a season. *Med Sci Sports Exerc, 43*(3), 388–397.

Dartt, D. A., & Willcox, M. D. (2013). Complexity of the tear film: Importance in homeostasis and dysfunction during disease. *Exp Eye Res, 117*, 1–3.

da Silveira, E. G., Prato, L. S., Pilati, S. F. M., & Arthur, R. A. (2023). Comparison of oral cavity protein abundance among caries-free and caries-affected individuals-a systematic review and meta-analysis. *Front Oral Health, 4*, 1265817.

Davison, G., Allgrove, J., & Gleeson, M. (2009). Salivary antimicrobial peptides (LL-37 and alpha-defensins HNP1–3), antimicrobial and IgA responses to prolonged exercise. *Eur J Appl Physiol, 106*(2), 277–284.

Davison, G., & Diment, B. C. (2010). Bovine colostrum supplementation attenuates the decrease of salivary lysozyme and enhances the recovery of neutrophil function after prolonged exercise. *Br J Nutr, 103*(10), 1425–1432.

Davidson, W. J., Verity, W. S., Traves, S. L., Leigh, R., Ford, G. T., & Eves, N. D. (2012). Effect of incremental exercise on airway and systemic inflammation in patients with COPD. *J Appl Physiol, 112*(12), 2049–2056.

Dawes, C. (2008). Salivary flow patterns and the health of hard and soft oral tissues. *J Am Dent Assoc, 139*(Suppl), 18S–24S.

de Jong, H. K., van der Poll, T., & Wiersinga, W. J. (2010). The systemic pro-inflammatory response in sepsis. *J Innate Immun, 2*(5), 422–430.

Del Giudice, M., & Gangestad, S. W. (2018). Rethinking IL-6 and CRP: Why they are more than inflammatory biomarkers, and why it matters. *Brain Behav Immun, 70*, 61–75.

Denguezli-Bouzgarrou, M., Ben Saad, H., Ben Chiekh, I., Gaied, S., Tabka, Z., & Zbidi, A. (2007). Role of lung inflammatory mediators as a cause of training-induced lung function changes in runners. *Sci Sports, 22*(1), 35–42.

De Smet, K., & Contreras, R. (2005). Human antimicrobial peptides: Defensins, cathelicidins and histatins. *Biotechnol Lett, 27*(18), 1337–1347.

Docherty, S., Harley, R., McAuley, J. J., Crowe, L. A. N., Pedret, C., Kirwan, P. D.,... Millar, N. L. (2022). The effect of exercise on cytokines: implications for musculoskeletal health: a narrative review. *BMC Sports Sci Med Rehabil, 14*(1), 5.

do Nascimento, E. S., Sampaio, L. M., Peixoto-Souza, F. S., Dias, F. D., Gomes, E. L., Greiffo, F. R.,... Costa, D. (2015). Home-based pulmonary rehabilitation improves clinical features and systemic inflammation in chronic obstructive pulmonary disease patients. *Int J Chron Obstruct Pulmon Dis, 10*, 645–653.

Fábián, T. K., Hermann, P., Beck, A., Fejérdy, P., & Fábián, G. (2012). Salivary defense proteins: Their network and role in innate and acquired oral immunity. *Int J Mol Sci, 13*(4), 4295–4320.

Fahlman, M. M., & Engels, H. J. (2005). Mucosal IgA and URTI in American college football players: A year longitudinal study. *Med Sci Sports Exerc, 37*(3), 374–380.

Fedewa, M. V., Hathaway, E. D., & Ward-Ritacco, C. L. (2017). Effect of exercise training on C reactive protein: A systematic review and meta-analysis of randomised and non-randomised controlled trials. *Br J Sports Med, 51*(8), 670–676.

Fischer, C. P. (2006). Interleukin-6 in acute exercise and training: what is the biological relevance? *Exerc Immunol Rev, 12*, 6–33.

França-Pinto, A., Mendes, F. A., de Carvalho-Pinto, R. M., Agondi, R. C., Cukier, A., Stelmach, R.,... Carvalho, C. R. (2015). Aerobic training decreases bronchial hyperresponsiveness and systemic inflammation in patients with moderate or severe asthma: a randomized controlled trial. *Thorax, 70*(8), 732–739.

Freitas, P. D., Ferreira, P. G., Silva, A. G., Stelmach, R., Carvalho-Pinto, R. M., Fernandes, F. L.,... Carvalho, C. R. (2017). The role of exercise in a weight-loss program on clinical control in obese adults with asthma. A randomized controlled trial. *Am J Respir Crit Care Med, 195*(1), 32–42.

Fullard, R.J., Snyder, C. (1990). Protein levels in nonstimulated and stimulated tears of normal human subjects. *Invest Ophthalmol Vis Sci, 31*(6), 1119–26.

Gill, S. K., Hankey, J., Wright, A., Marczak, S., Hemming, K., Allerton, D. M.,... Costa, R. J. (2015). The impact of a 24-h ultra-marathon on circulatory endotoxin and cytokine profile. *Int J Sports Med, 36*(8), 688–695.

Gill, S. K., Teixeira, A. M., Rosado, F., Hankey, J., Wright, A., Marczak, S.,... Costa, R. J. (2014). The impact of a 24-h ultra-marathon on salivary antimicrobial protein responses. *Int J Sports Med, 35*(11), 966–971.

Gillum, T. L., Kuennen, M., Gourley, C., Schneider, S., Dokladny, K., & Moseley, P. (2013). Salivary antimicrobial protein response to prolonged running. *Biol Sport, 30*(1), 3–8.

Gleeson, M. (2000). Mucosal immunity and respiratory illness in elite athletes. *Int J Sports Med, 21*(Suppl 1), S33–43.

Gleeson, M., Bishop, N., Oliveira, M., McCauley, T., Tauler, P., & Muhamad, A. S. (2012). Respiratory infection risk in athletes: association with antigen-stimulated IL-10 production and salivary IgA secretion. *Scand J Med Sci Sports, 22*(3), 410–417.

Gleeson, M., McDonald, W. A., Pyne, D. B., Cripps, A. W., Francis, J. L., Fricker, P. A.,... Clancy, R. L. (1999). Salivary IgA levels and infection risk in elite swimmers. *Med Sci Sports Exerc, 31*(1), 67–73.

Gleeson, M., & Pyne, D. B. (2000). Special feature for the Olympics: Effects of exercise on the immune system: Exercise effects on mucosal immunity. *Immunol Cell Biol, 78*(5), 536–544.

Gleeson, M., Pyne, D. B., Elkington, L. J., Hall, S. T., Attia, J. R., Oldmeadow, C.,... Callister, R. (2017). Developing a multi-component immune model for evaluating the risk of respiratory illness in athletes. *Exerc Immunol Rev, 23*, 52–64.

Gomes, E. C., Stone, V., & Florida-James, G. (2011). Impact of heat and pollution on oxidative stress and CC16 secretion after 8 km run. *Eur J Appl Physiol, 111*(9), 2089–2097.

Granger, D. A., Fortunato, C. K., Beltzer, E. K., Virag, M., Bright, M. A., & Out, D. (2012). Focus on methodology: Salivary bioscience and research on adolescence: An integrated perspective. *J Adolesc, 35*(4), 1081–1095.

Gunga, H. C., Machotta, A., Schobersberger, W., Mittermayr, M., Kirsch, K., Koralewski, E.,... Rocker, L. (2002). Neopterin, IgG, IgA, IgM, and plasma volume changes during long-distance running. *Pteri-dines, 13*, 15–20.

Hall, M. W., Wellappuli, N. C., Huang, R. C., Wu, K., Lam, D. K., Glogauer, M.,... Senadheera, D. B. (2023). Suspension of oral hygiene practices highlights key bacterial shifts in saliva, tongue, and tooth plaque during gingival inflammation and resolution. *ISME Commun, 3*(1), 23.

Hallstrand, T. S., Moody, M. W., Aitken, M. L., & Henderson, W. R., Jr. (2005). Airway immunopathology of asthma with exercise-induced bronchoconstriction. *J Allergy Clin Immunol, 116*(3), 586–593.

Halson, S. L., Lancaster, G. I., Jeukendrup, A. E., & Gleeson, M. (2003). Immunological responses to overreaching in cyclists. *Med Sci Sports Exerc, 35*(5), 854–861.

Hanson, P. G., & Flaherty, D. K. (1981). Immunological responses to training in conditioned runners. *Clin Sci (Lond), 60*(2), 225–228.

Hanstock, H. G., Edwards, J. P., & Walsh, N. P. (2019). Tear lactoferrin and lysozyme as clinically relevant biomarkers of mucosal immune competence. *Front Immunol, 10*, 1178.

Hanstock, H. G., Walsh, N. P., Edwards, J. P., Fortes, M. B., Cosby, S. L., Nugent, A.,... Yong, X. H. (2016). Tear fluid SIgA as a noninvasive biomarker of mucosal immunity and common cold risk. *Med Sci Sports Exerc, 48*(3), 569–577.

Harmon, A. G., Hibel, L. C., Rumyantseva, O., & Granger, D. A. (2007). Measuring salivary cortisol in studies of child development: watch out--what goes in may not come out of saliva collection devices. *Dev Psychobiol, 49*(5), 495–500.

He, C. S., Tsai, M. L., Ko, M. H., Chang, C. K., & Fang, S. H. (2010). Relationships among salivary im-munoglobulin A, lactoferrin, and cortisol in basketball players during a basketball season. *Eur J Appl Physiol, 110*(5), 989–995.

Heaney, J. L. J., Killer, S. C., Svendsen, I. S., Gleeson, M., & Campbell, J. P. (2018). Intensified training increases salivary free light chains in trained cyclists: Indication that training volume increases oral inflammation. *Physiol Behav, 188*, 181–187.

Hendley, J. O. (1998). Epidemiology, pathogenesis, and treatment of the common cold. *Semin Pediatr Infect Dis, 9*(1), 50–55.

Hendley, J. O., & Gwaltney, J. M. Jr. (1998). Mechanisms of transmission of rhinovirus infections. *Epide-miol Rev, 10,* 243–58.

Hiscock, N., Chan, M. H., Bisucci, T., Darby, I. A., & Febbraio, M. A. (2004). Skeletal myocytes are a source of interleukin-6 mRNA expression and protein release during contraction: Evidence of fiber type specificity. *FASEB J, 18*(9), 992–994.

Hostrup, M., Hansen, E. S. H., Rasmussen, S. M., Jessen, S., & Backer, V. (2023). Asthma and exercise-induced bronchoconstriction in athletes: Diagnosis, treatment, and anti-doping challenges. *Scand J Med Sci Sports* (in press).

Hucklebridge, F., Clow, A., & Evans, P. (1998). The relationship between salivary secretory immunoglob-ulin A and cortisol: Neuroendocrine response to awakening and the diurnal cycle. *Int J Psychophysiol, 31*(1), 69–76.

Iorgulescu, G. (2009). Saliva between normal and pathological. Important factors in determining systemic and oral health. *J Med Life, 2*(3), 303–307.

Jenkins, A. R., Holden, N. S., Gibbons, L. P., & Jones, A. W. (2020). Clinical outcomes and inflammatory responses of the frequent exacerbator in pulmonary rehabilitation: A prospective Cohort study. *COPD*, *17*(3), 253–260.

Johansen, F. E., Braathen, R., & Brandtzaeg, P. (2001). The J chain is essential for polymeric Ig receptor-mediated epithelial transport of IgA. *J Immunol*, 167(9), 5185–5192.

Jollès, P., & Jollès, J. (1984). What's new in lysozyme research? Always a model system, today as yesterday. *Mol Cell Biochem*, *63*(2), 165–189.

Jones, A. W., Mironas, A., Mur, L. A. J., Beckmann, M., Thatcher, R., & Davison, G. (2023). Vitamin D status modulates innate immune responses and metabolomic profiles following acute prolonged cycling. *Eur J Nutr*, *62*(7), 2977–2990.

Kang, J. H., & Kho, H. S. (2019). Blood contamination in salivary diagnostics: Current methods and their limitations. *Clin Chem Lab Med*, *57*(8), 1115–1124.

Karacabey, K., Peker, İ., Saygın, Ö., Cıloglu, F., Ozmerdivenli, R., & Bulut, V. (2005a). Effects of acute aerobic and anaerobic exercise on humoral immune factors in elite athletes. *Biotechnol Biotechnol Equip*, *19*, 175–180.

Karacabey, K., Saygin, O., Ozmerdivenli, R., Zorba, E., Godekmerdan, A., & Bulut, V. (2005b). The effects of exercise on the immune system and stress hormones in sportswomen. *Neuro Endocrinol Lett*, *26*, 361–366.

Keller, C., Steensberg, A., Pilegaard, H., Osada, T., Saltin, B., Pedersen, B. K., & Neufer, P. D. (2001). Transcriptional activation of the IL-6 gene in human contracting skeletal muscle: Influence of muscle glycogen content. *FASEB J*, *15*(14), 2748–2750.

Kimura, F., Aizawa, K., Tanabe, K., Shimizu, K., Kon, M., Lee, H.,... Kono, I. (2008). A rat model of saliva secretory immunoglobulin: a suppression caused by intense exercise. *Scand J Med Sci Sports*, *18*(3), 367–372.

Kiwata, J., Anouseyan, R., Desharnais, R., Cornwell, A., Khodiguian, N., & Porter, E. (2014). Effects of aerobic exercise on lipid-effector molecules of the innate immune response. *Med Sci Sports Exerc*, *46*(3), 506–512.

Klentrou, P., Cieslak, T., MacNeil, M., Vintinner, A., & Plyley, M. (2002). Effect of moderate exercise on salivary immunoglobulin A and infection risk in humans. *Eur J Appl Physiol*, *87*(2), 153–158.

Koibuchi, E., & Suzuki, Y. (2014). Exercise upregulates salivary amylase in humans (Review). *Exp Ther Med*, *7*(4), 773–777.

Korsrud, F. R., & Brandtzaeg, P. (1980). Quantitative immunohistochemistry of immunoglobulin- and J-chain-producing cells in human parotid and submandibular salivary glands. *Immunology*, *39*(2), 129–140.

Kudsk, K. A. (2002). Current aspects of mucosal immunology and its influence by nutrition. *Am J Surg*, *183*(4), 390–398.

Kunz, H., Bishop, N. C., Spielmann, G., Pistillo, M., Reed, J., Ograjsek, T.,... Simpson, R. J. (2015). Fitness level impacts salivary antimicrobial protein responses to a single bout of cycling exercise. *Eur J Appl Physiol*, *115*(5), 1015–1027.

Kurimoto, Y., Saruta, J., To, M., Yamamoto, Y., Kimura, K., & Tsukinoki, K. (2016). Voluntary exercise increases IgA concentration and polymeric Ig receptor expression in the rat submandibular gland. *Biosci Biotechnol Biochem*, *80*(12), 2490–2496.

Kurowski, M., Jurczyk, J., Olszewska-Ziąber, A., Jarzębska, M., Krysztofiak, H., & Kowalski, M. L. (2018). A similar pro/anti-inflammatory cytokine balance is present in the airways of competitive athletes and non-exercising asthmatics. *Adv Med Sci*, *63*(1), 79–86.

Lee, B., Kim, Y., Kim, Y. M., Jung, J., Kim, T., Lee, S. Y.,... Ryu, J. H. (2019). Anti-oxidant and anti-inflammatory effects of aquatic exercise in allergic airway inflammation in mice. *Front Physiol*, *10*, 1227.

Legrand, D., Elass, E., Pierce, A., & Mazurier, J. (2004). Lactoferrin and host defence: An overview of its immuno-modulating and anti-inflammatory properties. *Biometals*, *17*(3), 225–229.

Leicht, C. A., Goosey-Tolfrey, V. L., & Bishop, N. C. (2018). Exercise intensity and its impact on relationships between salivary immunoglobulin A, saliva flow rate, and plasma cortisol concentration. *Eur J Appl Physiol*, *118*(6), 1179–1187.

Lindh, E. (1975). Increased resistance of Immunoglobulin A dimers to proteolytic degradation after binding of secretory component. *J Immunol*, *114*(1 Pt 2), 284–286.

Lomax, M. (2016). Airway dysfunction in elite swimmers: Prevalence, impact, and challenges. *Open Access J Sports Med*, *7*, 55–63.

Mackinnon, L. T., Ginn, E., & Seymour, G. J. (1993). Decreased salivary Immunoglobulin A secretion rate after intense interval exercise in elite kayakers. *Eur J Appl Physiol*, *67*, 180–184.

Mackinnon, L. T., & Hooper, S. (1994). Mucosal (secretory) immune system responses to exercise of varying intensity and during overtraining. *Int J Sports Med*, *15*(Suppl 3), S179–S183.

Macpherson, A. J., Geuking, M. B., & McCoy, K. D. (2012). Homeland security: IgA immunity at the frontiers of the body. *Trends Immunol*, *33*(4), 160–167.

Margaritelis, N. V., Theodorou, A. A., Chatzinikolaou, P. N., Kyparos, A., Nikolaidis, M. G., & Paschalis, V. (2021). Eccentric exercise per se does not affect muscle damage biomarkers: Early and late phase adaptations. *Eur J Appl Physiol*, *121*(2), 549–559.

Martin, N., Lindley, M. R., Hargadon, B., Monteiro, W. R., & Pavord, I. D. (2012). Airway dysfunction and inflammation in pool- and non-pool-based elite athletes. *Med Sci Sports Exerc*, *44*(8), 1433–1439.

Martinez-Gomez, D., Eisenmann, J. C., Wärnberg, J., Gomez-Martinez, S., Veses, A., Veiga, O. L.,... Marcos, A. (2010). Associations of physical activity, cardiorespiratory fitness, and fatness with low-grade inflammation in adolescents: The AFINOS Study. *Int J Obes (Lond)*, *34*(10), 1501–1507.

Martinez-Gomez, D., Gomez-Martinez, S., Ruiz, J. R., Diaz, L. E., Ortega, F. B., Widhalm, K.,... Marcos, A. (2012). Objectively-measured and self-reported physical activity and fitness in relation to inflammatory markers in European adolescents: The HELENA Study. *Atherosclerosis*, *221*(1), 260–267.

McClelland, D. B., Shearman, D. J., & Van Furth, R. (1976). Synthesis of immunoglobulin and secretory component by gastrointestinal mucosa in patients with hypogammaglobulinaemia or IgA deficiency. *Clin Exp Immunol*, *25*(1), 103–111.

McDowell, S. L., Hughes, R. A., Hughes, R. J., Housh, T. J., & Johnson, G. O. (1992). The effect of exercise training on salivary immunoglobulin A and cortisol responses to maximal exercise. *Int J Sports Med*, *13*(8), 577–580.

McKune, A. J., Smith, L. L., Semple, S. J., & Wadee, A. A. (2005). Influence of ultra-endurance exercise on immunoglobulin isotypes and subclasses. *Br J Sports Med*, *39*(9), 665–670.

Miller, E. G., Sethi, P., Nowson, C. A., Dunstan, D. W., & Daly, R. M. (2017). Effects of progressive resistance training and weight loss versus weight loss alone on inflammatory and endothelial biomarkers in older adults with type 2 diabetes. *Eur J Appl Physiol*, *117*(8), 1669–1678.

Min, H. J., Min, S. J., Kang, H., & Kim, K. S. (2020). Differential nasal expression of heat shock proteins 27 and 70 by aerobic exercise: A preliminary study. *Int J Med Sci*, *17*(5), 640–646.

Moraes, H., Aoki, M. S., Freitas, C. G., Arruda, A., Drago, G., & Moreira, A. (2017). SIgA response and incidence of upper respiratory tract infections during intensified training in youth basketball players. *Biol Sport*, *34*(1), 49–55.

Morgans, R., Orme, P., Anderson, L., Drust, B., & Morton, J. P. (2014). An intensive winter fixture schedule induces a transient fall in salivary IgA in English premier league soccer players. *Res Sports Med*, *22*(4), 346–354.

Moy, M. L., Teylan, M., Weston, N. A., Gagnon, D. R., Danilack, V. A., & Garshick, E. (2014). Daily step count is associated with plasma C-reactive protein and IL-6 in a US cohort with COPD. *Chest*, *145*(3), 542–550.

Müns, G., Rubinstein, I., & Singer, P. (1996). Neutrophil chemotactic activity is increased in nasal secretions of long-distance runners. *Int J Sports Med*, *17*(1), 56–59.

Murphy, R. C., Lai, Y., Nolin, J. D., Aguillon Prada, R. A., Chakrabarti, A., Novotny, M. V.,... Hallstrand, T. S. (2021). Exercise-induced alterations in phospholipid hydrolysis, airway surfactant, and eicosanoids and their role in airway hyperresponsiveness in asthma. *Am J Physiol Lung Cell Mol Physiol*, *320*(5), L705–L714.

Nash, D., Hughes, M. G., Butcher, L., Aicheler, R., Smith, P., Cullen, T., & Webb, R. (2023). IL-6 signalling in acute exercise and chronic training: Potential consequences for health and athletic performance. *Scand J Med Sci Sports*, *33*(1), 4–19.

Navazesh, M. (1993). Methods for collecting saliva. *Ann N Y Acad Sci*, *694*, 72–77.

Navazesh, M., & Christensen, C. M. (1982). A comparison of whole mouth resting and stimulated salivary measurement procedures. *J Dent Res*, *61*(10), 1158–1162.

Needleman, I., Ashley, P., Petrie, A., Fortune, F., Turner, W., Jones, J.,... Porter, S. (2013). Oral health and impact on performance of athletes participating in the London 2012 Olympic Games: A cross-sectional study. *Br J Sports Med*, *47*(16), 1054–1058.

Nehlsen-Cannarella, S. L., Nieman, D. C., Balk-Lamberton, A. J., Markoff, P. A., Chritton, D. B., Guse-witch, G.,... Lee, J. W. (1991a). The effects of moderate exercise training on immune response. *Med Sci Sports Exerc*, *23*(1), 64–70.

Nehlsen-Cannarella, S. L., Nieman, D. C., Jessen, J., Chang, L., Gusewitch, G., Blix, G. G., & Ashley, E. (1991b). The effects of acute moderate exercise on lymphocyte function and serum immunoglobulin levels. *Int J Sports Med*, *12*(4), 391–398.

Netea, M. G., van der Meer, J. W., van Deuren, M., & Kullberg, B. J. (2003). Proinflammatory cytokines and sepsis syndrome: not enough, or too much of a good thing? *Trends Immunol*, *24*(5), 254–258.

Neville, V., Gleeson, M., & Folland, J. P. (2008). Salivary IgA as a risk factor for upper respiratory infections in elite professional athletes. *Med Sci Sports Exerc*, *40*(7), 1228–1236.

Nieman, D. C., Dumke, C. I., Henson, D. A., McAnulty, S. R., McAnulty, L. S., Lind, R. H., & Morrow, J. D. (2003). Immune and oxidative changes during and following the Western States Endurance Run. *Int J Sports Med*, *24*(7), 541–547.

Nieman, D. C., Henson, D. A., Smith, L. L. L., Utter, A. C., Vinci, D. M., Davis, J. M., Kaminsky, D. E., & Shute, M. (2001). Cytokine changes after a marathon race. *J Appl Physiol (1985)*, *91*(1), 109–114.

Nieman, D. C., Henson, D. A., Fagoaga, O. R., Utter, A. C., Vinci, D. M., Davis, J. M., & Nehlsen-Cannarella, S. L. (2002). Change in salivary IgA following a competitive marathon race. *Int J Sports Med*, *23*(1), 69–75.

Nieman, D. C., Tan, S. A., Lee, J. W., & Berk, L. S. (1989). Complement and immunoglobulin levels in athletes and sedentary controls. *Int J Sports Med*, *10*(2), 124–128.

Northoff, H., Berg, A. (1991). Immunologic mediators as parameters of the reaction to strenuous exercise. *Int J Sports Med*, *12*(Suppl 1), S9–S15.

Nunes, J. A., Crewther, B. T., Ugrinowitsch, C., Tricoli, V., Viveiros, L., de Rose, D. Jr., & Aoki, M. S. (2011). Salivary hormone and immune responses to three resistance exercise schemes in elite female athletes. *J Strength Cond Res*, *25*(8), 2322–2327.

Oliver, S. J., Laing, S. J., Wilson, S., Bilzon, J. L., Walters, R., & Walsh, N. P. (2007). Salivary Immunoglobulin A response at rest and after exercise following a 48 h period of fluid and/or energy restriction. *Br J Nutr*, *97*(6), 1109–1116.

Ostrowski, K., Rohde, T., Zacho, M., Asp, S., & Pedersen, B. K. (1998). Evidence that interleukin-6 is produced in human skeletal muscle during prolonged running. *J Physiol*, *508*(3), 949–953.

Ostrowski, K., Rohde, T., Asp, S., Schjerling, P., & Pedersen, B. K. (1999). Pro- and anti-inflammatory cytokine balance in strenuous exercise in humans. *J Physiol*, *515*(1), 287–291.

O'Sullivan, N. L., & Montgomery, P. C. (2015). Ocular mucosal immunity. In J. Mestecky, W. Strober, M. W. Russell, B. L. Kelsall, H. Cheroutre, B. N. Lambrecht (Eds) *Mucosal Immunology*, (pp. 1873–1897). London: Academic Press..

Palmer, F. M., Nieman, D. C., Henson, D. A., McAnulty, S. R., McAnulty, L., Swick, N. S., Utter, A. C., Vinci, D. M., & Morrow, J. D. (2003). Influence of vitamin C supplementation on oxidative and salivary IgA changes following an ultramarathon. *Eur J Appl Physiol*, *89*(1), 100–107.

Parsons, J. P., Baran, C. P., Phillips, G., Jarjoura, D., Kaeding, C., Bringardner, B., Wadley, G., Marsh, C. B., & Mastronarde, J. G. (2008). Airway inflammation in exercise-induced bronchospasm occurring in athletes without asthma. *J Asthma*, *45*(5), 363–367.

Peake, J. M., Della Gatta, P., Suzuki, K., & Nieman, D. C. (2015). Cytokine expression and secretion by skeletal muscle cells: Regulatory mechanisms and exercise effects. *Exerc Immunol Rev*, *21*, 8–25.

Pedersen, B. K. (2009). The diseasome of physical inactivity--and the role of myokines in muscle--fat cross talk. *J Physiol*, *587*(Pt 23), 5559–5568.

Pedersen, B. K. (2017). Anti-inflammatory effects of exercise: Role in diabetes and cardiovascular disease. *Eur J Clin Invest*, *47*(8), 600–611.

Pedersen, L., Lund, T. K., Barnes, P. J., Kharitonov, S. A., & Backer, V. (2008). Airway responsiveness and inflammation in adolescent elite swimmers. *J Allergy Clin Immunol*, *122*(2), 322–327, 327.e1.

Perkins, E., & Davison, G. (2022). Epstein-Barr Virus (EBV) DNA as a Potential Marker of in vivo immunity in professional footballers. *Res Q Exerc Sport*, *93*(4), 861–868.

Petibois, C., Cazorla, G., & Déléris, G. (2003). The biological and metabolic adaptations to 12 months training in elite rowers. *Int J Sports Med*, *24*(1), 36–42.

Phillips, C. M., Dillon, C. B., & Perry, I. J. (2017). Does replacing sedentary behaviour with light or moderate to vigorous physical activity modulate inflammatory status in adults? *Int J Behav Nutr Phys Act*, *14*(1), 138.

Price, O. J., Sewry, N., Schwellnus, M., Backer, V., Reier-Nilsen, T., Bougault, V., Pedersen, L., Chenuel, B., Larsson, K., & Hull, J. H. (2022a). Prevalence of lower airway dysfunction in athletes: A systematic review and meta-analysis by a subgroup of the IOC consensus group on 'acute respiratory illness in the athlete'. *Br J Sports Med*, *56*(4), 213–222.

Price, O. J., Walsted, E. S., Bonini, M., Brannan, J. D., Bougault, V., Carlsen, K. H., Couto, M., Kippelen, P., Moreira, A., Pite, H., Rukhadze, M., & Hull, J. H. (2022b). Diagnosis and management of allergy and respiratory disorders in sport: An EAACI task force position paper. *Allergy*, *77*(10), 2909–2923.

Proctor, G. B., & Carpenter, G. H. (2007). Regulation of salivary gland function by autonomic nerves. *Auton Neurosci*, *133*(1), 3–18.

Proctor, G. B., Garrett, J. R., Carpenter, G. H., & Ebersole, L. E. (2003). Salivary secretion of immunoglobulin A by submandibular glands in response to autonomimetic infusions in anaesthetised rats. *J Neuroimmunol*, *136*(1–2), 17–24.

Rapson, A., Collman, E., Faustini, S., Yonel, Z., Chapple, I. L., Drayson, M. T., … Heaney, J. L. J. (2020). Free light chains as an emerging biomarker in saliva: Biological variability and comparisons with salivary IgA and steroid hormones. *Brain Behav Immun*, *83*, 78–86.

Robson-Ansley, P., Barwood, M., Canavan, J., Hack, S., Eglin, C., Davey, S., … Ansley, L. (2009). The effect of repeated endurance exercise on IL-6 and sIL-6R and their relationship with sensations of fatigue at rest. *Cytokine*, *45*(2), 111–116.

Rocha, E. M., Alves, M., Rios, J. D., & Dartt, D. A. (2008). The aging lacrimal gland: Changes in structure and function. *Ocul Surf*, *6*(4), 162–174.

Rothschild-Rodriguez, D., Causer, A. J., Brown, F. F., Collier-Bain, H. D., Moore, S., Murray, J., … Campbell, J. P. (2022). The effects of exercise on complement system proteins in humans: A systematic scoping review. *Exerc Immunol Rev*, *28*, 1–35.

Ruiz, J. R., Ortega, F. B., Warnberg, J., & Sjöström, M. (2007). Associations of low-grade inflammation with physical activity, fitness and fatness in prepubertal children; the European Youth Heart Study. *Int J Obes (Lond)*, *31*(10), 1545–1551.

Ruiz, J. R., Ortega, F. B., Wärnberg, J., Moreno, L. A., Carrero, J. J., Gonzalez-Gross, M., … Sjöström, M. (2008). Inflammatory proteins and muscle strength in adolescents: The Avena study. *Arch Pediatr Adolesc Med*, *162*(5), 462–468.

Sari-Sarraf, V., Reilly, T., & Doran, D. A. (2006). Salivary IgA response to intermittent and continuous exercise. *Int J Sports Med*, *27*(11), 849–855.

Sari-Sarraf, V., Reilly, T., & Doran, D. A. (2007). The effects of single and repeated bouts of soccer-specific exercise on salivary IgA. *Arch Oral Biol*, *52*(6), 526–532.

Scannapieco, F. A., Solomon, L., & Wadenya, R. O. (1994). Emergence in human dental plaque and host distribution of amylase-binding streptococci. *J Dent Res*, *73*(10), 1627–1635.

Semple, S. J., Smith, L. L., McKune, A. J., Hoyos, J., Mokgethwa, B., San Juan, A. F.,... Wadee, A. A. (2006). Serum concentrations of C reactive protein, alpha1 antitrypsin, and complement (C3, C4, C1 esterase inhibitor) before and during the Vuelta a España. *Br J Sports Med*, *40*(2), 124–127.

Seys, S. F., Hox, V., Van Gerven, L., Dilissen, E., Marijsse, G., Peeters, E.,... Bullens, D. M. (2015). Damage-associated molecular pattern and innate cytokine release in the airways of competitive swimmers. *Allergy*, *70*(2), 187–194.

Shephard, R. J. (2001). Sepsis and mechanisms of inflammatory response: Is exercise a good model? *Br J Sports Med*, *35*(4), 223–230.

Shephard, R. J. (2002). Cytokine responses to physical activity, with particular reference to IL-6: Sources, actions, and clinical implications. *Crit Rev Immunol*, *22*(3), 165–182.

Silva, R. A., Almeida, F. M., Olivo, C. R., Saraiva-Romanholo, B. M., Martins, M. A., & Carvalho, C. R. (2016). Exercise reverses OVA-induced inhibition of glucocorticoid receptor and increases anti-inflammatory cytokines in asthma. *Scand J Med Sci Sports*, *26*(1), 82–92.

Silva, R. A., Vieira, R. P., Duarte, A. C., Lopes, F. D., Perini, A., Mauad, T.,... Carvalho, C. R. (2010). Aerobic training reverses airway inflammation and remodelling in an asthma murine model. *Eur Respir J*, *35*(5), 994–1002.

Singh, P. K., Tack, B. F., McCray P. B., Jr., & Welsh, M. J. (2000). Synergistic and additive killing by antimicrobial factors found in human airway surface liquid. *Am J Physiol Lung Cell Mol Physiol*, *279*(5), L799–805.

Škrgat, S., Korošec, P., Kern, I., Šilar, M., Šelb, J., Fležar, M., & Marčun, R. (2018). Systemic and airway oxidative stress in competitive swimmers. *Respir Med*, *137*, 129–133.

Smith, J. K., Chi, D. S., Krish, G., Reynolds, S., & Cambron, G. (1990). Effect of exercise on complement activity. *Ann Allergy*, *65*(4), 304–310.

Stahl, U., Willcox, M., & Stapleton, F. (2012). Osmolality and tear film dynamics. *Clin Exp Optom*, *95*(1), 3–11.

Starkie, R. L., Angus, D.J., Rolland, J., Hargreaves, M., & Febbraio, M. A. (2000). Effect of prolonged, submaximal exercise and carbohydrate ingestion on monocyte intracellular cytokine production in humans. *J Physiol, 528*(3), 647–55.

Starkie, R. L., Arkinstall, M. J., Koukoulas, I., Hawley, J. A., & Febbraio, M. A. (2001). Carbohydrate ingestion attenuates the increase in plasma interleukin-6, but not skeletal muscle interleukin-6 mRNA, during exercise in humans. *J Physiol, 533*(2), 585–91.

Starkie, R., Ostrowski, S. R., Jauffred, S., Febbraio, M., & Pedersen, B. K. (2003). Exercise and IL-6 infusion inhibit endotoxin-induced TNF-alpha production in humans. *FASEB J*, *17*(8), 884–886.

Stavnezer, J. (1996). Immunoglobulin class switching. *Curr Opin Immunol*, *8*(2), 199–205.

Steensberg, A., Fischer, C. P., Keller, C., Møller, K., & Pedersen, B. K. (2003). IL-6 enhances plasma IL-1ra, IL-10, and cortisol in humans. *Am J Physiol Endocrinol Metab*, *285*(2), E433–437.

Steensberg, A., van Hall, G., Osada, T., Sacchetti, M., Saltin, B., & Pedersen, B. K. (2000). Production of interleukin-6 in contracting human skeletal muscles can account for the exercise-induced increase in plasma interleukin-6. *J Physiol*, *529*(Pt 1), 237–242.

Strugnell, R. A., & Wijburg, O. L. (2010). The role of secretory antibodies in infection immunity. *Nat Rev Microbiol*, *8*(9), 656–667.

Suki, W. N., & Massry, S. G (Eds.) (1998). *Suki and Massry's Therapy of Renal Diseases and Related Disorders*. Kluer Academic Publishers.

Surda, P., Walker, A., Putala, M., & Siarnik, P. (2017). Prevalence of rhinitis in athletes: Systematic review. *Int J Otolaryngol*, *2017*, 8098426.

Teeuw, W., Bosch, J. A., Veerman, E. C., & Amerongen, A. V. (2004). Neuroendocrine regulation of salivary IgA synthesis and secretion: Implications for oral health. *Biol Chem*, *385*(12), 1137–1146.

Tiollier, E., Gomez-Merino, D., Burnat, P., Jouanin, J. C., Bourrilhon, C., Filaire, E.,... Chennaoui, M. (2005a). Intense training: Mucosal immunity and incidence of respiratory infections. *Eur J Appl Physiol*, *93*(4), 421–428.

Tiollier, E., Schmitt, L., Burnat, P., Fouillot, J. P., Robach, P., Filaire, E.,... Richalet, J. P. (2005b). Living high-training low altitude training: effects on mucosal immunity. *Eur J Appl Physiol*, *94*(3), 298–304.

Tomasi, T. B., Trudeau, F. B., Czerwinski, D., & Erredge, S. (1982). Immune parameters in athletes before and after strenuous exercise. *J Clin Immunol*, *2*(3), 173–178.

Travis, S. M., Singh, P. K., & Welsh, M. J. (2001). Antimicrobial peptides and proteins in the innate defense of the airway surface. *Curr Opin Immunol*, *13*(1), 89–95.

Turner, S. E. G., Loosemore, M., Shah, A., Kelleher, P., & Hull, J. H. (2021). Salivary IgA as a potential biomarker in the evaluation of respiratory tract infection risk in athletes. *J Allergy Clin Immunol Pract*, *9*(1), 151–159.

Ullum, H., Haahr, P. M., Diamant, M., Palmø, J., Halkjaer-Kristensen, J., & Pedersen, B. K. (1994). Bicycle exercise enhances plasma IL-6 but does not change IL-1 alpha, IL-1 beta, IL-6, or TNF-alpha pre-mRNA in BMNC. *J Appl Physiol, 77*(1), 93–97.

Underdown, B. J., & Dorrington, K. J. (1974). Studies on the structural and conformational basis for the relative resistance of serum and secretory immunoglobulin A to proteolysis. *J Immunol, 112*(3), 949–959.

Vaisberg, M., Suguri, V. M., Gregorio, L. C., Lopes, J. D., & Bachi, A. L. (2013). Cytokine kinetics in nasal mucosa and sera: New insights in understanding upper-airway disease of marathon runners. *Exerc Immunol Rev, 19*, 49–59.

van der Strate, B. W., Beljaars, Molema, G., Harmsen, M. C., & Meijer, D. K. (2001). Antiviral activities of lactoferrin. *Antiviral Res, 52*(3), 225–239.

Vieira, R. P., Claudino, R. C., Duarte, A. C., Santos, A. B., Perini, A., Faria Neto, H. C.,... Carvalho, C. R. (2007). Aerobic exercise decreases chronic allergic lung inflammation and airway remodeling in mice. *Am J Respir Crit Care Med, 176*(9), 871–877.

Vieira, R. P., Silva, R. A., Oliveira-Junior, M. C., Greiffo, F. R., Ligeiro-Oliveira, A. P., Martins, M. A.,... Carvalho, C. R. (2014). Exercise deactivates leukocytes in asthma. *Int J Sports Med, 35*(7), 629–635.

Vieira, R. P., Toledo, A. C., Ferreira, S. C., Santos, A. B., Medeiros, M. C., Hage, M.,... Carvalho, C. R. (2011). Airway epithelium mediates the anti-inflammatory effects of exercise on asthma. *Respir Physiol Neurobiol, 175*(3), 383–389.

Walsh, N. P., Bishop, N. C., Blackwell, J., Wierzbicki, S. G., & Montague, J. C. (2002). Salivary IgA response to prolonged exercise in a cold environment in trained cyclists. *Med Sci Sports Exerc, 34*(10), 1632–1637.

Walsh, N. P., Blannin, A. K., Clark, A. M., Cook, L., Robson, P. J., & Gleeson, M. (1999). The effects of high-intensity intermittent exercise on saliva IgA, total protein, and alpha-amylase. *J Sports Sci, 17*(2), 129–134.

Walsh, N. P., Gleeson, M., Shephard, R. J., Gleeson, M., Woods, J. A., Bishop, N. C.,... Simon, P. (2011). Position statement. Part one: Immune function and exercise. *Exerc Immunol Rev, 17*, 6–63.

Walsh, N. P., Montague, J. C., Callow, N., & Rowlands, A. V. (2004). Saliva flow rate, total protein concentration and osmolality as potential markers of whole body hydration status during progressive acute dehydration in humans. *Arch Oral Biol, 49*(2), 149–54.

Wang, C. H., Chou, P. C., Joa, W. C., Chen, L. F., Sheng, T. F., Ho, S. C.,... Kuo HP. (2014). Mobile-phone-based home exercise training program decreases systemic inflammation in COPD: a pilot study. *BMC Pulm Med, 14*, 142.

Wang, X., Wang, Z., & Tang, D. (2021). Aerobic Exercise Alleviates Inflammation, Oxidative Stress, and Apoptosis in Mice with Chronic Obstructive Pulmonary Disease. *Int J Chron Obstruct Pulmon Dis, 16*, 1369–1379.

Ward, P. P., Paz, E., & Conneely, O. M. (2005). Multifunctional roles of lactoferrin: A critical overview. *Cell Mol Life Sci, 62*(22), 2540–2548.

Weight, L. M., Alexander, D., & Jacobs, P. (1991). Strenuous exercise analogous to the acute-phase response? Clin Sci, 81, 677–83.

Weiler, J. M., Brannan, J. D., Randolph, C. C., Hallstrand, T. S., Parsons, J., Silvers, W.,... Wallace, D. (2016). Exercise-induced bronchoconstriction update-2016. *J Allergy Clin Immunol, 138*(5), 1292–1295.

West, N.P., Pyne, D.B., Renshaw, G., & Cripps, A.W. (2006). Antimicrobial peptides and proteins, exercise and innate mucosal immunity. *FEMS Immunol Med Microbiol, 48*(3), 293–304.

Wira, C. R., & Rossoll, R. M. (1991). Glucocorticoid regulation of the humoral immune system. Dexamethasone stimulation of secretory component in serum, saliva, and bile. *Endocrinology, 128*(2), 835–42.

6 Systems immunity and exercise

*Hawley E. Kunz, Brian A. Irving, J. Philip Karl,
and Heather Quiriarte*

Introduction

Cells of the immune system are found in every organ and tissue in the body, providing defence against pathogens and cancer and aiding in wound healing and tissue repair. The extensive crosstalk between the immune system and the body's tissues is vital for maintaining tissue homeostasis. While the effects of exercise on immune cells and proteins in blood and saliva have greatly enhanced our understanding of the effects of exercise on the immune system, the effect of exercise on tissue-resident immune cells has emerged as an essential area of investigation. Some of the most well-studied adaptations to exercise are those observed in adipose tissue and skeletal muscle. Not only does exercise affect leucocytes within these tissues, but the essential roles these tissue-resident immune cells play in mediating the adaptations to exercise is becoming increasingly apparent. In addition, the effects of exercise on the gut microbiome and its capacity to mediate immunologic and physiologic responses to exercise are emerging areas of research. The gut microbiome and the immune system are closely intertwined, with the microbiome shaping the immune system and the immune system shaping the microbiome. This chapter will focus on the effects of exercise on the immune cells of the adipose tissue and skeletal muscle, interactions between exercise, the gut microbiome, and the gastrointestinal immune system, and the critical role of the immune system in adaptations to exercise within these biological systems.

Adipose tissue

Adipose tissue comprises 30–40% of total body mass in healthy weight [i.e., body mass index (BMI) of 18.5–25 kg/m^2] women, 15–25% of total body mass in men of healthy weight, and can exceed 50% of total body mass in individuals with obesity. In addition to its essential role in energy and fat storage, adipose tissue is a highly metabolically active tissue that regulates energy balance, protects various organs, and is a vital endocrine organ. Most adipose tissue in adults is white adipose tissue (WAT), composed of large, unilocular, lipid-laden white adipocytes specialised for energy storage and mobilisation. In specific regions of humans' necks, shoulders, posterior thoraxes, and abdomens, small brown adipose tissue populations exist, comprising 0.2–3.0% of total adipose tissue mass. Brown adipocytes are multilocular, rich in mitochondria, promote thermogenesis, and share the same progenitors as skeletal muscle. WAT includes [1] the subcutaneous adipose tissue (SAT) that accounts for >80% of fat and resides beneath the skin with primary depots in the abdominal, gluteal, and femoral regions; [2] visceral adipose tissue (VAT), which is found intraabdominally and is associated with the internal organs; and [3] intermuscular adipose tissue (IMAT), adipose tissue located between muscle fibres. Due to its

DOI: 10.4324/9781003256991-6

accessibility and abundance, SAT is the most studied adipose depot and is the emphasis of this section. Within WAT, adipocytes represent ~90% of total adipose tissue volume but constitute only 20–40% of the adipose tissue cellular content. In addition to adipocytes, adipose tissue contains an extracellular matrix, vasculature, and a stromal vascular fraction (SVF), which is a heterogeneous population of cells that includes endothelial cells, fibroblasts, preadipocytes, stem cells, pericytes, and leucocytes.

Adipose tissue leucocytes

Leucocyte populations residing within the SVF of WAT play essential roles in health and disease. All major leucocyte cell types can be found in WAT (Table 6.1), and these cells serve both immune and non-immune functions [see Trim and Lynch (2021) for an in-depth review of the functions of adipose leucocytes]. Adipose harbors many immune cell populations vital for regulating adipocyte homeostasis, maintaining adipose tissue health, regulating adipogenesis and adipocyte metabolism, promoting adipose tissue angiogenesis, and helping facilitate whole-body immune responses. Adipose-resident leucocytes also lie at the nexus of immunometabolism, the complex interplay between organismal metabolism and the immune system.

Table 6.1 Adipose tissue leucocyte populations and their functions. Adapted from Trim and Lynch (2021)

Cell Type	Functions
Macrophages (Russo & Lumeng, 2018; Trim & Lynch, 2021; Kolb & Zhang, 2020)	• Most abundant immune cells in adipose tissue • Highly heterogeneous, often with disparate functions depending on polarisation • Recycle fatty acids released from adipocytes • Scavenge and form crown-like structures that engulf dead/dying adipocytes • Promote adipose tissue angiogenesis • Regulate adipogenesis • Promote adipose tissue maintenance and homeostasis • Secrete anti-inflammatory cytokines in lean tissue • Increase in number and shift to a pro-inflammatory phenotype with obesity • Involved in the pathophysiology of metabolic diseases and endocrine dysregulation • Exacerbate other diseases through direct and indirect mechanisms (e.g. adipose-resident macrophages contribute to breast cancer directly via cytokine secretion and indirectly by increasing adipocyte oestrogen production)
Dendritic Cells (Macdougall & Longhi, 2019)	• Involved in immunoregulation within the adipose tissue • Recruit regulatory T cells • Limit pro-inflammatory signalling • Contribute to adipogenesis and insulin sensitivity
Monocytes (Weisberg et al., 2003)	• Generate populations of macrophages and DCs
Eosinophils (Vohralik et al., 2020)	• Produce IL-4 and IL-13 which support anti-inflammatory macrophages • Release IL-4 which enhances fatty acid mobilisation and adipocyte maturation • Decrease in number with obesity and ageing

(Continued)

Table 6.1 (Continued)

Cell Type	Functions
Neutrophils (Watanabe et al., 2019)	• Promote adipose inflammation by inducing IL-1β production by adipocytes and recruiting macrophages • Increase in number with a high-fat diet
Mast Cells (Elieh Ali Komi et al., 2020)	• Promote differentiation of preadipocytes • Remodel adipose tissue extracellular matrix • Support adipose tissue angiogenesis • Exacerbate inflammation through the release of TNFα and other cytokines and through the recruitment of leucocytes • Increase in number with obesity and exacerbate the metabolic syndrome
NK Cells (Lee et al., 2016)	• Increase in number with high-fat diet and obesity • Regulate the number of adipose tissue macrophages and promote inflammation and insulin resistance through the production of TNFα and other cytokines
Cytotoxic CD8+ T cells (Wang & Wu, 2018; Han et al., 2017)	• Adipose tissue serves as a reservoir for memory CD8+ T cells • Along with macrophages, surround dead/dying adipocytes • Increase in number and exacerbate adipose tissue inflammation in obesity
Helper CD4+ T cells (Wang & Wu, 2018)	• Th1 cells increase in number, produce higher amounts of IFNγ, and are associated with metabolic dysfunction with obesity • Th17 cells increase in number and promote inflammation and insulin resistance through the release of IL-17 in obesity • Th2 cells produce the anti-inflammatory cytokines IL-4, IL-5, and IL-13 that support anti-inflammatory macrophages
Regulatory T cells (Becker et al., 2017; Wang & Wu, 2018)	• Control adipose tissue inflammation • Maintain adipose tissue and systemic metabolic homeostasis • Produce IL-4 and IL-13 • Decline in number with obesity • Increase in number during ageing
γδ T cells (Hu et al., 2020; Kohlgruber et al., 2018)	• Produce IL-17A which regulates adaptive thermogenesis and regulatory T cell accumulation • Promote adipose sympathetic innervation
iNKT cells (Lynch et al., 2015)	• Produce IL-2 which promote adipose tissue regulatory T cell proliferation and function • Support an anti-inflammatory macrophage phenotype via the production of IL-2 and IL-10 • Promote adipocyte homeostasis, metabolic health, and glycaemic control • Decrease in number and become pathogenic with obesity
Innate Lymphoid Type 1 Cells (Boulenouar et al., 2017; Wang et al., 2019)	• Promote inflammation and fibrogenesis in obesity • Regulate adipose tissue macrophage homeostasis
Innate Lymphoid Type 2 Cells (Molofsky et al., 2013)	• Produce IL-4, IL-5, and IL-13 • Control adipose inflammation by supporting eosinophil, anti-inflammatory macrophage, and regulatory T cell numbers
B Cells (Camell et al., 2019; Frasca & Blomberg 2017)	• B1 cells produce IgM and IL-10 and are primarily anti-inflammatory • B2 cells outnumber B1 cells and are pro-inflammatory, secreting IgG and inflammatory cytokines • Impair lipolysis and thermogenesis • Increase in number with ageing in an Nlrp3 inflammasome-dependent manner • Recruited to adipose tissue during obesity and promote T-cell activation and inflammation

Individual exercise bouts and adipose tissue leucocytes

During a single bout of exercise, adipose tissue is highly active, serving as an important energy source. While an exercise-induced increase in adipose tissue lipolysis is well-established, the effects of individual exercise bouts on the cells of the SVF have begun to be investigated more recently. Exercise modality, intensity, participant characteristics, and the timing of sample collection are likely important factors when studying the effects of exercise on immune cells within the adipose tissue. In overweight-to-obese adults, 60 minutes of endurance exercise increased gene expression of the angiogenic regulator *VEGFA* 1 hour post-exercise, while gene expression of inflammation- and immune-related genes (*TNFA, IL-1B, NLRP3, MCP1, CD206*) was not affected (Van Pelt, Guth, & Horowitz, 2017). However, 4 hours after an individual bout of exercise in young males (BMI 22.8±0.6 kg/m^2), adipose mRNA expression of inflammation, and monocyte- and macrophage-associated genes was upregulated, indicating potential immune cell infiltration into adipose tissue. Interestingly, exercise training attenuated this response (Fabre et al., 2018). Exercise-induced immune alterations in adipose tissue likely represent a transient response. No alterations were observed in the overall proportions of immune cells 12 hours after a 60-minute bout of endurance exercise performed by regularly exercising adults with obesity (BMI 33 ± 3 kg/m^2), although the proportion of preadipocytes in the SVF was lower at this time point (Ludzki et al., 2020). Additional work is required to fully understand the effects of individual exercise bouts on the immune cells of the adipose tissue and the time course of these changes.

The adipose tissue secretome is also affected by individual bouts of exercise, and the expression of genes encoding multiple cytokines and adipokines is altered in the hours after exercise (Frydelund-Larsen et al., 2007). A single bout of high-intensity exercise triggers the release of extracellular vesicles (EVs) into circulation, some of which originate in adipose tissue (Vanderboom et al., 2021). Adipose-derived EVs facilitate intercellular and inter-organ crosstalk, carrying cytokines, adipokines, lipids, proteins, and nucleic acids to target cells locally and distally (Z. Huang & Xu, 2021). Acute alterations in adipose secretory signalling and the transient recruitment of immune cells following a single exercise session likely play an important role in the long-term adipose tissue adaptations associated with exercise training.

Exercise training and adipose tissue leucocytes

Most research on the effects of exercise training on adipose-resident leucocytes has been conducted in the context of obesity and suggests that aerobic exercise training supports an anti-inflammatory environment and improves adipose–immune dynamics. In individuals with obesity, adipose tissue is characterised by adipocyte hypertrophy, adipogenesis, hypoxia, fibrosis, and a pro-inflammatory profile. Adipose macrophages switch from anti-inflammatory (M2) to pro-inflammatory (M1), and the proportions of CD8+ T cells, neutrophils, γδ T cells, mast cells, and B cells increase with concomitant reductions in regulatory T cells, iNK T cells, and type 2 lymphoid cells. These shifts in immune cell populations have important implications for adipose tissue, whole-body inflammation, and metabolic health.

The mechanisms by which exercise training may improve the adipose tissue–immune cell profile and combat the negative adaptations associated with obesity are multifactorial. Winn, Cottam, Wasserman, and Hasty (2021) recently proposed four mechanisms by which aerobic exercise training may have beneficial effects on the immune profile of adipose tissue: (1) remodelling and reducing adipose tissue mass; (2) decreasing the recruitment and accumulation of immune cells; (3) shifting the immune profile from pro-inflammatory to anti-inflammatory; and (4) altering the adipose tissue secretome (Figure 6.1).

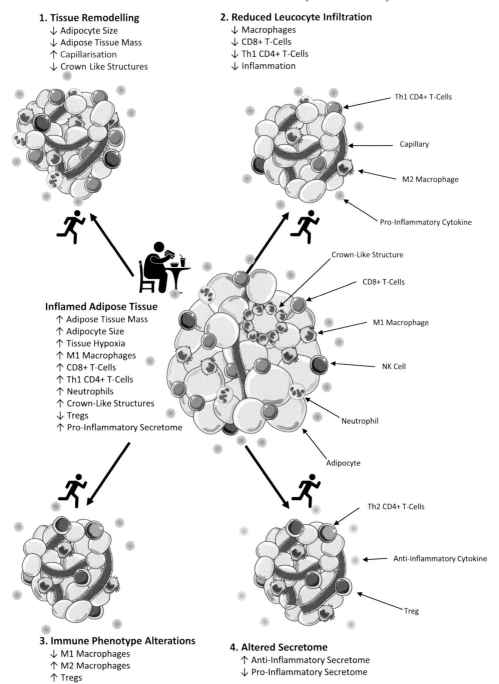

Figure 6.1 Effects of exercise training on adipose tissue.

Legend: "The increased energy balance associated with a sedentary lifestyle and excess energy intake is associated with profound changes in adipose tissue, which include increased adipose tissue mass and adipose tissue inflammation, defined by increased tissue hypoxia, an infiltration of pro-inflammatory leukocytes, and a pro-inflammatory secretome. The beneficial effects of exercise training on the adipose tissue include 1) adipose tissue remodeling; 2) reduced accumulation of immune cells; 3) a shift from a pro-inflammatory profile to an anti-inflammatory immune profile; and 4) alterations in the adipose tissue secretome. Figure generated with Shutterstock Vector ID 105093671 and images from BioIcons and Servier Medical Art, provided by Servier, licensed under a Creative Commons Attribution 3.0 unported license."

Adipose tissue remodelling in response to exercise training

Adipose tissue is an important energy source during exercise, and regular physical activity can reduce adipose tissue mass by creating a negative energy balance. In addition to reducing the overall burden of adipose–immune interactions, reductions in adipose mass and adipocyte size also result in significant tissue remodelling that fosters a healthy immune environment. In mice fed a high-fat diet, exercise training increased adipose tissue capillary density, reducing tissue hypoxia and improving nutrient delivery, likely contributing to reduced adipose inflammation. In these mice, exercise training also reduced the number of crown-like structures—pro-inflammatory complexes of macrophages surrounding dying adipocytes (Kolahdouzi, Talebi-Garakani, Hamidian, & Safarzade, 2019). Adipose-resident leucocytes may also play a role in this remodelling, as they have been shown to support angiogenesis and extracellular matrix remodelling within the adipose tissue (Elieh Ali Komi et al., 2020).

Exercise training reduces adipose–leucocyte infiltration

Exercise training may reduce the recruitment, retention, and accumulation of inflammatory leucocytes in adipose tissue. In a cohort of men who were overweight, 12 weeks of combined aerobic and endurance training downregulated the expression of transcripts associated with macrophages, inflammation, and immune pathways (Lee et al., 2019). Similarly, in mice, exercise training reversed high-fat diet-induced elevations in pro-inflammatory macrophages, CD8+ T cells, and inflammatory cytokine gene expression in adipose tissue (Bradley, Jeon, Liu, & Maratos-Flier, 2008; Kawanishi, Mizokami, Yano, & Suzuki, 2013). This apparent reduction in adipose inflammation may result from decreased infiltration of immune cells into adipose tissue. The chemokine RANTES and its cognate receptor CCR5 promote inflammatory immune cell chemotaxis and survival, and their levels are elevated in the adipose tissue of individuals with obesity compared with lean. Three months of combined aerobic and resistance training reduced the expression of both RANTES and CCR5 in the adipose tissue of participants with obesity, with concomitant reductions in adipose TNFα and IL-6 levels (Baturcam et al., 2014). In high-fat-fed mice, exercise training downregulated mRNA expression of multiple chemoattractant proteins [ICAM1, monocyte chemoattractant proteins (MCP) 1 and 2, and macrophage inflammatory proteins (MIP) 1α and 1β] in adipose tissue, and inhibited the infiltration of monocytes, macrophages, and neutrophils into the adipose tissue (Kawanishi et al., 2013; Kawanishi, Niihara, Mizokami, Yada, & Suzuki, 2015; Kawanishi, Yano, Yokogawa, & Suzuki, 2010). In humans with obesity, exercise training reduces the proportion of circulating monocytes and progenitor cells expressing TLR4 and CCR2 (Gleeson, McFarlin, & Flynn, 2006; Niemiro et al., 2018). CCR2 is a homing receptor for sites of inflammation, and TLR4 drives inflammatory macrophage differentiation. The exercise-induced reduction in cells expressing these markers suggests that exercise may reduce the infiltration of inflammatory leucocytes into the adipose tissue.

Exercise training induced immune phenotype alterations

As the most abundant leukocyte in adipose tissue, macrophages play an important role in the inflammatory profile of the tissue. Broadly characterised, macrophages can be polarised towards an M1 pro-inflammatory phenotype or an M2 anti-inflammatory phenotype. In the inflamed adipose tissue associated with obesity, the M1 polarisation state, characterised by a predominance of classical macrophages and high TNFα and IL-6 production, prevails (Gleeson et al., 2011). In high-fat-fed mice, exercise training facilitates the phenotypic switching from M1 to

M2 polarisation. Protein expression of the M1 marker CD68 is decreased, while expression of the M2 markers CD206 and CD163 is increased and expression of TLR4 and TNFα is attenuated (Kawanishi et al., 2010; Kolahdouzi et al., 2019).

Exercise training and the adipose tissue secretome

Adipose tissue acts as a paracrine and endocrine organ, secreting factors that have essential local and systemic effects. More than 1,700 proteins have been identified in conditioned media generated from WAT, and 441 of these proteins were characterised as secreted proteins (Deshmukh et al., 2019). Within adipose tissue, excess free fatty acids, adipokines (e.g., leptin), and secreted inflammatory mediators (e.g., TNFα, IL-6, IL-1β, MCP-1) exacerbate adipose inflammation by recruiting and activating immune cells. These inflammatory mediators also have systemic effects and are thought to contribute to impaired myogenesis (O'Leary et al., 2018) and whole-body insulin resistance (Hotamisligil, Shargill, & Spiegelman, 1993). Exercise training appears to have beneficial effects on the adipose tissue secretome. Circulating concentrations of adiponectin, secreted frizzled-related protein-5 (SFRP5), and omentin, adipokines with anti-inflammatory and insulin-sensitising properties, are elevated following aerobic exercise training (Babaei & Hoseini, 2022). In mice, the transplantation of SAT from exercise-trained mice into sedentary recipient mice improved glucose tolerance and skeletal muscle glucose uptake in the recipient mice (Stanford et al., 2015), demonstrating the beneficial effects of exercise-trained adipose tissue on distal tissues. Adipose-derived adipokines, cytokines, and EVs may mediate these effects. In mice, exercise reversed high-fat diet-induced elevations in inflammatory cytokine gene expression in adipose tissue (Bradley et al., 2008), and the expression of 62 genes associated with secreted proteins was significantly altered in the adipose tissue of dysglycaemic men, following 12 weeks of exercise (Lee et al., 2019).

Adipose tissue–immune cells are recognised as important regulators of metabolic health, and interventions that improve the adipose–leukocyte milieu may have therapeutic benefits. However, because obesity and weight gain alter the adipose–leukocyte milieu, it is important to distinguish the effects of exercise-induced weight loss on adipose leucocytes from the effects of exercise in and of itself. Recently, 36 adults with obesity (BMI 33±3 kg/m²) underwent 12 weeks of either moderate-intensity endurance training or high-intensity interval training, but their nutritional intake was controlled such that body weight did not change. Even without weight loss, both endurance and high-intensity interval training reduced adipocyte size and increased adipose tissue capillarisation (Ahn et al., 2022). Similarly, during weight loss interventions, the method by which energy balance is altered (i.e., energy restriction by reduced energy caloric intake or enhanced energy expenditure through exercise) appears to affect the adipose tissue response to weight loss. Individuals who reduced caloric intake by ~18% for 12 weeks reduced their total adipose tissue mass by 8.9% and exhibited enhanced lipolysis and increased adipose tissue gene expression of pro-inflammatory macrophage markers. A separate cohort of 11 adults underwent 12 weeks of combined endurance and resistance training, increasing energy expenditure by ~17%. These participants reduced adipose tissue mass by 10.9%, but the changes in adipose tissue gene expression indicated reduced inflammation, with decreased expression of genes associated with macrophages and T cells (Lee et al., 2016). These studies suggest that the benefits of exercise on adipose tissue are not only the result of exercise-induced weight loss.

Researchers face numerous challenges in trying to understand the effects of exercise training on the adipose tissue–immune profile. Timing of the post-exercise training sample is critical in determining if changes in adipose immunity represent the cumulative effects of successive individual exercise bouts or if changes occur only transiently in response to exercise bouts. In one study, adipose samples collected one day after the final training session of a 12-week intervention, inflammatory signalling pathways were activated. However, four days following the last exercise session, the expression of key factors involved in inflammatory signalling had returned to pre-training levels (Ahn et al., 2022). In addition, adipose tissue is distributed throughout the body in distinct depots, and the composition and function of these different depots can vary considerably and may differ in their responses to exercise. Even different subcutaneous depots have exhibited unique responses to exercise training. In a cohort of South African women with obesity, gene expression of markers of inflammation was elevated in gluteal, but not abdominal, SAT after 12 weeks of combined endurance and resistance training (Nono Nankam et al., 2020). As the most accessible depot, SAT has been studied most extensively. However, the effects of exercise on other adipose tissue depots are of interest. VAT is considered an immunogenic tissue and is thought to significantly contribute to chronic inflammation in obesity and ageing. In adult and old mice, exercise training appears to promote an anti-inflammatory phenotype in VAT (Ziegler et al., 2019), but additional work is required in humans. Given its proximity to skeletal muscle and potential paracrine effects, IMAT may also be an important target of exercise training (Sparks, Goodpaster, & Bergman, 2021).

Skeletal muscle

Skeletal muscle is essential for maintaining physical function and metabolic health across the lifespan and is an exquisitely malleable tissue. The advantageous skeletal muscle adaptations to exercise are well documented and include the stimulation of protein synthesis, hypertrophy, increased strength, enhanced regenerative capacity, improved oxidative capacity, and heightened insulin sensitivity (Distefano & Goodpaster, 2018; Irving et al., 2015; Phillips, Tipton, Aarsland, Wolf, & Wolfe, 1997). In contrast, noxious stimuli such as skeletal muscle disuse induced by reduced physical activity, short-term bed rest, and limb immobilisation can lead to maladaptive responses (Brook et al., 2022; Glover et al., 2008; Standley et al., 2020). The skeletal muscle's many adaptive and maladaptive responses to stimuli are mechanistically regulated in part by both acute and chronic inflammatory processes (Peake, Neubauer, Della Gatta, & Nosaka, 2017). The immune system is essential in the adaptive skeletal muscle response to exercise bouts.

Skeletal muscle leucocytes

Skeletal muscle harbors tissue-resident immune cells and recruits immune cells in response to physiological stimuli. Anti- and pro-inflammatory macrophages, monocytes, neutrophils, eosinophils, T cells, NK cells, B cells, and regulatory T cells have all been identified within the skeletal muscle. Macrophages comprise most skeletal muscle leucocytes, residing primarily near blood vessels and in the connective tissue encasing the muscle. Through both acute and long-term processes, leucocytes mechanistically regulate many of the skeletal muscle's adaptive responses to exercise (Peake et al., 2017).

Individual exercise bouts and skeletal muscle leucocytes

Individual exercise bouts result in bioenergetic and mechanical stress that can induce mild to severe *sterile* skeletal muscle damage (Furrer, Eisele, Schmidt, Beer, & Handschin, 2017). Several cell types are essential for the adequate repair and remodelling of the skeletal muscle following

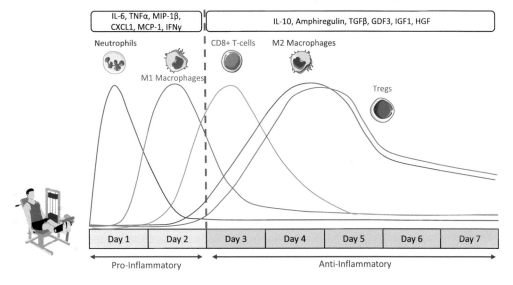

Figure 6.2 Immune response to acute, muscle-damaging exercise.

Legend: "Acute exercise can induce sterile muscle damage that initiates a coordinated immune response to aid in muscle repair and regeneration. Following the initial damage, neutrophils rapidly infiltrate the skeletal muscle and release cytokines that recruit monocytes/macrophages and CD8+ T cells to the tissue. The initial influx of macrophages and T cells are predominantly pro-inflammatory, but as muscle repair continues, the pro-inflammatory environment shifts to an anti-inflammatory environment via a coordinated process that involved regulatory T cells and cytokines released from the infiltrating leukocytes. CXCL, chemokine (C-X-C motif) ligand; GDF, grow/differentiation factor; IGF, insulin-like growth factor; HGF, hepatocyte growth factor; IFN, interferon; IL, interleukin; MCP, monocyte chemoattractant protein; MIP, macrophage inflammatory protein; TGF, transforming growth factor; TNF, tumor necrosis factor Figure adapted with permissions from Tidball (2017). Figure generated with Shutterstock Vector ID: 2169477623 and images from BioIcons and Servier Medical Art, provided by Servier, licensed under a Creative Commons Attribution 3.0 unported license."

damage, including satellite cells, fibro-adipogenic progenitor cells, and leucocytes (Tidball, 2017). Eccentric exercise is a classic model for inducing skeletal muscle damage. In response to eccentric exercise, damaged muscle fibres recruit and activate resident and peripheral innate (neutrophils, macrophages, monocytes) and adaptive (T cells) leucocytes to coordinate muscle repair and regeneration (Figure 6.2).

Neutrophil response to individual exercise bouts

Neutrophils are rapidly recruited to the muscle following exercise, where they phagocytose cellular debris. As the number of neutrophils in the skeletal muscle peaks 12–24 hours after damage, one of the primary roles of neutrophils following exercise appears to be initiating the coordinated muscle immune response. By releasing cytokines and reactive oxygen species, neutrophils create a pro-inflammatory environment that attracts monocytes/macrophages to the injured tissue and promotes M1 polarisation (Kawanishi, Mizokami, Niihara, Yada, & Suzuki, 2016).

Monocyte and macrophage response to individual exercise bouts

The roles of monocytes and macrophages in skeletal muscle repair and regeneration in response to individual exercise bouts and regular exercise training are only beginning to be fully appreciated. Still, these cells are well known for their role in repair and regeneration following cytotoxic injury (Chazaud, 2020). In response to muscle tissue injury, monocytes exhibiting a classical CD14^{++} CD16$^-$ pro-inflammatory phenotype are recruited to the skeletal muscle (Soehnlein & Lindbom, 2010). The monocytes are quickly transformed into pro-inflammatory macrophages (M1) designed to remove cellular debris via phagocytosis and release pro-inflammatory cytokines, including IL-1β, TNFα, and IFNγ (Furrer et al., 2017; Peake et al., 2017). Following this initial pro-inflammatory phase, M1 macrophages transition to anti-inflammatory (M2) macrophages, which release factors that promote tissue repair (Jensen et al., 2020). The infiltration of non-classical CD14^{low}CD16$^+$ monocytes post-injury also facilitates this switch, converting to M2 macrophages and improving skeletal muscle regeneration (San Emeterio, Olingy, Chu, & Botchwey, 2017).

The concept that M1 and M2 macrophages play a role in skeletal muscle tissue repair after exercise bouts is widely accepted, but the mechanisms by which these macrophage subtypes contribute to this critical adaptive response remain incompletely understood. M1 macrophages help clear cellular debris, likely shed from contracting muscle fibres and surrounding extracellular matrix. In addition, the secretion of pro-inflammatory cytokines and insulin-like growth factor 1 (IGF1) by M1 macrophages is reported to increase the proliferation of satellite cells and myoblasts (muscle stem cells and the myogenic progenitor cells they give rise to). Preclinical models that knock out pro-inflammatory cytokines (TNFα, IL-6, IFNγ) have all shown some degree of reduced skeletal muscle regenerative capacity (Collins & Grounds, 2001; Serrano, Baeza-Raja, Perdiguero, Jardi, & Munoz-Cinoves, 2008; Walton et al., 2019).

As the repair process continues, infiltrating macrophages undergo a phenotypic transition from M1 to M2 macrophages. A seminal article by Arnold et al. (2007) elegantly demonstrated that the phagocytosis of cellular debris led to the conversion of M1 to M2 macrophages. These M2 macrophages begin to secrete anti-inflammatory and pro-myogenic cytokines (Arnold et al., 2007; Minari & Thomatieli-Santos, 2022), such as the transforming growth factor β (TGFβ) family member GDF3, which facilitates recruitment and fusion of primary myoblasts to form myotubes (Varga et al., 2016). In addition, co-culturing myoblasts with M2 macrophages supported the fusion of myoblasts to form myotubes (Arnold et al., 2007). Thus, in response to exercise bouts, macrophages appear to play a role in stoking the flames of inflammation but also help extinguish the fire while initiating the rebuilding process.

T cell response to individual exercise bouts

Emerging data suggest that T cells also play a role in skeletal muscle repair and regeneration in response to exercise-induced muscle injury (Deyhle et al., 2020; Deyhle et al., 2015). CD4+ and CD8+ T cells are recruited to the muscle in response to injury and appear in elevated numbers during repair and regeneration. Muscle-resident CD8+ T cells appear to be early responders following muscle damage and play a critical role in coordinating inflammation and muscle regeneration. Genetic deletion of *CD8a* significantly reduced the infiltration of macrophages into injured muscles and impaired muscle regeneration (Tidball, 2017). Regulatory T cells are also gaining recognition as potentially important players in the coordinated response to muscle damage. Recruited to the muscle days after injury, regulatory T cells facilitate the M1 to M2 phenotypic switch and release amphiregulin, which enhances the expression of multiple myogenic factors that promote muscle differentiation. Indeed, in mice, regulatory T cells depletion impaired the M1 to M2 macrophage phenotypic switch and attenuated skeletal muscle repair (Burzyn et al., 2013).

Soluble factors coordinate the skeletal muscle immune response to individual exercise bouts

Elevations in circulating cytokines (e.g., IL-6, TNFα, MIP-1β) following individual bouts of exercise are well documented (Pedersen & Toft, 2000) and likely orchestrate the recruitment of immune cells to sites of damage. IL-6 released from immune and muscle cells has been proposed as the essential regulator of the immune response to exercise (Furrer et al., 2017). Through distinct signalling pathways, IL-6 can help orchestrate pro- and anti-inflammatory processes. IL-6 binds to a receptor complex comprising the IL-6 receptor α (IL-6Rα) and glycoprotein (gp) 130. In the classical (*cis*)-signalling pathway, IL-6 released from contracting muscle cells binds to membrane-bound IL-6Rα, promoting anti-inflammatory processes. In the *trans*-signalling pathway, IL-6 released from immune cells binds to soluble IL-6Rα, which then dimerises with membrane-bound gp130 and activates pro-inflammatory signalling cascades (Nara & Watanabe, 2021; Pullamsetti, Seeger, & Savai, 2018). In addition, IL-6 helps drive skeletal muscle hypertrophy through its ability to activate satellite cells (Serrano et al., 2008). However, IL-6 receptor blockade was found to have little effect on leucocyte recruitment during exercise (Bay et al., 2020), suggesting that there are likely redundant signalling pathways to initiate the inflammatory responses to exercise that facilitate tissue repair and remodelling.

The recruitment of neutrophils from circulation to exercised muscle is likely coordinated, at least partly, by chemoattractant signalling from cytokines released from exercising muscle (Yamada et al., 2002). Muscle-resident macrophages release CXC chemokine ligand 1 (CXCL1) and MCP-1, potent neutrophil chemoattractant proteins (Tidball, 2017). *In vitro*, human myotubes (the multinucleated precursors to muscle fibres) exhibit increasing chemotaxis indices when subjected to progressively higher degrees of mechanical strain, even without overt cellular injury (Tsivitse, Mylona, Peterson, Gunning, & Pizza, 2005).

Following closely behind the infiltration of neutrophils, M1 macrophages infiltrate the tissue. In the days following exercise, these M1 macrophages undergo phenotypic conversion from M1 to M2 macrophages in a highly coordinated process resulting from crosstalk between exercised muscle and immune cells. The initial influx of monocytes and macrophages is facilitated by chemotactic signals released during exercise (e.g., IL-6) and from infiltrating neutrophils. These recruited monocytes and macrophages enter a microenvironment enriched with pro-inflammatory proteins released by neutrophils (e.g., TNFα and IFNγ), promoting M1 polarisation (Tidball, 2017). Multiple signals coordinate the phenotypic switch from M1 to M2 macrophages. Regulatory T cells recruited to the tissue secrete IL-10 and amphiregulin, and the infiltrating macrophages begin to secrete IL-10 and IGF1, promoting phenotypic conversion to the M2 phenotype (Burzyn et al., 2013). The phenotypic switch of M1 to M2 macrophages is also initiated by the phagocytosis of cellular debris and exposure to local changes in cellular metabolites (Arnold et al., 2007; Jensen et al., 2020; Rogeri et al., 2020). A recent study suggests lactate released from contracting skeletal muscle fibres promotes the conversion of M1 to M2 macrophages, stimulating angiogenesis and muscle tissue repair (Zhang et al., 2020).

In addition, recent data suggest that skeletal muscle activation of peroxisome proliferator-activated receptor γ co-activator 1α (PGC-1α) plays a role in the immune response. PGC-1α is a well-defined exercise-inducible transcription factor that regulates mitochondrial biogenesis, helping to orchestrate skeletal muscle remodelling. PGC-1α stimulates the recruitment of monocytes to the muscle and promotes their secretion of MCP-1. In turn, MCP-1 helps activate local endothelial cells, pericytes, and smooth muscle cells to support angiogenesis (Rowe et al., 2014). It also appears to regulate myokine expression. Muscle-specific overexpression of PGC-1α promoted M2-type polarisation, increased levels of anti-inflammatory cytokines, and enhanced the activity of tissue-resident macrophages, likely facilitating the remodeling of skeletal muscle tissue in response to exercise (Furrer et al., 2017).

Exercise training and skeletal muscle leucocytes

By promoting tissue repair and regeneration, mitochondrial biogenesis, and extracellular matrix remodeling, leucocytes play a vital role in skeletal muscle adaptations to exercise. However, the effects of exercise training on muscle-resident leukocyte phenotypes and functions have not been fully elucidated. The local leukocyte milieu appears to play an essential role in protein turnover and maintenance of muscle mass. For example, IGF1 secreted by M2 macrophages supports muscle preservation, while monocyte infiltration and the activation of resident M1 macrophages are thought to be a driving force behind muscle atrophy, resulting from disuse or disease (Kawanishi, Funakoshi, & Machida, 2018). Exercise training likely alters the phenotype and function of muscle-resident leucocytes. Twelve weeks of aerobic cycle training was recently shown to increase the number of M2 macrophages in the skeletal muscle, and this was associated with increased gene expression of growth factors (HGF, IGF1) and extracellular matrix proteins (Walton et al., 2019).

As members of the adaptive immune system, T cells are often thought to promote skeletal muscle adaptations that protect the muscle from repeated bouts of exercise (RBE). RBE are associated with adaptations to the muscle extracellular matrix and structural integrity that protect the muscle from subsequent exercise-induced injury. Using immunohistochemistry, Deyhle et al. (2015) showed that $CD8^+$ T cells are recruited to the skeletal muscle to a greater extent following a repeated muscle-lengthening exercise than the initial (naive) bout of exercise. Their data suggested that $CD8^+$ T cells were likely sensitized by the initial bout of exercise (performed 28 days before the RBE). However, in a follow-up study, Deyhle et al. (2020) used flow cytometry to further characterise the role of T cells in the RBE and found the increase in T cells following the RBE was not as robust. Further, the adoptive transfer of RBE-exposed T cells failed to provide any protection from exercise-induced muscle damage. Thus, the role of T cells in the adaptive response to RBE is still under debate.

Ageing, obesity, and physical inactivity have also been shown to affect muscle macrophage phenotypes and function at rest and in response to exercise (Jensen et al., 2020; Przybyla et al., 2006; Reidy et al., 2019; Sorensen et al., 2019). Dysregulation of the skeletal muscle leukocyte milieu and the chronic inflammation associated with ageing are postulated to contribute to the blunted hypertrophic responses to exercise often observed in older adults (Lang, Frost, Nairn, MacLean, & Vary, 2002). However, while the skeletal muscle immune response to an individual bout of exercise in non-exercising older males is characterised by predominantly pro-inflammatory signatures, the response of lifelong exercisers resembled that of young, healthy adults (Lavin et al., 2020). Therefore, exercise training may counteract the negative impact of ageing on the skeletal muscle leucocyte profile. Indeed, recent work suggests that 12 weeks of endurance exercise can lead to elevations in the proportion of M2 macrophages per muscle fibre in healthy adults, and these elevations correlate with training-induced muscle hypertrophy (Walton et al., 2019). These results support the notion that endurance training promotes an anti-inflammatory environment that may facilitate muscle growth. Current data also suggest that phenotypic changes in the immune cell profiles within the skeletal muscle may help regulate skeletal muscle hypertrophy in response to regular resistance training (Long et al., 2022).

Skeletal muscle-adipose-immune crosstalk

Exercise impacts virtually every tissue and system in the body, and the exercise-induced release of cytokines, adipokines, and myokines facilitates the crosstalk between these systems. Tissue crosstalk was proposed as a mechanism behind exercise adaptations over 60 years ago (Kao &

Ray, 1954), but it was decades before this concept was explored and recognised as an important driver of the systemic effects of exercise. The immune system is a target for the signals sent from adipose tissue and skeletal muscle and it also contributes to exercise-induced alterations in the adipose and skeletal muscle secretome.

Much of the adipose tissue secretome is derived from the adipose-resident immune cells. Pro-inflammatory adipose-resident macrophages secrete pro-inflammatory cytokines that are thought to contribute to metabolic derangements, impaired insulin signalling and lipid and glucose metabolism in the liver and skeletal muscle, and dysregulated skeletal muscle mito-chondrial function (Kojta, Chacinska, & Blachnio-Zabielska, 2020; Valerio et al., 2006). These metabolic derangements are exacerbated by obesity and physical inactivity and attenuated by exercise training. Therefore, many of the whole-body benefits of exercise may stem from its beneficial effects on adipose–leukocyte interactions and the adipose tissue secretome. Regular physical activity increases circulating adiponectin (Becic, Studenik, & Hoffmann, 2018), an adipokine with insulin-sensitising and anti-inflammatory effects.

Skeletal muscle also serves as a key regulator of the immune system. More than 300 potential myokines have been identified as part of the skeletal muscle secretome, many of which have immune-modulating properties (Giudice & Taylor, 2017). The myokines released from exercis-ing muscles have endocrine effects that may reduce inflammation and alleviate symptoms of diseases associated with sedentary behaviour (Pedersen & Febbraio, 2012). The type 1 helper cell (Th1) cytokine IL-15 is highly expressed in skeletal muscle and has a role in the prolifera-tion, activation, and distribution of NK cells and B cells and aids in the survival of naive T cells (Conlon et al., 2015). IL-7, a recently identified myokine, plays an important role in develop-ing immature lymphocytes (Duggal, Pollock, Lazarus, Harridge, & Lord, 2018; Haugen et al., 2010). Elevated circulating concentrations of IL-6 are associated with inflammation, obesity, and insulin resistance. However, transient increases in IL-6 concentrations, which may occur as IL-6 is released from contracting muscle during exercise, have been shown to promote the production of anti-inflammatory cytokines IL-1 receptor antagonist (IL-1ra) and IL-10, inhibit-ing IL-1β signal transduction and TNFα production, respectively (Steensberg, Fischer, Keller, Moller, & Pedersen, 2003).

Exercise-stimulated intercellular crosstalk may also occur through the secretion and uptake of EVs and exosomes. A type of EV considered critical for intercellular communication, ex-osomes, are endosomal in origin, range in size from ~40 to 160 nm, have lipid bilayers, and are secreted by nearly all cell types under normal physiological conditions and disease states (Kal-luri & LeBleu, 2020). Exosomes contain diverse cargo, including proteins, lipids, RNAs, and DNA, that can induce local and systemic changes (He, Zheng, Luo, & Wang, 2018). By transfer-ring their cargo, exosomes can affect the recipient cell's physiological properties and facilitate cellular adaptations. A single bout of exercise triggers the release of EVs and exosomes into cir-culation, some exhibiting protein signatures indicative of adipose tissue and skeletal muscle ori-gins (Vanderboom et al., 2021) and some carrying myokines and other signalling proteins with effects ranging from signal transduction to immune cell proliferation (Kowal et al., 2016). These exosomes are cleared from circulation during the early recovery phase from exercise (Fruhbeis, Helmig, Tug, Simon, & Kramer-Albers, 2015), suggesting that the exosomes released during exercise are quickly taken in by recipient cells, where they perform myriad functions.

The combination of exosomes, cytokines, myokines, and adipokines released during ex-ercise, collectively called exerkines, have important health implications. Likely through its anti-inflammatory effects on circulation and tissues, physical activity reduces the risk of multi-ple types of cancer and controls cancer progression (Moore et al., 2016). Exercise-conditioned serum, which contains the exerkines secreted during exercise, has consistently been shown to inhibit tumor formation and growth, suggesting the molecular factors secreted during exercise may have therapeutic potential in cancer treatment (Hojman, Gehl, Christensen, & Pedersen, 2018).

Many beneficial adaptations to exercise likely derive from the combined effects of exercise on multiple tissues. The skeletal muscle transcriptional response to exercise differs from that of adipose tissue. In overweight or obese adults, eight weeks of endurance training elicited transcriptional changes in the skeletal muscle associated with improved oxidation and substrate storage. In adipose tissue, lipogenesis, insulin signalling, and lipid storage transcripts were downregulated after an exercise bout (Dreher et al., 2023). While not surprising given the independent roles these tissues play during exercise, the distinct effects of exercise on skeletal muscle and adipose tissue highlight exercise's systemic, tissue-dependent, and synergistic effects.

Gastrointestinal immune system and the gut microbiome

Gastrointestinal (GI) tract contains the body's most extensive collection of lymphoid tissues and leucocytes. Within the GI immune system, multiple intestinal epithelial cell types, the distribution of which varies along the length of the GI tract, provide physical and chemical barriers against compounds within the GI lumen and mucosa. Goblet cells produce mucus that coats the intestinal mucosa, providing a physical deterrent against microorganisms and antigens. An additional line of defence is provided by an immune barrier comprising antibodies [e.g., secretory immunoglobulin A (IgA)], anti-microbial peptides [e.g., islet-derived protein (REG)-IIIγ], and glycoproteins. Paneth cells residing within villous crypts of the small intestine protect the epithelium by secreting various anti-microbial peptides, including REG-IIIγ, α-defensins, and lysozymes (Maynard, Elson, Hatton, & Weaver, 2012; Mowat & Agace, 2014). Microfold (M) cells are specialized for the phagocytosis and transcytosis of antigens and microbes across the epithelial barrier. These epithelial components of the innate immune system are complemented by leucocytes that reside in gut-associated lymphoid tissues (GALT; Peyer's patches, isolated lymphoid follicles, and mesenteric lymph nodes) and the lamina propria. Predominant leukocyte populations include dendritic cells, macrophages, regulatory T cells, Th17 cells, Th1 cells, NK cells, and IgA$^+$ B cells (Mowat & Agace, 2014). Like intestinal epithelial cells, the densities of these populations vary along the GI tract. Collectively, they sample and process antigens and respond to tissue damage and encroaching microbes in part by releasing a suite of pro- and anti-inflammatory cytokines (e.g., TGFβ, IFNγ, IL-4, IL-6, IL-1β, IL-10, IL-12, IL-13, IL-17, IL-22, IL-23) and various chemokines. The result is a carefully orchestrated and regulated immune response that balances immune tolerance with activation against potential pathogens (Ruff, Greiling, & Kriegel, 2020).

The gut microbiome is a diverse and dense ecosystem of bacteria, archaea, fungi, and other microorganisms within the GI tract. Most of the community resides in the colon, where the slow transit of nutrients, access to mucus and various cellular debris, immuno-tolerance, luminal pH, and minimal to non-existent oxygen levels provide an environment conducive to the survival of trillions of microbes (Donaldson, Lee, & Mazmanian, 2016). The gut microbiome contains an estimated 100 times more genes than the human genome (Qin et al., 2010). That genetic capacity increases biochemical complexity within human physiology and provides myriad functions and capabilities the human body would otherwise not have. For example, the gut microbiome metabolises non-digestible dietary components, produces essential vitamins and unique anti-microbial compounds, and shapes the development, function, and activity of biological systems throughout the body. Though the gut microbiome is influenced to some extent by genetics, environmental factors, including diet, medications, environmental exposures, disease, and exercise habits, are more influential (Falony et al., 2016; Karl et al., 2018; Rothschild et al., 2018). In addition, some microbiome features remain stable over months to years, while others are dynamic and malleable over hours and days. The result is a community in constant flux toward and away from an individualized basal state. That flux modulates gut-brain signalling via the enteric nervous system and vagus nerve (Cryan et al., 2019), influences the types and concentrations of compounds that provide metabolic substrates or act as signalling molecules throughout the body (Koppel & Balskus, 2016), and alters systemic immune function (Ruff et al., 2020). Conversely, the gut microbiome constantly responds to alterations within the GI environment resulting from changes in physiology, xenobiotic exposures, and activity within the GI immune system (Zheng, Liwinski, & Elinav, 2020).

The GI immune system and gut microbiome continuously engage in bi-directional interactions. The leucocytes within the GALT, intestinal epithelium, and lamina propria survey and respond to individual constituents of the gut microbiome through the recognition of cellular structures and microbial metabolites. Cellular structures such as lipopolysaccharide (LPS), flagellin, and polysaccharide A are known as microbe-associated molecular patterns (MAMPs). These MAMPs activate pattern recognition receptors (PRRs), including toll-like receptors, C-type lectin receptors, and nucleotide-binding oligomerization domain (NOD)-like receptors that are expressed on epithelial cells throughout the GI tract (Maynard et al., 2012; Mowat & Agace, 2014). Similarly, microbial metabolites such as short-chain fatty acids (SCFA), secondary bile acids, and tryptophan-derived compounds bind receptors on intestinal epithelial and immune cells (Zheng et al., 2020). Activation of PRRs, epithelial cells, and immune cells within the GI tract initiates signalling cascades, resulting in the production and secretion of mucus, anti-microbial compounds, and cytokines that direct an immune response designed to maintain homeostasis within both the GI tract and the gut microbiome (Maynard et al., 2012; Ruff et al., 2020; Zheng et al., 2020). In this way, the gut microbiome directs GI immune responses and is integral to the development and function of the GI immune system.

Perturbations in the gut microbiome or immunodeficiency can transiently or more permanently skew the bi-directional interactions with the GI immune system and gut microbiome toward persistent inflammation and immune activation in the GI tract and, subsequently, in peripheral tissues throughout the body, including adipose tissues and the skeletal muscle. Conversely, there is increasing evidence that the gut microbiome can be altered in ways that are considered health-promoting, and that the gut microbiome likely mediates some of the health benefits achieved through various lifestyle interventions. Within that context, interactions between exercise and the gut microbiome are increasingly being explored (Hughes, 2019; Mohr et al., 2020).

Effects of exercise training on GI immunity and the gut microbiome

Exercise influences bi-directional interactions between the GI immune system and the gut microbiome through multiple pathways (Figure 6.3). Exercise, especially when prolonged or strenuous (e.g., $\geq\sim70\%$ VO_2 max), activates the autonomic nervous system, increases body temperature, and reduces splanchnic and mesenteric blood flow to redistribute oxygen to working muscles. Tissue hyperthermia, hypoperfusion, and subsequent reperfusion alter GI barrier function and the GI environment, thereby influencing activity within the GI immune system. Though these effects of exercise bouts are transient, RBE do have more sustained effects on the GI immune system. In mice, exercise training can promote lymphocyte turnover, increase antioxidant and anti-inflammatory capacity, inhibit pro-inflammatory cytokine expression in intestinal leucocytes, increase luminal IgA concentrations, and improve intestinal immune function (Allen

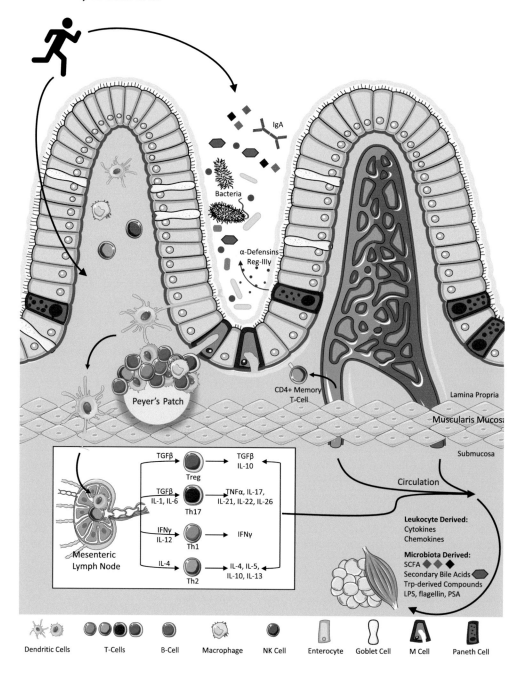

Figure 6.3 Exercise, the gastrointestinal immune system and the gut microbiome.

Legend: "The gastrointestinal immune system and gut microbiome are in constant bi-directional communication. Gut microbes and their metabolites induce responses within both innate and adaptive components of the gastrointestinal immune system while the gastrointestinal immune system acts as both a physical and biochemical barrier to the microbiome. Exercise training impacts these bi-directional interactions through effects on gastrointestinal physiology, barrier function and immunity. Exercise-mediated effects on the gastrointestinal immune system, the gut microbiome and their interactions influence biological systems throughout the body. Ig, immunoglobulin; IL, interleukin; IFN, interferon; LPS, lipopolysaccharide; PSA, polysaccharide A; SCFA, short-chain fatty acid; TGF, transforming growth factor; TNF, tumor necrosis factor.; Treg, regulatory T cell. Figure generated with images from BioIcons and Servier Medical Art, provided by Servier, licensed under a Creative Commons Attribution 3.0 unported license."

et al., 2017; Hoffman-Goetz, Pervaiz, Packer, & Guan, 2010; Packer & Hoffman-Goetz, 2012; Viloria et al., 2011).

Exercise training may also impact the gut microbiome independent of any effects on the GI immune system. Exercise decreases GI transit time and influences nutrient absorption. Those effects, along with diet composition, impact the availability of metabolic substrates to the gut microbiome, which is an important modulator of community composition and metabolic activity (Zmora, Suez, & Elinav, 2019). Changes in GI transit time may also alter colonic bile acid concentrations and composition. Bile acids have anti-microbial effects and are transformed by the gut microbiome into secondary bile acids with signalling functions within the GI tract and throughout the body (Cai, Sun, & Gonzalez, 2022).

The effect of exercise training on the gut microbiome is an active and rapidly growing area of research. Research to date indicates that exercise training alters the gut microbiome and although the effects observed are often inconsistent across studies and, in some cases, contradictory, several trends have emerged. In observational studies, the gut microbiome of athletes has been shown to differ from that of less active individuals by being more diverse, having an increased capacity for carbohydrate and branched-chain amino acid metabolism and SCFA production, and having higher proportions of certain taxa generally considered to be health-promoting, such as *Faecalibacterium* and *Akkermansia muciniphilia* (Mohr et al., 2020). However, to what extent those differences are due to exercise training, residual effects of individual exercise bouts or other factors such as diet is unclear. In experimental studies, exercise training appears to decrease the relative abundance of Proteobacteria, a phylum of LPS-producing bacteria often implicated in the development of inflammation and associated diseases (Huttenhower, Kostic, & Xavier, 2014). Exercise training also appears to increase the relative abundances of *Lactobacillus* and *Bifidobacterium*, two bacterial genera widely recognised as beneficial, and which include strains that enhance immune function and improve GI physiology (Gibson et al., 2017). Further, exercise training may increase the relative abundance of *Akkermansia muciniphilia*, a mucus-degrading bacteria with anti-inflammatory effects thought to enhance GI barrier function and metabolic health. Increased community diversity, an ecological characteristic generally regarded as beneficial, has also been observed following exercise training, as has an increased capacity within the community to produce SCFA. SCFA are produced from bacterial fermentation of carbohydrates primarily and are widely regarded as beneficial. Health benefits of the most abundant SCFA—acetate, propionate, and butyrate—include reducing colonic pH and inflammation, stimulating epithelial cell growth, enhancing GI immunity, and deterring pathogen colonization (Macfarlane & Macfarlane, 2012). Collectively, these data suggest that the effects of exercise training on the gut microbiome are largely beneficial. However, results do vary across studies (Hughes, 2019). That variability may be due, in part, to participant characteristics (e.g., species, training status, body composition, diet, and lifestyle habits) and differential effects of training intensity, duration and mode on GI immunity and physiology, and, subsequently, the gut microbiome.

Effects of the gut microbiome on exercise performance and adaptation to exercise training

Research has also focused on determining the extent to which the gut microbiome mediates physiologic and behavioural responses to exercise. Animal studies in which rodents are born and raised in sterile conditions (known as germ-free animals) or are treated with antibiotics that deplete the commensal gut microbiome facilitate studying the gut microbiome's role in physiologic and immunologic responses to exercise. Responses of germ-free or antibiotic-treated rodents to exercise interventions can then be compared to those of rodents that have a gut microbiome. Fecal microbiome transplantation experiments can also be conducted wherein the gut microbiome from exercised animals or sedentary controls (the donors) are transplanted into germ-free or antibiotic-treated recipients. Recapitulation of exercise-induced changes in the physiology, immunology, behaviour, or other outcomes seen in exercised donor animals in the recipient animals suggests that the gut microbiome causes those changes. Although limitations exist and translation to humans requires caution (Walter, Armet, Finlay, & Shanahan, 2020), these and other models have provided evidence that the gut microbiome may be one factor mediating exercise performance and adaptations to exercise. For example, reduced muscle strength and endurance capacity are seen in antibiotic-treated and germ-free mice, but this can be prevented with fecal microbiome transplants and attenuated with SCFA supplementation (W. C. Huang, Chen, Chuang, Chiu, & Huang, 2019; Lahiri et al., 2019; Nay et al., 2019; Okamoto et al., 2019). In one notable study, *Veillonella*, a genus of lactate-metabolising commensal gut bacteria, and bacterial genes within the metabolic pathway through which lactate is converted to the SCFA propionate were found to be enriched in fecal samples of endurance athletes. A subsequent series of experiments suggested a mechanism whereby *Veillonella* converted lactate produced during exercise into propionate, which was hypothesized to provide a glycogen-sparing energy source during exercise. In support of this, feeding mice a *Veillonella* strain isolated from one athlete improved aerobic endurance performance (Scheiman et al., 2019).

Exercise-induced changes in the gut microbiome and SCFA profiles have also been shown to mediate some of the health benefits of exercise in animal models. For example, a more favourable inflammatory profile in the colon, reduced leukocyte infiltration, improved gut morphology, increased capacity for butyrate production, and an attenuated response to experimentally induced colitis were observed in germ-free mice receiving transplants from exercise-trained relative to sedentary donor animals (Allen et al., 2017). In one randomised controlled trial, changes in the gut microbiomes and metabolites produced by gut microbes, such as SCFA and the neurotransmitter gamma-amino butyric, differed in individuals with insulin resistance who saw improvements in insulin sensitivity (responders) relative to those who did not (non-responders), following a 12-week exercise training intervention. Mice receiving fecal transplants from the responder cohort, but not the non-responder cohort, demonstrated improvements in glucose metabolism and insulin sensitivity when fed a high-fat diet, implicating the gut microbiome as one factor contributing to inter-individual variability in the metabolic response to exercise training (Liu et al., 2020). To what extent the gut microbiome may also influence inter-individual variability in immunologic responses to exercise is unknown.

In summary, current evidence suggests that exercise training may alter the gut microbiome in several potentially health-promoting ways. These effects are mediated through effects on GI physiology and the GI immune system. Conversely, the gut microbiome is likely one factor influencing responses and adaptations to exercise training within multiple biological systems that impact health and exercise performance, including the GI immune system.

Chapter summary

Just as physical activity impacts nearly every system in the body, the cells of the immune system infiltrate every body tissue. The role of the immune system in mediating the beneficial effects of exercise on the body is becoming increasingly recognised, as are the effects of exercise on tissue-resident immune cells. The complex interplay between exercise, the immune system, skeletal muscle, adipose tissue, the gastrointestinal tract, and the gut microbiome and the means of communication and crosstalk between these systems (Figure 6.4) are exciting areas of investigation.

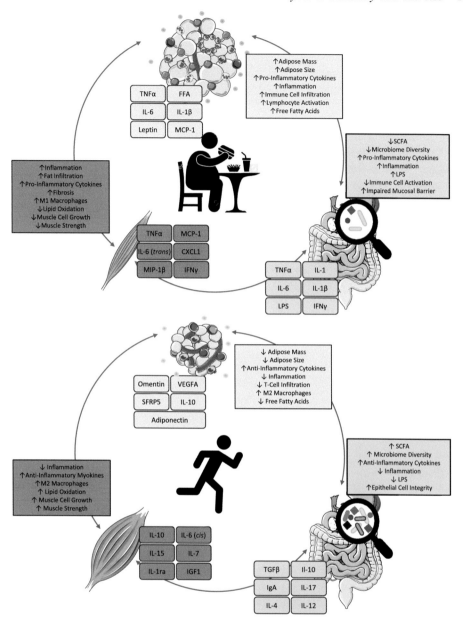

Figure 6.4 Effects of obesity and exercise training on the immuno-phenotype and secretome of adipose tissue, the gastrointestinal (GI) tract, and the skeletal muscle.

Legend: "Obesity and a sedentary lifestyle have a profound impact on the cellular composition, morphology, phenotype, and secretome of the GI tract, the adipose tissue, and the skeletal muscle. Many of the tissue-specific alterations associated with obesity are characterized by increased immune infiltration and pro-inflammatory immune profiles. The secretion of pro-inflammatory proteins, cytokines, chemokines, adipokines, and myokines from these tissues facilitates intra- and inter-tissue crosstalk, resulting in both local and systemic effects. In contrast, exercise training appears to promote an anti-inflammatory phenotype in the GI tract, adipose tissue, and skeletal muscle. Exercise-induced changes in the secretome of these tissues likely contribute to the systemic anti-inflammatory effects of exercise. CXCL, chemokine (C-X-C motif) ligand FFA, free fatty acid; Ig, immunoglobulin; IGF, insulin-like growth factor; IFN, interferon; IL, interleukin; LPS, lipopolysaccharide; MCP, monocyte chemoattractant protein; MIP, macrophage inflammatory protein; SCFA, short-chain fatty acids; TGF, transforming growth factor; TNF, tumor necrosis factor. Figure generated with Shutterstock Vector ID 105093671 and images from BioIcons and Servier Medical Art, provided by Servier, licensed under a Creative Commons Attribution 3.0 unported license."

Key points

- Leucocytes reside and play key roles in all tissues of the body, including the adipose tissue, the skeletal muscle, and the GI system.
- Regular physical activity appears to improve the adipose–leucocyte profile through remodeling tissue, reducing leukocyte infiltration, promoting an anti-inflammatory phenotype, and altering the adipose tissue secretome.
- Skeletal muscle leucocytes coordinate an adaptive response to individual bouts of exercise that facilitate muscle repair and regeneration, comprising an acute inflammatory response followed by an anti-inflammatory, pro-regenerative profile.
- Adipose tissue, skeletal muscle, and the leucocytes within these tissues facilitate intercellular crosstalk during exercise through the secretion of cytokines, adipokines, myokines, EVs, and exosomes.
- Exercise training improves the gut microbiome through its effects on GI physiology and the leucocytes residing within the GI tract. The gut microbiome, in turn, may mediate some of the immunologic and physiologic effects of exercise.

References

Ahn, C., Ryan, B. J., Schleh, M. W., Varshney, P., Ludzki, A. C., Gillen, J. B., … Horowitz, J. F. (2022). Exercise training remodels subcutaneous adipose tissue in adults with obesity even without weight loss. *J Physiol, 600*(9), 2127–2146.

Allen, J. M., Mailing, L. J., Cohrs, J., Salmonson, C., Fryer, J. D., Nehra, V., … Woods, J. A. (2017). Exercise training-induced modification of the gut microbiota persists after microbiota colonization and attenuates the response to chemically-induced colitis in gnotobiotic mice. *Gut Microbes, 9(2)*, 115–130.

Arnold, L., Henry, A., Poron, F., Baba-Amer, Y., van Rooijen, N., Plonquet, A., … Chazaud, B. (2007). Inflammatory monocytes recruited after skeletal muscle injury switch into antiinflammatory macrophages to support myogenesis. *J Exp Med, 204*(5), 1057–1069.

Babaei, P., & Hoseini, R. (2022). Exercise training modulates adipokine dysregulations in metabolic syndrome. *Sports Med Health Sci, 4*(1), 18–28.

Baturcam, E., Abubaker, J., Tiss, A., Abu-Farha, M., Khadir, A., Al-Ghimlas, F., … Dehbi, M. (2014). Physical exercise reduces the expression of RANTES and its CCR5 receptor in the adipose tissue of obese humans. *Mediators Inflamm, 2014*, 627150.

Bay, M. L., Heywood, S., Wedell-Neergaard, A. S., Schauer, T., Lehrskov, L. L., Christensen, R. H., … Ellingsgaard, H. (2020). Human immune cell mobilization during exercise: effect of IL-6 receptor blockade. *Exp Physiol, 105*(12), 2086–2098.

Becic, T., Studenik, C., & Hoffmann, G. (2018). Exercise increases adiponectin and reduces leptin levels in prediabetic and diabetic individuals: Systematic review and meta-analysis of randomized controlled trials. *Med Sci (Basel), 6*(4), 97.

Becker, M., Levings, M. K., & Daniel, C. (2017). Adipose-tissue regulatory T cells: Critical players in adipose-immune crosstalk. *Eur J Immunol, 47*(11), 1867–1874.

Boulenouar, S., Michelet, X., Duquette, D., Alvarez, D., Hogan, A. E., Dold, C., … Lynch, L. (2017). Adipose type one innate lymphoid cells regulate macrophage homeostasis through targeted cytotoxicity. *Immunity, 46*(2), 273–286.

Bradley, R. L., Jeon, J. Y., Liu, F. F., & Maratos-Flier, E. (2008). Voluntary exercise improves insulin sensitivity and adipose tissue inflammation in diet-induced obese mice. *Am J Physiol Endocrinol Metab, 295*(3), E586–594.

Brook, M. S., Stokes, T., Gorissen, S. H. M., Bass, J. J., McGlory, C., Cegielski, J., … Atherton, P. J. (2022). Declines in muscle protein synthesis account for short-term muscle disuse atrophy in humans in the absence of increased muscle protein breakdown. *J Cachexia Sarcopenia Muscle, 13(4)*, 2005–2016.

Burzyn, D., Kuswanto, W., Kolodin, D., Shadrach, J. L., Cerletti, M., Jang, Y., … Mathis, D. (2013). A special population of regulatory T cells potentiates muscle repair. *Cell, 155*(6), 1282–1295.

Cai, J., Sun, L., & Gonzalez, F. J. (2022). Gut microbiota-derived bile acids in intestinal immunity, inflammation, and tumorigenesis. *Cell Host Microbe, 30*(3), 289–300.

Camell, C. D., Gunther, P., Lee, A., Goldberg, E. L., Spadaro, O., Youm, Y. H., … Dixit, V. D. (2019). Aging induces an Nlrp3 inflammasome-dependent expansion of adipose B cells that impairs metabolic homeostasis. *Cell Metab, 30*(6), 1024–1039, e1026.

Chazaud, B. (2020). Inflammation and skeletal muscle regeneration: Leave it to the macrophages! *Trends Immunol, 41*(6), 481–492.

Collins, R. A., & Grounds, M. D. (2001). The role of tumor necrosis factor-alpha (TNF-alpha) in skeletal muscle regeneration. Studies in TNF-alpha(-/-) and TNF-alpha(-/-)/LT-alpha(-/-) mice. *J Histochem Cytochem, 49*(8), 989–1001.

Conlon, K. C., Lugli, E., Welles, H. C., Rosenberg, S. A., Fojo, A. T., Morris, J. C., … Waldmann, T. A. (2015). Redistribution, hyperproliferation, activation of natural killer cells and CD8 T cells, and cytokine production during first-in-human clinical trial of recombinant human interleukin-15 in patients with cancer. *J Clin Oncol, 33*(1), 74–82.

Cryan, J. F., O'Riordan, K. J., Cowan, C. S. M., Sandhu, K. V., Bastiaanssen, T. F. S., Boehme, M., … Dinan, T. G. (2019). The microbiota-Gut-brain axis. *Physiol Rev, 99*(4), 1877–2013.

Deshmukh, A. S., Peijs, L., Beaudry, J. L., Jespersen, N. Z., Nielsen, C. H., Ma, T., … Scheele, C. (2019). Proteomics-based comparative mapping of the secretomes of human brown and white adipocytes reveals EPDR1 as a novel Batokine. *Cell Metab, 30*(5), 963–975, e967.

Deyhle, M. R., Carlisle, M., Sorensen, J. R., Hafen, P. S., Jesperson, K., Ahmadi, M., … Hyldahl, R. D. (2020). Accumulation of skeletal muscle t cells and the repeated Bout effect in rats. *Med Sci Sports Exerc, 52*(6), 1280–1293.

Deyhle, M. R., Gier, A. M., Evans, K. C., Eggett, D. L., Nelson, W. B., Parcell, A. C., & Hyldahl, R. D. (2015). Skeletal muscle inflammation following repeated Bouts of lengthening contractions in humans. *Front Physiol, 6*, 424.

Distefano, G., & Goodpaster, B. H. (2018). Effects of exercise and aging on skeletal muscle. *Cold Spring Harb Perspect Med, 8*(3), a029785.

Donaldson, G. P., Lee, S. M., & Mazmanian, S. K. (2016). Gut biogeography of the bacterial microbiota. *Nat Rev Microbiol, 14*(1), 20–32.

Dreher, S. I., Irmler, M., Pivovarova-Ramich, O., Kessler, K., Jurchott, K., Sticht, C., … Moller, A. (2023). Acute and long-term exercise adaptation of adipose tissue and skeletal muscle in humans: A matched transcriptomics approach after 8-week training-intervention. *Int J Obes (Lond), 47*(4), 313–324.

Duggal, N. A., Pollock, R. D., Lazarus, N. R., Harridge, S., & Lord, J. M. (2018). Major features of immunesenescence, including reduced thymic output, are ameliorated by high levels of physical activity in adulthood. *Aging Cell, 17*(2), e12750.

Elieh Ali Komi, D., Shafaghat, F., & Christian, M. (2020). Crosstalk between mast cells and adipocytes in physiologic and pathologic conditions. *Clin Rev Allergy Immunol, 58*(3), 388–400.

Fabre, O., Ingerslev, L. R., Garde, C., Donkin, I., Simar, D., & Barres, R. (2018). Exercise training alters the genomic response to acute exercise in human adipose tissue. *Epigenomics, 10*(8), 1033–1050.

Falony, G., Joossens, M., Vieira-Silva, S., Wang, J., Darzi, Y., Faust, K., … Raes, J. (2016). Population-level analysis of gut microbiome variation. *Science, 352*(6285), 560–564.

Frasca, D., & Blomberg, B. B. (2017). Adipose tissue inflammation induces B cell inflammation and decreases B cell function in aging. *Front Immunol, 8*, 1003.

Fruhbeis, C., Helmig, S., Tug, S., Simon, P., & Kramer-Albers, E. M. (2015). Physical exercise induces rapid release of small extracellular vesicles into the circulation. *J Extracell Vesicles, 4*, 28239.

Frydelund-Larsen, L., Akerstrom, T., Nielsen, S., Keller, P., Keller, C., & Pedersen, B. K. (2007). Visfatin mRNA expression in human subcutaneous adipose tissue is regulated by exercise. *Am J Physiol Endocrinol Metab, 292*(1), E24–31.

Furrer, R., Eisele, P. S., Schmidt, A., Beer, M., & Handschin, C. (2017). Paracrine cross-talk between skeletal muscle and macrophages in exercise by PGC-1 alpha-controlled BNP. *Scientific Reports, 7*, 40789.

Gibson, G. R., Hutkins, R., Sanders, M. E., Prescott, S. L., Reimer, R. A., Salminen, S. J., … Reid, G. (2017). Expert consensus document: The International Scientific Association for Probiotics and

Prebiotics (ISAPP) consensus statement on the definition and scope of prebiotics. *Nat Rev Gastroenterol Hepatol, 14*(8), 491–502.

Giudice, J., & Taylor, J. M. (2017). Muscle as a paracrine and endocrine organ. *Curr Opin Pharmacol, 34,* 49–55.

Gleeson, M., Bishop, N. C., Stensel, D. J., Lindley, M. R., Mastana, S. S., & Nimmo, M. A. (2011). The anti-inflammatory effects of exercise: mechanisms and implications for the prevention and treatment of disease. *Nat Rev Immunol, 11*(9), 607–615.

Gleeson, M., McFarlin, B., & Flynn, M. (2006). Exercise and Toll-like receptors. *Exerc Immunol Rev, 12,* 34–53.

Glover, E. I., Phillips, S. M., Oates, B. R., Tang, J. E., Tarnopolsky, M. A., Selby, A., … Rennie, M. J. (2008). Immobilization induces anabolic resistance in human myofibrillar protein synthesis with low and high dose amino acid infusion. *J Physiol, 586*(24), 6049–6061.

Han, S. J., Glatman Zaretsky, A., Andrade-Oliveira, V., Collins, N., Dzutsev, A., Shaik, J., … Belkaid, Y. (2017). White adipose tissue is a reservoir for memory T cells and promotes protective memory responses to infection. *Immunity, 47*(6), 1154–1168, e1156.

Haugen, F., Norheim, F., Lian, H., Wensaas, A. J., Dueland, S., Berg, O., … Drevon, C. A. (2010). IL-7 is expressed and secreted by human skeletal muscle cells. *Am J Physiol Cell Physiol, 298*(4), C807–816.

He, C., Zheng, S., Luo, Y., & Wang, B. (2018). Exosome theranostics: Biology and translational medicine. *Theranostics, 8*(1), 237–255.

Hoffman-Goetz, L., Pervaiz, N., Packer, N., & Guan, J. (2010). Freewheel training decreases pro- and increases anti-inflammatory cytokine expression in mouse intestinal lymphocytes. *Brain Behav Immun, 24*(7), 1105–1115.

Hojman, P., Gehl, J., Christensen, J. F., & Pedersen, B. K. (2018). Molecular mechanisms linking exercise to cancer prevention and treatment. *Cell Metab, 27*(1), 10–21.

Hotamisligil, G. S., Shargill, N. S., & Spiegelman, B. M. (1993). Adipose expression of tumor necrosis factor-alpha: direct role in obesity-linked insulin resistance. *Science, 259*(5091), 87–91.

Hu, B., Jin, C., Zeng, X., Resch, J. M., Jedrychowski, M. P., Yang, Z., … Spiegelman, B. M. (2020). Gammadelta T cells and adipocyte IL-17RC control fat innervation and thermogenesis. *Nature, 578*(7796), 610–614.

Huang, W. C., Chen, Y. H., Chuang, H. L., Chiu, C. C., & Huang, C. C. (2019). Investigation of the effects of microbiota on exercise physiological adaption, performance, and energy utilization using a gnotobiotic animal model. *Front Microbiol, 10,* 1906.

Huang, Z., & Xu, A. (2021). Adipose extracellular vesicles in intercellular and inter-organ crosstalk in metabolic health and diseases. *Front Immunol, 12,* 608680.

Hughes, R. L. (2019). A review of the role of the Gut microbiome in personalized sports nutrition. *Front Nutr, 6,* 191.

Huttenhower, C., Kostic, A. D., & Xavier, R. J. (2014). Inflammatory bowel disease as a model for translating the microbiome. *Immunity, 40*(6), 843–854.

Irving, B. A., Lanza, I. R., Henderson, G. C., Rao, R. R., Spiegelman, B. M., & Sreekumaran Nair, K. (2015). Combined training enhances skeletal muscle mitochondrial oxidative capacity independent of age. *J Clin Endocrinol Metab, 100*(4), 1654–1663.

Jensen, S. M., Bechshoft, C. J. L., Heisterberg, M. F., Schjerling, P., Andersen, J. L., Kjaer, M., & Mackey, A. L. (2020). Macrophage subpopulations and the acute inflammatory response of elderly human skeletal muscle to physiological resistance exercise. *Front Physiol, 11,* 811.

Kalluri, R., & LeBleu, V. S. (2020). The biology, function, and biomedical applications of exosomes. *Science, 367*(6478).

Kao, F. F., & Ray, L. H. (1954). Regulation of cardiac output in anesthetized dogs during induced muscular work. *Am J Physiol, 179*(2), 255–260.

Karl, J. P., Hatch, A. M., Arcidiacono, S. M., Pearce, S. C., Pantoja-Feliciano, I. G., Doherty, L. A., & Soares, J. W. (2018). Effects of psychological, environmental and physical stressors on the Gut microbiota. *Front Microbiol, 9,* 2013.

Kawanishi, N., Funakoshi, T., & Machida, S. (2018). Time-course study of macrophage infiltration and inflammation in cast immobilization-induced atrophied muscle of mice. *Muscle Nerve, 57*(6), 1006–1013.

Kawanishi, N., Mizokami, T., Yano, H., & Suzuki, K. (2013). Exercise attenuates M1 macrophages and CD8+ T cells in the adipose tissue of obese mice. *Med Sci Sports Exerc, 45*(9), 1684–1693.

Kawanishi, N., Mizokami, T., Niihara, H., Yada, K., & Suzuki, K. (2016). Neutrophil depletion attenuates muscle injury after exhaustive exercise. *Med Sci Sports Exerc, 48*(10), 1917–1924.

Kawanishi, N., Niihara, H., Mizokami, T., Yada, K., & Suzuki, K. (2015). Exercise training attenuates neutrophil infiltration and elastase expression in adipose tissue of high-fat-diet-induced obese mice. *Physiol Rep, 3*(9).

Kawanishi, N., Yano, H., Yokogawa, Y., & Suzuki, K. (2010). Exercise training inhibits inflammation in adipose tissue via both suppression of macrophage infiltration and acceleration of phenotypic switching from M1 to M2 macrophages in high-fat-diet-induced obese mice. *Exerc Immunol Rev, 16*, 105–118.

Kohlgruber, A. C., Gal-Oz, S. T., LaMarche, N. M., Shimazaki, M., Duquette, D., Koay, H. F., ... Lynch, L. (2018). gammadelta T cells producing interleukin-17A regulate adipose regulatory T cell homeostasis and thermogenesis. *Nat Immunol, 19*(5), 464–474.

Kojta, I., Chacinska, M., & Blachnio-Zabielska, A. (2020). Obesity, bioactive lipids, and adipose tissue inflammation in insulin resistance. *Nutrients, 12*(5), 1305.

Kolahdouzi, S., Talebi-Garakani, E., Hamidian, G., & Safarzade, A. (2019). Exercise training prevents high-fat diet-induced adipose tissue remodeling by promoting capillary density and macrophage polarization. *Life Sci, 220*, 32–43.

Kolb, R., & Zhang, W. (2020). Obesity and breast cancer: A case of inflamed adipose tissue. *Cancers (Basel), 12*(6), 1686.

Koppel, N., & Balskus, E. P. (2016). Exploring and understanding the biochemical diversity of the human microbiota. *Cell Chem Biol, 23*(1), 18–30.

Kowal, J., Arras, G., Colombo, M., Jouve, M., Morath, J. P., Primdal-Bengtson, B., ... Thery, C. (2016). Proteomic comparison defines novel markers to characterize heterogeneous populations of extracellular vesicle subtypes. *Proc Natl Acad Sci U S A, 113*(8), E968–977.

Lahiri, S., Kim, H., Garcia-Perez, I., Reza, M. M., Martin, K. A., Kundu, P., ... Pettersson, S. (2019). The gut microbiota influences skeletal muscle mass and function in mice. *Sci Transl Med, 11*(502), eaan5662.

Lang, C. H., Frost, R. A., Nairn, A. C., MacLean, D. A., & Vary, T. C. (2002). TNF-alpha impairs heart and skeletal muscle protein synthesis by altering translation initiation. *Am J Physiol Endocrinol Metab, 282*(2), E336–347.

Lavin, K. M., Perkins, R. K., Jemiolo, B., Raue, U., Trappe, S. W., & Trappe, T. A. (2020). Effects of aging and lifelong aerobic exercise on basal and exercise-induced inflammation. *J Appl Physiol (1985), 128*(1), 87–99.

Lee, B. C., Kim, M. S., Pae, M., Yamamoto, Y., Eberle, D., Shimada, T., ... Lee, J. (2016). Adipose Natural killer cells regulate adipose tissue macrophages to promote insulin resistance in obesity. *Cell Metab, 23*(4), 685–698.

Lee, S., Norheim, F., Langleite, T. M., Noreng, H. J., Storas, T. H., Afman, L. A., ... NutriTech, C. (2016). Effect of energy restriction and physical exercise intervention on phenotypic flexibility as examined by transcriptomics analyses of mRNA from adipose tissue and whole body magnetic resonance imaging. *Physiol Rep, 4*(21), e13019.

Lee, S., Norheim, F., Langleite, T. M., Gulseth, H. L., Birkeland, K. I., & Drevon, C. A. (2019). Effects of long-term exercise on plasma adipokine levels and inflammation-related gene expression in subcutaneous adipose tissue in sedentary dysglycaemic, overweight men and sedentary normoglycaemic men of healthy weight. *Diabetologia, 62*(6), 1048–1064.

Liu, Y., Wang, Y., Ni, Y., Cheung, C. K. Y., Lam, K. S. L., Wang, Y., ... Xu, A. (2020). Gut microbiome fermentation determines the efficacy of exercise for diabetes prevention. *Cell Metab, 31*(1), 77–91, e75.

Long, D. E., Peck, B. D., Lavin, K. M., Dungan, C. M., Kosmac, K., Tuggle, S. C., ... Peterson, C. A. (2022). Skeletal muscle properties show collagen organization and immune cell content are associated with resistance exercise response heterogeneity in older persons. *J Appl Physiol (1985), 132*(6), 1432–1447.

Ludzki, A. C., Krueger, E. M., Baldwin, T. C., Schleh, M. W., Porsche, C. E., Ryan, B. J., ... Horowitz, J. F. (2020). Acute aerobic exercise remodels the adipose tissue progenitor cell phenotype in obese adults. *Front Physiol, 11*, 903.

Lynch, L., Michelet, X., Zhang, S., Brennan, P. J., Moseman, A., Lester, C., … Brenner, M. B. (2015). Regulatory iNKT cells lack expression of the transcription factor PLZF and control the homeostasis of T(reg) cells and macrophages in adipose tissue. *Nat Immunol, 16*(1), 85–95.

Macdougall, C. E., & Longhi, M. P. (2019). Adipose tissue dendritic cells in steady-state. *Immunology, 156*(3), 228–234.

Macfarlane, G. T., & Macfarlane, S. (2012). Bacteria, colonic fermentation, and gastrointestinal health. *J AOAC Int, 95*(1), 50–60.

Maynard, C. L., Elson, C. O., Hatton, R. D., & Weaver, C. T. (2012). Reciprocal interactions of the intestinal microbiota and immune system. *Nature, 489*(7415), 231–241.

Minari, A. L. A., & Thomatieli-Santos, R. V. (2022). From skeletal muscle damage and regeneration to the hypertrophy induced by exercise: what is the role of different macrophage subsets? *Am J Physiol Regul Integr Comp Physiol, 322*(1), R41–R54. Doi:10.1152/ajpregu.00038.2021

Mohr, A. E., Jager, R., Carpenter, K. C., Kerksick, C. M., Purpura, M., Townsend, J. R., … Antonio, J. (2020). The athletic gut microbiota. *J Int Soc Sports Nutr, 17*(1), 24.

Molofsky, A. B., Nussbaum, J. C., Liang, H. E., Van Dyken, S. J., Cheng, L. E., Mohapatra, A., … Locksley, R. M. (2013). Innate lymphoid type 2 cells sustain visceral adipose tissue eosinophils and alternatively activated macrophages. *J Exp Med, 210*(3), 535–549.

Moore, S. C., Lee, I. M., Weiderpass, E., Campbell, P. T., Sampson, J. N., Kitahara, C. M., … Patel, A. V. (2016). Association of leisure-time physical activity with risk of 26 types of cancer in 1.44 million adults. *JAMA Intern Med, 176*(6), 816–825.

Mowat, A. M., & Agace, W. W. (2014). Regional specialization within the intestinal immune system. *Nat Rev Immunol, 14*(10), 667–685.

Nara, H., & Watanabe, R. (2021). Anti-inflammatory effect of muscle-derived Interleukin-6 and Its involvement in lipid metabolism. *Int J Mol Sci, 22*(18), 9889.

Nay, K., Jollet, M., Goustard, B., Baati, N., Vernus, B., Pontones, M., … Koechlin-Ramonatxo, C. (2019). Gut bacteria are critical for optimal muscle function: a potential link with glucose homeostasis. *Am J Physiol Endocrinol Metab, 317*(1), E158–e171.

Niemiro, G. M., Allen, J. M., Mailing, L. J., Khan, N. A., Holscher, H. D., Woods, J. A., & De Lisio, M. (2018). Effects of endurance exercise training on inflammatory circulating progenitor cell content in lean and obese adults. *J Physiol, 596*(14), 2811–2822.

Nono Nankam, P. A., Mendham, A. E., De Smidt, M. F., Keswell, D., Olsson, T., Bluher, M., & Goedecke, J. H. (2020). Changes in systemic and subcutaneous adipose tissue inflammation and oxidative stress in response to exercise training in obese black African women. *J Physiol, 598*(3), 503–515.

O'Leary, M. F., Wallace, G. R., Davis, E. T., Murphy, D. P., Nicholson, T., Bennett, A. J., … Jones, S. W. (2018). Obese subcutaneous adipose tissue impairs human myogenesis, particularly in old skeletal muscle, via resistin-mediated activation of NFkappaB. *Sci Rep, 8*(1), 15360.

Okamoto, T., Morino, K., Ugi, S., Nakagawa, F., Lemecha, M., Ida, S., … Maegawa, H. (2019). Microbiome potentiates endurance exercise through intestinal acetate production. *Am J Physiol Endocrinol Metab, 316*(5), E956–E966.

Packer, N., & Hoffman-Goetz, L. (2012). Exercise training reduces inflammatory mediators in the intestinal tract of healthy older adult mice. *Can J Aging, 31*(2), 161–171.

Peake, J. M., Neubauer, O., Della Gatta, P. A., & Nosaka, K. (2017). Muscle damage and inflammation during recovery from exercise. *J Appl Physiol (1985), 122*(3), 559–570.

Pedersen, B. K., & Febbraio, M. A. (2012). Muscles, exercise and obesity: Skeletal muscle as a secretory organ. *Nat Rev Endocrinol, 8*(8), 457–465.

Pedersen, B. K., & Toft, A. D. (2000). Effects of exercise on lymphocytes and cytokines. *Br J Sports Med, 34*(4), 246–251.

Phillips, S. M., Tipton, K. D., Aarsland, A., Wolf, S. E., & Wolfe, R. R. (1997). Mixed muscle protein synthesis and breakdown after resistance exercise in humans. *Am J Physiol, 273*(1 Pt 1), E99–107.

Przybyla, B., Gurley, C., Harvey, J. F., Bearden, E., Kortebein, P., Evans, W. J., … Dennis, R. A. (2006). Aging alters macrophage properties in human skeletal muscle both at rest and in response to acute resistance exercise. *Exp Gerontol, 41*(3), 320–327.

Pullamsetti, S. S., Seeger, W., & Savai, R. (2018). Classical IL-6 signaling: a promising therapeutic target for pulmonary arterial hypertension. *J Clin Invest, 128*(5), 1720–1723.

Qin, J., Li, R., Raes, J., Arumugam, M., Burgdorf, K. S., Manichanh, C., ... Ehrlich, S. D. (2010). A human gut microbial gene catalogue established by metagenomic sequencing. *Nature, 464*(7285), 59.

Reidy, P. T., McKenzie, A. I., Mahmassani, Z. S., Petrocelli, J. J., Nelson, D. B., Lindsay, C. C., ... Drummond, M. J. (2019). Aging impairs mouse skeletal muscle macrophage polarization and muscle-specific abundance during recovery from disuse. *Am J Physiol Endocrinol Metab, 317*(1), E85–E98.

Rogeri, P. S., Gasparini, S. O., Martins, G. L., Costa, L. K. F., Araujo, C. C., Lugaresi, R., ... Lancha, A. H. (2020). Crosstalk between skeletal muscle and immune system: Which roles do IL-6 and glutamine play? *Front Physiol, 11,* 582258.

Rothschild, D., Weissbrod, O., Barkan, E., Kurilshikov, A., Korem, T., Zeevi, D., ... Segal, E. (2018). Environment dominates over host genetics in shaping human gut microbiota. *Nature, 555*(7695), 210–215.

Rowe, G. C., Raghuram, S., Jang, C., Nagy, J. A., Patten, I. S., Goyal, A., ... Arany, Z. (2014). PGC-1alpha induces SPP1 to activate macrophages and orchestrate functional angiogenesis in skeletal muscle. *Circ Res, 115*(5), 504–517.

Ruff, W. E., Greiling, T. M., & Kriegel, M. A. (2020). Host-microbiota interactions in immune-mediated diseases. *Nat Rev Microbiol, 18*(9), 521–538.

Russo, L., & Lumeng, C. N. (2018). Properties and functions of adipose tissue macrophages in obesity. *Immunology, 155*(4), 407–417.

San Emeterio, C. L., Olingy, C. E., Chu, Y., & Botchwey, E. A. (2017). Selective recruitment of nonclassical monocytes promotes skeletal muscle repair. *Biomaterials, 117,* 32–43.

Scheiman, J., Luber, J. M., Chavkin, T. A., MacDonald, T., Tung, A., Pham, L. D., ... Kostic, A. D. (2019). Meta-omics analysis of elite athletes identifies a performance-enhancing microbe that functions via lactate metabolism. *Nat Med, 25*(7), 1104–1109.

Serrano, A. L., Baeza-Raja, B., Perdiguero, E., Jardi, M., & Munoz-Cinoves, P. (2008). Interleukin-6 is an essential regulator of satellite cell-mediated skeletal muscle hypertrophy. *Cell Metabolism, 7*(1), 33–44.

Soehnlein, O., & Lindbom, L. (2010). Phagocyte partnership during the onset and resolution of inflammation. *Nat Rev Immunol, 10*(6), 427–439.

Sorensen, J. R., Kaluhiokalani, J. P., Hafen, P. S., Deyhle, M. R., Parcell, A. C., & Hyldahl, R. D. (2019). An altered response in macrophage phenotype following damage in aged human skeletal muscle: Implications for skeletal muscle repair. *FASEB J, 33*(9), 10353–10368.

Sparks, L. M., Goodpaster, B. H., & Bergman, B. C. (2021). The metabolic significance of intermuscular adipose tissue: Is IMAT a friend or a foe to metabolic health? *Diabetes, 70*(11), 2457–2467.

Standley, R. A., Distefano, G., Trevino, M. B., Chen, E., Narain, N. R., Greenwood, B., ... Goodpaster, B. H. (2020). Skeletal muscle energetics and mitochondrial function are impaired following 10 days of bed rest in older adults. *J Gerontol A Biol Sci Med Sci, 75*(9), 1744–1753.

Stanford, K. I., Middelbeek, R. J., Townsend, K. L., Lee, M. Y., Takahashi, H., So, K., ... Goodyear, L. J. (2015). A novel role for subcutaneous adipose tissue in exercise-induced improvements in glucose homeostasis. *Diabetes, 64*(6), 2002–2014.

Steensberg, A., Fischer, C. P., Keller, C., Moller, K., & Pedersen, B. K. (2003). IL-6 enhances plasma IL-1ra, IL-10, and cortisol in humans. *Am J Physiol Endocrinol Metab, 285*(2), E433–437.

Tidball, J. G. (2017). Regulation of muscle growth and regeneration by the immune system. *Nat Rev Immunol, 17*(3), 165–178.

Trim, W. V., & Lynch, L. (2021). Immune and non-immune functions of adipose tissue leukocytes. *Nat Rev Immunol, 22(6),* 371–386.

Tsivitse, S. K., Mylona, E., Peterson, J. M., Gunning, W. T., & Pizza, F. X. (2005). Mechanical loading and injury induce human myotubes to release neutrophil chemoattractants. *Am J Physiol Cell Physiol, 288*(3), C721–729.

Valerio, A., Cardile, A., Cozzi, V., Bracale, R., Tedesco, L., Pisconti, A., ... Nisoli, E. (2006). TNF-alpha downregulates eNOS expression and mitochondrial biogenesis in fat and muscle of obese rodents. *J Clin Invest, 116*(10), 2791–2798.

Vanderboom, P. M., Dasari, S., Ruegsegger, G. N., Pataky, M. W., Lucien, F., Heppelmann, C. J., … Nair, K. S. (2021). A size-exclusion-based approach for purifying extracellular vesicles from human plasma. *Cell Rep Methods, 1*(3), 100055.

Van Pelt, D. W., Guth, L. M., & Horowitz, J. F. (2017). Aerobic exercise elevates markers of angiogenesis and macrophage IL-6 gene expression in the subcutaneous adipose tissue of overweight-to-obese adults. *J Appl Physiol (1985), 123*(5), 1150–1159.

Varga, T., Mounier, R., Patsalos, A., Gogolak, P., Peloquin, M., Horvath, A., … Nagy, L. (2016). Macrophage PPARgamma, a lipid activated transcription factor controls the growth factor GDF3 and skeletal muscle regeneration. *Immunity, 45*(5), 1038–1051.

Viloria, M., Lara-Padilla, E., Campos-Rodriguez, R., Jarillo-Luna, A., Reyna-Garfias, H., Lopez-Sanchez, P., … Garcia-Latorre, E. (2011). Effect of moderate exercise on IgA levels and lymphocyte count in mouse intestine. *Immunol Invest, 40*(6), 640–656.

Vohralik, E. J., Psaila, A. M., Knights, A. J., & Quinlan, K. G. R. (2020). EoTHINophils: Eosinophils as key players in adipose tissue homeostasis. *Clin Exp Pharmacol Physiol, 47*(8), 1495–1505.

Walter, J., Armet, A. M., Finlay, B. B., & Shanahan, F. (2020). Establishing or exaggerating causality for the gut microbiome: Lessons from human microbiota-associated rodents. *Cell, 180*(2), 221–232.

Walton, R. G., Kosmac, K., Mula, J., Fry, C. S., Peck, B. D., Groshong, J. S., … Peterson, C. A. (2019). Human skeletal muscle macrophages increase following cycle training and are associated with adaptations that may facilitate growth. *Sci Rep, 9*(1), 969.

Wang, H., Shen, L., Sun, X., Liu, F., Feng, W., Jiang, C., … Bi, Y. (2019). Adipose group 1 innate lymphoid cells promote adipose tissue fibrosis and diabetes in obesity. *Nat Commun, 10*(1), 3254.

Wang, Q., & Wu, H. (2018). T cells in adipose tissue: Critical players in immunometabolism. *Front Immunol, 9*, 2509.

Watanabe, Y., Nagai, Y., Honda, H., Okamoto, N., Yanagibashi, T., Ogasawara, M., … Takatsu, K. (2019). Bidirectional crosstalk between neutrophils and adipocytes promotes adipose tissue inflammation. *FASEB J, 33*(11), 11821–11835.

Weisberg, S. P., McCann, D., Desai, M., Rosenbaum, M., Leibel, R. L., & Ferrante, A. W., Jr. (2003). Obesity is associated with macrophage accumulation in adipose tissue. *J Clin Invest, 112*(12), 1796–1808.

Winn, N. C., Cottam, M. A., Wasserman, D. H., & Hasty, A. H. (2021). Exercise and adipose tissue immunity: Outrunning inflammation. *Obesity (Silver Spring), 29*(5), 790–801.

Yamada, M., Suzuki, K., Kudo, S., Totsuka, M., Nakaji, S., & Sugawara, K. (2002). Raised plasma G-CSF and IL-6 after exercise may play a role in neutrophil mobilization into the circulation. *J App Physiol, 92*(5), 1789–1794.

Zhang, J., Muri, J., Fitzgerald, G., Gorski, T., Gianni-Barrera, R., Masschelein, E., … De Bock, K. (2020). Endothelial lactate controls muscle regeneration from ischemia by inducing M2-like macrophage polarization. *Cell Metab, 31*(6), 1136–1153, e1137.

Zheng, D., Liwinski, T., & Elinav, E. (2020). Interaction between microbiota and immunity in health and disease. *Cell Res, 30*(6), 492–506.

Ziegler, A. K., Damgaard, A., Mackey, A. L., Schjerling, P., Magnusson, P., Olesen, A. T., … Scheele, C. (2019). An anti-inflammatory phenotype in visceral adipose tissue of old lean mice, augmented by exercise. *Sci Rep, 9*(1), 12069.

Zmora, N., Suez, J., & Elinav, E. (2019). You are what you eat: Diet, health and the gut microbiota. *Nat Rev Gastroenterol Hepatol, 16*(1), 35–56.

7 Exercise and immune competency

In vivo research

Kate M. Edwards, Sarah C. Marvin, and Marian L. Kohut

Introduction

It is clear from epidemiology evidence that exercise has an array of health impacts, including altering infection susceptibility (see Chapter 9). This is thought to be a result of exercise altering the functional competency of the immune system. When attempting to understand how exercise impacts immunity, it is necessary from a research perspective – given the profound complexity of an immune response – to take different methodological approaches. One constructivist approach is to analyse separate components of the immune response in an isolated and controlled manner. For example, using highly controlled *in vitro* assays in the laboratory to measure the impact of exercise on cellular responses to immune challenge (e.g., viral antigens). However, it is difficult to bridge the findings from these isolated *in vitro* experiments with outcomes from more holistic studies in the real-world – for example, findings from epidemiology studies showing that an active lifestyle is associated with reduced incidence of a particular disease. Given this, it is necessary to conduct translational research to bridge this gap.

In the context of exercise immunology, the principle translational approach that is commonly used is the controlled administration of an immune challenge to a host *in vivo* (e.g., an animal or a human) and subsequently measuring the 'outcome' of the immune response. For example, administering an antigen (e.g., via vaccination) and then determining the antibody response to that antigen in the days and/or weeks thereafter. Other approaches include the assessment of local tissue responses to immune challenge, such as wound healing or delayed-type-hypersensitivity reactions in the skin, or exposure to a pathogen.

Importantly, each method has advantages and disadvantages, which provide further nuance or limitations in the interpretation of the findings observed. Generally, *in vivo* measures (i.e., performed within a living organism) are insightful as they permit the study of a functioning immune system, and thereby utilise a representative model with capability to generate results that are clinically relevant. However, it is difficult to control all variables which may confound or contribute to outcomes, and these studies are often costly and raise complex ethical considerations. These translational approaches, which evaluate responses to immunological challenge are conducted interchangeably in animals and humans. There is considerable debate surrounding the use of humans and animals in immunological research. Whilst evidence from human models is the most clinically applicable, their use in research carries significant ethical burden centred around the risk of harm coupled with lack of personal benefit for participants. As such, animal models are used extensively when comparable human research is difficult experimentally, logistically, and ethically. Along with the inherent ethical implications in using animals in research, animal models have some limitations as the pathology of the infectious disease and immune responses may differ between humans and animals, although the extent of

DOI: 10.4324/9781003256991-7

this difference varies by disease. There are newer "humanized" mouse models in which human cells or tissues are engrafted into immunodeficient mice (Lai & Chen, 2018), and as a result, the immune response better reflects that of humans for a given infectious disease. However, these models have not been used in the exercise immunology field to date. Another consideration in animal studies is the mode of exercise selected and species. Typically, mice or rats are used in animal studies, and the most common exercise modes are treadmill running, voluntary wheel running, or swimming. However, the ways in which these different forms of exercise are administered may create additional stress for the animal, with activation of the hypothalamic–pituitary–adrenal (HPA) axis, which may impact the immune response to infection. Guidelines to minimise stress in animal exercise should be followed ("Resource Book for the Design of Animal Exercise Protocols", 2006). Thus, animals have advantages and disadvantages for use, but with careful selection of animal disease model and exercise mode, the findings can have important translational relevance. In fact, the results obtained from both translational exercise immunology research – from both animals and humans – usually have direct clinical implications. Indeed, such studies not only tell us about the effects of exercise on the immune system in a living model, but they inspire us to investigate exercise as an intervention to be harnessed for real-world impact.

A summary of the more common translational *in vivo* models that are used to examine the immune response to challenge will be provided next. In doing so, a summary of the effects of exercise – both individual bouts of exercise and regular exercise – on immune outcomes measured in those models will also be provided. Methods of immune challenge in the local tissues will be provided first, and thereafter, an overview of immune challenge models at the systemic – i.e., whole organism – level will be provided.

In vivo methodologies

Commonly used tissue-level methods include delayed-type hypersensitivity and wound-healing techniques where the model challenge is isolated to a single location (e.g. skin), and commonly used system-level methods involve assessing the whole organism response to vaccination or infection challenge. These will be discussed, in turn, next.

Delayed-type hypersensitivity

Delayed-type hypersensitivity (DTH) reactions generally refer to a cell-mediated immuno-inflammatory response that develops 24–72 hours following exposure to sensitisation to an antigen. A recall response is usually measured, whereby the sensitising agent used is a previously encountered antigen – yet, novel agents that induce a primary DTH response also exist. The term originates from the skin test used to diagnose tuberculosis, and functions to differentiate between delayed cell-mediated responses and more immediate antibody-mediated responses. This reaction is generally studied on the skin due to its accessibility. Where animal models are utilised, the ear pinnae and footpads of smaller animals such as rodents are typically used.

The DTH response consists of two phases

1 Sensitisation phase: initial contact with an antigen is made and a cell-mediated (T-cell) primary response is activated.
2 Effector/challenge phase: subsequent exposure to the antigen results in a local inflammatory response.

A characteristic DTH response presents as skin indurations of 2–20 mm in diameter accompanied by redness at the site due to increased blood flow in superficial arteries (Córdoba et al., 2008). The magnitude of this response is quantified by the diameter and thickness of indurations, as well as the degree of swelling present (Burlingham et al., 2005).

The DTH response reflects the interaction of memory T cells, antigen-presenting cells, effector T cells, and cytokines to coordinate an effective immune response upon exposure to a previously encountered antigen. This test can also be used clinically as an index of immune compromise, whereby a lack of DTH response to recall antigens is considered evidence of anergy (Burlingham et al., 2005). However, DTH reactions cannot predict the potential for a response to antigens that have not yet been encountered – it simply reflects prior immunologic sensitisation.

Models using DTH often apply an intervention at time of primary or secondary exposure to the antigen, to examine the effect on the holistic cell-mediated immune response. For example, in 1997, Dhabhar and McEwen used the ear pinnae of rats to measure DTH reactions to 2,4-dinitro-1-fluorobenzene (DNFB) following acute and chronic stress. Acute stress in the form of shaking and restraint applied for 2 hours prior to antigenic challenge significantly enhanced the DTH response. The strength of this reaction was further enhanced by increasing either the duration to 5 hours or the intensity of applied stress. Contrastingly, when observing the impact of chronic stress, the DTH reaction was suppressed when people were repeatedly exposed to stress for 6 hours daily over a three- to five-week period prior to testing. When examining stress models, the magnitude and direction of effect on DTH reactions are highly dependent on the type and degree of stress applied. Animal research has shown the acute administration of low-dose corticosterone and epinephrine can enhance DTH reactions, whilst high-dose corticosterone and low-dose dexamethasone suppress this response (Dhabhar, 2000).

Effects of an individual bout of exercise on DTH reactions

When focusing on an individual bout of exercise performed immediately prior to the sensitisation phase, the effects on the DTH response are mixed and appear to be largely driven by the combination of exercise intensity and duration. Nevertheless, a transient suppression of the DTH response with prolonged and strenuous exercise has been observed many times in the literature and across various antigens (Bruunsgaard et al., 1997; Harper Smith et al., 2011; Davison et al., 2016; Jones et al., 2019). For example, a seminal study by Bruunsgaard et al. (1997) assessed DTH responses to an array of antigens applied to the skin after prolonged exercise, including: tetanus and diphtheria toxoids, streptococcus, tuberculin, and proteus bacterial antigens, and candida and trichophyton fungal antigens. A skin test score was used to assess responses and it was found that DTH responses were generally impaired in those who conducted prolonged exercise compared with controls. A possible mechanistic explanation for this may be due to altered numbers of circulating T-Helper 1 (Th1) cells which are integral in inducing DTH reactions. The cytokines produced by these cells (e.g. IL-2, IFN-γ) signal macrophages to an area after encounter with a recall antigen, thus supporting cell-mediated inflammatory reactions. However, a bout of prolonged and strenuous exercise is associated with a decrease in the percentage of circulating Th1 cells (Steensberg et al., 2001), due to the redistributive effects of exercise on immune cells, and in turn thus acutely suppressing the DTH reaction in the skin.

Interestingly, a 2015 randomised-controlled trial (Diment et al., 2015) investigated how exercise affected the immune response prior to sensitisation with the hapten diphenylcyclopropenone (DPCP), which is thought to induce a primary – rather than recall – response. It was found that individuals who completed high-intensity short-duration exercise (30 minutes at 80%

VO_2 peak) 20 minutes prior to sensitisation with DPCP demonstrated similar skin reactions at the challenge phase to those that performed moderate-intensity prolonged-duration exercise (2 hours at 60% VO_2 peak). Skin reactions to these exercise bouts were comparable to that of the control group (i.e. rest/no exercise). This finding supports other studies showing that a lower exercise dose produces little to no effect on DTH responsiveness (Nieman & Wentz, 2019; Bruunsgaard et al., 1997). Contrastingly, the skin reaction in the study by Diment and colleagues was significantly impaired among those that completed high-intensity prolonged-duration exercise (2 hours at 80% VO_2 peak). Interestingly, eliciting a positive response would have required a dose of DPCP 4.4 times higher than the dose used with controls, thus showing the extent to which the skin reaction was temporarily impaired. The mechanism responsible for the apparent suppression of responses to DPCP is not well understood, in part because the molecular effects of DPCP in the skin are still being understood (Gulati et al., 2014). However, the mechanism may be similar to that involved in the suppression of recall responses, whereby exercise invokes a redistribution of immune cells to different tissue sites, thus leading to functional 'impairment' in other sites, such as the skin.

Effects of exercise training on DTH reactions

The impact of regular exercise on DTH responses has not been studied extensively. The effects of regular exercise on DTH reactions in humans have more often been studied in older individuals as a means of determining whether an exercise intervention or even simply being habitually physically active can offset some of the immune function abnormalities seen with ageing. DTH reactions to antigens including keyhole-limpet haemocyanin (KLH) and a purified protein derivative of tuberculin have been studied in older participants who regularly exercise (Okutsu et al., 2008; Smith et al., 2004). In one such study, physically active or sedentary, younger (20–35 years old) or older (60–79 years old) men received KLH immunisations. As expected, it was found that KLH reactivity was mostly reduced in ageing, yet the physically active older group had significantly higher DTH responses – and anti-KLH IgM and IgG responses – compared with the sedentary older group (Smith et al., 2004).

In a randomised-controlled trial (Chin A Paw et al., 2000), the effects of exercise training on DTH responsivity to seven antigens were assessed: tetanus and diphtheria toxoid, streptococcus, tuberculin, candida, proteus, and trichophyton. It was found that DTH responsiveness declined over a 17-week period among those who did not exercise; however, DTH responsiveness remained unchanged in those conducting moderate-intensity resistance training. As mentioned earlier, the DTH reaction is driven by Th cells. During ageing, there is a decline in naive T cells and an accumulation of oligoclonal memory T cells, resulting in a compromised ability to deal with novel antigens. As outlined in Chapter 8, regular exercise may help to preserve this ratio, thus augmenting the ability of T cells to respond to – and thereafter maintain responsivity to – novel antigens to produce DTH reactions.

Wound healing

Wound healing is a multi-phase process that begins within minutes following an injury and can last for several weeks or months to years. Each phase requires distinct contributions from the immune system. The major phases of wound healing are:

Haemostasis: there is a cessation of bleeding, usually within minutes of injury. Platelets release signalling molecules that attract immune cells to the site. Inflammation: occurs within several hours to several days following injury. Neutrophils and monocytes are recruited to the site, and resident immune cells including macrophages are activated. Macrophage phenotype switching is critical for the phases of healing, with pro-inflammatory M1 macrophages switching towards M2 phenotypes during later stages of healing. Proliferation: injured cells are removed, mediated in part by the macrophages. The body starts to rebuild the damaged tissues. This involves the proliferation of epithelial cells, migration of keratinocytes, angiogenesis, and granulation tissue formation. Remodelling: scar development occurs in this phase. Collagen fibres are re-organised and remodelled, increasing the strength of the wound. Overtime, the wound becomes less visible. This phase may last for several months to years. Wounds can be classified as acute or chronic. Acute wounds relate to an incision, excision, burn, or suction (blister). Chronic wounds are developed as a consequence of health conditions such as diabetes or malnutrition. These are often challenging to study as they rely on the development of models that can reproducibly replicate the chronicity associated with impaired healing (Stephens et al., 2013).

Wound-healing models are used to enhance understanding of the tissue repair process and test new protocols or treatments for wound healing. These studies are intricate due to the complexity of the healing process and the multifactorial nature of the wound environment. There are several invasive and non-invasive techniques used to quantify the progressive changes in wounds during healing. The most appropriate assessment will vary according to the wound, and it is determined by healing pathophysiology, factors delaying healing, and optimal wound-bed conditions for therapeutic effectiveness (Masson-Meyers et al., 2020). *In vivo* wound models are typically favoured as they permit the study of a functioning immune system and provide a more realistic representation of the wound environment. Since rodents are easy to house and maintain, they are typically used despite distinct differences in skin structure and physiology as well as wound-healing processes. These differences must be considered when assessing the translational value of these models (Nauta et al., 2013).

Effects of individual exercise bouts on wound healing

Short-term stress and individual bouts of exercise both have the potential to augment wound healing through an improved immune response. An exercise bout is known to increase blood flow, enhance collagen synthesis, improve oxygenation, modulate growth factors, and promote angiogenesis. Each of these are essential components of wound healing. Furthermore, exercise triggers the release of pro-inflammatory cytokines such as IL-15 which aid in the early stages of healing. As such, it is speculated that each bout of exercise plays a vital role in the acceleration of wound healing (Wong et al., 2019). Similar to the DTH model, the timing of an exercise bout following wounding allows researchers to better understand the unique contribution of exercise to an individual phase of wound healing.

The available evidence is predominantly from animal research. There is variety in the types of exercise used among rodent models, with the length of each bout ranging from 30 to 60 minutes. The duration of interventions and timing around wounding also varies, with studies commencing exercise 3 days prior to wounding and continuing for between 3 and 10 days post-wounding. Despite key differences, these models have established that moderate-intensity exercise commenced prior to wounding and continued in the days following accelerated healing (Kawanishi et al., 2022; Pence et al., 2012; Keylock et al., 2008). However, this may be influenced by factors, such as age, sex, and obesity, and these effects may be associated with the role of macrophage phenotype switching. In obesity and in ageing, M1 phenotypes are associated with chronic inflammation, while exercise is found to induce accelerated switching to M2 phenotypes and that acceleration is associated with increased rates of healing (Kawanishi et al., 2010; Kawanishi et al., 2022). Overall, exercise has been shown to exert its greatest influence on the healing rate in the early stages of the healing process (days 1–5 post-wounding), corresponding to the inflammatory phase.

Evidence for acute wounds in humans is limited and mostly confined to surgical wounds, likely due to ethical considerations surrounding deliberate wounding in human models. A recent study among postpartum women demonstrated that Kegel exercise in the first week following childbirth accelerates wound healing (Gustirini et al., 2020). The exercise group was instructed to contract the muscles as if holding urine for 5 seconds and then relax for 5 seconds, repeating five times and completing the set five times per day (25 contractions per day), after 7 days perineal wound assessment indicated healing was accelerated by exercise, 80% of the treatment groups had healed by the 8th day compared with 33% of the control group. Contrastingly, earlier research in post-operative wounds did not find any measurable improvement in wound healing at one week following total hip replacement with early post-operative exercise; however, the authors proposed that the intensity of exercise was insufficient to induce significant changes at the wound site (Whitney & Parkman, 2004). Further research in human models is necessary to better understand wound-healing responses to acute exercise.

Effects of exercise training on wound healing

Regular exercise training applied before wound healing can be used to study the influence of exercise on all subsequent phases of the wound healing process. The evidence available demonstrates that regular exercise results in faster wound healing, and typically this has been shown in aged animals or older humans, or in the context of chronic disease. One compelling randomised-controlled trial used an acute punch biopsy wound model and incorporated three months of cycling training at 70% maximum heart rate (cycling 30 min plus 15 min walking/jogging or arm cranking, 3 × per week). Participants underwent an acute wound procedure (3.5 mm on the back of the non-dominant upper arm) after one month of training and continued the regular exercise programme for a further two months. Wounds were healed completely in the participants in the exercise group after 29 days, significantly faster than the 39 days observed in the control group (Emery et al., 2005). In the chronic wound-healing setting, the effects of exercise on wounds in type II diabetes have been investigated. Type II diabetes is a chronic health condition that commonly includes wounds, indeed diabetic foot ulcers affect up to 30% of people with diabetes. These wounds are complicated by the hyperglycaemic and hypoxic wound environments which result in increased inflammatory signals and inhibit the switching of macrophages to M2 phenotypes. Exercise is also challenging in that weight-bearing has a negative impact on diabetic foot ulcer healing. However, exercise has been demonstrated to be remarkably effective in increasing healing rates. In one study of supervised cycling (three times a week for 12 weeks progressively increasing intensity from 60% to 85% maximum heart rate, and increasing duration up to 50 min), 61% of exercise participants had achieved wound closure, compared with 3% in the control group (Nwankwo et al., 2014).

In terms of optimising the wound-healing responses from exercise, one animal study has compared the effect of exercise intensity on wound healing. Mice ran for 45 min 5 times a week for 8 weeks prior to the wounding procedure; results demonstrated that compared with controls who trained only 5 min per session, moderate-intensity training animals had better wound closure and enhanced re-epithelialisation seven days after wounding, while high-intensity training mice had worse wound closure but enhanced re-epithelialisation, and severe-intensity training mice had worse wound closure (Zogaib & Monte-Alto-Costa, 2011). Thus, it is likely that an optimal dose of exercise exists, and further research is needed to understand the possible mechanisms at play.

Vaccination

Vaccination traditionally refers to the injection of a dead or attenuated pathogen to induce an adaptive immune response and protect against disease. Advances in vaccine research have allowed for the development of many types of vaccines, including messenger RNA (mRNA) vaccines, subunit, recombinant, polysaccharide, and conjugate vaccines, toxoid vaccines, and viral vector vaccines. The immune response to vaccination involves the coordination of various cells, including antigen-presenting cells, T cells, and B cells. Responses are most commonly quantified by antibody titre or concentration. However, the assessment of immune responses to vaccination may vary by vaccine type as viral vector vaccines and mRNA vaccines induce significant T-cell immunity along with the more commonly measured serum antibody. The time course of antibody production and qualitative response is different for an initial (primary) vaccination as compared with a secondary immunisation which can be a booster vaccine. The earliest immunoglobin (IgM) peaks approximately five-days following primary vaccination. High-affinity IgG and IgA antibodies then reach a peak approximately 28 days post-vaccination. Upon subsequent exposure to the antigen (secondary immunisation or exposure to infectious pathogen), these antibody responses are more rapid and of greater magnitude and are primarily IgG antibodies in serum. Many studies assess antibody quantity; however, the functional capacity of antibody may also be assessed by determining neutralising antibody which represents the ability of the antibody to prevent infection. Immunological memory is essential for the success of vaccines as both memory T cells and B cells remember the antigen and produce a rapid and robust immune response on secondary exposure to antigens. Therefore, vaccination models may be used to evaluate B-cell antibody production and/or T-cell mediated immunity as well as an assessment of primary or secondary recall response.

The vaccination model provides a means of examining associations between behavioural or psychosocial variables and the *in vivo* response of the immune system to antigenic challenge, in contrast to the more localised tissue models of wound healing and DTH, the vaccination model is seen as a whole organism system model. The ethical considerations weigh favourably for this model; low risk of injury or illness, especially when using therapeutically approved vaccines and benefit to participants in providing the protection from disease the vaccine affords. A higher antibody response to vaccination is considered a general indicator of a well-functioning immune system. More specifically, it is likely associated with greater protection against the particular disease in question. T cell mediated immunity to vaccination may also have an important role in protection, and the degree of protection from T cells may vary by pathogen or specific population receiving the vaccine (Wherry & Barouch, 2022). For example, in older adults, T cell immunity has been shown to serve as a better correlate of protection than serum antibody (McElhaney et al., 2006). Therefore, the strength of the immune response to vaccination (either antibody or T cell immunity) has direct clinical relevance, both in terms of maximising protection and as a model of the response to naturalistic infection (Burns et al., 2007).

Effects of individual bouts of exercise on vaccination

Vaccine studies, with their safety and acceptability for participants, as well as the clinical utility of intervention-induced changes in response have been the most commonly used methodology for examining the effect of exercise on immune response to challenge. Since 2006, at least 17 studies (randomised-controlled or controlled trials) have reported the effects of exercise bouts on the response to vaccinations in human studies. This data includes different vaccines (most commonly influenza, but also including human papillomavirus, meningococcal, pneumococcal, tetanus, diphtheria, and, recently, COVID-19 vaccines) and different populations (most data found in young healthy adults, with athletes, older adults, and young adolescents also included). Although many (approximately 65%) of these studies individually show positive effects in some aspect, Cochrane meta-analyses on more homogeneous subsets of studies showed no overall effect of exercise on responses to influenza vaccines (Grande et al., 2016). However, a more recent individual participant data meta-analysis showed greater antibody responses when vaccinated after an individual bout of exercise (Bohn-Goldbaum et al., 2022). Nevertheless, there is insufficient data to understand why this happens, due to differences in protocol and design. For example, approximately half of the studies have used aerobic exercise and half, resistance exercise, and intervention intensity and duration also vary considerably. Most studies have used moderate-intensity protocols in healthy populations; three studies have used athletes and high-intensity acute exercise bouts, with two finding greater antibody responses with vaccination immediately after marathon/ironman completions.

There has been some success in testing theories that may explain why exercise has variable efficacy. One such hypothesis posits that the enhancements seen with an individual bout of exercise are mostly seen in the responses that are the weakest – suggesting a ceiling effect that exercise has little effect on boosting strong responses. For example, younger adults – in whom immune responses are often strong, and thus at the 'ceiling' – may not benefit from exercise-induced enhancement of immune responses. This was tested in a randomised-controlled trial which randomised participants to either a 'strong' dose (using a full dose of vaccine) or a 'weak' dose (using a half dose vaccine). It was found that those who completed the exercise intervention had significantly larger antibody responses than those who rested in the 'weak' response group, but responses were similar (between exercisers and controls) in the 'strong' response group (Edwards et al., 2012). Separately, it has also been shown that the timing and duration of the exercise intervention are very important. For example, Hallam and colleagues recently showed that 90 minutes (but not 45 minutes) of low- to moderate-intensity aerobic exercise completed after inoculation enhanced responses to influenza and COVID-19 vaccinations (Hallam et al., 2022). Interestingly, they also noted that in older participants (> 44 years) in whom the control response was weaker than younger counterparts, the enhancement was greater, in line with Edwards' reduced dose findings. This trial was notably different in intervention timing, as almost all other trials in this research area have administered exercise before the injection.

To date, the majority of studies have recruited young healthy adults; only three studies have included older adults with all three finding no effect of exercise, including the post-vaccination (45 min) protocol. Despite the hypothesis that exercise enhancements might be most effective when control responses are weak, these studies recruiting older adults have found no effect of exercise. Nevertheless, it may be that individual bouts of exercise are insufficient to enhance responses in older adults, due to declining immune competency in ageing, termed immunosenescence. Indeed, it may be that longer-term exposure to exercise training, which is known to alter features of immunosenescence (Chapter 8) may instead be needed to augment immune responses to vaccination in older adults. Nevertheless, future studies will no doubt teach us more about the potential for designing exercise tasks to improve vaccine-induced protection in at-risk populations and/or for allowing dose-sparing regimens for public health efficacy.

Effects of exercise training on vaccination

Different approaches have been used to assess the impact of exercise training on vaccine responses in human studies. Many studies have employed a cross-sectional comparison approach between individuals based on their level of cardiorespiratory fitness or based upon reporting of physical activity. The majority of cross-sectional studies have examined the antibody response to influenza vaccination. Given that influenza vaccination is typically administered annually, it is relatively straightforward to find participants planning to receive the vaccine. Very recent studies have evaluated antibody responses to COVID-19 vaccination. It is expected that future studies will also focus on the immune response to COVID-19 vaccine as it is likely to be administered frequently. Across many studies, higher levels of physical activity or greater cardiorespiratory fitness are associated with either higher antibody concentrations and/or slower decline of antibody over time in response to influenza or COVID-19 vaccine. These findings generally hold across populations including older adults (Bachi et al., 2013; De Araújo et al., 2015; Keylock et al., 2007; Kohut et al., 2002; Schuler et al., 2003; Silva et al., 2023), as well as younger elite athletes who showed higher neutralising antibody after influenza vaccination (Ledo et al., 2020). Although influenza vaccination does not likely represent a primary immune response given the likelihood of previous infection or prior vaccination, there is evidence to suggest that higher physical activity is also associated with enhanced antibody response to a novel antigen (Smith et al., 2004). The benefits of physical activity on vaccine response also extend to other populations, as two recent studies demonstrate that in immunocompromised individuals, greater physical activity was associated with higher antibody titre, higher neutralising antibody, and a slower decline in serum antibody over time following COVID-19 vaccination (Gualano et al., 2022a; Gualano et al., 2022b). Maintenance of serum antibody levels to pneumococcal serotypes, whether due to natural infection or vaccination, was also higher in older adults with greater physical activity (Whittaker et al., 2023). It is worth noting that whereas most vaccine studies have evaluated antibody response, Wong et al. recently found that greater physical activity in older adults was associated with greater vaccine-specific IgG-secreting B cells assessed by ELISpot assay (Wong et al., 2019). Evidence also shows an increased T cell response to vaccination in more highly fit or active young adults (Ledo et al., 2020) and older adults (Smith et al., 2004). With newer vaccine formulations that induce significant T cell-mediated immunity, it is expected that the impact of exercise on protective T cell function will be studied more thoroughly in the near future. Although antibody response serves as a correlate of vaccine efficacy, the ability of a vaccine to truly protect against severe disease requiring hospitalisation or mortality could be considered the most important efficacy endpoint. In order to assess vaccine efficacy in this manner, a large number of participants are required. Yet, promising findings from a recent study demonstrated that greater physical activity was linked to greater vaccine efficacy in terms of reduced risk of COVID-19-related hospital admission using a test negative case–control study design (Collie et al., 2023).

In contrast to cross-sectional comparisons, several randomised-controlled trials have been performed in which participants were assigned to an exercise intervention or control treatment, and vaccine responses were measured before, after, or during the intervention. Regular moderate-intensity aerobic exercise and Taiji and Qigong exercises practiced over several months increased antibody response to the influenza vaccine or resulted in higher antibody levels over time (Felisimo et al., 2021; Kohut et al., 2004; Woods et al., 2009; Yang et al., 2007). When vaccines contained a novel antigen to assess primary antibody response, regular exercise training resulted in a similar benefit of increased vaccine-specific IgM and IgG1 antibody (Grant et al., 2008). Evidence also supports the possibility that T cell immune response to vaccination may also be improved by regular exercise training (Irwin et al., 2007; Kohut et al., 2005). It is worth noting that not all studies have consistently found improved immune response to vaccination (Long et al., 2013; Hayney et al., 2014). While the reasons for the inconsistent findings are unclear, longer-term interventions may be necessary, as positive findings have been observed with five or more months of exercise. A specific intensity of exercise may also be required for

significant improvements. Future studies will be needed to determine whether a specific duration of training or intensity of exercise is necessary to improve immune response to immunisation, but most evidence suggests that regular exercise has positive benefits.

Infectious challenge models

Studies involving immune challenge with an infectious pathogen have important clinical relevance, as protection from infection or a reduction in the severity of infection are desirable
health outcomes. These models can be viewed as the closest to translation as they involve
exposure to a pathogen in a controlled setting in order to derive information about the pathogenesis of disease and evaluate preventative or therapeutic approaches. This model permits
observation of the life cycle of a virus in its entirety (Lambkin-Williams et al., 2018). For
ethical reasons, these studies are typically conducted in animal models rather than human
participants. However, a few studies exist in which humans were randomly assigned to exercise treatment during an infection or after virus administration. The use of an animal infection
model can allow the researcher to address many questions with respect to the role of exercise.
For example, measures of morbidity or mortality have obvious clinical relevance and provide evidence concerning the possibility that exercise may protect against infectious disease.
Other measures, such as pathogen load, immune response across different tissues, or specific
immune functions, may also be studied, which improves the overall understanding of how
exercise may be protective. The type of pathogen can be varied, such that it is possible to learn
whether exercise is protective against broad pathogen categories, (i.e., bacteria, viruses, fungal, or parasites), or if the effects of exercise are specific for each specific pathogen. The route
of pathogen entry (respiratory, gastrointestinal, genitourinary) can also be modified to better
understand if exercise impacts the first line of host defence in a tissue-specific manner. The
timing at which immune responses or other markers of infection are measured in relation to
the time of infection can allow one to identify exercise effects on specific immune parameters
that may be critical at a certain stage of disease. For example, if pathogen load (number of
viral particles, bacterial counts, etc.) is measured within the first couple of days of infection,
any exercise-related effects likely reflect innate immunity host defences. However, if assessed
later in the course of infection, pathogen load may reflect the ability of the adaptive immune
system to eliminate the pathogen, which is important in clearing infection. The effects of exercise on infectious disease have generally been studied as either exercise bouts, administered
just before or during infection, or the infectious disease challenge follows a period of exercise
training. In models where an individual bout of exercise is examined, many researchers have
administered heavy, prolonged, intense exercise to determine whether very stressful exercise
could have negative effects at or near the time of infection. In contrast, exercise-training studies have generally examined moderate-intensity exercise over some period (weeks/months)
after which animals are infected with a pathogen to determine if regular exercise could confer
protective benefits.

Effects of individual exercise bouts on infectious challenge

In general, if strenuous or prolonged exercise is administered immediately prior to infection or
during the first several days of infection, exercise is detrimental to the host. Although exercise
protocols vary, the typical amount of exercise resulting in negative consequences is between
75 minutes and 3 or more hours of vigorous exercise. These findings have been observed

with viruses that were administered through different routes of administration, including respiratory, intraperitoneal, intracerebral, and genitourinary tract. Morbidity and mortality were significantly increased or trended towards an increase when vigorous exercise occurred immediately prior to infection or for several days after infection across multiple pathogens, including herpes simplex virus type 1 (HSV-1) (Brown et al., 2004; Davis et al., 1997), influenza (Ilbäck et al., 1984; Lowder et al., 2005; Murphy et al., 2008), coxsackievirus (Gatmaitan et al., 1970; Kiel et al., 1989), or HSV-2 (Adachi et al., 2021). The amount of exercise that resulted in increased severity of infection (morbidity or mortality) was substantial. For example, in one study involving intranasal infection Influenza A PR8/34 (H1N1) infection, mice completed 2.5 hours of treadmill running at 8–12 metres/minute daily starting on the first day of infection and continuing through day three post-infection. This type of exercise increased the morbidity of infection (Lowder et al., 2005). In another study of intranasal infection with Influenza A/Aichi2/68 (H3N2), mice completed 50 minutes of forced swimming exercise also starting on the day of infection and continuing for six days. Lethality from the virus was greater in mice that completed the vigorous swimming exercise (Ihlback et al., 1984). Although both studies noted increase severity of disease or mortality from influenza infection by an intranasal route, it is possible that effects of exhaustive exercise are pathogen-dependent or route-dependent, given that the investigators using the forced swim exercise model did not find increased mortality from infection with the bacterial pathogen *Francisella tularensis* delivered by the intraperitoneal route (Ilbäck et al., 1984). Other findings suggest that even a single session of exhaustive exercise consisting of 2.5–3.5 hours of treadmill running at gradually increasing speeds until voluntary exhaustion, performed just prior pathogen challenge with mouse adapted HSV-1 virus (intranasal route), also increased mortality and morbidity from infection (Davis et al., 1997). Immune alterations associated with this type of exercise and infection are reduced antigen-specific IFN-γ and IL-2, decreased ability to produce IL-12 and impaired NK cytotoxicity, all of which are important responses against viral challenge. The immunosuppressive responses persisted two days post-exercise but were no longer present at one-week post-exercise (Kohut et al., 2001). It is worth noting that in one study, by altering the timing of vigorous exercise in relation to pathogen challenge (8 hours post-exercise in contrast to 18 hours), the negative effects of exercise were attenuated (Adachi et al., 2021). In addition to higher viral loads with vigorous exercise, other negative consequences included increased severity of myocarditis from coxsackie virus infection (Ilbäck et al., 1989; Kiel et al., 1989). Another potential confounder with respect to the potential immunosuppressive effects of vigorous exercise are other environmental, physical, or psychological stresses that may accompany the exercise. As a classic example, one study completed many years ago found that non-human primates exercised to fatigue by swimming exercise showed increased severity of polio virus infection (Levinson et al., 1945). Yet, a water immersion comparison group included in the same study also showed increased severity of infection leading investigators to consider the possibility that chilling due to water could contribute to disease severity. However, in testing the effect of chilling alone, animals were restrained in the water (Levinson et al., 1945) and restraint stress is well-recognised in promoting greater severity of infection across pathogens (Padgett et al., 1998; Campbell et al., 2001; Belay et al., 2017). Therefore, although animal models have advantages with control over exposure to other potential stressors, unlike human studies, it is extremely important to consider the mode of exercise and the extent to which additional stressors may confound the effects of exercise alone.

In contrast to negative consequences, some research shows that light- or moderate-intensity exercise (30–45 minutes) may minimise mortality or morbidity. Two models have been used to address this question, (1) exercise applied either once or for several days prior to infection, or (2) exercise applied within the first few days of infection. With respect to the first scenario, a single session of light to moderate treadmill running for 45 minutes just before viral challenge (Davis et al., 1997; Sim et al., 2009), or 6 days of 60-minute moderate-intensity exercise, including a final exercise session before virus (Murphy et al., 2004) reduced morbidity or mortality of HSV-1 or influenza infection, including a reduction of lung viral load and inflammatory cytokines or chemokines (Sim et al., 2009). When exercise was administered at a light intensity (estimated 55–65% of VO_2 max) for 30 minutes on the day of the influenza infection and for three subsequent days, mortality was decreased, a trend towards reduced viral load, and a shift in the lung cytokine and chemokine profile along with reduced lung cellularity, suggestive of reduced inflammation were found (Lowder et al., 2005; Lowder et al., 2006). One interpretation of the existing data is that light or moderate exercise stress in the context of infection may have some benefits that are likely pathogen-dependent, whereas exhaustive exercise exacerbates morbidity and increases the risk of mortality. Together, these findings support the J-curve model in which moderate exercise may attenuate severity of infection whereas exhaustive exercise may increase likelihood of a more severe disease outcome (Nieman, 1994) (see Chapter 9). However, the negative effects of stressful exercise may depend on the timing at which exercise is administered in relation to infectious challenge (Adachi et al., 2021) and may further vary by type of pathogen. In addition, potential confounders including other stressors (psychosocial stress, environmental stress) may influence the risk or severity of disease. With a better understanding of mechanisms, it may be possible in the future to predict whether a certain type of exercise during a given type of infection may be beneficial or detrimental.

Human studies are very limited, yet in one interesting study, participants were infected with a type of cold virus (rhinovirus) and either exercised moderately over 10 days (every other day for 40 minutes) or did not exercise (Weidner et al., 1998). There were no differences in symptom scores between the two groups, suggesting that exercise did not increase nor attenuate symptoms. Other measures of viral load or immune response were not measured in the study, so it is difficult to conclude whether this type of moderate exercise had immune benefits, but there did not appear to be a worse outcome from infection. It is also difficult to extrapolate from these findings to other more severe infections. Based on case study reports, one concern of exercise during certain infections is the risk of myocarditis, which arose during the COVID-19 pandemic. Although this question remains under investigation, there are reports of myocarditis or cardiac involvement in athletes infected with COVID-19, but thus far, it appears that athletes do not experience higher rates of myocarditis or cardiac complications (Daniels et al., 2021; Moulson et al., 2021; Symanski et al., 2022). In older literature, participants who were already infected with hepatitis were assigned to exercise or no exercise at a time point at which symptoms were minimal or in the early recovery phase. Even though exercise was strenuous in one of these studies, there were no differences in the duration of infection, recovery, or measures of liver function (Repsher & Freebern, 1969; Edlung, 1971). However, the findings from these studies may also be complicated by confounding factors as one study evaluated members of the military who had been in the Vietnam War and who likely experience other stressors, and participants in the second study included narcotics users. One additional limiting factor is that infection was not controlled as participants were already infected with hepatitis and may have been at different stages in the disease course, which may limit the conclusions that can be drawn. In summary, data concerning acute infection and exercise in humans is largely limited to case reports, and some evidence suggests that strenuous exercise during certain types of infections may lead to complications, including myocarditis (Halle et al., 2021).

Effects of exercise training on infectious challenge

Epidemiological studies in humans provide some evidence that regular moderate exercise may be associated with fewer or less severe infections, but evidence learned from randomised-controlled trials in which an infectious pathogen is administered instead relies on animal models. The potential protective effect of exercise has been evaluated in animal models of viral, bacterial, and parasite infections. Across most studies, animals undergo a moderate exercise-training programme ranging from two weeks to several months, after which time animals are infected with a pathogen. In most cases, animals rest for several hours to a couple of days after the last exercise session before being infected with a pathogen to minimise any short-term acute effects of an exercise bout. Across multiple studies involving a range of pathogens, findings suggest that exercise reduces disease severity and often limits pathogen load. For example, morbidity or mortality from influenza infection is attenuated by exercise training, and lower amounts of virus are found in the lungs of exercised mice (Ilbäck et al., 1984; Sim et al., 2009; Warren et al., 2013). In two studies, mice ran on a treadmill at a moderate intensity for 45 minutes, 5 days per week for 2–3 months. After the last exercise session, mice rested for 24 hours and were then infected with influenza virus. In both studies, exercise reduced infection morbidity and viral load in the lungs up to day 5 post-infection. Immune alterations induced by exercise training were suggestive of reduced inflammation including changes in cytokines or chemokine profiles, cell infiltration, and tissue damage assessed by histopathology (Sim et al., 2009; Warren et al., 2013). Furthermore, the same protective effects of exercise against influenza infection were found in a diet-induced obesity model in which high-fat diet-treated mice ran on a treadmill for 8 weeks, 5 days per week, 45 minutes per session at approximately 65–80% of VO_2 max (Warren et al., 2015). Interestingly, exercise was protective, but the mechanism differed in high-fat diet-induced obese mice as compared with non-obese mice. It is recognised that obesity has negative consequences for several respiratory diseases including influenza and COVID-19, and the findings that exercise may attenuate the consequences of obesity may have relevance for a significant portion of the population that may be overweight or obese.

Concerning bacterial pneumonia infection (*Streptococcus pneumoniae* and *Pseudomonas aeruginosa*), exercise training also attenuated morbidity and reduced bacterial load or colonisation, with these protective benefits present in aged mice as well (Olivo et al., 2014; Stravinskas Durigon et al., 2018). In these studies, mice ran on a treadmill 5 days per week, 60 minutes per day, at approximately 50% of estimated maximal capacity for 4 or 5 weeks. Mice were rested after the last exercise session for either 24 or 72 hours before infection with the bacterial respiratory pathogens. Exercise training by voluntary wheel running for less than three weeks also conferred greater survival in mice infected with another type of bacteria, *Salmonella typhimurium* (Cannon & Kluger, 1984). Parasite infection models have also been tested to determine whether exercise training would attenuate the severity of infection. The parasites that have been studied are. *Toxoplasma gondii*, which causes the disease toxoplasmosis, or *Trypanosoma cruzi*, which causes Chagas disease. The results indicate that survival and recovery from these parasite infections are improved by exercise training. Furthermore, exercise reduces parasite load (Bortolini et al., 2016; Chao et al., 1992; Moreira et al., 2014; Lucchetti et al., 2017; Schebeleski-Soares et al., 2009). Tissue damage from these infections was also attenuated by exercise. Taken together, the evidence from animal studies in which pathogen exposure dose and total amount of exercise can be controlled is quite strong. Multiple findings show exercise training induced enhanced protection against viruses, bacteria, or parasites. Interestingly, exercise is beneficial across different types of pathogens, given that the type of host immune defences may vary widely by pathogen type.

Mechanisms of exercise-induced changes to infectious challenge

Animal models of exercise and infection provide the opportunity to measure immune response in tissues that are directly affected by the pathogen or contribute to immune host defence. In this section, the discussion is confined to experiments in which an infectious pathogen was administered *in vivo*, followed by an assessment of immune response. With respect to individual bouts of vigorous exercise that result in increased morbidity or mortality from infection, studies are limited, but suggest that host defences may be compromised in a pathogen-specific manner and may also show evidence of more general immunosuppression. For example, in an animal model of respiratory viral infection with HSV, alveolar antiviral resistance function was impaired by strenuous exercise, along with reduced production of HSV-specific Th1 cytokines as measured in the spleen (Davis et al., 1997; Kohut et al., 2001). The macrophage antiviral resistance contributed to protection, and when a nutritional supplement (oat-β glucan) was provided to mice before infection, alveolar macrophage viral resistance was restored, leading to improved morbidity/mortality from infection (Murphy et al., 2009). In addition, the exercise-associated increase in catecholamines was responsible for the impairment of macrophage antiviral function (Kohut et al., 1998). An impairment of NK cell cytotoxicity was also found, persisting for at least two days post-infection, and may have also contributed to impaired immunity (Kohut et al., 2001). In a separate model of infection in which strenuous, prolonged exercise exacerbated HSV-2 genitourinary infection, plasmacytoid dendritic cells (pDCs) trafficking to bone marrow driven by exercise-induced glucocorticoids was responsible for the immunosuppressive effects of exercise. These studies suggest that neuroendocrine factors may contribute to acute exercise-induced alterations of immune response after pathogen challenge. Other potential mechanisms likely to be investigated in the near future include exercise-induced changes in metabolism, as activation of immune cells involves metabolic reprogramming which alters metabolic requirements.

In exercise-training studies, several patterns have emerged regarding the potential mechanisms by which exercise confers protection in response to infectious disease. One consistent finding is an anti-inflammatory effect, particularly at the site of infection. With respiratory tract infection, either due to influenza virus or bacterial pneumonia pathogens, there is reduced recruitment of immune cells to the site of infection, including neutrophils and lymphocytes (Lowder et al., 2006; Sim et al., 2009; Olivo et al., 2014; Warren et al., 2015; Stravinskas Durigon et al., 2018). These same studies also show a decrease in pro-inflammatory cytokines at the site of infection and a decrease in chemokines, which is likely linked to the decrease in immune-cell recruitment. Some studies show that exercise causes a shift in the cytokine profile that may be associated with improved host defence, as in the case of parasite infection (Bortolini et al., 2016) or may limit inflammation (Lowder et al., 2006). In the normal course of infection, inflammatory responses are important in the early phase of host defence, and so it is interesting that at these early time points post-infection, an exercise-related reduction of inflammatory response appears to be protective. One potential explanation for this seemingly contradictory finding is that exercise may limit pathogen load, reducing requirements for potent immune activation. Evidence supporting this possibility comes from a study showing that exercise training increases the early production of type I interferon and interferon-inducible gene expression, which are mechanisms that contribute to a reduction of viral replication (Warren et al., 2015). In the later phase of infection, a prolonged inflammatory response could be detrimental, delaying recovery from infection and contributing to tissue damage. Yet many of the aforementioned studies mentioned above also show an exercise-induced reduction of inflammatory cytokines or cells in the later phase of infection, accompanied by a trend towards less tissue damage, indicating the beneficial effects of anti-inflammatory processes. Several other immune functions that may contribute to enhanced host defence include increased ciliary beat frequency that can improve mucociliary clearance of influenza virus (in obese animals only) (Warren et al., 2015) and increased antioxidant defence or decreased lipid peroxidation in tissues affected by bacterial or parasite pathogens (Olivo et al., 2014; Stravinskas Durigon et al., 2018; Novaes et al., 2016) which may also limit tissue damage.

With respect to adaptive immunity, antigen-specific responses, including assessment of T cells and serum antibody, have been evaluated after exercise training and infection. Findings show that exercise training decreased the total number of influenza-specific CD8+ T cells in the lungs, but the functional capacity of the cells tended to increase in both obese and non-obese animals (as measured by influenza-specific IFN-γ) (Warren et al., 2015). Serum antibody was increased by exercise training prior to viral infection (HSV, influenza) Warren et al., 2015; Kohut et al., 2001). Other studies also report enhanced antibody responses to antigens after training (see review by Suzuki, 2015). It is important to note that exercise training has also been linked with lower antibody response compared with control that persisted for six months post-infection (Warren et al., 2013). However, these findings may be attributed to a reduction in lung viral load, which limited the requirement for a potent antibody response, given that when the same pathogen was delivered via an intraperitoneal site, in which viral replication did not occur, the exercise-related reduction of antibody was not present. In addition, the amount of antibody produced in exercised mice was protective when mice were challenged with the same pathogen months later. These studies highlight the importance of using *in vivo* infection models, where an increase or decrease of an immune response can be put in the context of overall disease outcome. Therefore, even though reductionist approaches provide important mechanistic infor-mation, it is also essential to test the potential mechanisms in a "whole body" systems-based approach in which a clear clinical outcome can be established.

Summary

In summary, an array of different translational models have been employed in exercise immunol-ogy with the aim of investigating how exercise affects immune competency *in vivo* (Table 7.1). These methods have included the assessment of local tissue-level responses, via DTH responses

Table 7.1 Summary of the effects of single bouts of exercise and exercise training on *in vivo* models of immune response to challenge

	Effects of single bouts of exercise	*Effects of exercise training*
Delayed-Type-Hypersensitivity	↓ Response reduced (found in prolonged, strenuous bouts)	↑ response enhanced (found in aged population)
Wound healing	↑ Increased rate of healing (moderate intensity, effects greatest in early stages of healing)	↑ Increased rate of healing (greatest benefit with moderate-intensity exercise)
Vaccination	↑ ↔ Increased response (mixed results, suggestions for greater effects within overall weaker responses, and exercise immediately after injection)	↑ Increased response (greater peak antibody responses and slower response decline, greater T-cell responses, greater efficacy shown by reduced hospitalisation)
Infectious challenge	↑ ↔ Increased protection (morbidity and mortality reduced by light/moderate-intensity exercise) ↓ Reduced protection (morbidity and mortality increased by strenuous/prolonged exercise)	↑ ↔ Increased protection (reduced disease severity with light/moderate-intensity exercise training)

and wound healing, and also more systemic responses, such as vaccination and challenge via an infectious pathogen. These approaches have been conducted in both animals and humans. Each method has provided unique insight into the role of exercise in altering immune resistance to infection. Across the models summarised here, the pattern of effect emerges that exercise training almost always results in improved immune function, especially when moderate-intensity exercise training is examined. The effects of individual bouts of exercise are more mixed, with reduced immune function found with strenuous, prolonged exercise in some models and improved immune responses with moderate-intensity exercise. Nevertheless, each method adopted has numerous advantages and disadvantages and as such, it remains unclear how exercise alters immune competency *in vivo*, particularly in humans.

Key points

- Studies in exercise immunology have investigated how exercise affects the functional competency of the immune system *in vivo* (i.e., in a living organism, such as a human or an animal) using approaches that include wound healing or methods of immune challenge (e.g., vaccination or experimental infection).
- DTH reactions generally refer to a cell-mediated immuno-inflammatory response that develops 24–72 hours following exposure to an antigen and this model has been used to assess *in vivo* immunity in exercise immunology.
- An individual bout of exercise performed immediately prior to sensitisation with an antigen can lead to a transient suppression of the DTH response, which is speculated as being driven by changes to the profile of Th cells.
- The DTH response is impaired with ageing, but exercise-training, particularly among older adults, has been linked with stronger DTH responses or at least maintenance of this response. It has been speculated that these observations are due to the effects of exercise on the profile of Th cells.
- Wound healing comprises major phases, including haemostasis, inflammation, proliferation, and remodelling, and has been used in human and animal studies examining individual bouts of exercise (30–60 minutes) and taking place 3 days prior to wounding and continuing for between 3 and 10 days post-wounding.
- Generally, bouts of moderate-intensity exercise commenced prior to wounding and continued in the days following accelerated healing, with the greatest effects in the early stages of the healing process (days 1–5 post-wounding), corresponding to the inflammatory phase.
- Human and animal studies show that exercise training results in faster wound healing, and typically this has been shown in older animals or older humans, or in the context of chronic disease.
- Vaccination involves the coordination of antigen-presenting T cells and B cells. Responses are most commonly quantified by antibody titres (or concentration) but sometimes also B and T cell function. Studies have examined many different vaccines, including influenza, human papillomavirus, meningococcal, pneumococcal, tetanus, diphtheria, and, recently, COVID-19.
- In the context of individual exercise bouts, around half of all studies show positive effects, and although there is some indication that exercise might help most with "weak" responses (e.g., that might be exhibited by older people with chronic disease), results are mixed, partly due to a variety of study designs. Overall, individual bouts of exercise appear not to improve vaccine responses.
- Cross-sectional studies have shown that high levels of physical activity or high cardiorespiratory fitness are associated with stronger responses to vaccines, including antibody

concentration, B cell function and T cell function, and a slower decline in antibody concentration. These findings are supported by several randomised and controlled trials of exercise training, but not all.

- Infectious challenge models involve the administration of a pathogen, and although some human studies in exercise immunology exist, most are limited to animal models, with outcomes such as morbidity, mortality, and pathogen-specific immune responses by various cells of the immune system.
- If strenuous and prolonged exercise (e.g., 1–3 hours) is undertaken immediately prior to infectious challenge or during the first few days of infection, morbidity and mortality are increased, accompanied by impaired T cell and NK cell responses. If light- or moderate-intensity and shorter duration exercise (e.g. 30–45 minutes) is undertaken in the hours or days prior to infectious challenge, morbidity and mortality are reduced.
- Exercise training for several weeks or months prior to infectious challenge reduces disease severity, limits pathogen load, and reduces overall morbidity and mortality – which has been shown with a variety of pathogens, including viruses (e.g., influenza) bacteria (e.g., *Salmonella typhimurium*), and parasites (e.g., *Toxoplasma gondii*).

References

Adachi, A., Honda, T., Dainichi, T., Egawa, G., Yamamoto, Y., Nomura, T., … Kabashima, K. (2021). Prolonged high-intensity exercise induces fluctuating immune responses to herpes simplex virus infection via glucocorticoids. *J Allergy Clin Immun*o, l, 1588, e7.

Bachi, A. L., Suguri, V. M., Ramos, L. R., Mariano, M., Vaisberg, M., & Lopes, J. D. (2013). Increased production of autoantibodies and specific antibodies in response to influenza virus vaccination in physically active older individuals. *Results Immunol*, 3, 10–16.

Belay, T., Woart, A., & Graffeo, V. (2017). Effect of cold water-induced stress on immune response, pathology and fertility in mice during Chlamydia muridarum genital infection. *Pathog Dis*, 045.

Bohn-Goldbaum, E., Owen, K. B., Lee, V. Y. J., Booy, R., & Edwards, K. M. (2022). Physical activity and acute exercise benefit influenza vaccination response: A systematic review with individual participant data meta-analysis. *PLOS ONE*, 17(6), e0268625.

Bortolini, M. J., Silva, M. V., Alonso, F. M., Medeiros, L. A., Carvalho, F. R., Costa, L. F., … Mineo, J. (2016). Strength and aerobic physical exercises are able to increase survival of toxoplasma gondii-infected C57BL/6 mice by interfering in the IFN-γ expression. *Front Physiol, 7*, 641.

Brown, A. S., Davis, J. M., Murphy, E. A., Carmichael, M. D., Ghaffar, A., & Mayer, E. P. (2004). Gender differences in viral infection after repeated exercise stress. *Med Sci Sports Exerc*, 36(8), 1290–1295.

Bruunsgaard, H., Hartkopp, A., Mohr, T., Konradsen, H., Heron, I., Mordhorst, C. H., & Pedersen, B. K. (1997). In vivo cell-mediated immunity and vaccination response following prolonged, intense exercise. *Med Sci Sports Exerc*, 29(9), 1176–1181.

Burlingham, W. J., Jankowska-Gan, E., Vanbuskirk, A. M., Pelletier, R. P., & Orosz, C. G. (2005). Chapter 35—Delayed type hypersensitivity responses. In M. T. Lotze & A. W. Thomson (Eds.), *Measuring Immunity* (pp. 407–418). London: Academic Press.

Burns, V. E., Phillips, A. C., & Edwards, K. M. (2007). Vaccination. In G. Fink (Ed), *Encyclopedia of Stress* (pp. 813–817). London: Elsevier.

Campbell, T., Meagher, M. W., Sieve, A., Scott, B., Storts, R., Welsh, T. H., & Welsh, C. J. (2001). The effects of restraint stress on the neuropathogenesis of Theiler's virus infection: I. Acute Disease. *Brain Behav Immun*, September, 15(3), 235–254.

Cannon, J. G., & Kluger, M. J. (1984). Exercise enhances survival rate in mice infected with Salmonella typhimurium. *Proc Soc Exp Biol Med*, April, 175(4), 518–521.

Chao, C. C., Strgar, F., Tsang, M., & Peterson, P. K. (1992). Effects of swimming exercise on the pathogenesis of acute murine Toxoplasma gondii Me49 infection. *Clin Immunol Immunopathol*, February, 62(2), 220–226.

Chin A Paw, M. J. M., De Jong, N., Pallast, E. G. M., Kloek, G. C., Schouten, E. G., & Kok, A. F. J. (2000). Immunity in frail elderly: A randomized controlled trial of exercise and enriched foods. *Med Sci Sports Exerc*, 32(12), 2005–2011.

Collie, S., Saggers, R. T., Bandini, R., Steenkamp, L., Champion, J., Gray, G., … Patricios, J. (2023). Association between regular physical activity and the protective effect of vaccination against SARS-CoV-2 in a South African case-control study. *Br J Sports Med*, February, 57(4), 205–211.

Córdoba, F., Wieczorek, G., Preussing, E., & Bigaud, M. (2008). Modeling Of delayed type hypersensitivity (DTH. In the non-human primate (NHP). *Drug Discovery Today: Disease Models,* 5, 63–71.

Daniels, C. J., Rajpal, S., & Greenshields, J. T. (2021). Prevalence of clinical and subclinical myocarditis in competitive athletes with recent SARS-CoV-2 infection: Results from the big ten COVID-19 cardiac registry. *JAMA Cardiol, 6(9),* 1078–1087.

Davis, J. M., Kohut, M. L., Colbert, L. H., Jackson, D. A., Ghaffar, A., & Mayer, E. P. (1997). Exercise, alveolar macrophage function, and susceptibility to respiratory infection. *J Appl Physiol*, 83(5), 1461–1466.

Davison, G., Kehaya, C., Diment, B. C., & Walsh, N. P. (2016). Carbohydrate supplementation does not blunt the prolonged exercise-induced reduction of in vivo immunity. *Eur J Nut*, 55(4), 1583–1593.

De Araújo, A. L., Silva, L. C. R., Fernandes, J. R., Matias, M. D. S. T., Boas, L. S., Machado, C. M., … Benard, G. (2015). Elderly men with moderate and intense training lifestyle present sustained higher antibody responses to influenza vaccine. *AGE*, 37(6), 105.

Dhabhar, F. S. (2000). Acute stress enhances while chronic stress suppreses skin immunity. The role of stress hormones and leukocyte trafficking. *Ann N Y Acad Sci*, 917, 876–893.

Dhabhar, F. S., & McEwen, B. S. (1997). Acute stress enhances while chronic stress suppresses cell-mediated immunity in vivo: A potential role for leukocyte trafficking. *Brain, Behav Imm*, 11(4), 286–306.

Diment, B. C., Fortes, M. B., Edwards, J. P., Hanstock, H. G., Ward, M. D., Dunstall, H. M., … Walsh, N. P. (2015). Exercise intensity and duration effects on in vivo immunity. *Med Sci Sports Exerc*, 47(7), 1390–1398.

Edlung, A. (1971). The effect of defined physical exercise in the early convalescence of viral hepatitis. *Scand J Infect Dis,* 3(3), 189–196.

Edwards, K. M., Pung, M. A., Tomfohr, L. M., Ziegler, M. G., Campbell, J. P., Drayson, M. T., & Mills, P. J. (2012). Acute exercise enhancement of pneumococcal vaccination response: A randomised controlled trial of weaker and stronger immune response. *Vaccine*, 30(45), 6389–6395.

Emery, C. F., Kiecolt-Glaser, J. K., Glaser, R., Malarkey, W. B., & Frid, D. J. (2005). Exercise Accelerates Wound Healing Among Healthy Older Adults: A Preliminary Investigation. *The J Gerontol Series A: Biologi Sci Med Sci*, 60(11), 1432–1436.

Felisimo, E. S., Santos, J. M. B., Rossi, M., Santos, C. A. F., Durigon, E. L., Oliveira, D. B. L., … Bachi, A. L. L. (2021). Better Response to influenza virus vaccination in physically trained older adults is associated with reductions of cytomegalovirus-specific immunoglobulins as well as improvements in the inflammatory and CD8+ T-cell profiles. *Front Immunol*, 12 713–763.

Gatmaitan, B. G., Chason, J. L., & Lerner, A. M. (1970). Augmentation of the virulence of Murine Coxsackie-virus B-3 myocardiopathy by exercise. *J Exp Med*, 1, 131(6), 1121–1136.

Grande, A. J., Reid, H., Thomas, E. E., Nunan, D., & Foster, C. (2016). Exercise prior to influenza vaccination for limiting influenza incidence and its related complications in adults. *Cochrane Database Syst Rev*, 2016(8), CD011857.

Grant, R. W., Mariani, R. A., Vieira, V. J., Fleshner, M., Smith, T. P., Keylock, K. T., … Woods, J. A. (2008). Cardiovascular exercise intervention improves the primary antibody response to keyhole limpet hemocyanin (KLH) in previously sedentary older adults. *Brain Behav Immun*, August, 22(6), 923–932.

Gualano, B., Lemes, I. R., Silva, R. P., Pinto, A. J., Mazzolani, B. C., Smaira, F. I., … Bonfa, E. (2022a). Association between physical activity and immunogenicity of an inactivated virus vaccine against SARS-CoV-2 in patients with autoimmune rheumatic diseases. *Brain Behav Immun*, 101, 49–56.

Gualano, B., Lemes, Í. R., Silva, R. P., Pinto, A. J., Mazzolani, B. C., Smaira, F. I., … Bonfa, E. (2022b). Physical activity and antibody persistence 6 months after the second dose of CoronaVac in immunocompromised patients. *Scand J Med Sci Sports*, October, 32(10), 1510–1515.

Gulati, N., Suárez-Fariñas, M., Fuentes-Duculan, J., Gilleaudeau, P., Sullivan-Whalen, M., Correa Da Rosa, J., … Krueger, J. G. (2014). Molecular characterization of human skin response to diphencyprone at peak and resolution phases: Therapeutic insights. *J Invest Dermatol*, 134(10), 2531–2540.

Gustirini, R., Pratama, R. N., Maya, R. A. A., & Mardalena. (2020). The effectiveness of Kegel exercise for the acceleration of perineum wound healing on postpartum women. *Adv Health Sci Res*, 27, 400–402.

Hallam, J., Jones, T., Alley, J., & Kohut, M. L. (2022). Exercise after influenza or COVID-19 vaccination increases serum antibody without an increase in side effects. *Brain, Behav Immun*, 102, 1–10.

Halle, M., Binzenhöfer, L., Mahrholdt, H., Johannes Schindler, M., Esefeld, K., & Tschöpe, C. (2021). Myocarditis in athletes: A clinical perspective. *Eur J Prevent Cardiol*, 28(10), 1050–1057.

Harper Smith, A. D., Coakley, S. L., Ward, M. D., Macfarlane, A. W., Friedmann, P. S., & Walsh, N. P. (2011). Exercise-induced stress inhibits both the induction and elicitation phases of in vivo T-Cell-mediated immune responses in humans. *Brain, Behav Immun*, 25(6), 1136–1142.

Hayney, M. S., Coe, C. L., Muller, D., Obasi, C. N., Backonja, U., Ewers, T., & Barrett, B. (2014). Age and psychological influences on immune responses to trivalent inactivated influenza vaccine in the meditation or exercise for preventing acute respiratory infection (MEPARI) trial. *Hum Vaccin Immunother*, 10(1), 83–91.

Ilbäck, N. G., Fohlman, J., & Friman, G. (1989). Exercise in coxsackie B3 myocarditis: Effects on heart lymphocyte subpopulations and the inflammatory reaction. *Am Heart J*, June, 117(6), 1298–302.

Ilbäck, N. G., Friman, G., Beisel, W. R., Johnson, A. J., & Berendt, R. F. (1984). Modifying effects of exercise on clinical course and biochemical response of the myocardium in influenza and tularemia in mice. *Infect Immun*, August, 45(2), 498–504.

Irwin, M. R., Olmstead, R., & Oxman, M. N. (2007). Augmenting immune responses to varicella zoster virus in older adults: A randomized, controlled trial of Tai Chi. *J Am Geriatr Soc*, April, 55(4), 511–517.

Jones, A. W., March, D. S., Thatcher, R., Diment, B., Walsh, N. P., & Davison, G. (2019). The Effects of bovine colostrum supplementation on in vivo immunity following prolonged exercise: A randomised controlled trial. *Eur J Nutr*, 58(1), 335–344.

Kawanishi, M., Kami, K., Nishimura, Y., Minami, K., Senba, E., Umemoto, Y., … Tajima, F. (2022). Exercise-induced increase in M2 macrophages accelerates wound healing in young mice. *Physiol Rep*, 10(19), 15447.

Kawanishi, N., Yano, H., Yokogawa, Y., & Suzuki, K. (2010). Exercise training inhibits inflammation in adipose tissue via both suppression of macrophage infiltration and acceleration of phenotypic switching from M1 to M2 macrophages in high-fat-diet-induced obese mice. *Exerc Immunol Rev*, 16, 105–118.

Keylock, K., Todd, Lowder, T., Leifheit, K. A., Cook, M., Mariani, R. A., Ross, K., … Woods, J. A. (2007). Higher antibody, but not cell-mediated, responses to vaccination in high physically fit elderly. *J Appl Physiol*, 102(3), 1090–1098.

Keylock, K. T., Vieira, V. J., Wallig, M. A., Dipietro, L. A., Schrementi, M., & Woods, J. A. (2008). Exercise accelerates cutaneous wound healing and decreases wound inflammation in aged mice. *Am J Physiol. Regul, Integr Comp Physiol*, 294(1), 179–184.

Kiel, R. J., Smith, F. E., Chason, J., Khatib, R., & Reyes, M. P. (1989). Coxsackievirus B3 myocarditis in C3H/HeJ mice: Description of an inbred model and the effect of exercise on virulence. *Eur J Epidemiol*, September, 5(3), 348–350.

Kohut, M. L., Arntson, B. A., Lee, W., Rozeboom, K., Yoon, K. J., & Cunnick, J. E. (2004). Moderate exercise improves antibody response to influenza immunization in older adults. *Vaccine*, 22(17–18), 2298–2306.

Kohut, M. L., Boehm, G. W., & Moynihan, J. A. (2001). Prolonged exercise suppresses antigen-specific cytokine response to upper respiratory infection. *J App Physiol*, 90(2), 678–684.

Kohut, M. L., Cooper, M. M., Nickolaus, M. S., Russell, D. R., & Cunnick, J. E. (2002). Exercise and psychosocial factors modulate immunity to influenza vaccine in elderly individuals. *J Gerontol A Biol Sci Med Sci*, September, 57(9), M557–562.

Kohut, M. L., Davis, J. M., Jackson, D. A., Colbert, L. H., Strasner, A., Essig, D. A., … Mayer, E. P. (1998). The role of stress hormones in exercise-induced suppression of alveolar macrophage antiviral function. *J Neuroimmunol*, January, 81(1–2), 193–200.

Kohut, M. L., Lee, W., Martin, A., Arnston, B., Russell, D. W., Ekkekakis, P., … Cunnick, J. E. (2005). The exercise-induced enhancement of influenza immunity is mediated in part by improvements in psychosocial factors in older adults. *Brain Behav Immun*, July, 19(4), 357–366.

Lai, F., & Chen, Q. (2018). Humanized mouse models for the study of infection and pathogenesis of human viruses. *Viruses*, 10(11), 643.

Lambkin-Williams, R., Noulin, N., Mann, A., Catchpole, A., & Gilbert, A. S. (2018). The human viral challenge model: Accelerating the evaluation of respiratory antivirals, vaccines and novel diagnostics. *Respir Res*, 19(1), 123.

Ledo, A., Schub, D., Ziller, C., Enders, M., Stenger, T., Gärtner, B. C., … Sester, M. (2020). Elite athletes on regular training show more pronounced induction of vaccine-specific T-cells and antibodies after tetravalent influenza vaccination than controls. *Brain, Behav Immun*, 83, 135–145.

Levinson, S. O., Milzer, A., & Lewin, P. (1945). Effect of fatigue, chilling and mechanical trauma on resistance to experimental poliomyelitis. *Am J Hyg*, 42, 204–221.

Long, J. E. P., Ring, C. P., Bosch, J. A. P., Eves, F. P., Drayson, M. T. Mbc., PhD, C., … Burns, V. E. P. (2013). A life-style physical activity intervention and the antibody response to pneumococcal vaccination in women. *Psychosomatic Med*, 75(8), 774–782.

Lowder, T., Padgett, D. A., & Woods, J. A. (2005). Moderate exercise protects mice from death due to influenza virus. *Brain Behav Immun*, September, 19(5), 377–380.

Lowder, T., Padgett, D. A., & Woods, J. A. (2006). Moderate exercise early after influenza virus infection reduces the Th1 inflammatory response in lungs of mice. *Exerc Immunol Rev*, 12, 97–111.

Lucchetti, B. F. C., Zanluqui, N. G., Ataides Raquel, H., Lovo-Martins, M. I., Tatakihara, V. L. H., Oliveira Belém, M., … Martins-Pinge, M. C. (2017). Moderate treadmill exercise training improves cardiovascular and nitrergic response and resistance to trypanosoma cruzi infection in mice. *Front Physiol*, 8, 315.

Masson-Meyers, D. S., Andrade, T. A. M., Caetano, G. F., Guimaraes, F. R., Leite, M. N., Leite, S. N., & Frade, M. A. C. (2020). Experimental models and methods for cutaneous wound healing assessment. *Int J Exp Pathol*, 101(1–2), 21–37.

McElhaney, J. E., Xie, D., Hager, W. D., Barry, M. B., Wang, Y., Kleppinger, A., … Bleackley, R. C. (2006). T cell responses are better correlates of vaccine protection in the elderly. *J Immunol*, 15, 176(10), 6333–6339.

Moreira, N. M., Zanoni, J. N., Oliveira Dalálio, M. M., Almeida Araújo, E. J., Braga, C. F., & Araújo, S. M. (2014). Physical exercise protects myenteric neurons and reduces parasitemia in Trypanosoma cruzi infection. *Exp Parasitol, 141,* 68–74.

Moulson, N., Petek, B. J., & Drezner, J. A. (2021). SARS-CoV-2 cardiac involvement in young competitive athletes. *Circulation*, 144, 256–266.

Murphy, E. Angela, Davis, J. M., Carmichael, M. D., Mayer, E. P., & Ghaffar, A. (2009). Benefits of oat β-glucan and sucrose feedings on infection and macrophage antiviral resistance following exercise stress. *Am J Physiol-Regul, Integr Comp Physiol*, 297(4), R1188–R1194.

Murphy, E.A., Davis, J. M., Brown, A. S., Carmichael, M. D., Rooijen, N., Ghaffar, A., & Mayer, E. P. (2004). Role of lung macrophages on susceptibility to respiratory infection following short-term moderate exercise training. *Am J Physiol Regul Integr Comp Physiol*, December, 287(6), R1354–1358.

Murphy, E. A., Davis, J. M., Carmichael, M. D., Gangemi, J. D., Ghaffar, A., & Mayer, E. P. (2008). Exercise stress increases susceptibility to influenza infection. *Brain Behav Immun*, 22(8), 1152–1155.

Nauta, A. C., Gurtner, G. C., & Longaker, M. T. (2013). Adult stem cells in small animal wound healing models. In R. G. Gourdie & T. A. Myers (Eds.), *Wound Regeneration and Repair: Methods and Protocols* (pp. 81–98). Totowa, NJ: Humana Press.

Nieman, D. C. (1994). Exercise, infection, and immunity. *Int J Sports Med*, October, 15(Suppl 3), S131–141.

Nieman, D. C., & Wentz, L. M. (2019). The compelling link between physical activity and the body's defense system. *J Sport Health Sci*, 8(3), 201–217.

Novaes, R. D., Gonçalves, R. V., Penitente, A. R., Bozi, L. H., Neves, C. A., Maldonado, I. R., … Talvani, A. (2016). Modulation of inflammatory and oxidative status by exercise attenuates cardiac morphofunctional remodeling in experimental Chagas cardiomyopathy. *Life Sci*, 152, 210–219.

Nwankwo, M. J., Okoye, G. C., Victor, E. A., & Obinna, E. A. (2014). Effect of twelve weeks supervised aerobic exercise on ulcer healing and changes in selected biochemical profiles of diabetic foot ulcer subjects. *Int J Diabetes Res*, 3(3), 41–48.

Okutsu, M., Yoshida, Y., Zhang, X., Tamagawa, A., Ohkubo, T., Tsuji, I., & Nagatomi, R. (2008). Exercise training enhances in vivo tuberculosis purified protein derivative response in the elderly. *J App Physiol*, 104(6), 1690–1696.

Olivo, C. R., Miyaji, E. N., Oliveira, M. L. S., Almeida, F. M., Lourenço, J. D., Abreu, R. M., … Martins, M. A. (2014). Aerobic exercise attenuates pulmonary inflammation induced by Streptococcus pneumoniae. *J App Physiol*, 117(9), 998–1007.

Padgett, D. A., Sheridan, J. F., Dorne, J., Berntson, G. G., Candelora, J., & Glaser, R. (1998). Social stress and the reactivation of latent herpes simplex virus type 1. *Proc Natl Acad Sci U S A*, 9, 95(12), 7231–7235.

Paw, C. A., J., M., Jong, N., Pallast, E. G., Kloek, G. C., Schouten, E. G., & Kok, F. J. (2000). Immunity In frail elderly: A randomized controlled trial of exercise and enriched foods. *Med Sci Sports Exerc*, 32(12), 2005–2011.

Pence, B. D., DiPietro, L. A., & Woods, J. A. (2012). Exercise speeds cutaneous wound healing in high-fat diet-induced obese mice. *Med Sci Sports Exerc*, 44(10), 1846–1854.

Repsher, L. H., & Freebern, R. K. (1969). Effects of early and vigorous exercise on recovery from infectious hepatitis. *N Engl J Med*, 18, 281(25), 1393–1396.

Resource Book for the Design of Animal Exercise Protocols. (n.d.). Retrieved from https://www.physiology.org/career/policy-advocacy/animal-research/resource-book-for-the-design-of-animal-exercise-protocols?SSO=Y

Schebeleski-Soares, C., Occhi-Soares, R. C., Franzói-de-Moraes, S. M., Oliveira Dalálio, M. M., Almeida, F. N., Ornelas Toledo, M. J., & Araújo, S. M. (2009). Preinfection aerobic treadmill training improves resistance against Trypanosoma cruzi infection in mice. *Appl Physiol Nutr Metab*, August, 34(4), 659–665.

Schuler, P. B., Leblanc, P. A., & Marzilli, T. S. (2003). Effect of physical activity on the production of specific antibody in response to the 1998–99 influenza virus vaccine in older adults. *J Sports Med Phys Fitness*, September, 43(3), 404.

Silva, B. R., Monteiro, F. R., Cezário, K., Amaral, J. B. D., Paixão, V., Almeida, E. B., … Bachi, A. L. L. (2023). Older Adults who maintained a regular physical exercise routine before the pandemic show better immune response to vaccination for COVID-19. *Int J Environ Res Public Health*, 20(3), 1939.

Sim, Y. J., Yu, S., Yoon, K. J., Loiacono, C. M., & Kohut, M. L. (2009). Chronic exercise reduces illness severity, decreases viral load, and results in greater anti-inflammatory effects than acute exercise during influenza infection. *J Infect Dis*, 1, 200(9), 1434–1442.

Smith, T. P., Kennedy, S. L., & Fleshner, M. (2004). Influence of age and physical activity on the primary in vivo antibody and T cell-mediated responses in men. *J App Physiol*, 97(2), 491–498.

Steensberg, A., Toft, A. D., Bruunsgaard, H., Sandmand, M., Halkjaer-Kristensen, J., & Pedersen, B. K. (2001). Strenuous exercise decreases the percentage of Type 1 T cells in the circulation. *J App Physiol*, 91(4), 1708–1712.

Stephens, P., Caley, M., & Peake, M. (2013). Alternatives for animal wound model systems. In R. G. Gourdie & T. A. Myers (Eds.), *Wound Regeneration and Repair: Methods and Protocols* (pp. 177–201). Totowa, NJ: Humana Press.

Stravinskas Durigon, T., MacKenzie, B., Carneiro Oliveira-Junior, M., Santos-Dias, A., De Angelis, K., Malfitano, C., … Vieira, R. P. (2018). Aerobic exercise protects from pseudomonas aeruginosa-induced pneumonia in elderly mice. *J Innate Immun*, 10(4), 279–290.

Suzuki, K. (2015). Effects of exercise on antibody production. *World J Immunol*, 5(3), 160–166.

Symanski, J. D., Tso, J. V., Phelan, D. M., & Kim, J. H. (2022). Myocarditis in the athlete: A focus on COVID-19 sequelae. *Clin Sports Med*, July, 41(3), 455–472.

Warren, K. J., Olson, M. M., Thompson, N. J., Cahill, M. L., Wyatt, T. A., Yoon, K. J., … Kohut, M. L. (2015). Exercise improves host response to influenza viral infection in obese and non-obese mice through different mechanisms. *PLoS One*, 25, 10(6), e0129713.

Warren, K., Thompson, N., Wannemuehler, M., & Kohut, M. (2013). Antibody and CD8+ T cell memory response to influenza A/PR/8/34 infection is reduced in treadmill-exercised mice, yet still protective. *J App Physiol*, 114(10), 1413–1420.

Weidner, T. G., Cranston, T., Schurr, T., & Kaminsky, L. A. (1998). The effect of exercise training on the severity and duration of a viral upper respiratory illness. *Med Sci Sports Exerc*, 30(11), 1578–1583.

Wherry, E. J., & Barouch, D. H. (2022). T cell immunity to COVID-19 vaccines. *Science*, 19, 377(6608), 821–822.

Whitney, J. D., & Parkman, S. (2004). The effect of early postoperative physical activity on tissue oxygen and wound healing. *Biol Resr Nurs*, 6(2), 79–89.

Whittaker, A. C., Nys, L., Brindle, R. C., & Drayson, M. T. (2023). Physical activity and sleep relate to antibody maintenance following naturalistic infection and/or vaccination in older adults. *Brain Behav Immun Health, 32,* 100661.

Wong, G. C. L., Narang, V., Lu, Y., Camous, X., Nyunt, M. S. Z., Carre, C., … Larbi, A. (2019). Hallmarks of improved immunological responses in the vaccination of more physically active elderly females. *Exerc Immunol Rev*, 25, 20–33.

Wong, W., Crane, E. D., Kuo, Y., Kim, A., & Crane, J. D. (2019). The exercise cytokine interleukin-15 rescues slow wound healing in aged mice. *J Biol Chem*, 294(52), 20024–20038.

Woods, J. A., Keylock, K. T., Lowder, T., Vieira, V. J., Zelkovich, W., Dumich, S., … McAuley, E. (2009). Cardiovascular exercise training extends influenza vaccine seroprotection in sedentary older adults: The immune function intervention trial. *J Am Geriatr Soc*, December, 57(12), 2183–2191.

Yang, Y., Verkuilen, J., Rosengren, K. S., Mariani, R. A., Reed, M., Grubisich, S. A., & Woods, J. A. (2007). Effects of a Taiji and Qigong intervention on the antibody response to influenza vaccine in older adults. *Am J Chin Med*, 35(4), 597–607.

Zogaib, F. G., & Monte-Alto-Costa, A. (2011). Moderate intensity physical training accelerates healing of full-thickness wounds in mice. *Braz J Med Biol Res*, 44(10), 1025–1035.

8 Exercise immunology and immunosenescence

Guillaume Spielmann and Richard J. Simpson

Introduction

Over the past century, there has been a significant increase in life expectancy, with projections indicating that over 20% of the U.S. population will be 65 years or older by the year 2030. Despite this remarkable progress, the average 'healthspan', representing the portion of life spent in good health, is trailing behind. Consequently, there is a rising prevalence of age-related conditions, including cancer, Alzheimer's disease, osteoarthritis, osteoporosis, and infections. Exercise and physical activity enhance overall health and well-being in older adults. One reason for this preventive or delaying effect on diseases is that exercise preserves the ageing immune system. The immune system undergoes significant changes with age, contributing to increased infection rates, low-grade inflammation, heightened disease susceptibility like cancer, and diminished vaccine responses in older populations. Referred to as immunosenescence, this process involves impairments in both innate and adaptive immunity. These impairments include reduced cytotoxicity of effector lymphocytes, diminished quality and quantity of antibodies from B cells, thymic atrophy, a higher frequency of highly differentiated T cell subsets with low proliferative capacity and elevated inflammatory responses, and compromised migratory and antimicrobial function in neutrophils and monocytes. This can manifest in poor control of latent viral infections (e.g. CMV or *Cytomegalovirus* which leads to sub-clinical reactivation, or VZV or *Varicella Zoster Virus* which causes Shingles), and is indicative of systemic immune system deterioration, potentially driving T cell exhaustion. Middle-aged and older adults who are physically active and/or have high levels of cardiorespiratory fitness often exhibit blunted hallmark features of immunosenescence. In contrast, detrimental age-related changes in immunity can manifest prematurely in younger, inactive, and/or "individuals with obesity". This chapter describes well-known age-related changes to innate and adaptive immunity, evaluates the effects of exercise and/or physical activity on these changes, and discusses potential mechanisms by which exercise can preserve the ageing immune system. Additionally, it explores the evidence supporting the idea that exercise may have immune-rejuvenating effects, such as the reversal of immunosenescence.

Age-related changes in innate immunity

Innate immunity is considered the first line of defence against pathogens. It involves nonspecific mechanisms like physical barriers (e.g., skin), soluble proteins (e.g. complement, cytokines, acute-phase proteins), and immune cells (e.g., macrophages, neutrophils) that quickly recognise and attack invaders. This rapid response provides immediate protection for the host while the adaptive immune response, with its more specific defences, is mobilised. Ageing can impact the

DOI: 10.4324/9781003256991-8

complement system, a crucial component of the innate immune system responsible for recognis-ing and eliminating pathogens. Ageing may lead to a decline in complement activity, affecting the ability to recognise and clear pathogens. This diminished function may contribute to an in-creased susceptibility to infections and a potential dysregulation of immune responses in older adults. Moreover, ageing adversely affects haematopoiesis, the process of blood cell formation, and lymphoid progenitor cells crucial for immune development. With ageing, there is a notable decline in the activity and production of CD34+ haematopoietic stem cells (Della Bella et al., 2007), leading to reduced production of various blood cells, including immune cells, and an altered differentiation pattern that may favour myeloid cell production over lymphoid cells, impacting the overall immune response. The peripheral cellular functions in the innate immune system, notably those of neutrophils, monocytes/macrophages, dendritic cells, and NK cells, also diminish with ageing.

Neutrophils

Neutrophils are the most abundant circulating leukocyte (10^{11} produced each day) and are pri-marily responsible for engulfing and destroying bacteria and other pathogens, have a key role in host defence against infections. Neutrophils migrate daily from the bone marrow to the blood as mature terminally differentiated cells. The lack of cellular division in the circulation, and exceptionally short half-life (0.5–5 days) resulted in a long-standing perception that neutrophils are homogeneous and lack functional diversity. However, this perception is incorrect, as neutro-phils are now regarded as a complex, heterogeneous, multi-functional population with multiple roles in the pathophysiology of ageing and disease (Adrover, Nicolas-Avila, & Hidalgo, 2016; Ng, Ostuni, & Hidalgo, 2019). New stages of neutrophil development, distinct from monocyte development, have revealed neutrophil specific mature, progenitor, and precursor cells (Evrard et al., 2018; Kim et al., 2017; Zhu et al., 2018). In older adults, the phenotype of circulating neutrophil shifts from cells expressing high levels of CXCR2, CD16, and CD62L, while aged neutrophils lose expression of CXCR2 and CD16, and gain CD11b, CD11c, CD49d, ICAM-1, and TLR4As (Rosales, 2018). Interestingly, the number of total circulating neutrophils does not appear to decrease with age, but they instead adopt a more mature and immunosuppres-sive CD16^{bright}/CD62L^{dim} phenotype characterised by impaired oxidative burst generation and phagocytic activities (Sauce et al., 2017). Further, chronic inflammatory conditions such as inflamm-ageing (discussed in more detail later in this chapter) promote the accumulation of dys-functional neutrophils with extended lifespans (>5 days) but with decreased chemotaxis, speed, and accuracy during migration (Sapey et al., 2014). The impaired ability of neutrophils to mi-grate to infection sites in older adults is not the only age-associated burden to the innate immune system. Indeed, while neutrophils found in younger healthy adults release net-like structures composed of DNA-histone complexes and proteins (NETs) to trap pathogens and facilitate their removal, neutrophils in older adults exhibit impaired NET formation, further compromising resolution of pathogenic challenges (Hazeldine et al., 2014).

Monocytes and macrophages

Circulating monocytes make up around 10% of leucocytes in blood and have a crucial role in phagocytosis, antigen presentation, tissue repair, and inflammation. Historically, monocytes have been divided into three main sub-populations based on their expression of CD14 and CD16 and effector functions, namely "pro-inflammatory" classical (CD14^{bright}/CD16−), "pa-trolling" or "non-classical" (CD14^{dim}/CD16+) and intermediate (CD14+/CD16+) monocytes which secrete vast amounts of pro-inflammatory cytokines IL-1β and TNF-α when stimulated (Gunther & Schultze, 2019). Ageing does not appear to be associated with significant changes in the numbers of circulating monocytes, but rather in the frequency of each specific subset (Seidler, Zimmermann, Bartneck, Trautwein, & Tacke, 2010). Specifically, both non-classical and intermediate monocytes with increased TNF-α-producing capacities (Cao et al., 2022) accumulate with ageing, at the expense of CD14^{bright}CD16− classical monocytes. Moreover, monocyte expression of co-stimulatory molecules (CD80/CD86) and production of cytokines decrease with ageing, particularly in the context of TLR signalling (Busse et al., 2015). In addition, monocytes isolated from older adults exhibit impaired expression of tumour necrosis factor receptor-associated factor 3 (TRAF3), a key upstream adaptor protein involved in type I IFN production (Molony et al., 2017). This leads to inadequate production of IFN-α, which in turn increases the susceptibility of older adults to respiratory viruses such as Influenza, Respiratory Syncytial Virus, and SARS-CoV-2.

Once circulating monocytes leave the peripheral blood compartment to enter tissues, they differentiate into macrophages, which specialise in phagocytosis, engulfing and digesting cellular debris, pathogens, and other foreign substances. They have a vital role in both innate and adaptive immune responses, contributing to tissue repair and regulation of immune reactions. In response to bacterial LPS, TNF-α, or IFN-γ stimulation, tissue-residing macrophages will be polarised into "classically-activated" M1 macrophages and produce pro-inflammatory cytokines when further stimulated. On the other hand, IL-4 and IL-13 stimulation drives macrophages towards more heterogenous phenotypes of "alternatively-activated" and "reparative" M2 macrophages which include M2a, M2b, M2c, and M2d phenotypes. While ageing impacts macrophage phenotype and function in a tissue-specific manner (Davies, Jenkins, Allen, & Taylor, 2013), it is generally agreed that muscle- and adipose-tissue residing macrophages shift towards an M1-like pro-inflammatory phenotype in older adults (Becker et al., 2018).

NK cells

NK cells belong to the family of Innate Lymphoid Cells I (ILC1) (Zhang & Huang, 2017), and have a pivotal role in immune-surveillance. They are capable of recognising and directly attacking infected or abnormal cells, including cancerous or senescent cells, without prior sensitisation. Although NK cells were initially assumed not to be affected by ageing, mounting evidence points towards a decline in NK cell function with chronological ageing (Nikolich-Zugich, 2018). Specifically, among total NK cells, there is an age-associated decrease in the proportion of CD56^{bright} NK cells and increased CD56^{dim} subsets, which leads to a substantial reduction in the ratio of CD56^{bright} to CD56^{dim} cells with ageing (Béziat, Descours, Parizot, Debré, & Vieillard, 2010; Le Garff-Tavernier et al., 2010; Solana, Campos, Pera, & Tarazona, 2014). Novel techniques such as single-cell RNA-seq have highlighted that immunosenescence is also associated with the accumulation of highly differentiated mature CD56^{dim} CD57+CD16+, and impaired killing of tumour cells (Borrego et al., 1999; Solana et al., 2014). The mechanisms behind the age-related alterations in NK cell function remain unclear; however, current evidence points towards the remodelling of NK cell receptor expression; with increases in inhibitory receptors (Sikorski, Swat, Brzezinska, Wroblewski, & Bisikirska, 1993) and decreased expression of activating receptors (NKG2D) (Almeida-Oliveira et al., 2011). Furthermore, circulating NK cells isolated from otherwise healthy older adults tend to express fewer cytotoxic granules perforin and granzymes than those from young adults, along with impaired IFN-γ production (Rukavina et al., 1998). Taken together, the impaired responsiveness of NK cells in older adults increases viral infection and cancer risk (Gounder et al., 2018; Ogata et al., 2001).

Dendritic cells

Dendritic cells are a type of immune cell that have critical functions in both the innate and adaptive immune system. They specialise in capturing, processing, and presenting antigens to T cells, initiating and coordinating immune responses. Dendritic cells are crucial for the recognition of pathogens and the activation of specific immune defences against infections. Like macrophages, dendritic cells have traditionally considered tissue-residing immune cells. However, it is now clear that these antigen-presenting cells (APCs) also include cells that migrate between various tissue and the lymphatic system to maximise their initiation and regulation of the adaptive immune response against pathogenic threats. Dendritic cells can be divided into three subtypes, plasmacytoid dendritic cells (pDCs) which produce large amounts of Type I and III interferons in response to TLR stimulation, myeloid dendritic cells (mDCs) which coordinate the adaptive immunity and monocyte-derived dendritic cells (MoDCs). Studies have reported lower numbers of circulating pDCs in older adults, while the number of mDCs and MoDC tend to remain stable (Panda et al., 2010). However, while the number of MoDC does not change with ageing, they have been reported to produce larger amounts of pro-inflammatory cytokines TNF-α, CXCL-10, and IL-6, even when stimulated with lower amount of LPS (Agrawal et al., 2007). Furthermore, dendritic cells isolated from older adults display a concomitant decreased ability to produce anti-inflammatory cytokines such as IL-10, with an increased sensitivity to self-antigens (Agrawal, Gollapudi, Gupta, & Agrawal, 2013). Taken together, these age-associated functional changes promote inflamm-ageing, limit tissue remodelling, and severely threaten tolerance against self-antigens.

Age-related changes in adaptive immunity

Adaptive immunity is the branch of the immune system that provides specific and targeted defence against pathogens. It involves the recognition of specific antigens, the development of immunological memory, and the ability to mount tailored responses to combat infections or other foreign substances. This system includes T cells, B cells, antibodies, and immune-memory mechanisms. Considering the interplay between the innate and adaptive arms of the immune system, it is not surprising that any age-induced alterations in innate immunity would also orchestrate decrements in adaptive immunity. A series of elegant studies conducted by Wikby and colleagues (Wikby, Johansson, Ferguson, & Olsson, 1994; Wikby et al., 2002) longitudinally followed around 100 Swedish octogenarians and nonagenarians for nine years and six years, respectively, and provided a comprehensive characterisation of the biological and cellular hallmarks of adaptive immunosenescence (Wikby, Mansson, Johansson, Strindhall, & Nilsson, 2008). From these studies, the concept of an Immune Risk Profile (IRP) (Brzezinska, Koziorowska, Wirpsza, & Zawadzki, 1977) was developed, which was shown to predict morbidity and mortality in older adults (Pawelec & Gouttefangeas, 2006; Wikby et al., 2008). The IRP is characterised by a combination of factors such as impaired T cell proliferation in response to mitogens, and accumulations of late-stage differentiated effector memory T cells, a decline in B cells, and an increase in markers of systemic inflammation, and a CD4:CD8 ratio that is less than 1.0. CMV serostatus was later also included in the IRP, and a defining characteristic of this profile is the diminished ability to maintain previously acquired viral infections in a latent state. This is evident with CMV, a widespread herpesvirus infecting approximately 50% of the U.S. population (with similar epidemiology across other western countries, but a higher prevalence in developing countries). CMV prevalence increases with age, and the virus persists throughout the lifetime (Bate, Dollard, & Cannon, 2010). CMV is prone to sporadic reactivation, linking it to T cell clonal exhaustion, suboptimal vaccine responses, and inflammation, thereby emerging as an independent predictor of all-cause morbidity and mortality among octogenarians and nonagenarians. The triggers for CMV reactivation, as with other herpes viruses, are multifaceted – thought to involve several mechanistic factors, such as inflammation, oxidative stress, and adrenergic activity – and also linked with physical factors (e.g., fatigue, ageing), socioeconomic factors (e.g., income, education, job status), lifestyle factors (e.g., smoking, physical activity), and psychosocial factors (e.g., adversity, anxiety, depression). The IRP is indicative of immune-ageing and is linked to higher susceptibility to infections, chronic diseases, and mortality in older adults. In this section, we will summarise the age-associated changes that occur within the B cell and T cell compartments, and provide an update on the evidence supporting the IRP in older adults.

B cells

B cells are responsible for producing antibodies that specifically target and neutralise pathogens such as bacteria and viruses. They also have a key role in immune memory, providing long-term protection against previously encountered pathogens. The number of circulating B cells decreases with age (Frasca, Diaz, Romero, D'Eramo, & Blomberg, 2017; Frasca et al., 2008), potentially due to bone marrow microenvironmental changes that promote haematopoietic stem cells to be biased towards myeloid cell production rather than a lymphoid cell lineage (Cho, Sieburg, & Muller-Sieburg, 2008). This is further supported by the "Immunos study", which identified sex-differences in CD19+ B cell numbers in age-matched adults (Breuil et al., 2010). In this study, older women diagnosed with osteoporosis had fewer circulating B cells than their age-matched counterparts with healthy bone mineral density. In addition, their B cells secreted a greater quantity of granulocyte macrophage colony-stimulating factor (GM-CSF), which likely continues to fuel the aforementioned myeloid bias among haematopoietic stem cells. In addition, stromal cells from the bone marrow of older adults are known to produce less IL-7 than cells from younger adults (Stephan, Reilly, & Witte, 1998), thus further dampening pro-B cell development and overall B-lymphopoiesis.

While total circulating B cell numbers decline with age, some B cell subsets are more affected by immunosenescence than others. Indeed, while switched memory B cells (CD19+/IgD−/CD27+) (Frasca, Diaz, Romero, D'Eramo et al., 2017) and IgM memory (CD19+/IgD+/IgM+) (Martin, Wu, Kipling, & Dunn-Walters, 2015) decrease with age, the proportion of naïve and double-negative B cells (CD19+/IgD−/CD27−) significantly increase (Frasca, Diaz, Romero, & Blomberg, 2017). This is of special clinical relevance among older adults since naive B cells are less able to mount rapid antibody responses against pathogenic exposure than switched memory B cells. Of note, double-negative B cells have been implicated in the aetiology of various inflammatory conditions such as multiple sclerosis (Claes et al., 2016), systemic lupus erythematosus (Jenks et al., 2018), and even obesity (Frasca, Romero, Garcia, Diaz, & Blomberg, 2021) as large producers of pro-inflammatory cytokines. It is thus likely that their accumulation in older adults further amplifies inflamm-ageing.

Advancing age is associated with reduced antibody responses to vaccination, especially during primary exposure to a vaccine (Weinberger et al., 2018). This is likely multifactorial, but it is now clear that age leads to reductions in the frequencies of B1 cells (Rodriguez-Zhurbenko, Quach, Hopkins, Rothstein, & Hernandez, 2019) known for spontaneously secreting IgM and IgA immunoglobulins as part of innate-like immunity (Holodick, Rodriguez-Zhurbenko, & Hernandez, 2017) and B2 cells (McKenna, Washington, Aquino, Picker, & Kroft, 2001) involved in the production of high-affinity antibodies. In addition to changes in B cell numbers and function, age-associated decrements in T cell function are also likely to contribute to this reduced response to vaccination (see Chapter 7 for how exercise can potentially improve vaccine responses in older adults).

T cells

T cells have diverse roles, including directly attacking infected cells or coordinating the function of other immune cells. There are various subsets of T cells that work synergistically to mount effective immune responses, providing defence against infections and contributing to immune memory. The CD4+ 'helper' T cell coordinates immune responses by releasing cytokines as signalling molecules, and activating B cells and cytotoxic CD8+ T cells. Indeed, CD8+ cytotoxic T cells directly attack and destroy infected or tumour cells by releasing toxic substances. Regulatory T cells have a crucial role in regulating immune tolerance by suppressing excessive immune responses, preventing autoimmunity and maintaining immune system balance.

Older adults possess similar numbers of circulating T cells as young adults; however, there are vast differences in the subsets of both CD4+ and CD8+ T cells, with older adults having greater proportions of cells with a late-stage differentiation phenotype (CD28−/CD27−/CCR7−/KLRG1+/PDCD1+/CD57+/CD45RA+). In young adults, circulating CD4+ helper T cells are more abundant than CD8+ cytotoxic T cells with a CD4:CD8 ratio between 1.5 and 2.5 generally accepted to be normal in healthy individuals. Conversely, the numbers of memory and highly differentiated CD8+ T cells increase with ageing, leading to an inverted CD4:CD8 ratio below 1.0. The progressive accumulation of senescent T cells is thought to be a natural consequence of (1) multiple exposures to external pathogens and persistent viral infections throughout the lifespan and (2) physiological thymic involution, compromising naive T cell generation. Persistent antigenic stimuli promote repeated and excessive rounds of T cell division, leading to T cell cycle arrest and immunosenescence (Spaulding, Guo, & Effros, 1999). In that state, T cells will no longer clonally expand upon further antigenic stimulation, but will still retain effector cell properties (i.e. recognising and killing virally infected cells). Specifically, highly differentiated and senescent CD8+ T cells circulating in the blood of older adults, express large amount of Granzyme K, Granzyme B, and Granulysin (Zheng et al., 2020). Studies highlighting CMV immunodominance, especially in older adults (Khan, Cobbold, Keenan, & Moss, 2002), propose that frequent bouts of CMV reactivation may be the driving force behind the expansion of highly differentiated T cells, promoting premature senescence and limiting the amount of available "immune space" for naive and less-differentiated cells to populate the immune compartment (Simpson, 2011).

In addition to promoting impaired immune response to pathogens, senescent T cells exhibit aberrant pro-inflammatory cytokines production, and increased resistance to apoptosis, thus limiting their clearance (Spaulding et al., 1999). This phenotype, also termed senescence-associated secretory phenotype (SASP) was originally defined in senescent fibroblasts as cells that secreted a variety of pro-inflammatory cytokines (Coppe et al., 2008) and produced elevated levels of reactive oxygen species (Passos et al., 2010). Similarly, senescent CD4+ and CD8+ T cells secrete large amounts of inflammatory cytokines and contain high levels of cytotoxic proteins; however, unlike fibroblast SASP, T cells secrete high levels of TNF-α, IL-18, and CCL16 instead of IL-6 and IL-1β (Callender et al., 2018; Henson et al., 2014).

Ageing also affects discrete subsets of CD4+ helper T cells. Indeed, both animal and human studies have shown that the balance between Th1 and Th2 responses is comprised in older adults, as ageing leads to Th2-type dominance (Mirza, Pollock, Hoelzinger, Dominguez, & Lustgarten, 2011). Similarly, ageing is associated with an increased number of Th17 and Th22 cells, known for promoting inflammation by releasing IL-6, IL-17A, IL-17F, IL-21, and IL-23 (Cosmi et al., 2008) and IL-22, IL-13, IL-26, and TNF-α respectively (Chen et al., 2018). Specifically, naive and memory Th17 cells in humans have been shown to be affected by ageing, with older adults producing more IL-17 from naive CD4+ T cells but less IL-17 from memory CD4 T cells compared with young adults (Lee et al., 2011). In addition, peripheral activated regulatory CD4+/CD25+/FOXP3+/CD45RA− T cells have an important role in the maintenance and regulation of the immune system by suppressing excessive immune responses to both self and innocuous antigens through curbing the activation, proliferation, and functions of a variety of cells, increase with ageing (Raynor et al., 2013).

Inflamm-ageing as a driver of age-related disease

Chronic low-grade inflammation is considered to underly many age-related diseases including cancer (Allavena, Garlanda, Borrello, Sica, & Mantovani, 2008; Mantovani & Pierotti, 2008), neurodegenerative diseases (Giunta et al., 2008), stroke (Christensen, Boysen, & Johannesen, 2004), and metabolic diseases such as type 2 diabetes (Pedersen et al., 2003). Inflamm-ageing is a term coined to describe the chronic, low-grade inflammation that occurs with ageing. It combines "inflammation" and "ageing" and is characterised by a state of increased inflammatory markers in the absence of infection or overt injury. Inflamm-ageing reflects a shift in the balance of the immune response, with a persistent pro-inflammatory state that may disrupt normal cellular functions and tissue homeostasis over time. It is defined by increased concentrations of circulating inflammatory mediators, acute-phase proteins and cytokines such as C-reactive protein, IL-1β, IL-6, IL-15, and TNF-α (Bartlett et al., 2012; Bruunsgaard et al., 1999; Ershler et al., 1993; Forsey et al., 2003) accompanied by a decrease in anti-inflammatory cytokines such as IL-2 and IL-10. Inflamm-ageing is primarily associated with changes in the innate immune system, specifically involving chronic activation of macrophages and neutrophils. The dysregulation of innate immune responses, along with factors like cellular senescence, mitochondrial dysfunction, and the accumulation of damage-associated molecular patterns, contribute to inflamm-ageing. While adaptive immunity is not the primary driver of inflamm-ageing, there can be interactions and crosstalk between innate and adaptive immunity, and age-related changes in the adaptive immune system, such as thymic involution and reduced T-cell diversity, may indirectly influence the inflammatory environment. While largely considered detrimental, it should also be noted that inflamm-ageing has been suggested to help clear pathogens and malignant cells, while also assisting with tissue repair (Fulop et al., 2017). This is supported by a Japanese longitudinal study of 1,554 semi-supercentenarians, which showed that low-grade inflammation was the single most important predictor of longevity, over age, sex, and biomarkers of cellular senescence such as leukocyte telomere length (Arai et al., 2015).

Regardless of whether inflamm-ageing is beneficial or detrimental to the health of older adults, its aetiology is likely multifactorial, with several potential mechanistic underpinnings. First, the level of inflammation in older adults appears to be partially dependent on genetic pre-disposition, with studies showing that polymorphisms in the promoter of certain cytokines genes, including IL-6 (Rea et al., 2003) and IL-10 (Lio et al., 2004; Lio et al., 2002) lead to an inflated cumulative inflammatory response (Franceschi et al., 2007). However, genetic pre-disposition to inflamm-ageing only partially explains the vast disparities in immune health observed between healthy and frail immunocompromised older adults. The balance between adipose tissue and muscle mass has a crucial role in shaping the inflammatory status of the ageing body. Excessive adipose tissue, especially visceral fat, releases pro-inflammatory molecules, contributing to systemic inflammation. Conversely, skeletal muscle, when active, produces anti-inflammatory myokines, and maintaining or increasing muscle mass through exercise is associated with reduced inflammation (see Chapter 6). While CMV is considered to be a driver of T cell ageing and exhaustion, a longitudinal study found that hallmark features of inflamm-ageing such as CRP, TNF-α, and IL-6 increased over a decade, with concomitant reductions in the anti-inflammatory cytokine IL-10, despite no change in CMV serostatus or viral antibody titres. This indicates that latent CMV infection is unlikely to be a major driver of inflamm-ageing, which is not surprising given that CMV seropositivity can be as high as 80% in adults over the age of 60 years (Fowler et al., 2022).

Immune responses to exercise in older adults

Physical activity involves any skeletal muscle movement requiring energy expenditure, encompassing daily life activities. Individual bouts of exercise and exercise training are subtypes of physical activity, which can be a structured, repetitive regimen with the goal of enhancing or sustaining physical fitness, often measured by factors such as cardiorespiratory fitness. While these terms are sometimes used interchangeably, both exercise training and physical activity can significantly impact the ageing immune system. Exercise is theorised to enhance immune competence through four major pathways: (1) improved immune-surveillance and; (2) immune-organs crosstalk; (3) immune system maintenance and remodelling; and (4) resolution of inflammation (Figure 8.1). Both individual bouts and regular exercise training contribute to this response, with effects being direct (influencing immune cells and molecules) or indirect (altering physiological states impacting immunity and inflammation, (e.g., reducing visceral

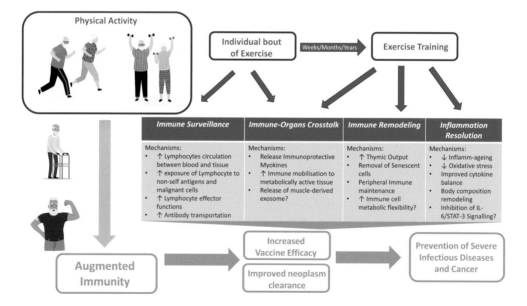

Figure 8.1 Theoretical framework of the exercise-mediated improvements in immune function in older adults.

Legend: "Chronological aging is associated with decrements in immune function (immunosenescence). Individual bouts of aerobic and/or resistance exercise improve immune surveillance and crosstalk between the immune system and other physiological systems (muscle, adipose tissue, gut microbiome etc..). When repeated regularly, the cumulative effects of individual bouts of exercise promote immune remodelling and resolution of inflamm-ageing in older adults. Taken together, regular exercise improves body composition, vaccine efficacy and tumour control, thus reducing the risk of severe infectious disease and cancer."

adiposity or moderating the biological stress response). Furthermore, the mode (e.g., aerobic or resistance exercise) and specific exercise prescriptions, including frequency, intensity, and duration are likely to be crucial factors influencing how the immune system responds to periods of physical activity and exercise training. In this section, we delineate distinctions in immune responses to individual bouts of exercise between the young and older adults and how this may contribute to alterations in immune-surveillance. Additionally, we explore how exercise training can potentially preserve or reshape immunity while addressing inflammatory cascades linked to inflamm-ageing.

Ageing and the immune response to individual bouts of exercise

Immune-surveillance is the inherent defence mechanism that perpetually monitors and identifies abnormal cells or pathogens (e.g., viruses, bacteria). This vigilant process relies on the continuous recirculation of immune cells and antibodies, which, upon detecting anomalies, collaborate to neutralise the threat. A single session of moderate-to-vigorous intensity exercise triggers the biological stress response, heightening the activity of the sympathetic nervous system and the hypothalamic–pituitary–adrenal axis. This results in increased haemodynamic shear stress, various cytokines, catecholamines, and neurotransmitters, leading to an immediate influx of leucocytes into the bloodstream (Simpson, Kunz, Agha, & Graff, 2015; Walsh et al., 2011). Notably, lymphocytes, especially those with high effector functions like CD8+ T cells and NK cells, respond robustly to exercise, along with APCs such as monocytes and dendritic cells. After exercise cessation, these mobilised cells swiftly exit the bloodstream to survey secondary lymphoid organs (e.g., lymph nodes, spleen, tonsils) and peripheral tissues, including the lower and upper respiratory tracts. While the immune response to each exercise session is thought to enhance immune-surveillance by increasing the chance of interaction between immune cells and potential threats, conclusive experimental evidence – especially in humans – is still pending. Nevertheless, discernible differences in this exercise-induced immune response between young and older adults may influence immune-surveillance dynamics and are described in detail below.

Cellular response to individual bouts of exercise

The transient increase in innate and adaptive immune cell numbers in peripheral blood in response to a single bout of exercise is well-documented in young and older adults. A seminal study from Joseph Cannon in the early 1990s compared the neutrophil response to downhill running between younger ($n = 9$; 22–29 years) and older ($n = 12$; 55–74 years) men. While the number of circulating neutrophils immediately post-exercise increased in both age groups, the peak exercise-induced neutrophil increase was 62% lower in the older group than in the younger group (Cannon et al., 1990). Interestingly, however, neutrophil function following exercise, as determined by mitogen-induced superoxide release, remained identical between the age groups. These results were later confirmed by the same research group, where older adults appeared to exhibit much lower exercise-induced neutrophil mobilisation compared with younger adults in response to eccentric contractions (45 min of downhill running at 78% maximum heart rate) (Cannon, Fiatarone, Fielding, & Evans, 1994).

Similarly, it has been hypothesised that monocytes from physically active older adults may exhibit better functions than those from sedentary age-matched adults, due to frequent and recurrent exposure to exercise-induced heightened shear stress. This is supported by evidence highlighting that exercise responsiveness of monocytes is maintained with age and physical activity status. A study by Minuzzi et al. (2019), comparing the monocyte response to a single bout of maximal cycling exercise between young active adults ($n = 9$, age: 31.8 ± 3.0 years, VO_2 max: 46.8 ± 6.1 ml kg min^{-1}), middle-aged sedentary adults ($n = 10$, age: 54.2 ± 5.9 years, VO_2 max: 29.3 ± 4.1 ml kg min^{-1}) and master athletes ($n = 20$, age: 53.1 ± 8.8 years, VO_2 max: 40.3 ± 11.2 ml kg min^{-1}) showed similar increases in the numbers of circulating monocytes immediately post-exercise across the groups (Minuzzi et al., 2019). This is likely to be of clinical relevance since circulating monocytes will eventually extravasate into tissues, including the muscle microenvironment, where they will differentiate into macrophages and promote muscle repair, all of which are believed to be impaired in sedentary older adults (Peake, Della Gatta, & Cameron-Smith, 2010).

Individual bouts of exercise also lead to robust changes in NK cell numbers and function in adults of all ages. In contrast to neutrophils and monocytes which displayed differential sensitivity to exercise based on age, NK cell activity following a bout of maximal exercise function has been shown to be similar between young (21–39 years) and older (>64 years) adults (Fiatarone et al., 1989). This finding is further supported by a meta-analysis that consolidated the results from 12 studies on the effects of exercise bouts on NK cell function (Rumpf et al., 2021). While only two studies from this meta-analysis included older adults, they showed similar mobilisation of NK cells with improved NK cell cytotoxicity between young and old, immediately following resistance exercise (Flynn et al., 1999; McFarlin, Flynn, Phillips, Stewart, & Timmerman, 2005). Similarly, Evans and colleagues recently observed similar NK cell mobilisation to exercise bouts in healthy older women (59 ± 5 years, $n = 9$) (Evans et al., 2015). However, since the purpose of this study was to evaluate changes in NK cell responsiveness to a bout of exercise in breast cancer survivors, they did not measure NK cell cytotoxicity in the healthy control group, and only in breast cancer survivors. In this specific population, they saw no change in NK cell cytotoxicity in response to a bout of exercise, replicating data from others (van der Pompe, Bernards, Kavelaars, & Heijnen, 2001).

As discussed throughout this chapter, immune responses to stressors, such as exercise are diminished in older adults. This is especially documented amongst T cell subsets, with multiple reports showing blunted T cell mobilisation following moderate–vigorous and maximal exercise in older adults (Ceddia et al., 1999; Simpson et al., 2008; Spielmann, Bollard et al., 2014). However, all T cell subsets are not equally affected with advancing age. For example, it has been proposed that exercise can shift the balance back towards Th1-type responses following an intensity-dependent pattern (Malm, 2004). Many groups have reported that moderate exercise tends to lead to a shift towards Th1-type cytokines responses, whereas strenuous or exhaustive exercise causes Th2-type dominance (Ostrowski, Rohde, Asp, Schjerling, & Pedersen, 1999; Pedersen & Bruunsgaard, 2003). Using older mice, Kohut et al. found that moderate exercise training increased antigen-specific cell production of Th1 cytokines but not Th2 cytokines (Kohut, Boehm, & Moynihan, 2001). This effect was not found in younger mice, indicating that age may interact with exercise in Th1/Th2 shifts.

Advancing age is associated with an accumulation of terminally differentiated T cells with a narrower T cell receptor (TCR) repertoire diversity. As T cells advance through repeated activation and differentiation stages, they upregulate high-affinity β2 adrenergic receptors (Slota, Shi, Chen, Bevans, & Weng, 2015). This is especially true amongst CD8+ T cells, known for expressing more β2 adrenergic receptors than CD4+ T cells (Sanders, 2012). In this context, the release of catecholamines in response to an exercise bout leads to a preferential mobilisation of senescent T cells in blood, which can be inhibited by exposure to β2 antagonists (Graff et al., 2018). Consequently, it is only logical to see similar mobilisation of senescent T cells between young and older adults, if not only due to the shear increase in circulating senescent T cells in older participants. This finding has been reproduced across many exercise modalities, including cycling (Minuzzi et al., 2018; M. Ross et al., 2018), running (Simpson et al., 2010), swimming (Theall et al., 2020), resistance exercise (Graff et al., 2022) and futsal (Cury-Boaventura et al., 2018). Interestingly, it has been suggested that while chronological ageing may not affect the mobilisation of senescent T cells in response to exercise bouts, latent infection with CMV may have a greater role in this rapid ingress and egress of senescent T cells in peripheral blood post-exercise. Indeed, in a study comparing the mobilisation of senescent T cells in response to 30-min cycling at 80% peak power between young ($n = 16$ men, 28.6 ± 4.7 years, 50% CMV seropositive) and old ($n = 16$ men, 55.4 ± 4.0 years, 50% CMV seropositive), showed that CMV infection prevented the age-associated decline in CD8+ T-cell mobilisation following exercise, due to the inflated senescent T-cell pool in the CMV-seropositive participants (Spielmann, Bol-lard et al., 2014). It can be concluded that at least some of the changes in T-cell exercise respon-siveness observed in older adults can be explained by infection history, and years of senescent T-cell pool inflation as a direct consequence of repeated viral reactivations. This appears to be confirmed by a recent study showing that a single bout of exercise, at least in young adults, leads to a preferential mobilisation of the most dominant T cell clone, as determined by deep TCR-β chain sequencing and tandem single-cell RNA sequencing following a graded cycling exercise up to 80% VO_2 max (Zuniga et al., 2023).

A few researchers have examined the responses of other subtypes of T cells such as regulatory T cells and angiogenic T cells to exercise in humans. Wilson et al. found a significant increase in CD4+CD25+FOXP3 cells following a single bout of exercise in elite adolescent swimmers (Wilson, Zaldivar, Schwindt, Wang-Rodriguez, & Cooper, 2009). This was later confirmed by a study from Minuzzi et al. (2017), showing that master athletes (53.2 ± 9.1 years) who cycled to exhaustion exhibited increased numbers and proportions of CD4+/CD25+/CD127[low] regulatory T cells (Minuzzi et al., 2017). When it comes to angiogenic T cells, a study conducted by Ross et al. (2018) in 18 healthy men aged 18–75 years showed that not only do these cells decrease with ageing at rest, mobilisation into the blood in response to 30 minutes of cycling at 70% VO_2 max was much smaller among older adults than among young adults (Ross et al., 2018). Considering the crucial role played by angiogenic T cells in the formation of new blood vessels and the repair of damaged endothelium, this blunted response to exercise may at least partially explain the increased risk in vascular dysfunction observed in ageing.

Soluble responses to individual bouts of exercise

The effects of individual bouts of exercise on the release of cytokines in young adults are well-documented (Pedersen, Ostrowski, Rohde, & Bruunsgaard, 1998); however, fewer studies have been conducted on soluble responses to exercise in older adults. A study conducted on 20 heathy sedentary adults (66.9 ± 6.1 years) showed that moderate-intensity treadmill walking (3.5 km/h at 12% incline) for 5 minutes leads to significant increases in circulating TNF-α but not IL-6 (Signorelli et al., 2003). Contrarily, Reihmane et al. showed that 45 minutes of dynamic knee-extensor exercise at low intensity ($19.5W \pm 0.9W$) was sufficient to show increases in IL-6, but not TNF-α in 15 sedentary men of similar age (68.1 ± 1.1 years) (Reihmane, Gram, Vigelso, Wulff Helge, & Dela, 2016). These seemingly contradictory results could potentially be ex-plained by differences in fitness status or exercise intensity rather than age *per se*. Indeed, a study comparing the cytokine response to moderate- or high-intensity individual bouts of exercise of 16 sedentary (VO_2 peak: 22.6 ± 2.8 mL kg^{-1} min^{-1}, 72 ± 6 years) and 14 trained older adults (VO_2 peak: 37.4 ± 5.9 mL kg^{-1} min^{-1}, 69 ± 5 years) found that IL-6 and IL-10 increased post-exercise but not TNF-α, regardless of exercise intensity or fitness level (Windsor et al., 2018). This is supported by a recent study that showed greater exercise-induced changes in circulating IL-6,

IL-15, IL-7, and IL-1RA in older physically active adults than their age-matched sedentary counterparts in response to 30 minutes of cycling at 50% VO_2 peak (MacNeil, Tarnopolsky, & Crane, 2021). Interestingly, they showed no difference in circulating TNF-α following exercise. Exercise bouts also lead to the release of thymoprotective myokines IL-7 and IL-15, both at the mRNA (Nielsen et al., 2007) and protein levels (Haugen et al., 2010; Tamura et al., 2011) regardless of age (Crane et al., 2015). Considering that these myokines have been implicated in the reduction of thymic involution and hallmarks of premature immune-ageing, this suggests that the cumulative immune responses to isolated bouts of exercise throughout a structured aerobic training programme likely lead to the improved immune function of older active individuals.

Systemic immune response to individual bouts of exercise

Taken together, changes in cellular and soluble immunity in response to individual bouts of exercise in older adults lead to alterations in systemic immunity. Specifically, alterations in vaccination responses following single bouts of exercise have been demonstrated in older adults. However, not all studies are consistent. For example, a single bout of exercise does not appear to improve B cell function among older adults as demonstrated by antibody response to vaccine challenges. Long et al. (2012) reported no age- or exercise effects on the antibody response of young ($n = 60$, 21.3 ± 2.8 years) or older ($n = 60$, 57.9 ± 4.4 years) adults to pneumococcal or influenza vaccines administered after a brisk 45-min walk at >55% of participants' age-predicted maximum heart rate (Long et al., 2012). This was supported by a report from Edwards et al. (2015), where a single bout of resistance exercise at 60% of one repetition maximum was not sufficient to improve antibody responses to an influenza vaccine in 46 healthy adults aged 73 ± 7 years (Edwards et al., 2015). When Ranadive et al. (2014) vaccinated 59 volunteers aged 55–75 years (23 men/32 women) against influenza following 40 minutes of moderate-intensity aerobic exercise at an intensity of 55–65% of their maximum heart rate, only the women who exercised showed improvements in antibody titres against H1N1. A potential explanation for the discrepancies in findings can be found in the differences in exercise duration between studies. Indeed, a recent study from Hallam et al., compared with resting control, assessed whether cycling at 60–70% of maximum heart rate for 45 or 90 minutes, prior to being vaccinated with monovalent Influenza A/California/7/09 H1N1 vaccine, impacted the antibody response of young and old adults (Hallam, Jones, Alley, & Kohut, 2022). Interestingly, while 45 minutes of exercise was insufficient to improve the antibody response to the H1N1 vaccine, older adults who exercised for 90 minutes exhibited greater responses to the vaccine than the other groups. Of note, this study was also the first study to show that 90 minutes of moderate-intensity exercise improved the response to a COVID-19 vaccine (see Chapter 7 for more details of how exercise affects vaccine responses).

Ageing and the immune response to exercise training

Cellular responses to regular exercise

As described above, individual bouts of exercise have profound effects on both innate and adaptive arms of the immune system in older adults. While the changes in circulating cytokines, immune cell numbers, and subset distribution and functions in response to individual bouts of exercise are transient, repeating these bouts frequently over extended periods of time leads to a profound remodelling of the immune system in older adults, and facilitates inflammation resolution.

A recent cross-sectional study in 211 older healthy adults (67 ± 5 years) showed that neutrophils isolated from older adults who tended to walk more than 10,000 steps per day were able to migrate towards the chemoattractant IL-8 with greater accuracy than neutrophils from relatively sedentary older adults who walked up to 5,000 steps per day (Bartlett et al., 2016). Perhaps more importantly, the same group showed that a simple 10-week long exercise intervention was sufficient to improve neutrophil phagocytosis and oxidative burst along with monocyte phagocytosis and the percentage of monocytes producing an oxidative burst (Bartlett et al., 2017). This interesting study compared the efficacy of two different exercise interventions, 30–45 minutes of moderate-intensity continuous cycling at 70% of maximum heart rate or 18–25 minutes of high-intensity-interval-training (HIIT) on a cycle ergometer with bouts of between 15 and 60 seconds at >90% of maximum heart rate. Both interventions led to similar improvements in neutrophil and monocyte functions. Another cross-sectional study highlighted similar positive associations between regular exercise (more than 1 hour at least twice a week for over three years) and greater neutrophil phagocytosis in older exercisers (age 65.1 ± 4.5 years) than in age-matched sedentary controls (66.3 ± 3.2 years) (Yan et al., 2001).

Age-associated decrements in NK cell function also appear to be at least partly curtailed with regular exercise. Crist et al. (1989) compared the impact of 16 weeks of aerobic exercise (20 minutes, three times per week at 50% of Heart Rate Reserve) on the ability of NK cells from 22 older women (72 ± 1 years) to kill myelogenous leukaemia cells K562 *in vitro* (Crist, Mackinnon, Thompson, Atterbom, & Egan, 1989). Specific lysis (%) of K562 cells by isolated NK cells increased in both the exercising group and the non-exercising control; however, the increase in NK cell function was the greatest in the exercising women. On the other hand, certain studies exploring the effects of exercise have found apparently contradictory results. When assessing changes in NK-cell cytotoxicity of older adults in response to supervised moderate aerobic exercise (three times a week) for six months in men (Woods et al., 1999), the authors highlighted improvements in the exercise group, compared with the non-exercising control group. However, in a study conducted by Nieman et al. (1993), 12 weeks of walking for 30 minutes at 60% of Heart Rate Reserve in previously sedentary older women (67–85 years) (Nieman et al., 1993) was not sufficient to improve NK cell function. While these results could suggest that the intervention was not long enough to alter NK cell function in this population, it is crucial to note that the older women who were categorised as highly active in this study (aerobic exercisers for at least 7 hours a week for the past five years) exhibited the highest NK cell killing capacity of K562 cells, compared with their age-matched sedentary counterparts (Nieman et al., 1993) at baseline. If 12 weeks is too short to see improvements in NK-cell cytotoxicity, slightly longer interventions of 15 weeks of cycling at 70–75% VO_2 peak (three times per week) have resulted in increased NK cell cytotoxicity in post-menopausal breast cancer survivors (65–85 years) (Fairey et al., 2005).

Considering the benefits of exercise training on neutrophils and NK cells in older adults, it is not surprising that other cell types such as B cells and T cells are also impacted. In this context, a study by Papp et al. (2021) followed 29 older women (67.0 ± 3.7 years) who took part in a six-week long functional gymnastic intervention (Papp et al., 2021). At the end of the intervention, the frequency of circulating double-negative (DN) B cells (CD19+/IgD−/CD27−) decreased (4.8 ± 2.0% vs. 4.0 ± 1.8%; p = 0.0298), while the percentages of un-switched memory B cells (CD19+/IgD+/CD27+) increased (11.6 ± 6.6% vs. 12.3 ± 6.5%; p = 0.0246). Considering that DN B cells have been implicated with the propagation of inflamm-ageing, the authors concluded that the decrease in DN B cells, with the concomitant increase in un-switched memory B cells indicates that the exercise intervention led to overall improvements in B cell competency in this population. Conversely, cross-sectional studies comparing circulating B cell subsets between healthy young adults ($n = 55$, 20–36 years), with sedentary older ($n = 75$, 57–80 years) and master cyclists aged 55–79 years ($n = 125$), have reported contradictory findings (Duggal, Pollock, Lazarus, Harridge, & Lord, 2018). Indeed, a study from Duggal et al. (2018) showed that while lifelong exercisers have fewer circulating naive B cells (CD19+/ IgD+/CD27−) than young adults, they have more of these cells than their age-matched sedentary counterparts (Duggal et al., 2018). Furthermore, contrary to the results obtained by Papp et al. (2021), active older adults had more DN B cells than sedentary older participants and fewer un-switched memory B cells. Lifelong exercisers also exhibited greater frequencies of B-regulatory cells than their age-matched sedentary counterparts (Duggal et al., 2018). While this study has some limitations, including its cross-sectional nature, it suggests that regular exercise appears to be beneficial for dampening the age-associated decline in B cell function and enhancing the regulation of immune responses in older adults.

Sedentary behaviour and other modifiable lifestyle factors are associated with premature accumulation of senescent T cells, regardless of age. Indeed, smoking, excessive adiposity and sedentary lifestyles appear to accelerate telomere shortening in peripheral blood mononuclear cells, suggesting a consequential increase in dysfunctional senescent T cells accumulation, to the detriment of competent naive and central-memory T cells in young adults (Muezzinler, Zaineddin, & Brenner, 2014; Valdes et al., 2005). Furthermore, physical inactivity and the associated excess-body mass in young adolescents have also been linked with the premature accumulation of senescent T cells to the detriment of naive cells (Spielmann, Johnston, O'Connor, Foreyt, & Simpson, 2014). This association between body composition and accumulation of dysfunctional senescent T cells in children aged 10–13 years is surprising as these features of immunosenescence are classically observed in much older adults. Consequently, it could be hypothesised that a lifetime of sedentary behaviour may be a more impactful contributor to decline in immune competency than the passing of time alone. This is supported by cross-sectional studies highlighting that physically active older men and women are more likely to exhibit reductions in biomarkers associated with immunosenescence compared to their sedentary counterparts (Simpson & Guy, 2010), including longer leukocyte telomere lengths, enhanced *in vitro* T cell responses to mitogens (Ferguson, Wikby, Maxson, Olsson, & Johansson, 1995), elevated *in vivo* immune responses to vaccines and recall antigens (Effros et al., 1994; Goronzy et al., 2001). Regular physical activity and greater aerobic fitness are strongly associated with reduced proportions of senescent and higher proportions of naive cells, particularly within the CD8+ T cell compartment in healthy adult men aged 18–61 years (Spielmann et al., 2011). Since the aforementioned association between aerobic fitness and numbers of senescent T cells in men withstood adjustment for age and CMV serostatus, the importance of physical activity may be potentially greater than chronological age and infection history on immunosenescence. This concept is further supported by studies conducted on master athletes, and lifelong exercisers who exhibit greater recent thymic emigrants (RTE) and naive T cells, along with reduced accumulation of highly differentiated memory EMRA T cells (CD45RA+/CCR7−) (Duggal et al., 2018), and in some reports, senescent CD4+ and CD8+ T cells (Minuzzi et al., 2018) than their age-matched sedentary counterparts. As discussed previously, lower hallmarks of T cell senescence observed in active older adults appear to be at least partially mediated reductions in age-associated thymic involution in those that exercise regularly due to the frequent release of the thymoprotective myokines IL-7 and IL-15 (Bartlett & Duggal, 2020). In master athletes, both myokines are found in greater concentrations in circulation at rest, than in sedentary older adults (Duggal et al., 2018), advocating for a role of repeated bouts of exercise on delaying the onset of immunosenescence.

Researchers have also suggested that regular physical activity exerts moderating effects on immunosenescence and the reduction associated inflamm-ageing by facilitating the destruction of excess viral-specific T cell clones via apoptosis, freeing up "immune space" for naive T cells to occupy and expand the antigenic T cell repertoire. Recently, Niemiro et al. (2022) conducted a randomised 12-week exercise intervention study in 24 post-menopausal women to compare the effects of high-intensity interval training (HITT) with moderate-intensity continuous training (MICT) on T-cell rejuvenation. Following the intervention, HITT led to reductions in CD4+ naive T cells, CD4+ RTE, and CD4:CD8 ratio, while MICT led to increased total T cells, CD4+ naive T cells, CD4+ CM T cells, and CD4+ RTE (Niemiro et al., 2022). The authors also found a near significant inverse relationship between improvements in aerobic fitness and the percentage of EMRA CD8+ T cells in both groups, from which they concluded that remodelling of the T cell compartment is likely dependent on significant improvements in fitness. This is of particular clinical relevance, since prior studies have identified improvements in neutrophil function in the aged in response to HITT (Bartlett et al., 2017), while in the Niemiro study HIIT reduced naive CD4+ T cell numbers by ~38% as opposed to MICT which tended to increase naive CD4+ T cells by ~20%. These results advocate for the need of personalised exercise intervention to improve specific components of the immune system in older adults, based on their personalised needs. Similarly, resistance training appears to elicit significant changes in the proportions of senescent T cells, with a study showing that a six-week strength endurance training intervention in CMV-seropositive women aged 61–79 years ($n = 100$) is sufficient to reduce circulating CD28−/CD57+ senescent CD8+ T cells by up to 44%, while also increasing naive CD8+ T cells (Cao Dinh et al., 2019). Similar findings have been shown in a study of 29 healthy older women (67.0 ± 3.7 years) who engaged in functional exercise sessions twice a week (lower and upper body exercises for 1 hour per session) for six weeks (Papp et al., 2021). While fewer changes were identified among CD4+ T cells at the end of the intervention, the six weeks of functional training led to reductions in EMRA CD8+ T cells and to concomitant increases in naive CD4+ T cells (Papp et al., 2021). Finally, regular physical activity appears to improve the function of unconventional T cells such as MAIT cells. For example, one study has shown that 12 weeks of combined aerobic and resistance training, restores some of the age-associated decline in MAIT cells' responsiveness to acute stressors such as a bout of exercise (Bates et al., 2021).

Soluble responses to regular exercise

Regular aerobic exercise moderates the level of circulating pro-inflammatory cytokines (Petersen & Pedersen, 2005) such as IL-6 and TNF-α and lowers LPS-stimulated *in vitro* secretion of IL-6, TNF-α, and IL-1β (Phillips, Flynn, McFarlin, Stewart, & Timmerman, 2010). These results were corroborated by a recent study examining the impact of physical activity and sedentary behaviour on over 3,000 older men (Parsons et al., 2017). In this population-based study, Parsons et al. highlighted an inverse relationship between moderate-level, low-level, and total level of physical activity and circulating IL-6 and CRP, suggesting that older men who were physically active were less likely to be affected by inflamm-ageing than their sedentary counterparts (Parsons et al., 2017). In addition to regular physical activity, six months of moderate-intensity aerobic training in older adults, was sufficient to reduce inflamm-ageing by promoting anti-inflammatory IL-4, IL-10, and TGF-β1 cytokine production in mitogen-stimulated immune cells, providing further evidence of therapeutic properties offered by physical activity (Smith, Dykes, Douglas, Krishnaswamy, & Berk, 1999). This reduction in inflamm-ageing is even clearer in highly fit older populations. For example, Minuzzi et al. (2019) showed that master athletes, as defined by individuals aged over 40 years old and who had a minimum of 20 years of regular participation in competitive sports and remained conditioned ($n = 20$, 53.1 ± 8.8 years), had higher circulating levels of anti-inflammatory IL-10, IL-4, and IL-1ra than their sedentary adults of the same age (Minuzzi et al., 2019). This is in support of previous findings from Duggal and colleagues, in which older male cyclists able to cycle 100 km in under 6.5 hours and older female cyclists able to cycle 60 km in under 5.5 hours, exhibited lower serum levels of the inflammatory thymosuppresive cytokines IL-6 than their age- and sex-matched sedentary counterparts.

Systemic immune response to regular exercise

Sedentary behaviour and physical inactivity in older adults are associated with poor vaccine efficacy, while older adults who meet or exceed the physical activity guidelines exhibit greater vaccine efficacy (de Araujo et al., 2015; Kohut, Cooper, Nickolaus, Russell, & Cunnick, 2002), suggesting that the beneficial impact of exercise training extends to B cell function. A study conducted of 112 older adult participants over the age of 75 years, assessed the relationship between physical activity, as determined by accelerometer-measured step counts and immune response to influenza vaccination (Wong et al., 2019). The authors divided the participants in quartiles based on sex and daily step counts, and classified women as sedentary (<10,927 steps/day, n = 28) or highly active (>18,509 steps/day, n = 28), and males as sedentary (<7174 steps/day, n = 17) or highly active (>14,770 steps/day, n = 17). While there was no difference in the proportion of circulating B cells or plasmablasts in sedentary and active men and women prior to vaccination, fitter participants had a greater increase in total B cells, plasmablasts and vaccine-specific B cells as early as seven days post-vaccination. Furthermore, B cells isolated from fitter participants were able to secrete more vaccine-specific IgG, as measured by Enzyme-Linked ImmunoSpot assays, than their sedentary counterparts, regardless of sex. This data confirms prior reports that a 10-month cardiovascular exercise intervention increased seroprotection against influenza. This study conducted by Woods et al. (2009) included 144 older adults (control: 70.1 ± 5.7 years, n = 70, exercisers: 69.6 ± 4.9 years, n = 74) who exercised under supervision for three days per week (walking, cycling, elliptical exercise, or stair climbing) at 60–70% VO_2 peak (Woods et al., 2009). The antibody response to influenza vaccination was similar for both exercisers and controls at week 3 and week 6 post-vaccination; however, exercise training led to a prolonged seroprotection against influenza, with antibody levels remaining stable at 24 weeks post-vaccination.

Summary – potential mechanisms of immune enhancement with exercise during ageing

The immune system undergoes significant changes with ageing, contributing to increased infection rates, low-grade inflammation, heightened disease susceptibility, and diminished vaccine responses. Throughout this chapter, we presented the beneficial effects of individual bouts of exercise, along with the effects of regular exercise training on multiple features of an ageing immune system. It is not clear, however, if exercise only functions to delay age-related declines in immunity, or if it can also help rejuvenate an already aged immune system. While cross-sectional studies provide limited insights due to the likelihood that physically active participants have maintained their activity levels over an extended period, recent randomised controlled trials in sedentary older adults are offering emerging evidence that exercise can indeed rejuvenate the declining immune system. Understanding the potential mechanisms behind a dual role for exercise in delaying and rejuvenating immunosenescence involves a multifaceted exploration. The preventive effects of exercise on immunosenescence encompass intricate processes, including influences on lymphoid progenitor cells in the bone marrow, the actions of muscle-derived cytokines, reductions in chronic low-grade inflammation, enhancements in antioxidant defences, improved microbiota composition, and direct modulation of diverse immune cell composition and function.

A critical aspect contributing to the protective effect of exercise is the reduction of chronic low-grade inflammation, a significant stressor on the immune system. Habitual physical activity, particularly when it leads to lower levels of body fat, proves advantageous as individuals age. Adipose tissue, prone to accumulation with age, houses macrophages and senescent cells that contribute to pro-inflammatory signals, intensifying inflamm-ageing and heightening susceptibility to age-related diseases (as discussed in Chapter 6). Skeletal muscle, recognised as an endocrine organ, releases myokines during exercise, including IL-6, conventionally viewed as pro-inflammatory, but exerting anti-inflammatory signals during exercise. This myokine-mediated anti-inflammatory environment is pivotal in supporting a healthier immune system during ageing. Myokines such as IL-7 and IL-15, expressed and released by exercising muscle, have essential roles in thymocyte development and lymphocyte proliferation, particularly for naive T cells. The decline in serum levels of these cytokines with age underscores the importance of active skeletal muscle in maintaining a healthy immune system, especially as physical inactivity and age-related sarcopenia limit muscle's immune regulatory function in old age.

Exercise also appears to protect against the reactivation of persistent viruses, such as CMV, known to burden the adaptive immune system. Higher cardiorespiratory fitness levels correlate inversely with latent viral control, particularly in people aged over 65 years. The mechanisms through which physical activity improves latent viral control involve potential triggers, including the redistribution of catecholamine-sensitive CD8+ T cells with viral antigen specificity and a highly differentiated phenotype. Stress reduction through exercise (see Chapter 15) most likely contributes to this protective effect, countering chronic stress which is linked to CMV reactivation and the induction of a senescent phenotype in peripheral T cells.

In terms of reversing immunosenescence, randomised control trials utilising exercise training to enhance immune responses to vaccination provide robust evidence. The concept of exercise increasing "immune space" emerged in the early 2010s as a theory explaining this phenomenon, suggesting that habitual exercise promotes the apoptosis or autophagy of senescent cells, allowing their replacement by new, fully functional cells, particularly T cells (Simpson & Guy, 2010). This process of exercise-induced cell death initiates triggers promoting the production of haematopoietic progenitor cells from the bone marrow, ultimately increasing the pool of fully functional naive T cells in the periphery. Muscle-derived cytokines, especially IL-7, are likely to have a key role in maintaining thymic mass and fostering the production of new naive T cells. While the precise mechanisms remain elusive, the notion of creating immune space through exercise stands as a compelling avenue in the rejuvenation of the age-weary immune system.

Key Points

- Chronological ageing is associated with functional declines in innate immunity.
- Hallmarks of immunosenescence include chronic low-grade inflammation (inflamm-ageing*)*, altered ability of innate immune cells to phagocytose pathogens, accumulation of terminally differentiated T cells and impaired vaccine responses.
- Lifestyle factors, including sedentary behaviour, accelerate the onset of immunosenescence and further aggravate its progression.

- Individual bouts of exercise stimulate immune cell mobilisation and surveillance in both young and old adults, while also inducing the release of thymoprotective myokines (IL-7 and IL-15), and anti-inflammatory cytokines IL-6 and IL-10 in older adults.
- Regular physical activity is associated with improved hallmarks of immunosenescence, including reduced inflamm-ageing, improved vaccination responses, neutrophil, and T cell proportions and effector functions.
- Future studies are needed to evaluate the therapeutic usage of different exercise modalities, intensity, and duration on the potential rejuvenation of the immune system in older adults.

References

Adrover, J. M., Nicolas-Avila, J. A., & Hidalgo, A. (2016). Aging: A temporal dimension for neutrophils. *Trends Immunol, 37*(5), 334–345.

Agrawal, A., Agrawal, S., Cao, J. N., Su, H., Osann, K., & Gupta, S. (2007). Altered innate immune functioning of dendritic cells in elderly humans: A role of phosphoinositide 3-kinase-signaling pathway. *J Immunol, 178*(11), 6912–6922.

Agrawal, S., Gollapudi, S., Gupta, S., & Agrawal, A. (2013). Dendritic cells from the elderly display an intrinsic defect in the production of IL-10 in response to lithium chloride. *Exp Gerontol, 48*(11), 1285–1292.

Allavena, P., Garlanda, C., Borrello, M. G., Sica, A., & Mantovani, A. (2008). Pathways connecting inflammation and cancer. *Curr Opin Genet Dev, 18*(1), 3–10.

Almeida-Oliveira, A., Smith-Carvalho, M., Porto, L. C., Cardoso-Oliveira, J., Ribeiro Ados, S., Falcao, R. R., … Diamond, H. R. (2011). Age-related changes in natural killer cell receptors from childhood through old age. *Hum Immunol, 72*(4), 319–329.

Arai, Y., Martin-Ruiz, C. M., Takayama, M., Abe, Y., Takebayashi, T., Koyasu, S., … von Zglinicki, T. (2015). Inflammation, but not telomere length, predicts successful ageing at extreme old age: A longitudinal study of semi-supercentenarians. *EBioMedicine, 2*(10), 1549–1558.

Bartlett, D. B., & Duggal, N. A. (2020). Moderate physical activity associated with a higher naive/memory T-cell ratio in healthy old individuals: Potential role of IL15. *Age Ageing, 49*(3), 368–373.

Bartlett, D. B., Firth, C. M., Phillips, A. C., Moss, P., Baylis, D., Syddall, H., … Lord, J. M. (2012). The age-related increase in low-grade systemic inflammation (Inflammaging) is not driven by cytomegalovirus infection. *Aging Cell, 11*(5), 912–915.

Bartlett, D. B., Fox, O., McNulty, C. L., Greenwood, H. L., Murphy, L., Sapey, E., … Lord, J. M. (2016). Habitual physical activity is associated with the maintenance of neutrophil migratory dynamics in healthy older adults. *Brain Behav Immun, 56*, 12–20.

Bartlett, D. B., Shepherd, S. O., Wilson, O. J., Adlan, A. M., Wagenmakers, A. J. M., Shaw, C. S., & Lord, J. M. (2017). Neutrophil and monocyte bactericidal responses to 10 weeks of low-volume high-intensity interval or moderate-intensity continuous training in sedentary adults. *Oxid Med Cell Longev, 2017*, 8148742.

Bate, S. L., Dollard, S. C., & Cannon, M. J. (2010). Cytomegalovirus seroprevalence in the United States: The national health and nutrition examination surveys, 1988–2004. *Clin Infect Dis, 50*(11), 1439–1447.

Bates, L. C., Hanson, E. D., Levitt, M. M., Richie, B., Erickson, E., Bartlett, D. B., & Phillips, M. D. (2021). Mucosal-associated invariant t cell response to acute exercise and exercise training in older obese women. *Sports (Basel), 9*(10).

Becker, L., Nguyen, L., Gill, J., Kulkarni, S., Pasricha, P. J., & Habtezion, A. (2018). Age-dependent shift in macrophage polarisation causes inflammation-mediated degeneration of enteric nervous system. *Gut, 67*(5), 827–836.

Béziat, V., Descours, B., Parizot, C., Debré, P., & Vieillard, V. (2010). NK cell terminal differentiation: Correlated stepwise decrease of NKG2A and acquisition of KIRs. *PloS One, 5*(8), e11966–e11966.

Borrego, F., Alonso, M. C., Galiani, M. D., Carracedo, J., Ramirez, R., Ostos, B., … Solana, R. (1999). NK phenotypic markers and IL2 response in NK cells from elderly people. *Exp Gerontol, 34*(2), 253–265.

Breuil, V., Ticchioni, M., Testa, J., Roux, C. H., Ferrari, P., Breittmayer, J. P., … Carle, G. F. (2010). Immune changes in post-menopausal osteoporosis: The Immunos study. *Osteoporos Int, 21*(5), 805–814.

Bruunsgaard, H., Andersen-Ranberg, K., Jeune, B., Pedersen, A. N., Skinhoj, P., & Pedersen, B. K. (1999). A high plasma concentration of TNF-alpha is associated with dementia in centenarians. *J Gerontol A Biol Sci Med Sci, 54*(7), M357–364.

Brzezinska, H., Koziorowska, B., Wirpsza, R., & Zawadzki, R. (1977). [Tissue damage in the larynx in children following endotracheal intubation. Clinical and pathomorphological studies]. *Otolaryngol Pol, 31*(3), 347–351.

Busse, S., Steiner, J., Alter, J., Dobrowolny, H., Mawrin, C., Bogerts, B., … Busse, M. (2015). Expression of HLA-DR, CD80, and CD86 in healthy aging and Alzheimer's disease. *J Alzheimers Dis, 47*(1), 177–184.

Callender, L. A., Carroll, E. C., Beal, R. W. J., Chambers, E. S., Nourshargh, S., Akbar, A. N., & Henson, S. M. (2018). Human CD8(+) EMRA T cells display a senescence-associated secretory phenotype regulated by p38 MAPK. *Aging Cell, 17*(1), e12675.

Cannon, J. G., Fiatarone, M. A., Fielding, R. A., & Evans, W. J. (1994). Aging and stress-induced changes in complement activation and neutrophil mobilization. *J Appl Physiol (1985), 76*(6), 2616–2620.

Cannon, J. G., Orencole, S. F., Fielding, R. A., Meydani, M., Meydani, S. N., Fiatarone, M. A., … Evans, W. J. (1990). Acute phase response in exercise: Interaction of age and vitamin E on neutrophils and muscle enzyme release. *Am J Physiol, 259*(6 Pt 2), R1214–1219.

Cao Dinh, H., Bautmans, I., Beyer, I., Onyema, O. O., Liberman, K., De Dobbeleer, L., … Njemini, R. (2019). Six weeks of strength endurance training decreases circulating senescence-prone T-lymphocytes in cytomegalovirus seropositive but not seronegative older women. *Immun Ageing, 16*, 17.

Cao, Y., Fan, Y., Li, F., Hao, Y., Kong, Y., Chen, C., … Zeng, H. (2022). Phenotypic and functional alterations of monocyte subsets with aging. *Immun Ageing, 19*(1), 63.

Ceddia, M. A., Price, E. A., Kohlmeier, C. K., Evans, J. K., Lu, Q., McAuley, E., & Woods, J. A. (1999). Differential leukocytosis and lymphocyte mitogenic response to acute maximal exercise in the young and old. *Med Sci Sports Exerc, 31*(6), 829–836.

Chen, X., Wang, Y., Wang, J., Wen, J., Jia, X., Wang, X., & Zhang, H. (2018). Accumulation of T-helper 22 cells, interleukin-22 and myeloid-derived suppressor cells promotes gastric cancer progression in elderly patients. *Oncol Lett, 16*(1), 253–261.

Cho, R. H., Sieburg, H. B., & Muller-Sieburg, C. E. (2008). A new mechanism for the aging of hematopoietic stem cells: Aging changes the clonal composition of the stem cell compartment but not individual stem cells. *Blood, 111*(12), 5553–5561.

Christensen, H., Boysen, G., & Johannesen, H. H. (2004). Serum-cortisol reflects severity and mortality in acute stroke. *J Neurol Sci, 217*(2), 175–180.

Claes, N., Fraussen, J., Vanheusden, M., Hellings, N., Stinissen, P., Van Wijmeersch, B., … Somers, V. (2016). Age-associated B cells with proinflammatory characteristics are expanded in a proportion of multiple Sclerosis patients. *J Immunol, 197*(12), 4576–4583.

Coppe, J. P., Patil, C. K., Rodier, F., Sun, Y., Munoz, D. P., Goldstein, J., … Campisi, J. (2008). Senescence-associated secretory phenotypes reveal cell-nonautonomous functions of oncogenic RAS and the p53 tumor suppressor. *PLoS Biol, 6*(12), 2853–2868.

Cosmi, L., De Palma, R., Santarlasci, V., Maggi, L., Capone, M., Frosali, F., … Annunziato, F. (2008). Human interleukin 17-producing cells originate from a CD161+CD4+ T cell precursor. *J Exp Med, 205*(8), 1903–1916.

Crane, J. D., MacNeil, L. G., Lally, J. S., Ford, R. J., Bujak, A. L., Brar, I. K., … Tarnopolsky, M. A. (2015). Exercise-stimulated interleukin-15 is controlled by AMPK and regulates skin metabolism and aging. *Aging Cell, 14*(4), 625–634.

Crist, D. M., Mackinnon, L. T., Thompson, R. F., Atterbom, H. A., & Egan, P. A. (1989). Physical exercise increases natural cellular-mediated tumor cytotoxicity in elderly women. *Gerontology, 35*(2–3), 66–71.

Cury-Boaventura, M. F., Gorjao, R., de Moura, N. R., Santos, V. C., Bortolon, J. R., Murata, G. M., … Hatanaka, E. (2018). The effect of a competitive futsal match on T lymphocyte surface receptor signaling and functions. *Front Physiol, 9*, 202.

Davies, L. C., Jenkins, S. J., Allen, J. E., & Taylor, P. R. (2013). Tissue-resident macrophages. *Nat Immunol, 14*(10), 986–995.

de Araujo, A. L., Silva, L. C., Fernandes, J. R., Matias Mde, S., Boas, L. S., Machado, C. M., … Benard, G. (2015). Elderly men with moderate and intense training lifestyle present sustained higher antibody responses to influenza vaccine. *Age (Dordr), 37*(6), 105.

Della Bella, S., Bierti, L., Presicce, P., Arienti, R., Valenti, M., Saresella, M., … Villa, M. L. (2007). Peripheral blood dendritic cells and monocytes are differently regulated in the elderly. *Clinical Immunology, 122*(2), 220–228.

Duggal, N. A., Pollock, R. D., Lazarus, N. R., Harridge, S., & Lord, J. M. (2018). Major features of immunesenescence, including reduced thymic output, are ameliorated by high levels of physical activity in adulthood. *Aging Cell, 17*(2), e12750.

Edwards, K. M., Pascoe, A. R., Fiatarone-Singh, M. S., Singh, N. A., Kok, J., & Booy, R. (2015). A randomised controlled trial of resistance exercise prior to administration of influenza vaccination in older adults. *Brain Behav Immun, 49*, e24–e25.

Effros, R. B., Boucher, N., Porter, V., Zhu, X., Spaulding, C., Walford, R. L., … Schachter, F. (1994). Decline in CD28+ T cells in centenarians and in long-term T cell cultures: A possible cause for both in vivo and in vitro immunosenescence. *Exp Gerontol, 29*(6), 601–609.

Ershler, W. B., Sun, W. H., Binkley, N., Gravenstein, S., Volk, M. J., Kamoske, G., … Weindruch, R. (1993). Interleukin-6 and aging: Blood levels and mononuclear cell production increase with advancing age and in vitro production is modifiable by dietary restriction. *Lymphokine Cytokine Res, 12*(4), 225–230.

Evans, E. S., Hackney, A. C., McMurray, R. G., Randell, S. H., Muss, H. B., Deal, A. M., & Battaglini, C. L. (2015). Impact of acute intermittent exercise on natural killer cells in breast cancer survivors. *Integr Cancer Ther, 14*(5), 436–445.

Evrard, M., Kwok, I. W. H., Chong, S. Z., Teng, K. W. W., Becht, E., Chen, J., … Ng, L. G. (2018). Developmental analysis of bone marrow neutrophils reveals populations specialized in expansion, trafficking, and effector functions. *Immunity, 48*(2), 364–379.e368.

Fairey, A. S., Courneya, K. S., Field, C. J., Bell, G. J., Jones, L. W., & Mackey, J. R. (2005). Randomized controlled trial of exercise and blood immune function in postmenopausal breast cancer survivors. *J Appl Physiol (1985), 98*(4), 1534–1540.

Ferguson, F. G., Wikby, A., Maxson, P., Olsson, J., & Johansson, B. (1995). Immune parameters in a longitudinal study of a very old population of Swedish people: A comparison between survivors and nonsurvivors. *J Gerontol A Biol Sci Med Sci, 50*(6), B378–382.

Fiatarone, M. A., Morley, J. E., Bloom, E. T., Benton, D., Solomon, G. F., & Makinodan, T. (1989). The effect of exercise on natural killer cell activity in young and old subjects. *J Gerontol, 44*(2), M37–45.

Flynn, M. G., Fahlman, M., Braun, W. A., Lambert, C. P., Bouillon, L. E., Brolinson, P. G., & Armstrong, C. W. (1999). Effects of resistance training on selected indexes of immune function in elderly women. *J Appl Physiol (1985), 86*(6), 1905–1913.

Forsey, R. J., Thompson, J. M., Ernerudh, J., Hurst, T. L., Strindhall, J., Johansson, B., … Wikby, A. (2003). Plasma cytokine profiles in elderly humans. *Mech Ageing Dev, 124*(4), 487–493.

Fowler, K., Mucha, J., Neumann, M., Lewandowski, W., Kaczanowska, M., Grys, M., … Diaz-Decaro, J. (2022). A systematic literature review of the global seroprevalence of cytomegalovirus: Possible implications for treatment, screening, and vaccine development. *BMC Public Health, 22*(1), 1659.

Franceschi, C., Capri, M., Monti, D., Giunta, S., Olivieri, F., Sevini, F., … Salvioli, S. (2007). Inflammaging and anti-inflammaging: A systemic perspective on aging and longevity emerged from studies in humans. *Mech Ageing Dev, 128*(1), 92–105.

Frasca, D., Diaz, A., Romero, M., & Blomberg, B. B. (2017). Human peripheral late/exhausted memory B cells express a senescent-associated secretory phenotype and preferentially utilize metabolic signaling pathways. *Exp Gerontol, 87*(Pt A), 113–120.

Frasca, D., Diaz, A., Romero, M., D'Eramo, F., & Blomberg, B. B. (2017). Aging effects on T-bet expression in human B cell subsets. *Cell Immunol, 321*, 68–73.

Frasca, D., Landin, A. M., Lechner, S. C., Ryan, J. G., Schwartz, R., Riley, R. L., & Blomberg, B. B. (2008). Aging down-regulates the transcription factor E2A, activation-induced cytidine deaminase, and Ig class switch in human B cells. *J Immunol, 180*(8), 5283–5290.

Frasca, D., Romero, M., Garcia, D., Diaz, A., & Blomberg, B. B. (2021). Obesity Accelerates Age-Associated Defects in Human B Cells Through a Metabolic Reprogramming Induced by the Fatty Acid Palmitate. *Front Aging, 2*, 828697.

Fulop, T., Larbi, A., Dupuis, G., Le Page, A., Frost, E. H., Cohen, A. A., … Franceschi, C. (2017). Immunosenescence and inflamm-aging as two sides of the same coin: Friends or foes? *Front Immunol, 8*, 1960.

Giunta, B., Fernandez, F., Nikolic, W. V., Obregon, D., Rrapo, E., Town, T., & Tan, J. (2008). Inflammaging as a prodrome to Alzheimer's disease. *J Neuroinflammation, 5*, 51.

Goronzy, J. J., Fulbright, J. W., Crowson, C. S., Poland, G. A., O'Fallon, W. M., & Weyand, C. M. (2001). Value of immunological markers in predicting responsiveness to influenza vaccination in elderly individuals. *J Virol, 75*(24), 12182–12187.

Gounder, S. S., Abdullah, B. J. J., Radzuanb, N., Zain, F., Sait, N. B. M., Chua, C., & Subramani, B. (2018). Effect of aging on NK cell population and their proliferation at Ex vivo culture condition. *Anal Cell Pathol (Amst), 2018*, 7871814.

Graff, R. M., Jennings, K., LaVoy, E. C. P., Warren, V. E., Macdonald, B. W., Park, Y., & Markofski, M. M. (2022). T-cell counts in response to acute cardiorespiratory or resistance exercise in physically active or physically inactive older adults: A randomized crossover study. *J Appl Physiol (1985), 133*(1), 119–129.

Graff, R. M., Kunz, H. E., Agha, N. H., Baker, F. L., Laughlin, M., Bigley, A. B., … Simpson, R. J. (2018). beta(2)-Adrenergic receptor signaling mediates the preferential mobilization of differentiated subsets of CD8+ T-cells, NK-cells and non-classical monocytes in response to acute exercise in humans. *Brain Behav Immun, 74*, 143–153.

Gunther, P., & Schultze, J. L. (2019). Mind the map: Technology shapes the myeloid cell space. *Front Immunol, 10*, 2287.

Hallam, J., Jones, T., Alley, J., & Kohut, M. L. (2022). Exercise after influenza or COVID-19 vaccination increases serum antibody without an increase in side effects. *Brain Behav Immun, 102*, 1–10.

Haugen, F., Norheim, F., Lian, H., Wensaas, A. J., Dueland, S., Berg, O., … Drevon, C. A. (2010). IL-7 is expressed and secreted by human skeletal muscle cells. *Am J Physiol Cell Physiol, 298*(4), C807–816.

Hazeldine, J., Harris, P., Chapple, I. L., Grant, M., Greenwood, H., Livesey, A., … Lord, J. M. (2014). Impaired neutrophil extracellular trap formation: A novel defect in the innate immune system of aged individuals. *Aging Cell, 13*(4), 690–698.

Henson, S. M., Lanna, A., Riddell, N. E., Franzese, O., Macaulay, R., Griffiths, S. J., … Akbar, A. N. (2014). p38 signaling inhibits mTORC1-independent autophagy in senescent human CD8(+) T cells. *J Clin Invest, 124*(9), 4004–4016.

Holodick, N. E., Rodriguez-Zhurbenko, N., & Hernandez, A. M. (2017). Defining natural antibodies. *Front Immunol, 8*, 872.

Jenks, S. A., Cashman, K. S., Zumaquero, E., Marigorta, U. M., Patel, A. V., Wang, X., … Sanz, I. (2018). Distinct effector B cells induced by unregulated toll-like Receptor 7 contribute to pathogenic responses in systemic lupus erythematosus. *Immunity, 49*(4), 725–739, e726.

Khan, N., Cobbold, M., Keenan, R., & Moss, P. A. (2002). Comparative analysis of CD8+ T cell responses against human cytomegalovirus proteins pp65 and immediate early 1 shows similarities in precursor frequency, oligoclonality, and phenotype. *J Infect Dis, 185*(8), 1025–1034.

Kim, M. H., Yang, D., Kim, M., Kim, S. Y., Kim, D., & Kang, S. J. (2017). A late-lineage murine neutrophil precursor population exhibits dynamic changes during demand-adapted granulopoiesis. *Sci Rep, 7*, 39804.

Kohut, M. L., Boehm, G. W., & Moynihan, J. A. (2001). Moderate exercise is associated with enhanced antigen-specific cytokine, but not IgM antibody production in aged mice. *Mech Ageing Dev, 122*(11), 1135–1150.

Kohut, M. L., Cooper, M. M., Nickolaus, M. S., Russell, D. R., & Cunnick, J. E. (2002). Exercise and psychosocial factors modulate immunity to influenza vaccine in elderly individuals. *J Gerontol A Biol Sci Med Sci, 57*(9), M557–562.

Lee, J. S., Lee, W. W., Kim, S. H., Kang, Y., Lee, N., Shin, M. S., … Kang, I. (2011). Age-associated alteration in naive and memory Th17 cell response in humans. *Clin Immunol, 140*(1), 84–91.

Le Garff-Tavernier, M., Béziat, V., Decocq, J., Siguret, V., Gandjbakhch, F., Pautas, E., … Vieillard, V. (2010). Human NK cells display major phenotypic and functional changes over the life span. *Aging Cell, 9*(4), 527–535.

Lio, D., Candore, G., Crivello, A., Scola, L., Colonna-Romano, G., Cavallone, L., … Caruso, C. (2004). Opposite effects of interleukin 10 common gene polymorphisms in cardiovascular diseases and in successful ageing: Genetic background of male centenarians is protective against coronary heart disease. *J Med Genet, 41*(10), 790–794.

Lio, D., Scola, L., Crivello, A., Colonna-Romano, G., Candore, G., Bonafe, M., … Caruso, C. (2002). Gender-specific association between -1082 IL-10 promoter polymorphism and longevity. *Genes Immun, 3*(1), 30–33.

Long, J. E., Ring, C., Drayson, M., Bosch, J., Campbell, J. P., Bhabra, J., … Burns, V. E. (2012). Vaccination response following aerobic exercise: Can a brisk walk enhance antibody response to pneumococcal and influenza vaccinations? *Brain Behav Immun, 26*(4), 680–687.

MacNeil, L. G., Tarnopolsky, M. A., & Crane, J. D. (2021). Acute, exercise-induced alterations in cytokines and chemokines in the blood distinguish physically active and sedentary aging. *J Gerontol A Biol Sci Med Sci, 76*(5), 811–818.

Malm, C. (2004). Exercise immunology: The current state of man and mouse. *Sports Med, 34*(9), 555–566.

Mantovani, A., & Pierotti, M. A. (2008). Cancer and inflammation: A complex relationship. *Cancer Lett, 267*(2), 180–181.

Martin, V., Wu, Y. C., Kipling, D., & Dunn-Walters, D. K. (2015). Age-related aspects of human IgM(+) B cell heterogeneity. *Ann N Y Acad Sci, 1362*(1), 153–163.

McFarlin, B. K., Flynn, M. G., Phillips, M. D., Stewart, L. K., & Timmerman, K. L. (2005). Chronic resistance exercise training improves natural killer cell activity in older women. *J Gerontol A Biol Sci Med Sci, 60*(10), 1315–1318.

McKenna, R. W., Washington, L. T., Aquino, D. B., Picker, L. J., & Kroft, S. H. (2001). Immunophenotypic analysis of hematogones (B-lymphocyte precursors) in 662 consecutive bone marrow specimens by 4-color flow cytometry. *Blood, 98*(8), 2498–2507.

Minuzzi, L. G., Chupel, M. U., Rama, L., Rosado, F., Munoz, V. R., Gaspar, R. C., … Teixeira, A. M. (2019). Lifelong exercise practice and immunosenescence: Master athletes cytokine response to acute exercise. *Cytokine, 115*, 1–7.

Minuzzi, L. G., Rama, L., Bishop, N. C., Rosado, F., Martinho, A., Paiva, A., & Teixeira, A. M. (2017). Lifelong training improves anti-inflammatory environment and maintains the number of regulatory T cells in masters athletes. *Eur J Appl Physiol, 117*(6), 1131–1140.

Minuzzi, L. G., Rama, L., Chupel, M. U., Rosado, F., Dos Santos, J. V., Simpson, R., … Teixeira, A. M. (2018). Effects of lifelong training on senescence and mobilization of T lymphocytes in response to acute exercise. *Exerc Immunol Rev, 24*, 72–84.

Mirza, N., Pollock, K., Hoelzinger, D. B., Dominguez, A. L., & Lustgarten, J. (2011). Comparative kinetic analyses of gene profiles of naive CD4+ and CD8+ T cells from young and old animals reveal novel age-related alterations. *Aging Cell, 10*(5), 853–867.

Molony, R. D., Nguyen, J. T., Kong, Y., Montgomery, R. R., Shaw, A. C., & Iwasaki, A. (2017). Aging impairs both primary and secondary RIG-I signaling for interferon induction in human monocytes. *Sci Signal, 10*(509).

Muezzinler, A., Zaineddin, A. K., & Brenner, H. (2014). Body mass index and leukocyte telomere length in adults: A systematic review and meta-analysis. *Obes Rev, 15*(3), 192–201.

Ng, L. G., Ostuni, R., & Hidalgo, A. (2019). Heterogeneity of neutrophils. *Nature Reviews Immunology, 19*(4), 255–265.

Nielsen, A. R., Mounier, R., Plomgaard, P., Mortensen, O. H., Penkowa, M., Speerschneider, T., ... Pedersen, B. K. (2007). Expression of interleukin-15 in human skeletal muscle effect of exercise and muscle fibre type composition. *J Physiol, 584*(Pt 1), 305–312.

Nieman, D. C., Henson, D. A., Gusewitch, G., Warren, B. J., Dotson, R. C., Butterworth, D. E., & Nehlsen-Cannarella, S. L. (1993). Physical activity and immune function in elderly women. *Med Sci Sports Exerc, 25*(7), 823–831.

Niemiro, G. M., Coletta, A. M., Agha, N. H., Mylabathula, P. L., Baker, F. L., Brewster, A. M., ... Simpson, R. J. (2022). Salutary effects of moderate but not high intensity aerobic exercise training on the frequency of peripheral T-cells associated with immunosenescence in older women at high risk of breast cancer: A randomized controlled trial. *Immun Ageing, 19*(1), 17.

Nikolich-Zugich, J. (2018). The twilight of immunity: Emerging concepts in aging of the immune system. *Nat Immunol, 19*(1), 10–19.

Ogata, K., An, E., Shioi, Y., Nakamura, K., Luo, S., Yokose, N., ... Dan, K. (2001). Association between natural killer cell activity and infection in immunologically normal elderly people. *Clin Exp Immunol, 124*(3), 392–397.

Ostrowski, K., Rohde, T., Asp, S., Schjerling, P., & Pedersen, B. K. (1999). Pro- and anti-inflammatory cytokine balance in strenuous exercise in humans. *J Physiol, 515*(Pt 1), 287–291.

Panda, A., Qian, F., Mohanty, S., van Duin, D., Newman, F. K., Zhang, L., ... Shaw, A. C. (2010). Age-associated decrease in TLR function in primary human dendritic cells predicts influenza vaccine response. *J Immunol, 184*(5), 2518–2527.

Papp, G., Szabo, K., Jambor, I., Mile, M., Berki, A. R., Arany, A. C., ... Balogh, L. (2021). Regular exercise may restore certain age-related alterations of adaptive immunity and rebalance immune regulation. *Front Immunol, 12*, 639308.

Parsons, T. J., Sartini, C., Welsh, P., Sattar, N., Ash, S., Lennon, L. T., ... Jefferis, B. J. (2017). Physical activity, sedentary behavior, and inflammatory and hemostatic markers in men. *Med Sci Sports Exerc, 49*(3), 459–465.

Passos, J. F., Nelson, G., Wang, C., Richter, T., Simillion, C., Proctor, C. J., ... von Zglinicki, T. (2010). Feedback between p21 and reactive oxygen production is necessary for cell senescence. *Mol Syst Biol, 6*, 347.

Pawelec, G., & Gouttefangeas, C. (2006). T-cell dysregulation caused by chronic antigenic stress: The role of CMV in immunosenescence? *Aging Clin Exp Res, 18*(2), 171–173.

Peake, J., Della Gatta, P., & Cameron-Smith, D. (2010). Aging and its effects on inflammation in skeletal muscle at rest and following exercise-induced muscle injury. *Am J Physiol Regul Integr Comp Physiol, 298*(6), R1485–1495.

Pedersen, B. K., & Bruunsgaard, H. (2003). Possible beneficial role of exercise in modulating low-grade inflammation in the elderly. *Scand J Med Sci Sports, 13*(1), 56–62.

Pedersen, B. K., Ostrowski, K., Rohde, T., & Bruunsgaard, H. (1998). The cytokine response to strenuous exercise. *Can J Physiol Pharmacol, 76*(5), 505–511.

Pedersen, M., Bruunsgaard, H., Weis, N., Hendel, H. W., Andreassen, B. U., Eldrup, E., ... Pedersen, B. K. (2003). Circulating levels of TNF-alpha and IL-6-relation to truncal fat mass and muscle mass in healthy elderly individuals and in patients with type-2 diabetes. *Mech Ageing Dev, 124*(4), 495–502.

Petersen, A. M., & Pedersen, B. K. (2005). The anti-inflammatory effect of exercise. *J Appl Physiol, 98*(4), 1154–1162.

Phillips, M. D., Flynn, M. G., McFarlin, B. K., Stewart, L. K., & Timmerman, K. L. (2010). Resistance training at eight-repetition maximum reduces the inflammatory milieu in elderly women. *Med Sci Sports Exerc, 42*(2), 314–325.

Ranadive, S. M., Cook, M., Kappus, R. M., Yan, H., Lane, A. D., Woods, J. A ... Fernhall, B. (2014). Effect of acute aerobic exercise on vaccine efficacy in older adults. *Med Sci Sports Exerc, 46*(3), 455–461.

Raynor, J., Sholl, A., Plas, D. R., Bouillet, P., Chougnet, C. A., & Hildeman, D. A. (2013). IL-15 Fosters age-driven regulatory T cell accrual in the face of declining IL-2 levels. *Front Immunol, 4*, 161.

Rea, I. M., Ross, O. A., Armstrong, M., McNerlan, S., Alexander, D. H., Curran, M. D., & Middleton, D. (2003). Interleukin-6-gene C/G 174 polymorphism in nonagenarian and octogenarian subjects in the

BELFAST study. Reciprocal effects on IL-6, soluble IL-6 receptor and for IL-10 in serum and monocyte supernatants. *Mech Ageing Dev, 124*(4), 555–561.

Reihmane, D., Gram, M., Vigelso, A., Wulff Helge, J., & Dela, F. (2016). Exercise promotes IL-6 release from legs in older men with minor response to unilateral immobilization. *Eur J Sport Sci, 16*(8), 1039–1046.

Rodriguez-Zhurbenko, N., Quach, T. D., Hopkins, T. J., Rothstein, T. L., & Hernandez, A. M. (2019). Human B-1 Cells and B-1 cell antibodies change with advancing age. *Front Immunol, 10*, 483.

Rosales, C. (2018). Neutrophil: A cell with many roles in inflammation or several cell types? *Front Physiol, 9*, 113–113.

Ross, M., Ingram, L., Taylor, G., Malone, E., Simpson, R. J., West, D., & Florida-James, G. (2018). Older men display elevated levels of senescence-associated exercise-responsive CD28(null) angiogenic T cells compared with younger men. *Physiol Rep, 6*(12), e13697.

Ross, M. D., Malone, E. M., Simpson, R., Cranston, I., Ingram, L., Wright, G. P., … Florida-James, G. (2018). Lower resting and exercise-induced circulating angiogenic progenitors and angiogenic T cells in older men. *Am J Physiol Heart Circ Physiol, 314*(3), H392–H402.

Rukavina, D., Laskarin, G., Rubesa, G., Strbo, N., Bedenicki, I., Manestar, D., … Podack, E. R. (1998). Age-related decline of perforin expression in human cytotoxic T lymphocytes and natural killer cells. *Blood, 92*(7), 2410–2420.

Rumpf, C., Proschinger, S., Schenk, A., Bloch, W., Lampit, A., Javelle, F., & Zimmer, P. (2021). The effect of acute physical exercise on NK-Cell cytolytic activity: A systematic review and meta-analysis. *Sports Med, 51*(3), 519–530.

Sanders, V. M. (2012). The beta2-adrenergic receptor on T and B lymphocytes: Do we understand it yet? *Brain Behav Immun, 26*(2), 195–200.

Sapey, E., Greenwood, H., Walton, G., Mann, E., Love, A., Aaronson, N., … Lord, J. M. (2014). Phosphoinositide 3-kinase inhibition restores neutrophil accuracy in the elderly: Toward targeted treatments for immunosenescence. *Blood, 123*(2), 239–248.

Sauce, D., Dong, Y., Campillo-Gimenez, L., Casulli, S., Bayard, C., Autran, B., … Elbim, C. (2017). Reduced oxidative burst by primed neutrophils in the elderly individuals is associated with increased levels of the CD16bright/CD62Ldim immunosuppressive subset. *J Gerontol A Biol Sci Med Sci, 72*(2), 163–172.

Seidler, S., Zimmermann, H. W., Bartneck, M., Trautwein, C., & Tacke, F. (2010). Age-dependent alterations of monocyte subsets and monocyte-related chemokine pathways in healthy adults. *BMC Immunol, 11*, 30.

Signorelli, S. S., Mazzarino, M. C., Di Pino, L., Malaponte, G., Porto, C., Pennisi, G., … Virgilio, V. (2003). High circulating levels of cytokines (IL-6 and TNFalpha), adhesion molecules (VCAM-1 and ICAM-1) and selectins in patients with peripheral arterial disease at rest and after a treadmill test. *Vasc Med, 8*(1), 15–19.

Sikorski, A. F., Swat, W., Brzezinska, M., Wroblewski, Z., & Bisikirska, B. (1993). A protein cross-reacting with anti-spectrin antibodies is present in higher plant cells. *Z Naturforsch C, 48*(7–8), 580–583.

Simpson, R. J. (2011). Aging, persistent viral infections, and immunosenescence: Can exercise "make space"? *Exerc Sport Sci Rev, 39*(1), 23–33.

Simpson, R. J., Cosgrove, C., Ingram, L. A., Florida-James, G. D., Whyte, G. P., Pircher, H., & Guy, K. (2008). Senescent T-lymphocytes are mobilised into the peripheral blood compartment in young and older humans after exhaustive exercise. *Brain Behav Immun, 22*(4), 544–551.

Simpson, R. J., Cosgrove, C., Chee, M. M., McFarlin, B. K., Bartlett, D. B., Spielmann, G., … Shiels, P. G. (2010). Senescent phenotypes and telomere lengths of peripheral blood T-cells mobilized by acute exercise in humans. *Exerc Immunol Rev, 16*, 40–55.

Simpson, R. J., & Guy, K. (2010). Coupling aging immunity with a sedentary lifestyle: Has the damage already been done?--a mini-review. *Gerontology, 56*(5), 449–458.

Simpson, R. J., Kunz, H., Agha, N., & Graff, R. (2015). Exercise and the regulation of immune functions. *Prog Mol Biol Transl Sci, 135*, 355–380.

Slota, C., Shi, A., Chen, G., Bevans, M., & Weng, N. P. (2015). Norepinephrine preferentially modulates memory CD8 T cell function inducing inflammatory cytokine production and reducing proliferation in response to activation. *Brain Behav Immun, 46*, 168–179.

Smith, J. K., Dykes, R., Douglas, J. E., Krishnaswamy, G., & Berk, S. (1999). Long-term exercise and atherogenic activity of blood mononuclear cells in persons at risk of developing ischemic heart disease. *JAMA, 281*(18), 1722–1727.

Solana, R., Campos, C., Pera, A., & Tarazona, R. (2014). Shaping of NK cell subsets by aging. *Curr Opin Immunol, 29*, 56–61.

Spaulding, C., Guo, W., & Effros, R. B. (1999). Resistance to apoptosis in human CD8+ T cells that reach replicative senescence after multiple rounds of antigen-specific proliferation. *Exp Gerontol, 34*(5), 633–644.

Spielmann, G., Bollard, C. M., Bigley, A. B., Hanley, P. J., Blaney, J. W., LaVoy, E. C., … Simpson, R. J. (2014). The effects of age and latent cytomegalovirus infection on the redeployment of CD8+ T cell subsets in response to acute exercise in humans. *Brain Behav Immun, 39*, 142–151.

Spielmann, G., Johnston, C. A., O'Connor, D. P., Foreyt, J. P., & Simpson, R. J. (2014). Excess body mass is associated with T cell differentiation indicative of immune ageing in children. *Clin Exp Immunol, 176*(2), 246–254.

Spielmann, G., McFarlin, B. K., O'Connor, D. P., Smith, P. J., Pircher, H., & Simpson, R. J. (2011). Aerobic fitness is associated with lower proportions of senescent blood T-cells in man. *Brain Behav Immun, 25*(8), 1521–1529.

Stephan, R. P., Reilly, C. R., & Witte, P. L. (1998). Impaired ability of bone marrow stromal cells to support B-lymphopoiesis with age. *Blood, 91*(1), 75–88.

Tamura, Y., Watanabe, K., Kantani, T., Hayashi, J., Ishida, N., & Kaneki, M. (2011). Upregulation of circulating IL-15 by treadmill running in healthy individuals: Is IL-15 an endocrine mediator of the beneficial effects of endurance exercise? *Endocr J, 58*(3), 211–215.

Theall, B., Wang, H., Kuremsky, C. A., Cho, E., Hardin, K., Robelot, L., … Spielmann, G. (2020). Allostatic stress load and CMV serostatus impact immune response to maximal exercise in collegiate swimmers. *J Appl Physiol (1985), 128*(1), 178–188.

Valdes, A. M., Andrew, T., Gardner, J. P., Kimura, M., Oelsner, E., Cherkas, L. F., … Spector, T. D. (2005). Obesity, cigarette smoking, and telomere length in women. *Lancet, 366*(9486), 662–664.

van der Pompe, G., Bernards, N., Kavelaars, A., & Heijnen, C. (2001). An exploratory study into the effect of exhausting bicycle exercise on endocrine and immune responses in post-menopausal women: Relationships between vigour and plasma cortisol concentrations and lymphocyte proliferation following exercise. *Int J Sports Med, 22*(6), 447–453.

Walsh, N. P., Gleeson, M., Shephard, R. J., Gleeson, M., Woods, J. A., Bishop, N., … Hoffman-Goete, L. (2011). Position statement part one: Immune function and exercise. *Exerc Immunol Rev, 17*, 6–63.

Weinberger, B., Haks, M. C., de Paus, R. A., Ottenhoff, T. H. M., Bauer, T., & Grubeck-Loebenstein, B. (2018). Impaired immune response to primary but not to booster vaccination against Hepatitis B in older adults. *Front Immunol, 9*, 1035.

Wikby, A., Johansson, B., Ferguson, F., & Olsson, J. (1994). Age-related changes in immune parameters in a very old population of Swedish people: A longitudinal study. *Exp Gerontol, 29*(5), 531–541.

Wikby, A., Johansson, B., Olsson, J., Lofgren, S., Nilsson, B. O., & Ferguson, F. (2002). Expansions of peripheral blood CD8 T-lymphocyte subpopulations and an association with cytomegalovirus seropositivity in the elderly: The Swedish NONA immune study. *Exp Gerontol, 37*(2–3), 445–453.

Wikby, A., Mansson, I. A., Johansson, B., Strindhall, J., & Nilsson, S. E. (2008). The immune risk profile is associated with age and gender: Findings from three Swedish population studies of individuals 20–100 years of age. *Biogerontology, 9*(5), 299–308.

Wilson, L. D., Zaldivar, F. P., Schwindt, C. D., Wang-Rodriguez, J., & Cooper, D. M. (2009). Circulating T-regulatory cells, exercise and the elite adolescent swimmer. *Pediatr Exerc Sci, 21*(3), 305–317.

Windsor, M. T., Bailey, T. G., Perissiou, M., Meital, L., Golledge, J., Russell, F. D., & Askew, C. D. (2018). Cytokine responses to acute exercise in healthy older adults: The effect of cardiorespiratory fitness. *Front Physiol, 9*, 203.

Wong, G. C. L., Narang, V., Lu, Y., Camous, X., Nyunt, M. S. Z., Carre, C., … Larbi, A. (2019). Hallmarks of improved immunological responses in the vaccination of more physically active elderly females. *Exerc Immunol Rev, 25*, 20–33.

Woods, J. A., Ceddia, M. A., Wolters, B. W., Evans, J. K., Lu, Q., & McAuley, E. (1999). Effects of 6 months of moderate aerobic exercise training on immune function in the elderly. *Mech Ageing Dev, 109*(1), 1–19.

Woods, J. A., Keylock, K. T., Lowder, T., Vieira, V. J., Zelkovich, W., Dumich, S., … McAuley, E. (2009). Cardiovascular exercise training extends influenza vaccine seroprotection in sedentary older adults: The immune function intervention trial. *J Am Geriatr Soc, 57*(12), 2183–2191.

Yan, H., Kuroiwa, A., Tanaka, H., Shindo, M., Kiyonaga, A., & Nagayama, A. (2001). Effect of moderate exercise on immune senescence in men. *Eur J Appl Physiol, 86*(2), 105–111.

Zhang, Y., & Huang, B. (2017). The development and diversity of ILCs, NK cells and their relevance in health and diseases. *Adv Exp Med Biol, 1024*, 225–244.

Zheng, Y., Liu, X., Le, W., Xie, L., Li, H., Wen, W., … Su, W. (2020). A human circulating immune cell landscape in aging and COVID-19. *Protein Cell, 11*(10), 740–770.

Zhu, Y. P., Padgett, L., Dinh, H. Q., Marcovecchio, P., Blatchley, A., Wu, R., … Hedrick, C. C. (2018). Identification of an early unipotent neutrophil progenitor with pro-tumoral activity in mouse and human bone marrow. *Cell Rep, 24*(9), 2329–2341.e2328.

Zuniga, T. M., Baker, F. L., Smith, K. A., Batatinha, H., Lau, B., Burgess, S. C., … Simpson, R. J. (2023). Clonal Kinetics and single-cell transcriptional profiles of T cells mobilized to blood by acute exercise. *Med Sci Sports Exerc, 55*(6), 991–1002.

9 Exercise immunology and infectious disease

Emily C. LaVoy and David C. Nieman

Introduction

The effects of exercise on the immune system have been studied for centuries, but work published in the 1980s and 1990s in the then-nascent field of exercise immunology frequently focused on the potential link between prolonged, intensive exercise and increased risk for infectious illness (see Chapter 1). The relationship between infection risk and exercise was postulated to resemble the curve of the letter 'J' whereby: moderate exercise was thought to diminish infection risk relative to a sedentary lifestyle, while high-intensity, long-lasting exercise increased infection risk (Nieman 1994). However, exceptions to this relationship exist, and the contribution of possible exercise-induced immunosuppression to infection risk is not fully understood. More recently, the extent that which exercise induces immunosuppression such that the individual is more vulnerable to infection has been a subject of debate (Simpson et al. 2020). Exercise has broad physiological and psychological impacts, including short- and long-term effects on the immune system. Translating the clinical and practical consequences of these impacts, often measured *ex vivo* under artificial conditions, is challenging. Further, the physiological impact of exercise on the immune system must be considered in the context in which exercise training and competitive sport are often practised. This includes changes to immunity and to pathogen exposure associated with travel, altered nutrition and hydration, psychological stress, temperature extremes, and air pollution. As each of these factors may also increase one's risk for infection, it is difficult to isolate the unique impact of exercise on susceptibility to communicable disease. Nonetheless, it is meaningful and important to examine the effects of exercise on immune function and infection risk. An improved understanding of the relationship between exercise, immunity, and infectious disease will facilitate evidence-based recommendations to lower infection risk and related effects on athletic endeavours. This chapter discusses the literature regarding moderate and heavy exercise and infection risk in the general population, and in athletes, the effects of moderate and heavy exercise on the immune system, and whether the immune system of people in the general population and athletes differ. Exercise recommendations in the context of novel, emerging pathogens, including SARS-CoV-2, and the COVID-19 pandemic are also provided.

Exercise and risk of infection

This section will discuss evidence for and against the idea that exercise can alter infection risk. Most studies specifically examined risk for upper respiratory tract infection (URTI), and thus the effects of exercise on URTI risk will be the primary focus of this section. Interested readers are referred to recent work that has examined the relationship between physical activity and more broadly defined infectious disease. These investigations demonstrate the protective

DOI: 10.4324/9781003256991-9

effect of habitual physical activity generally against infectious disease morbidity and mortality (Hamer et al. 2019; Chastin et al. 2021). Further, this section is not intended to be exhaustive, but rather to provide an overview of the work that has been conducted in this area. Interested readers are also referred to in-depth reviews of and perspectives on the relationship between exercise and URTI risk (König et al. 2000; Campbell and Turner 2018; Nieman and Wentz 2019; Simpson et al. 2020).

Many studies do not perform laboratory diagnostics to define infection but use self-report of acute illness symptoms to indicate infection; these results are summarised here as effects of exercise on URTI symptoms to distinguish from studies that use laboratory-based diagnostics. The physiological effects of exercise differing in duration and intensity vary, as do effects on the immune system and infection risk. While the effects of exercise are likely best understood as a continuum, for the purposes of organisation, this chapter groups studies into two sections: those that investigated effects of moderate exercise and those that investigated heavy exercise (see Figure 9.1). Whether moderate exercise training alters infection risk is discussed first. Here, moderate exercise approximates that which may be practised most frequently by non-athletes and refers to exercise lasting one hour or less and typically performed at a presumed intensity below the first ventilatory threshold or lactate threshold, or around 65–75% maximum heart rate (American College of Sports Medicine 2000). Thereafter, we consider the effects of 'heavy' exercise on infection risk; 'heavy' exercise refers to exercise as practised by athletes or achieved during competition, and/or lasting greater than one hour and assumed to be performed at an intensity at or above the first ventilatory threshold or lactate threshold.

Relationships between URTI and moderate exercise

In comparison with a sedentary lifestyle, the regular practice of moderate exercise may reduce the incidence of URTI. Table 9.1 provides a summary of those studies highlighted here. A randomised controlled trial comparing older women assigned to walk 40 minutes, 5 times

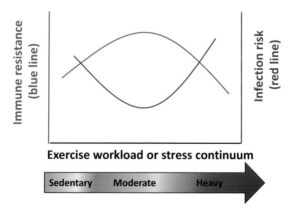

Figure 9.1 Model representing possible relationships between exercise workload, risk of upper respiratory symptom development, and immune resistance to infection.

Legend: "A model representing possible relationships between exercise workload, risk of upper respiratory symptom development, and immune resistance to infection. As exercise workload increases from sedentary to regular, moderate exercise, the literature generally supports a decrease in risk for URTI and an increase in immune resistance, possible due to augmented patterns of immunosurveillance. Less settled in the literature is whether the practice of long-duration high-intensity exercise (heavy exercise) increase the risk for URTI due to decreased immune resistance."

Table 9.1 Research investigating the relationships between moderate exercise and upper respiratory tract infection

Investigators; year published	Study population	Research design	Key findings
Randomised Controlled Trials			
Barrett et al. (2012)	149 adults (82% female; aged 59.3 ± 6.6 years)	Randomised to eight-week training in mindfulness meditation, moderate-intensity sustained exercise or observational control. Global illness severity assessed during a single cold and flu season	Exercise ($n = 47$) vs. control ($n = 51$): 26 episodes, 241 illness days, mean global severity = 248 vs 40 episodes, 453 illness days, and global severity = 35 ($p = .032$ for duration; $p = .16$ for mean global severity)
Barrett et al. (2018)	413 adults (76% female; aged 49.9 ± 11.8 years)	Randomised to eight-week behavioural training in mindfulness-based stress reduction, moderate-intensity sustained exercise (group sessions; home practice) or observational control. Acute respiratory illness symptoms surveyed weekly for nine months	Exercise vs. control: no difference in illness incidence rate (1.2 vs. 1.3) or mean days of duration (7.0 vs. 7.1). Regression model suggested lower global severity for exercise vs control ($p = .042$)
Chubak et al. (2006)	115 overweight and obese, sedentary, postmenopausal women (aged 60.7 ± 6.9 years)	Randomised to one-year moderate-intensity exercise 45 min/day × 5 days/week or 45 min/day × 1 day/week stretching control. Participants provided three-month recall of URTI symptoms quarterly	Illness incidence 30% in exercise vs. 48% in controls. Three-fold reduced risk of colds in exercise group vs. control in final three months
Klentrou et al. (2002)	20 sedentary adults (aged 25–50 years)	Randomised to 12 weeks of moderate-intensity exercise, 45 min/day × 3 days/week or control. URTI symptoms were self-recorded daily	Within exercise only, the number of days of influenza symptoms decreased and s-IgA concentration increased during intervention. Exercise vs. control: greater number of illness days in exercise in first six weeks (12.5 ± 3.1 vs. 7.7 ± 2.3)
Nieman, Nehlsen-Cannarella et al. (1990)	36 mildly obese sedentary women (aged 34.4 ± 1.1 years)	Randomised to 15-week moderate intensity 45 min/day × 5 days/week walking programme or observational control. URTI symptoms were self-recorded daily	Fewer days of URTI symptoms reported in walkers vs. controls (5.1 ± 1.2 vs. 10.8 ± 2.3 days). NK-cell cytotoxicity was greater in walkers, especially at week 6
Nieman et al. (1993)	32 sedentary women (aged 73.4 ± 1.2 years); 12 highly conditioned women (aged 72.5 ± 1.8 years)	Sedentary women randomised to 12-week moderate intensity 30–40 min/day × 5 days/week walking programme or stretching 45 min/day × 5 days/week. Highly conditioned women served as cross-sectional comparison. URTI symptoms were self-recorded daily	During intervention, URTI incidence was 8% in highly conditioned, 21% in walkers, and 50% in controls. Highly conditioned had greater NK-cell cytotoxicity and T-cell proliferation at baseline. No change in NK- or T-cell function in walkers

(Continued)

Table 9.1 (Continued)

Investigators; year published	Study population	Research design	Key findings
Nieman et al. (1998)	91 obese women (aged 45.6 ± 1.1 years)	Randomised to 12-week moderate intensity 45 min/day × 5 days/week walking programme or stretching 45 min/day × 4 days/week. One half of exercise and stretching groups additional reduced calorie consumption. URTI symptoms were self-recorded daily	Fewer days of URTI symptoms reported in walkers vs. controls (5.6 ± 0.9 vs. 9.4 ± 1.1 days). No group differences in immune measures
Sloan et al. (2013)	32 postmenopausal women (aged 54.1 ± 5.3 years)	Randomised to 16-week moderate intensity 30 min/day × 5 days/week at a walking pace to elicit 75% of max heart rate or 'continue with usual physical activity'	No group differences in the incidence of URTI symptoms or in symptom duration during incidences (control: 5.3 ± 1.5 days; exercise: 6.3 ± 2.2 days)
Silva et al. (2019)	26 older adults (73% female, aged 70 years [65–79 interquartile range])	Randomised to 36-week supervised exercise programme of 90 min/day × 3 days/week consisting of 30 min moderate intensity (60–80% max heart rate) aerobic exercise and 45-min resistance exercise, or 'continue with usual physical activity'	No difference in incidence of respiratory infections between groups (47% (95% CI: 23–70) in exercise, 44% (95% CI: 12–77) in control); no difference in severe or mild infections between groups
Observational Studies			
Ghilotti et al. (2018)	2038 adults (aged 26–64 years)	Baseline questionnaire on physical activity; URTI outcome was prospectively self-reported during a nine-month follow-up period	Middle and upper physical activity tertile did not differ from lowest tertile in URTI incidence (Estimated incidence rate ratios: low: referent; mid: 1.00 [95% CI 0.84–1.20]; upper: 1.01 [95% CI 0.84–1.21])
Mathews et al. (2002)	547 adults (49% female; aged 48.0 ± 12.4 years)	Participants followed for one year; interviewed for physical activity and illness symptoms every 90 days	29% reduced illness risk in upper vs. lower quartile of activity. Expenditure of 6–7 MET-h/day reduced illness risk by 20% in men and women. Risk reduction was most pronounced in the fall season
Nieman et al. (2011)	1002 adults (60% female; aged 18–85 years)	Participants followed for 12 weeks in winter and autumn seasons; baseline questionnaire on physical fitness levels; daily illness symptoms checklist	46% decrease in illness days in high vs. low physical fitness tertile. 43% decrease in those who reported ≥5 days/week aerobic activity vs. <1 day/week

each week for 12 weeks to women in a flexibility training group demonstrated lower incidence of URTI symptoms in walkers (Nieman et al. 1993). This study also provided comparisons with highly fit older women (exercised an average of 1.6 hours/day regularly over the prior 11 years); these highly fit older women displayed the lowest incidence of URTI symptoms (Nieman et al. 1993). Similarly, postmenopausal women assigned to 45 minutes moderate exercise five days per week reported significantly fewer colds during the year-long intervention compared with women assigned to flexibility training (Chubak et al. 2006). Another 12-week walking intervention among women with obesity also reduced the number of days of URTI symptoms amongst walkers relative to controls (Nieman et al. 1998). More broadly, 12 weeks of moderate exercise training reduced incidence of influenza-like symptoms amongst adults aged 25–50 years (Klentrou et al. 2002). Shorter interventions have also shown benefits. Adults aged >50 years randomised to an eight-week moderate-intensity exercise training intervention had less than half of the number of illness-associated missed days of work compared with adults in the observational control throughout the cold and influenza season (Barrett et al. 2012). Exercise training-induced improvements in cardiorespiratory fitness have been associated with decreased URTI incidence and may contribute to the trend in the results related above (Nieman, Nehlsen-Cannarella et al. 1990). A meta-analysis of randomised controlled trials examining moderate exercise training and URTI reported reduced incidence and duration of symptoms of the common cold (Lee et al. 2014).

A prospective study evaluating URTI events and self-reported physical activity at five intervals over one year found higher levels of physical activity were associated with reduced risk for URTI (Matthews et al. 2002). Similar results were reported by Nieman et al., where participants reporting five or more days of aerobic exercise per week had a 43% reduction in URTI symptoms in fall and winter months compared with sedentary participants (Nieman et al. 2011).

Not all studies support the effect of moderate exercise to lower risk for URTI symptoms, however. For example, a prospective cohort study tracking 2,038 adults aged 25–64 for nine months found no significant relationship between URTI incidence and self-reported physical activity, although the duration and severity of symptoms were not reported (Ghilotti et al. 2018). A 16-week home-based walking intervention did not improve URTI symptom incidence or duration among postmenopausal women, although the sample size was small ($N = 32$) (Sloan et al. 2013). An eight-month clinical trial in older adults reported no difference in incidence of self-report respiratory symptoms between those in the exercise group (three days/week of aerobic and resistance exercises) and the control group (Silva et al. 2019). Participants in an eight-week moderate-intensity exercise intervention demonstrated a trend ($p = .042$) for reduced illness severity over the 6-month follow-up period compared with those in the waitlist control (Barrett et al. 2018). But this result did not reach the *a priori* selected $p < .025$ threshold to reject the null hypothesis; the authors conclude that the trial was likely underpowered. Indeed, a Cochrane Database systematic review of 14 randomised and quasi-randomised controlled trials published between 1990 and 2018, examining exercise training and acute respiratory illness, did not find evidence that exercise reduced the number of self-reported illness episodes or the proportion of participants experiencing at least one episode during the study (Grande et al. 2020). The systematic review did report evidence for the capability of exercise to reduce symptom severity, though this evidence was classified as relatively weak. Differences between this meta-analysis and the aforementioned meta-analysis by Lee et al. may arise from differences in the inclusion criteria for exercise studies: Lee et al. only included studies that prescribed aerobic-type exercise for at least five days per week at an intensity of at least 60% of maximal heart rate (Lee et al. 2014). Grande et al. included a greater number of studies, including those that prescribed exercise just three days per week, as well as those that used other modalities (e.g., resistance training, Qigong). Thus, while contrary reports exist, current data generally support the idea that moderate exercise training can reduce risk for and symptoms of URTI. Importantly, there are no studies demonstrating an increased risk for symptoms of respiratory illness with moderate levels of exercise.

Relationships between URTI and heavy exercise in athletes

Elite athletes at the peak of competition are susceptible to symptoms of acute illness. Multiple surveys of athletes competing in the Summer and Winter Olympic Games of the last decade report that 7–9% of athletes present with illness to medical staff during the three-week competition period (Engebretsen et al. 2010; Soligard et al. 2017; Soligard et al. 2019). This is unsurprising given that these events involve a large gathering of people and travel, where exposure to opportunistic infection is higher. Nevertheless, these symptoms of acute illness can negatively impact athlete health and performance. Indeed, in Olympic and international-level competitions, illness interfered with training and competition ability in up to 10% of athletes (Gleeson and Pyne 2016; Soligard et al. 2019). Symptoms of URTI are the most common, accounting for 35–70% of illness presentations at sports medicine clinics or team physicians (Gleeson and Pyne 2016; Soligard et al. 2019). Gastrointestinal complaints are also common (Alonso et al. 2012; Soligard et al. 2017). Given the rates of infectious illness incidence and their potential to negatively impact performance, multiple studies have asked if heavy exercise increases athlete susceptibility to infectious illness, particularly URTI. Table 9.2 summarises the studies highlighted here.

A prospective study of URTI symptoms amongst participants in the 1982 Two Oceans Marathon in Cape Town versus matched non-running controls found 33% of runners had symptoms within two weeks of finishing the race compared with just 15% of live-in controls. Further, symptoms were more common amongst runners with faster finishing times (Peters and Bateman 1983). The same research group also found that 68% of runners developed URTI symptoms in the two weeks following the Comrades Ultramarathon, and symptom incidence was greatest amongst those who had the highest training volume (Peters et al. 1993). Similar results were reported by Nieman et al. among runners training for and racing the 1987 Los Angeles marathon. High volume (> 96 km/week) runners had a two-fold increased risk of reporting URTI in the two-month training period leading up to a marathon compared with runners training <32 km/week. Further, runners who completed the marathon race had a six-fold increased risk for URTI in the week following the race compared with those who trained but did not race (Nieman, Johanssen et al. 1990). Additional evidence for the relationship between training volume and infection risk was presented in a year-long prospective study of more than 500 runners, which observed that greater running mileage related to increased risk of URTI symptoms (Heath et al. 1991). More frequent self-reported illness symptoms relative to non-athlete controls have been reported among non-running athletic populations as well. For example, URTI symptoms have been positively related to training load amongst elite tennis players (Novas et al. 2003). Elite cross-country skiers competing in a cross-country skiing stage race were three times more likely to report illness than elite skiers not participating in the race (Svendsen et al. 2015). More recently, an observational study of Team Finland at the 2018 Winter Olympics reported URTI symptoms in 45% of athletes compared with 32% of staff members (Valtonen et al. 2019). It should be noted that in this later study, symptom duration did not differ between athletes or staff members.

Table 9.2 Research investigating the relationships between heavy exercise and upper respiratory tract infections

Investigators; year published	Study population	Research design	Key findings
Bayat et al. (2020)	20 elite female track and field athletes and 20 sedentary controls (aged 16–30 years)	Participants recorded daily training and URTI symptom logs during a 2.5-month winter training season	Mean number of URTI episodes did not differ between athletes and controls (1.0 ± 0.8 vs. 1.4 ± 0.8). Mean URTI duration did not differ between athletes and controls (5.4 ± 3.8 in athletes vs. 5.6 ± 3.0 in controls)
Ekblom et al. (2006)	1694 runners (340 females) completing a 42.2 km race	Participants provided three-week recall of URTI symptoms prior to race and a three-week recall after race	After race illness incidence (19%) did not differ from pre-race incidence (17%) overall. In runners without pre-race incidence, post-race incidence was 16%; runners with pre-race incidence reported 33% post-race incidence
Fahlman and Engels (2005)	75 male varsity college football athletes (20.5 ± 1.5 years), 25 male college student non-athletes (20.5 ± 1.6 years)	Participants recorded URTI symptoms in a weekly log over a 12-month period	A greater proportion of athletes reported URTI symptoms relative to controls in both fall and spring months; no difference in URTI duration
Fricker et al. (2005)	20 national and international male middle-distance and distance runners (aged 24.2 ± 3.1 years)	Participants recorded daily training and illness logs during a four-month Australian winter training season	No difference in mean weekly mileage, training intensity, or training load between ill and healthy runners; no relationship between training mileage, intensity, or load and number of illnesses
Gleeson et al. (1999)	26 elite swimmers (11 female; aged 16–24 years) and 12 swim team staff member controls (5 female; 19–41 years)	For seven months from fall to spring, participants recorded daily illness logs; suspected infections investigated by throat swab, saliva sampled monthly	Number of infections did not differ between swimmers and controls ($p = .26$). Baseline s-IgA negatively correlated with number of infections in swimmers and controls. s-IgA concentration declined over study in both swimmers and controls
Heath et al. (1991)	530 runners (16% female; aged 39.4 years)	Participants reported training log and URTI symptoms monthly for one year	Running >780 km/year increased risk of URTI
Hoffman and Krishnan (2014)	1212 active ultramarathon runners (32% female; aged 42.3 years)	Participants reported running experience and 12-month recall of illness, responses compared to data collected from general population by other studies	Runners reported 2.2 days of work/school missed and one day spent in bed in prior year due to illness or injury. This is 60% and 20%, respectively, of days reported in general population

(Continued)

Table 9.2 (Continued)

Investigators; year published	Study population	Research design	Key findings
Nieman et al. (1989)	294 runners	Participants reported two-month recall of URTI symptoms during training and one-week recall after 5-km, 10-km, and 21.1-km race	Training 42 km/week vs. 12 km/week associated with lower URTI risk. No effect of race on URTI
Nieman, Johanssen et al. (1990)	2311 Los Angeles marathon runners	Participants reported two-month recall of URTI symptoms before 42.2-km race and one week after race	Illness incidence 5.9× higher in runners who finished race vs. non-racing controls (13% vs. 2%). Runners training ³97 vs. <32 km/week at higher URTI risk
Novas et al. (2003)	17 nationally ranked female tennis players (aged 16.3 ± 2.0 years)	Participants recorded exercise duration and intensity and URTI symptoms daily for 12 weeks in Australian winter-spring. Saliva was collected once every two weeks	URTI symptoms increased with increasing training volume but did not relate to training intensity. s-IgA concentrations could not predict the development of URTI symptoms within the following seven days
Peters and Bateman (1983)	141 ultramarathon runners and 124 controls (aged 18–65 years)	Participants reported two-week recall of illness symptoms after 56-km race	Illness incidence 2× higher in runners after race vs. controls (33.3% vs. 15%) and was more common in faster runners
Peters et al. (1993)	84 runners and 73 non-runner controls	Runners randomised to receive daily vitamin C supplementation or placebo 21 days prior to race; participants reported two-week recall of illness symptoms after 90-km race	URTI incidence was greater in placebo runners vs. non-runner controls (68% vs. 49%)
Spence et al. (2007)	32 elite triathletes/cyclists (47% female), 31 recreational triathletes/cyclists (42% female), 20 sedentary controls (55% female) (aged 18–34 years)	Participants followed for five months in summer/autumn; reported daily illness symptoms	Illness incidence 4× greater in elite athletes and 2× greater in controls vs. recreational athletes. Greater number of illness days in elite athletes (311 days) and control (137 days) than recreational (92 days). Viral and bacterial infections were confirmed in 30% of illness episodes
Svendsen et al. (2015)	44 elite cross-country skiers (17 women; aged 24 ± 4 years)	Participants reported illness symptoms daily during and 10 days following Tour de Ski race	Illness incidence was 3× higher in skiers who raced the Tour de Ski vs. non-competing skiers (48% vs. 16%)

(Continued)

Table 9.2 (Continued)

Investigators; year published	Study population	Research design	Key findings
Svendsen et al. (2016)	39 elite cross-country skiers (17 female)	Retrospective cohort study examined daily training logs kept 2007–2015, and illness logs kept 2012–2015. Travel and competition also recorded	Only training variable identified in adjusted models as risk factor for URTI onset was training monotony. Competition and air travel were also significant predictors of reporting symptoms within seven days
Valtonen et al. (2019)	44 elite athletes, 68 Olympic team staff controls	Physician-recorded URTI symptoms daily during Winter Olympic Games (21 days). Samples assessed by molecular analyses to detect viruses	45% of athletes and 32% of controls experienced URTI symptoms. Viral aetiology of symptoms detected in 75% of athletes and 68% of controls

Whether there is a threshold of event duration (i.e., a marathon) or training volume that must be exceeded before an increase in illness symptoms – due to a corresponding decrease in immune resistance – is observed is unknown. However, data suggest a threshold likely exists. For example, in contrast to marathon competition and although likely raced at a high intensity, no increase in URTI episodes was reported in the week following running races of 5-km, 10-km, or half-marathon distance (Nieman et al. 1989). Further, URTI incidence over a five-month period was found to be lower among recreationally competitive triathletes and cyclists compared with both sedentary individuals and elite athletes; elite athlete training volume during this period was substantially higher than recreational athletes (Spence et al. 2007). On the other hand, increased rates of infectious illness symptoms following long-lasting exercise or high-volume training periods are not a universal finding. A prospective study over a seven-month time span found no difference in URTI between 26 elite swimmers and 12 controls; swimmers trained approximately 20–25 hours per week during this time (Gleeson et al. 1999). Additionally, no increase in infectious episodes in the three weeks following the 2000 Stockholm marathon was reported amongst all participants, nor did training volume in the six months before the race relate to infectious incidence (Ekblom et al. 2006). However, a third of finishers who had experienced an infection in the three weeks prior to the race also reported an infection after the race (Ekblom et al. 2006). More recently, a small study comparing elite female runners with matched sedentary controls in URTI symptoms during a winter season found no difference between groups in symptom incidence or duration (Bayat et al. 2020). A survey of more than 1200 active ultramarathon runners found that the runners reported fewer days of missed work or school due to illness than the general population (Hoffman and Krishnan 2014). Fricker et al. followed 20 international-level male distance runners over a four-month training period during the Australian winter and found no relationship between weekly training distance, intensity, or load and respiratory illness symptoms (Fricker et al. 2005). Similarly, training volume and intensity did not relate to the risk of reporting infectious symptoms amongst elite cross-country skiers assessed over an eight-year period (Svendsen et al. 2016). Collectively, data indicate that athletes may be at increased risk of developing symptoms of infectious illness, typically of the upper respiratory tract, following competition or a block of heavy training. However, not all studies support the notion of increased risk for infectious symptoms with long and strenuous training or acute exercise, indicating that developing illness due to training for or following a major athletic endeavour is not a forgone conclusion.

Possible exercise-induced mechanisms for altered immune resistance

During exercise, changes in haemodynamic forces, hormone and cytokine signalling, and body temperature result in a redistribution of immune cells from tissues into peripheral blood. This rapid mobilisation is related to the duration and intensity of the exercise. Consequently, moderate exercise may transiently improve immunosurveillance and, when practised on a regular basis, may translate to reduced incidence and severity of URTI. Conversely, acute changes to the immune system following a bout of heavy exercise may leave an individual with diminished pathogen defence and have been hypothesised to explain changes in infection risk.

The effects of moderate exercise on immunity

Moderate-intensity exercise, as well as vigorous exercise lasting less than one hour, generally appears to enhance cellular and mucosal immunity. As outlined in Chapter 4, leucocytes – especially natural killer (NK) cells and cytotoxic CD8+ T cells – are transiently mobilised from tissues into peripheral blood during exercise. Mobilised cells exhibit high tissue-migrating potential and cytotoxicity, and thus this mobilisation may provide a period of enhanced immunosurveillance (Martin et al. 2009; Walsh et al. 2011; Nieman and Wentz 2019). Salivary antimicrobial proteins also provide protection against pathogens and are thought to provide a first line of defence against URTI. For example, secretory immunoglobulin A (s-IgA) can bind to and opsonise pathogens including viruses. Although acute moderate-intensity exercise appears to have little effect on s-IgA, moderate exercise training can increase resting levels of s-IgA (Walsh et al. 2011). Whether the changes noted in individual cellular and mucosal immune parameters correspond to change in infection risk remains a matter of some debate (Walsh et al. 2011; Simpson et al. 2020), yet outlined next are studies that have investigated changes to immune parameters and infectious symptoms with moderate exercise training.

A randomised controlled trial examining the effects of 15 weeks of moderate exercise training on NK cell cytotoxicity and URTI symptoms in women found an increase in NK cell cytotoxicity, especially at week six of the intervention. The exercise group also had fewer days with URTI symptoms than controls (Nieman, Nehlsen-Cannarella et al. 1990). In contrast, no differences in NK- or T cell function were observed following 12 weeks of moderate exercise training in older women, despite reduced URTI symptoms (Nieman et al. 1993). Similarly, no differences in monocyte, granulocyte, or NK cell function were observed between women with obesity who completed a 12-week walking intervention and those in the control group, even though the walk group reported fewer days with URTI symptoms (Nieman et al. 1998). Yet, in a separate study, men and women aged 25–50 years who performed moderate exercise training for 12 weeks had a significant increase in s-IgA at the end of the intervention compared with those in a control group and s-IgA concentration was inversely related to the number of days of illness (Klentrou et al. 2002). Sloan et al. also report training-induced increases in s-IgA, although these did not relate to a decrease in URTI symptoms (Sloan et al. 2013).

As outlined in Chapter 7, several studies have explored whether moderate exercise enhances immune responses to vaccines. As the immune response to vaccination requires the coordination of multiple immune parameters and occurs *in vivo*, it is a useful paradigm to understand immune function. Multiple groups have examined the effects of cardiorespiratory and resistance exercise on the immune response to seasonal influenza vaccines. While some have reported improved antibody and/or cell-mediated immune responses to the influenza vaccine following exercise (Edwards et al. 2007; Campbell et al. 2010; Edwards et al. 2012), others report no difference between those who exercise and sedentary controls (Edwards and Booy 2013; Pascoe et al. 2014; Bohn-Goldbaum et al. 2020; Elzayat et al. 2021; Dinas et al. 2022). Sex differences may exist, for example increased humoral responses in women and increased cell-mediated responses in men (Edwards 2007). It is suggested that exercise training may be especially beneficial for populations that are generally less likely to respond to vaccination, such as older adults. A recent systematic review concluded that regular moderate aerobic exercise training can enhance immune responses to influenza and pneumonia vaccination in older adults (Song et al. 2020). The effects of moderate exercise training on antibody responses to a novel antigen such as keyhole limpet haemocyanin (KLH) have also been examined. The advantage of this model is that the novel stimulus removes the potential for lingering effects of prior vaccination or infection. Ten months of cardiorespiratory training in previously sedentary older adults increased primary antibody responses to KLH relative to controls (Grant et al. 2008). Together, moderate exercise appears capable of transiently increasing and/or improving immune parameters that, when repeated on a regular basis, may sum up to increase protection against infection.

The effects of long-lasting, heavy exercise on immunity

Heavy exercise training workloads and competition events are linked to immune dysfunction, inflammation, and oxidative stress. Neutrophil, NK cell, B cell, and T cell function, s-IgA output, monocyte expression of pathogen- and danger-signal-sensing proteins, macrophage expression of major histocompatibility complex (MHC) II, skin delayed-type hypersensitivity response, and other biomarkers of immune function can be altered for several hours to days during recovery from prolonged heavy exercise (Martin et al. 2009; Walsh et al. 2011; Nieman and Wentz 2019). Changes in immune cell phenotype and function may represent cell migration to other (non-blood) compartments (see Chapter 4) and s-IgA is a highly variable measure (see Chapter 5). Nevertheless, these immune changes have been proposed to leave the organism at increased risk for infection, especially when exhaustive exercise is repeated with insufficient recovery time between bouts. Data obtained using animal models link heavy exercise workloads with decreased lung macrophage viral resistance and reduced splenic lymphocyte function, as well as with increased mortality from infection (Martin et al. 2009). It should be noted, however, that samples for immune measures in humans are typically limited to saliva or peripheral blood; thus, little is known of the effects of exercise within human tissues (see Chapter 7). Furthermore, most functional measures have been performed *ex vivo* or *in vitro* and may not completely reflect immune function within the body (Gotlieb et al. 2015). Highlighted below are studies that have included measures of immune function and infection risk or *in vivo* immune measures following heavy exercise or training.

Secretory immunoglobulin A concentration has been frequently tested by investigators for its potential linkage to URTI symptoms in athletes (see Chapter 5). Increased training volume has been associated with decreased s-IgA in elite swimmers, and athletes with a low concentration or secretion rate of s-IgA demonstrate increased risk of self-reported URTI in some studies (Gleeson et al. 1999; Gleeson et al. 2012). A similar association between reduction in s-IgA and subsequent URTI has also been reported in elite yacht-racing athletes, collegiate soccer and football players, and professional soccer players (Putlur et al. 2004; Fahlman and Engels 2005; Neville et al. 2008; Perkins and Davison 2021). Conversely, while sensitive to training load over the course of

12 weeks of intense training, s-IgA did not relate to reports of URTI among tennis athletes (Novas et al. 2003). Salivary immunoglobulins, including s-IgA, also did not differ before and after a 15-week intensive training period amongst elite swimmers, nor did they relate to infectious episodes in this cohort (Pyne et al. 2001). In terms of immune cell function, neutrophil oxidative burst activity was found to be lower in elite swimmers during a 12-week intensive training period relative to sedentary controls; however, groups did not differ in URTI symptoms during this same time, suggesting that changes in neutrophil activity did not reach clinical significance (Pyne et al. 1995). Several immune parameters, including mitogen-induced lymphocyte proliferation, NK cell cytotoxicity, granulocyte and monocyte phagocytosis, oxidative burst activity, and plasma concentrations of interleukin (IL)-6, TNF-α, and IL-1ra also did not relate to the number of days with URTI symptoms in elite female rowers (Nieman et al. 2000).

As outlined in Chapter 7, Bruunsgaard et al. compared delayed-type hypersensitivity and vaccination responses between groups that completed a ½ Ironman triathlon immediately prior to treatment, non-exercising triathletes, and moderately trained men (Bruunsgaard et al. 1997). The triathlon was found to impair *in vivo* cell-mediated immunity, as a smaller cumulative response was noted in the group that first performed exercise. However, no differences in vaccine-induced antibody titres were noted between groups (Bruunsgaard et al. 1997). In contrast, a small study ($n = 4$ athletes) reported enhanced antibody production to a tetanus booster dose provided 30 minutes after a marathon-run relative to sedentary controls (Eskola et al. 1978). A more recent larger study also reported that elite athletes training an average of 14 hours per week exhibited more robust responses to the influenza vaccine, including both neutralising antibodies and multi-functional antigen-specific T cells (Ledo et al. 2020).

Differences in immune function between athletes and controls in the resting state

As frequent symptoms of illness are incompatible with elite exercise performance requirements, a 'survivor effect' for elite athletes has been suggested, where the immune systems of elite athletes can adapt and attenuate responses to greater workloads than the general population (Malm 2006). According to this reasoning, an update to the 'J-curve' has been proposed in which exercise workload and infection risk exhibit an 'S-curve' relationship, where elite athletes display heightened capacity to avoid negative effects of infection even with extreme physiological and psychological load associated with training and competition (Malm 2006). In support of this idea, a recent systematic review examining respiratory illness incidence in athletes across 124 studies found a higher incidence of URTI symptoms in non-elite athletes compared with elite athletes (8.7% vs 4.2%) (Derman, Badenhorst, Eken et al. 2022). Further, amongst elite cross-country skiers, those who won an Olympic or World Championship event reported shorter symptom durations compared with elite athletes who did not win a major event (Svendsen et al. 2016). In contrast, Spence et al. reported a higher rate of URTI among elite athletes than competitive recreational-level athletes (Spence et al. 2007).

Although the J-curve may suggest worse immune functioning in athletes relative to sedentary individuals following competition or heavy training periods, there is also evidence that exercise training can ultimately benefit the immune system. Differences in the immune systems of athletes and the general population in the resting state have been reported, generally for NK cell function (Nieman et al. 2000). For example, highly endurance-trained older women had greater NK cell cytotoxicity and greater T cell proliferative capacity relative to sedentary women (Nieman et al. 1993). These changes may in turn be due to phenotypic shifts arising from exercise participation, for example long-term exercise training appears to decrease the accumulation of senescent and exhausted T cells, and increase neutrophil phagocytic activity, and maintain leukocyte telomere length (Simpson et al. 2012; Nieman and Wentz 2019; Donovan et al. 2021). Even moderate exercise training may yield resting immune differences, such as an increase in CD4+ T cell number (Chastin et al. 2021).

Additional factors contributing to infectious symptom prevalence in athletes

Are acute respiratory illness symptoms caused by infection?

It has been suggested that URTI symptoms are not always infectious in origin, but rather result from airway inflammation or allergy (Spence et al. 2007; Kippelen et al. 2012; Robson-Ansley et al. 2012; Sjöström et al. 2019). Thus, athletes with allergic disease or who are exposed to environmental conditions likely to cause airway inflammation (e.g., cold, dry air, or high levels of air pollution) have a greater likelihood of experiencing symptoms of URTI. In support of this idea, predominant risk factors for reporting URTI symptoms in the 2012 Summer Olympics were mechanical and dehydration stresses generated within the airways, as well as the level of airborne pollutants, irritants, and allergens inhaled by the athlete under high ventilatory exercise conditions (Engebretsen et al. 2013). Reporting URTI symptoms was strongly associated with the presence of allergic disease in symptomatic 2010 London marathon runners (Robson-Ansley et al. 2012). Similarly, an investigation of acute respiratory symptoms in elite cyclists competing in the 21-stage Giro d'Italia reported allergies in four of the nine studied athletes before the race, and that respiratory symptoms during the race were related to environmental conditions of the race (Pollastri et al. 2021). The investigation by Spence et al. attempted to ascertain the pathogenic aetiology of reported cases of URTI in study participants but could only trace 30% of cases to a viral origin (Spence et al. 2007). Finally, no differences were found in the prevalence or load of ten common latent viruses between elite cross-country skiers after 10 months of high-performance training and healthy controls (Pyöriä et al. 2021).

Nevertheless, failure to identify a particular pathogen does not prove non-infectious aetiology. Rather, it could reflect a missed diagnosis. Negative test results for pathogen could be due to the suboptimal sampling technique, suboptimal timing, or an infection caused by an agent other than the viruses tested for in the assay. Most studies support that the self-diagnosis of acute respiratory illness is usually correct and that false positive reporting is uncommon (Barrett et al. 2009; Merk et al. 2013; Ghilotti et al. 2018). Certainly, some illnesses in athletes are infectious in origin. Molecular analyses were able to identify infectious aetiology in 75% of the symptomatic athletes from Team Finland at the 2018 Winter Olympics (Valtonen et al. 2019). Nearly 60% of elite athletes presenting with URTI symptoms to a sports medicine clinic had confirmed infectious illness (Cox et al. 2008). Not all infections may be acute, but rather may reflect a reactivation of a latent virus or exacerbation of chronic infection. An investigation into underlying medical conditions in athletes experiencing high incidence of infections found unresolved viral infections in 27% of the athletes, with evidence of Epstein–Barr virus (EBV) reaction in 22% (Reid et al. 2004). Gleeson et al. also concluded that EBV shedding contributed to URTI symptoms during a month-long intensive training period in elite swimmers (Gleeson et al. 2002). Similar conclusions were made in a study of collegiate rugby athletes during a month of intense training (Yamauchi et al. 2011).

Contribution of non-exercise factors to infectious illness susceptibility

Athletes training for and competing in endurance events face several factors that can increase susceptibility to infectious illness, including psychological stress, travel, and poor diet and sleep.

While illness that impedes performance or training will inconvenience an athlete regardless of its cause, it is important to understand the relative risk of each of these factors when designing effective approaches to reducing illness.

Athletes face additional physiological demands beyond those resulting from exercise. Many athletes travel for competition, which may lead to disrupted schedules, disrupted circadian rhythms if time zones are crossed, and disturbed sleep. Sleep deficiencies, such as shortened sleep or sleep of poor quality are associated with immune dysfunction (Besedovsky et al. 2019). A one-year retrospective study of 852 German endurance athletes showed that URTI risk was highest in those who also reported significant stress and sleep deprivation (König et al. 2000). Military recruits who reported sleeping less than six hours per night were four times more likely to be diagnosed with URTI (Wentz et al. 2018). Similar associations between insufficient sleep duration and URTI have also been reported in the general population (Robinson et al. 2021). Travel and increased training load may also lead to nutritional imbalance, which can impact immunity (Gleeson 2016). Low energy availability was identified as the leading variable associated with illness in Olympic-level athletes in the three months prior to the Summer Olympic Games (Drew et al. 2018). Further, athletic competition can induce psychological stress and anxiety (Rice et al. 2016). Both chronic and acute psychological stress have been linked to reduced immune function, including delayed wound healing, reactivation of latent viruses, lower response to vaccination, and increased severity of respiratory viral infection (Glaser and Kiecolt-Glaser 2005).

Travel to the competition itself also increase exposure to potential pathogens. Long-haul flights are a risk factor for the transmission of infectious disease in the general population (Leitmeyer and Adlhoch 2016). A recent systematic review identified international travel as a significant risk factor for acute respiratory illness in athletes (Derman, Badenhorst, Eken et al. 2022). Travelling across more than five time zones is associated with a two-to-three-fold increase in risk for illness among elite rugby players (Schwellnus et al. 2012). Further, 20% of the Team Finland athletes that went on to develop illness during the 2018 Winter Olympics were deemed to have developed the infection during the nine-hour flight to the Games (Valtonen et al. 2019). Similarly, air travel led to a nearly five-fold increased risk of infectious symptoms amongst elite cross-country skiers (Svendsen et al. 2016). Interestingly, this study also reported a nearly three-fold increased risk of infectious symptoms with competition, but no association of risk with training volume or intensity. Crowded conditions, such as in large marathon or mass-start races, also increase athlete exposure to potential pathogens. Indeed, increased risk of URTI has been reported in crowded, non-exercise situations (Choudhry et al. 2006; Coudeville et al. 2022). Thus, the external demands of competition, with accompanying psychological stress and increased exposure to pathogen within crowds, may elevate infection risk independent of exercise exertion.

Exercise in the context of novel, emerging pathogens including SARS-CoV-2

Vaccination and behavioural interventions designed to reduce virus transmission are important measures to reduce the disease burden of SARS-CoV-2/COVID-19. Exercise and physical activity also play an important role in preventing the worst effects of this virus and will likely be crucial against novel pathogens in future pandemics as well. Key risk factors for poor COVID-19 outcomes include advanced age, hypertension, type II diabetes mellitus, and obesity. These later three risk factors are well-known to be modifiable with exercise training (Nieman 2020). In the absence of chronic disease, physical inactivity is also a risk factor for severe COVID-19. Data from the UK Biobank demonstrated that physical inactivity was related to a

32% increased risk for COVID-19 hospitalisations (Hamer et al. 2019). Data from a large cohort of COVID-19 patients who visited Kaiser Permanente Southern California indicated physically inactive adults had approximately twice the risk for severe COVID-19 compared to adults who exercised ≥150 minutes per week during the previous two years (Sallis et al. 2021). Further, low exercise capacity has been associated with COVID-19 hospitalisation (Brawner et al. 2021). Although the nature of these data precludes assigning causation, these epidemiological studies support the idea that, as with other infectious diseases, the regular practice of moderate exercise can improve COVID-19 outcomes. Moderate exercise training may be especially important for older adults, given their increased risk for morbidity and mortality with respiratory illnesses such as COVID-19.

While frequent moderate exercise may be a prevention strategy for poor outcomes following infection with COVID-19 or other respiratory infections, some of the studies presented in this chapter suggest that exhaustive exercise bouts may instead lead to decreased viral defence. Although the direct clinical consequence of perturbations to immunity following heavy and prolonged exercise may be debated (Simpson et al. 2020), it is prudent to suggest caution during outbreaks of severe and novel infectious diseases. Indeed, animal data indicate that exercising to exhaustion before and during systemic infections increases morbidity and mortality rates (Martin et al. 2009; Nieman and Wentz 2019). In a call-to-action statement, the American College of Sports Medicine has recommended that individuals at high risk for SARS-CoV-2 exposure should refrain from exhaustive exercise, overreaching, and overtraining (Denay et al. 2020). Athletes returning to training and competition following infection should also be cautious. Prolonged and intense exercise sessions are not recommended until health is fully restored to avoid deterioration in immune health. Rather, a gradual resumption of exercise training is recommended with close monitoring for clinical deterioration (Denay et al. 2020).

Finally, a significant portion of people infected with SARS-CoV-2 go on to develop long COVID (Castanares-Zapatero et al. 2022). Typically characterised by the persistence of COVID-19 symptoms for more than four weeks, long COVID shares characteristics with myalgic encephalomyelitis and chronic fatigue syndrome (ME/CFS). Physical activity with a gradual increase in the total workload is recommended for ME/CFS. In these patients, emphasis must be given to decreasing fatigue, improving the overall quality of life, and avoiding symptom aggravation versus rapid improvements in fitness (Nieman 2021). These guidelines are likely prudent for patients with long COVID as well.

Summary

This chapter discussed the state of knowledge regarding moderate and heavy exercise and risk for acute respiratory infection in the general population and athletes, the effects of moderate and heavy exercise on the immune system, as well as immune differences between elite athletes and the general population. Although null findings exist, the literature generally supports the hypothesis that moderate exercise training can reduce the risk for and symptoms of URTI. Mechanistically, it may be that transient increases in immunosurveillance that occur with each bout of moderate exercise may translate to reduced incidence and severity of URTI when moderate exercise is repeated on a regular basis. Literature also supports the idea that practitioners of heavy, long-lasting exercise may have a greater risk of URTI. However, data countering this notion also exist, and thus the 'J-curve' hypothesis of infection risk will likely be a matter of debate for some time to come. Immune challenge trials, such as those used previously to study the effect of acute infection on athletic performance, would provide clarity (Friman et al. 1985; Lambkin-Williams et al. 2018). Another area for further study is the relative contribution of external (e.g., psychological stress, travel) versus exertional factors that may contribute to immune dysfunction in competitive athletes. An intriguing update to the J-curve is the 'S curve' hypothesis, which suggests that successful elite athletes have decreased risk for infectious illness allowing them to undertake greater training and competition loads than less successful, more illness-prone individuals. Finally, during widespread outbreaks of infectious pathogens such as SARS-CoV-2, it remains sensible to recommend the regular practice of moderate exercise. Following infection, however, athletes should take a cautious approach and make a slow return to long-duration heavy exercise.

References

Alonso J-M, Edouard P, Fischetto G, Adams B, Depiesse F, Mountjoy M. (2012). Determination of future prevention strategies in elite track and field: analysis of Daegu 2011 IAAF Championships injuries and illnesses surveillance. *Br J Sports Med.* 46(7):505–514.

American College of Sports Medicine, editor. (2000). *ACSM's Guidelines for Exercise Testing and Prescription.* 11th ed. Philadelphia: Lippincott Williams & Wilkins.

Barrett B, Brown RL, Mundt MP, Thomas GR, Barlow SK, Highstrom AD, Bahrainian M. (2009). Validation of a short form Wisconsin Upper Respiratory Symptom Survey (WURSS-21). *Health Qual Life Outcomes.* 7:76.

Barrett B, Hayney MS, Muller D, Rakel D, Ward A, Obasi CN, Brown R, …, Gern J. (2012). Meditation or exercise for preventing acute respiratory infection: A randomized controlled trial. *The Annals of Family Medicine* [Internet] [Accessed 2023 June 20] 10(4):337–346.

Barrett B, Hayney MS, Muller D, Rakel D, Brown R, Zgierska AE, …, Coe CL. (2018). Meditation or exercise for preventing acute respiratory infection (MEPARI-2): A randomized controlled trial. *PLoS One* 13(6):e0197778.

Bayat M, Asemani Y, Asemani S. (2020). Effect of exercise on upper respiratory tract infection in elite runners. *J Sports Med Phys Fitness.* 60(9):1269–1274.

Besedovsky L, Lange T, Haack M. (2019). The sleep-immune crosstalk in health and disease. *Physiol Rev.* 99(3):1325–1380.

Bohn-Goldbaum E, Pascoe A, Singh MF, Singh N, Kok J, Dwyer DE, …, Bohn-Boldbaum E. (2020). Acute exercise decreases vaccine reactions following influenza vaccination among older adults. *Brain Behav Immun.* 1:100009.

Brawner CA, Ehrman JK, Bole S, Kerrigan DJ, Parikh SS, Lewis BK,…, Keteyian SJ. (2021). Inverse relationship of maximal exercise capacity to hospitalization secondary to coronavirus disease 2019. *Mayo Clin Proc.* 96(1):32–39.

Bruunsgaard H, Hartkopp A, Mohr T, Konradsen H, Heron I, Mordhorst CH, Pedersen BK. (1997). In vivo cell-mediated immunity and vaccination response following prolonged, intense exercise. *Med Sci Sports Exerc.* 29(9):1176–1181.

Campbell JP, Edwards KM, Ring C, Drayson MT, Bosch JA, Inskip A, …, Burns VE. (2010). The effects of vaccine timing on the efficacy of an acute eccentric exercise intervention on the immune response to an influenza vaccine in young adults. *Brain Behav Immun.* 24(2):236–242.

Campbell JP, Turner JE. (2018). Debunking the myth of exercise-induced immune suppression: Redefining the impact of exercise on immunological health across the lifespan. *Front Immunol.* 9:648.

Castanares-Zapatero D, Chalon P, Kohn L, Dauvrin M, Detollenaere J, Maertens de Noordhout C, …, Van den Heede K. (2022). Pathophysiology and mechanism of long COVID: A comprehensive review. *Ann Med.* 54(1):1473–1487.

Chastin SFM, Abaraogu U, Bourgois JG, Dall PM, Darnborough J, Duncan E, …, Hamer M. (2021). Effects of regular physical activity on the immune system, vaccination and risk of community-acquired infectious disease in the general population: Systematic review and meta-analysis. *Sports Med.* 51(8):1673–1686.

Choudhry AJ, Al-Mudaimegh KS, Turkistani AM, Al-Hamdan NA. (2006). Hajj-associated acute respiratory infection among hajjis from Riyadh. *East Mediterr Health J.* 12(3–4):300–309.

Chubak J, McTiernan A, Sorensen B, Wener MH, Yasui Y, Velasquez M, …, Ulrich CM. (2006). Moderate-intensity exercise reduces the incidence of colds among postmenopausal women. *Am J Med.* 119(11):937–942.

Coudeville L, Amiche A, Rahman A, Arino J, Tang B, Jollivet O, …, Wu J. (2022). Disease transmission and mass gatherings: a case study on meningococcal infection during Hajj. *BMC Infect Dis.* 22(1):275.

Cox AJ, Gleeson M, Pyne DB, Callister R, Hopkins WG, Fricker PA. (2008). Clinical and laboratory evaluation of upper respiratory symptoms in elite athletes. *Clin J Sport Med.* 18(5):438–445.

Denay KL, Breslow RG, Turner MN, Nieman DC, Roberts WO, Best TM. (2020). ACSM call to action statement: COVID-19 considerations for sports and physical activity. *Curr Sports Med Rep.* 19(8):326–328.

Derman W, Badenhorst M, Eken M, Gomez-Ezeiza J, Fitzpatrick J, Gleeson M, …, Schwellnus M. (2022). Risk factors associated with acute respiratory illnesses in athletes: A systematic review by a subgroup of the IOC consensus on "acute respiratory illness in the athlete". *Br J Sports Med.* 56(11):639–650.

Derman W, Badenhorst M, Eken MM, Ezeiza-Gomez J, Fitzpatrick J, Gleeson M, Schwellnus M. (2022). Incidence of acute respiratory illnesses in athletes: A systematic review and meta-analysis by a subgroup of the IOC consensus on "acute respiratory illness in the athlete". *Br J Sports Med.* 56(11):630–638.

Dinas PC, Koutedakis Y, Ioannou LG, Metsios G, Kitas GD. (2022). Effects of exercise and physical activity levels on vaccination efficacy: A systematic review and meta-analysis. *Vaccines* 10(5):769.

Donovan T, Bain AL, Tu W, Pyne DB, Rao S. (2021). Influence of exercise on exhausted and senescent T cells: A systematic review. *Front Physiol.* 12:668327.

Drew M, Vlahovich N, Hughes D, Appaneal R, Burke LM, Lundy B, …, Lovell G. (2018). Prevalence of illness, poor mental health and sleep quality and low energy availability prior to the 2016 Summer Olympic Games. *Br J Sports Med.* 52(1):47–53.

Edwards KM, Booy R. (2013). Effects of exercise on vaccine-induced immune responses. *Hum Vaccin Immunother.* 9(4):907–910.

Edwards KM, Burns VE, Allen LM, McPhee JS, Bosch JA, Carroll D, …, Ring C. (2007). Eccentric exercise as an adjuvant to influenza vaccination in humans. *Brain Behav Immun.* 21(2):209–217.

Edwards KM, Pung MA, Tomfohr LM, Ziegler MG, Campbell JP, Drayson MT, Mills PJ. (2012). Acute exercise enhancement of pneumococcal vaccination response: A randomised controlled trial of weaker and stronger immune response. *Vaccine* 30(45):6389–6395.

Ekblom B, Ekblom O, Malm C. (2006). Infectious episodes before and after a marathon race. *Scand J Med Sci Sports.* 16(4):287–293.

Elzayat MT, Markofski MM, Simpson RJ, Laughlin M, LaVoy EC. (2021). No effect of acute eccentric resistance exercise on immune responses to Influenza vaccination in older adults: A randomized control trial. *Front Physiol.* 12:713183.

Engebretsen L, Soligard T, Steffen K, Alonso JM, Aubry M, Budgett R, Dvorak J, …, Mountjoy M. (2013). Sports injuries and illnesses during the London Summer Olympic Games 2012. *Br J Sports Med.* 47(7):407–414.

Engebretsen L, Steffen K, Alonso JM, Aubry M, Dvorak J, Junge A, Meeuwisse W, …, Wilkinson M. (2010). Sports injuries and illnesses during the Winter Olympic Games 2010. *Br J Sports Med.* 44(11):772–780.

Eskola J, Ruuskanen O, Soppi E, Viljanen MK, Järvinen M, Toivonen H, Kouvalainen K. (1978). Effect of sport stress on lymphocyte transformation and antibody formation. *Clin Exp Immunol.* 32(2):339.

Fahlman MM, Engels HJ. (2005). Mucosal IgA and URTI in American college football players: A year longitudinal study. *Med Sci Sports Exerc* 37(3):374–380.

Fricker PA, Pyne DB, Saunders PU, Cox AJ, Gleeson M, Telford RD. (2005). Influence of training loads on patterns of illness in elite distance runners. *Clin J Sport Med.* 15(4):246–252.

Friman G, Wright JE, Ilback NG, Beisel WR, White JD, Sharp DS, …, Vogel JA. (1985). Does fever or myalgia indicate reduced physical performance capacity in viral infections? *Acta Med.* 217(4):353–361.

Ghilotti F, Pesonen A-S, Raposo SE, Winell H, Nyrén O, Trolle Lagerros Y, Plymoth A. (2018). Physical activity, sleep and risk of respiratory infections: A Swedish cohort study. *PLoS One* 13(1):e0190270.

Glaser R, Kiecolt-Glaser JK. (2005). Stress-induced immune dysfunction: Implications for health. *Nat Rev Immunol.* 5(3):243–251.

Gleeson M. (2016). Immunological aspects of sport nutrition. *Immunol Cell Biol.* 94(2):117–123.

Gleeson M, Bishop N, Oliveira M, McCauley T, Tauler P, Muhamad AS. (2012). Respiratory infection risk in athletes: Association with antigen-stimulated IL-10 production and salivary IgA secretion. *Scand J Med Sci Sports.* 22(3):410–417.

Gleeson M, McDonald WA, Pyne DB, Cripps AW, Francis JL, Fricker PA, Clancy RL. (1999). Salivary IgA levels and infection risk in elite swimmers. *Med Sci Sports Exerc.* 31(1):67–73.

Gleeson M, Pyne DB, Austin JP, Lynn Francis J, Clancy RL, McDonald WA, Fricker PA. (2002). Epstein-Barr virus reactivation and upper-respiratory illness in elite swimmers. *Med Sci Sports Exerc.* 34(3):411–417.

Gleeson M, Pyne DB. (2016). Respiratory inflammation and infections in high-performance athletes. *Immunol Cell Biol.* 94(2):124–131.

Gotlieb N, Rosenne E, Matzner P, Shaashua L, Sorski L, Ben-Eliyahu S. (2015). The misleading nature of in vitro and ex vivo findings in studying the impact of stress hormones on NK cell cytotoxicity. *Brain Behav Immun.* 45:277–286.

Grande AJ, Keogh J, Silva V, Scott AM. (2020). Exercise versus no exercise for the occurrence, severity, and duration of acute respiratory infections. *Cochrane Database Syst Rev.* 4(4):CD010596.

Grant RW, Mariani RA, Vieira VJ, Fleshner M, Smith TP, Keylock KT, …, Woods JA. (2008). Cardiovascular exercise intervention improves the primary antibody response to keyhole limpet hemocyanin (KLH) in previously sedentary older adults. *Brain Behav Immun.* 22(6):923–932.

Hamer M, O'Donovan G, Stamatakis E. (2019). Lifestyle risk factors, obesity and infectious disease mortality in the general population: Linkage study of 97,844 adults from England and Scotland. *Prev Med.* 123:65–70.

Heath GW, Ford ES, Craven TE, Macera CA, Jackson KL, Pate RR. (1991). Exercise and the incidence of upper respiratory tract infections. *Med Sci Sports Exerc.* 23(2):152–157.

Hoffman MD, Krishnan E. (2014). Health and exercise-related medical issues among 1,212 ultramarathon runners: Baseline findings from the Ultrarunners Longitudinal TRAcking (ULTRA) Study. *PLoS One* 9(1):e83867.

Kippelen P, Fitch KD, Anderson SD, Bougault V, Boulet L-P, Rundell KW, …, McKenzie DC. (2012). Respiratory health of elite athletes - preventing airway injury: A critical review. *Br J Sports Med.* 46(7):471–476.

Klentrou P, Cieslak T, MacNeil M, Vintinner A, Plyley M. (2002). Effect of moderate exercise on salivary immunoglobulin A and infection risk in humans. *Eur J Appl Physiol.* 87(2):153–158.

König D, Grathwohl D, Weinstock C, Northoff H, Berg A. (2000). Upper respiratory tract infection in athletes: influence of lifestyle, type of sport, training effort, and immunostimulant intake. *Exerc Immunol Rev.* 6:102–120.

Lambkin-Williams R, Noulin N, Mann A, Catchpole A, Gilbert AS. (2018). The human viral challenge model: Accelerating the evaluation of respiratory antivirals, vaccines and novel diagnostics. *Respir Res.* 19(1):123.

Ledo A, Schub D, Ziller C, Enders M, Stenger T, Gärtner BC, …, Sester M. (2020). Elite athletes on regular training show more pronounced induction of vaccine-specific T-cells and antibodies after tetravalent influenza vaccination than controls. *Brain Behav Immun.* 83:135–145.

Lee HK, Hwang IH, Kim SY, Pyo SY. (2014). The effect of exercise on prevention of the common cold: A meta-analysis of randomized controlled trial studies. *Korean J Fam Med* 35(3):119–126.

Leitmeyer K, Adlhoch C. (2016). Review article: Influenza transmission on aircraft: A systematic literature review. *Epidemiology* 27(5):743–751.

Malm C. (2006). Susceptibility to infections in elite athletes: The S-curve. *Scand J Med Sci Sports.* 16(1):4–6.

Martin SA, Pence BD, Woods JA. (2009). Exercise and respiratory tract viral infections. *Exerc Sport Sci Rev.* 37(4):157–164.

Matthews CE, Ockene IS, Freedson PS, Rosal MC, Merriam PA, Hebert JR. (2002). Moderate to vigorous physical activity and risk of upper-respiratory tract infection. *Med Sci Sports Exerc.* 34(8):1242–1248.

Merk H, Kühlmann-Berenzon S, Bexelius C, Sandin S, Litton J-E, Linde A, Nyrén O. (2013). The validity of self-initiated, event-driven infectious disease reporting in general population cohorts. *PLoS One* 8(4):e61644.

Neville V, Gleeson M, Folland JP. (2008). Salivary IgA as a risk factor for upper respiratory infections in elite professional athletes. *Med Sci Sports Exerc.* 40(7):1228–1236.

Nieman DC. (1994). Exercise, infection, and immunity. *Int J Sports Med.* 15(Suppl 3):S131–141.

Nieman DC. (2020). Coronavirus disease-2019: A tocsin to our aging, unfit, corpulent, and immunodeficient society. *J Sport Health Sci.* 9(4):293–301.

Nieman DC. (2021). Exercise is medicine for immune function: Implication for COVID-19. *Curr Sports Med Rep.* 20(8):395–401.

Nieman DC, Henson DA, Gusewitch G, Warren BJ, Dotson RC, Butterworth DE, Nehlsen-Cannarella SL. (1993). Physical activity and immune function in elderly women. *Med Sci Sports Exerc.* 25(7):823–831.

Nieman DC, Henson DA, Austin MD, Sha W. (2011). Upper respiratory tract infection is reduced in physically fit and active adults. *Br J Sports Med.* 45(12):987–992.

Nieman DC, Johanssen LM, Lee JW. (1989). Infectious episodes in runners before and after a roadrace. *J Sports Med Phys Fitness.* 29(3):289–296.

Nieman DC, Johanssen LM, Lee JW, Arabatzis K. (1990). Infectious episodes in runners before and after the Los Angeles Marathon. *J Sports Med Phys Fitness.* 30(3):316–328.

Nieman DC, Nehlsen-Cannarella SL, Markoff PA, Balk-Lamberton AJ, Yang H, Chritton DB, …, Arabatzis K. (1990). The effects of moderate exercise training on natural killer cells and acute upper respiratory tract infections. *Int J Sports Med.* 11(6):467–473.

Nieman DC, Nehlsen-Cannarella SL, Henson DA, Koch AJ, Butterworth DE, Fagoaga OR, Utter A. (1998). Immune response to exercise training and/or energy restriction in obese women. *Med Sci Sports Exerc.* 30(5):679–686.

Nieman DC, Nehlsen-Cannarella SL, Fagoaga OR, Henson DA, Shannon M, Hjertman JM, …, Thorpe R. (2000). Immune function in female elite rowers and non-athletes. *Br J Sports Med.* 34(3):181–187.

Nieman DC, Wentz LM. (2019). The compelling link between physical activity and the body's defense system. *J Sport Health Sci.* 8(3):201–217.

Novas AMP, Rowbottom DG, Jenkins DG. (2003). Tennis, incidence of URTI and salivary IgA. *Int J Sports Med.* 24(3):223–229.

Pascoe AR, Fiatarone Singh MA, Edwards KM. (2014). The effects of exercise on vaccination responses: a review of chronic and acute exercise interventions in humans. *Brain Behav Immun.* 39:33–41.

Perkins E, Davison G. (2021). Epstein-Barr Virus (EBV) DNA as a Potential marker of in vivo immunity in professional footballers. *Res Q Exerc Sport.*:93(4):861–868.

Peters EM, Bateman ED. (1983). Ultramarathon running and upper respiratory tract infections. An epidemiological survey. *S Afr Med J.* 64(15):582–584.

Peters EM, Goetzsche JM, Grobbelaar B, Noakes TD. (1993). Vitamin C supplementation reduces the incidence of postrace symptoms of upper-respiratory-tract infection in ultramarathon runners. *Am J Clin Nutr.* 57(2):170–174.

Pollastri L, Macaluso C, Vinetti G, Tredici G, Lanfranconi F. (2021). Follow-up of acute respiratory disorders in cyclists competing in the 100th Giro d'Italia. *Int J Sports Med.* 42(3):234–240.

Putlur P, Foster C, Miskowski JA, Kane MK, Burton SE, Scheett TP, McGuigan MR. (2004). Alteration of immune function in women collegiate soccer players and college students. *J Sports Sci Med.* 3(4):234–243.

Pyne DB, Baker MS, Fricker PA, McDonald WA, Telford RD, Weidemann MJ. (1995). Effects of an intensive 12-wk training program by elite swimmers on neutrophil oxidative activity. *Med Sci Sports Exerc.* 27(4):536–542.

Pyne DB, McDonald WA, Gleeson M, Flanagan A, Clancy RL, Fricker PA. (2001). Mucosal immunity, respiratory illness, and competitive performance in elite swimmers. *Med Sci Sports Exerc.* 33(3):348–353.

Pyöriä L, Valtonen M, Luoto R, Grönroos W, Waris M, Heinonen OJ, …, Perdomo MF. (2021). Survey of viral reactivations in elite athletes: A case-control study. *Pathogens.* 10(6):666.

Reid VL, Gleeson M, Williams N, Clancy RL. (2004). Clinical investigation of athletes with persistent fatigue and/or recurrent infections. *Br J Sports Med.* 38(1):42–45.

Rice SM, Purcell R, De Silva S, Mawren D, McGorry PD, Parker AG. (2016). The mental health of elite athletes: A narrative systematic review. *Sports Med.* 46(9):1333–1353.

Robinson CH, Albury C, McCartney D, Fletcher B, Roberts N, Jury I, Lee J. (2021). The relationship between duration and quality of sleep and upper respiratory tract infections: a systematic review. *Fam Pract.* 38(6):802–810.

Robson-Ansley P, Howatson G, Tallent J, Mitcheson K, Walshe I, Toms C, …, Ansley L. (2012). Prevalence of allergy and upper respiratory tract symptoms in runners of the London marathon. *Med Sci Sports Exerc.* 44(6):999–1004.

Sallis R, Young DR, Tartof SY, Sallis JF, Sall J, Li Q, …, Cohen DA. (2021). Physical inactivity is associated with a higher risk for severe COVID-19 outcomes: a study in 48 440 adult patients. *Br J Sports Med.* 55(19):1099–1105.

Schwellnus MP, Derman WE, Jordaan E, Page T, Lambert MI, Readhead C, …, Kara S. (2012). Elite athletes travelling to international destinations >5 time zone differences from their home country have a 2–3-fold increased risk of illness. *Br J Sports Med.* 46(11):816–821.

Silva D, Arend E, Rocha SM, Rudnitskaya A, Delgado L, Moreira A, Carvalho J. (2019). The impact of exercise training on the lipid peroxidation metabolomic profile and respiratory infection risk in older adults. *Eur J Sport Sci.* 19(3):384–393.

Simpson RJ, Campbell JP, Gleeson M, Krüger K, Nieman DC, Pyne DB, Turner JE, Walsh NP. (2020). Can exercise affect immune function to increase susceptibility to infection? *Exerc Immunol Rev.* 26:8–22.

Simpson RJ, Lowder TW, Spielmann G, Bigley AB, LaVoy EC, Kunz H. (2012). Exercise and the aging immune system. *Ageing Res Rev.* 11(3):404-420.

Sjöström R, Söderström L, Klockmo C, Patrician A, Sandström T, Björklund G, …, Stenfors N. (2019). Qualitative identification and characterisation of self-reported symptoms arising in humans during experimental exposure to cold air. *Int J Circumpolar Health.* 78(1):1583528.

Sloan CA, Engels HJ, Fahlman MM, Yarandi HE, Davis JE. (2013). Effects of exercise on S-IGA and URS in postmenopausal women. *Int J Sports Med.* 34(1):81–86.

Soligard T, Palmer D, Steffen K, Lopes AD, Grant M-E, Kim D, …, Chang JY. (2019). Sports injury and illness incidence in the PyeongChang 2018 Olympic Winter Games: A prospective study of 2914 athletes from 92 countries. *Br J Sports Med.* 53(17):1085–1092.

Soligard T, Steffen K, Palmer D, Alonso JM, Bahr R, Lopes AD, …, Mountjoy M. (2017). Sports injury and illness incidence in the Rio de Janeiro 2016 Olympic Summer Games: A prospective study of 11274 athletes from 207 countries. *Br J Sports Med.* 51(17):1265–1271.

Song Y, Ren F, Sun D, Wang M, Baker JS, István B, Gu Y. (2020). Benefits of Exercise on Influenza or Pneumonia in older adults: A systematic review. *Int J Environ Res Public Health.* 17(8):2655.

Spence L, Brown WJ, Pyne DB, Nissen MD, Sloots TP, McCormack JG, …, Fricker PA. (2007). Incidence, etiology, and symptomatology of upper respiratory illness in elite athletes. *Med Sci Sports Exerc.* 39(4):577–586

Svendsen IS, Gleeson M, Haugen TA, Tønnessen E. (2015). Effect of an intense period of competition on race performance and self-reported illness in elite cross-country skiers. *Scand J Med Sci Sports.* 25(6):846–853.

Svendsen IS, Taylor IM, Tønnessen E, Bahr R, Gleeson M. (2016). Training-related and competition-related risk factors for respiratory tract and gastrointestinal infections in elite cross-country skiers. *Br J Sports Med.* 50(13):809–815.

Valtonen M, Waris M, Vuorinen T, Eerola E, Hakanen AJ, Mjosund K, …, Ruuskanen O. (2019). Common cold in Team Finland during 2018 Winter Olympic Games (PyeongChang): Epidemiology, diagnosis including molecular point-of-care testing (POCT) and treatment. *Br J Sports Med.* 53(17):1093–1098.

Walsh NP, Gleeson Michael, Shephard RJ, Gleeson Maree, Woods JA, Bishop NC, …, Hoffman-Goetz L, (2011). Position statement. Part one: Immune function and exercise. *Exerc Immunol Rev.* 17:6–63.

Wentz LM, Ward MD, Potter C, Oliver SJ, Jackson S, Izard RM, …, Walsh NP. (2018). Increased risk of upper respiratory infection in military recruits who report sleeping less than 6 h per night. *Mil Med.* 183(11–12):e699–e704.

Yamauchi R, Shimizu K, Kimura F, Takemura M, Suzuki K, Akama T, …, Akimoto T. (2011). Virus activation and immune function during intense training in rugby football players. *Int J Sports Med.* 32(5):393–398.

10 Exercise immunology and cancer

David B. Bartlett and Erik D. Hanson

Introduction

Cancer is newly diagnosed in over 19 million people worldwide yearly and is one of the leading causes of death in adults before age 70 (Sung et al., 2021). Earlier detection and new treatment options have significantly improved cancer prognosis. Cancer, cancer therapies, and existing and new comorbidities negatively impact patients' quality of life, physiologic reserve, and resilience to stressors. Such decrements are associated with an increased incidence of therapy failure, cancer recurrence, hospitalisations, cardiac abnormalities, and early mortality.

Physical activity and exercise are protective against the development of cancer (Matthews et al., 2020; Moore, Lee, Weiderpass et al., 2016). The risk of developing cancer is further lowered when physical activity and exercise participation are higher in a dose-response relationship. The World Health Organization guidelines suggest 150–300 minutes/week of moderate-intensity or 75–150 minutes/week of vigorous-intensity aerobic exercise (Bull et al., 2020). Additionally, the same physical activity guidelines apply to those diagnosed with cancer or survivors of cancer (Campbell et al., 2019; Ligibel et al., 2022).

A growing body of work suggests that exposure to exercise following a cancer diagnosis may lower cancer progression, reduce side effects and potentially reduce mortality (Friedenreich, Neilson, Farris, & Courneya, 2016). As such, many national healthcare providers promote integrative lifestyle approaches that may improve cancer prognosis alongside neoadjuvant, primary, and adjuvant therapies and into survivorship. Evidence suggests that exercise can improve cancer-related fatigue, quality of life, physical function, anxiety and depression, lymphedema, bone health, and sleep quality. However, the dose range that reduces cancer-specific or all-cause mortality remains unknown. This is likely due to the poorly understood molecular mechanisms by which exercise promotes anti-tumour effects and systemic benefits (K. A. Ashcraft, Peace, Betof, Dewhirst, & Jones, 2016; f, Dewhirst, & Jones, 2013). Regardless, it is becoming much more evident that exercise may be an adjunct therapy in cancer (Kathleen A. Ashcraft, Warner, Jones, & Dewhirst, 2019).

This book and other recent review articles have described how exercise and physical activity are generally associated with improved immune functions (Duggal, Niemiro, Harridge, Simpson, & Lord, 2019; Fiuza-Luces et al., 2023). However, cancer progression is often the result of immune system failure. Therefore, the role of exercise in promoting anti-cancer properties is most likely mediated in some way by the immune system. Many benefits are believed to operate through repeated bouts of exercise whereby immune cells, cytokines, chemokines, and growth factors are acutely altered in the body. Each bout promotes long-term adaptations. This chapter will explore the putative effects of exercise on reprogramming or maintaining the functions of the immune system during the cancer continuum (Figure 10.1). Before doing that, a brief

DOI: 10.4324/9781003256991-10

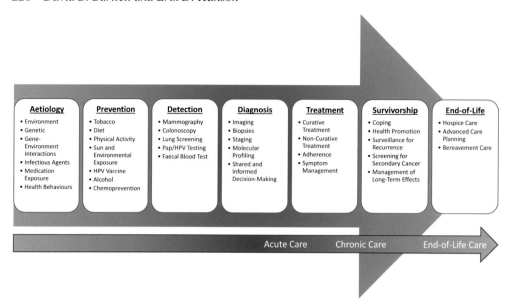

Figure 10.1 Cancer control continuum.

Legend: The cancer continuum describes the stages from cancer aetiology, where the biological mechanisms are determined, to prevention, detection, diagnosis, treatment, survivorship, and end-of-life.

overview of how the immune system responds to cancer is provided. Then, we first outline how exercise affects the immune system to prevent cancers. Secondly, the impact of exercise on people with solid or haematologic cancers is discussed. Thirdly, the impact of exercise on people who have completed cancer therapies (cancer survivors) is reviewed. Finally, a summary of these observations and potential key mechanisms is provided.

Immune responses to cancer: cancer immunoediting

To understand the impact of exercise on the immune system in cancer, it is necessary to introduce the most critical cells and factors in cancer (Figure 10.2), which have been comprehensively reviewed elsewhere (Gubin & Vesely, 2022; S. Liu, Sun, & Ren, 2023; Schreiber, Old, & Smyth, 2011; Vesely, Kershaw, Schreiber, & Smyth, 2011). Briefly, the immune system controls tumour progression and, surprisingly, can promote tumour progression through cells such as myeloid-derived suppressor cells (Shankaran et al., 2001). Cancer–immunity cycle involves critical steps to establish an effective anti-tumour response based on our fundamental understanding of general immune activation. These include NK cell recognition and elimination, tumour-antigen-processing, antigen presentation, T cell priming, T cell infiltration and recognition, and T cell effector responses.

Cancer immunoediting initiates after host cell transformation and malignant neoplasia development. In immunocompetent hosts, this often leads to the **elimination** of the transformed cells. During transformation, intrinsic tumour-suppressor mechanisms fail, causing an initial elevated level of 'danger signals', MHC-tumour-antigen expression and NK cell activating receptor expression. Following this, innate and adaptive immune cells, including CD8+, CD4+, and γδ T cells, NK cells, NKT cells, macrophages, and dendritic cells, act in concert to destroy transformed cells. During this phase, several important immune molecules are required. These

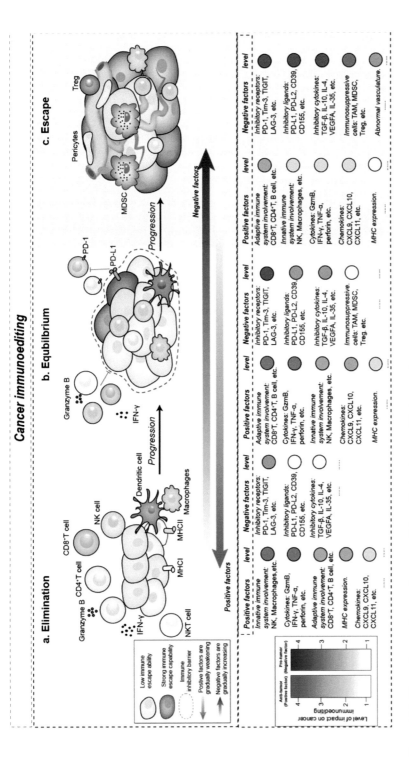

Figure 10.2 Three phases of cancer immunoediting: elimination, equilibrium, and escape.

Legend: (a) During the elimination phase, the innate and adaptive immune systems synergise to identify and eliminate malignant or transformed cancer cells before clinical detection. (b) During the equilibrium phase, a relative balance is established between the cancer cells and the immune system, with the immune system unable to eliminate the cancer cells and the cancer cells unable to evade immune surveillance. (c) During the escape phase, the immune system no longer restricts cancer cell growth and proliferation. The accumulation of rapidly proliferating cancer cells in combination with other stromal cells creates a more complex immunosuppressive microenvironment, thus further damaging the balance between cancer cells and the immune system. During the proceeding of cancer immunoediting, the ability of the immune system to monitor, recognise, and kill cancer cells is crucial in halting its progression. Factors that enhance this ability are positive, while those that enable cancer cells to evade immune recognition and killing are negative. The impact of these factors in the process of cancer immunoediting has been quantified and classified as strong expression (score 4), moderate intensity expression (score 3), weak expression (score 2), and pianissimo expression (score 1). TAMs (tumour-associated macrophages); MDSCs (myeloid-derived suppressor cells); MHC (major histocompatibility complex); GzmB (Granzyme B). Figure is taken unedited from Liu et al (2023) with permission.

include type I (i.e., IFNα, -β, ω) and type II (i.e., IFNγ) interferons that direct cancer-cell cytotoxicity and apoptosis inducers such as FasL, TRAIL, perforin, and granzymes that initiate cell death. The elimination phase culminates in a memory response against clones of any eliminated malignant cells.

Cancer may enter the **equilibrium** phase if the immune system fails to eliminate transformed cells. During the equilibrium phase, memory CD4+ and CD8+ T cells produce IFNγ in response to innate cell-produced IL-12. This maintains tumours in a state of functional dormancy by preventing the re-development of any malignant cell clones they have specificity against. During this phase, NK cells and molecules that initiate innate recognition and effector functions are required less. As such, the equilibrium phase is a function primarily of adaptive immunity. Equilibrium is typically the most prolonged phase of cancer immunoediting and likely extends throughout the host's life. Although the transition from escape to the equilibrium phase is progressive, many cancers will enter directly into the equilibrium phase or, under certain situations, directly into the **escape** phase.

Malignant cells that have acquired the ability to avoid immune identification or killing appear as progressive, observable tumours during the escape phase. The cancer-cell population can vary in reaction to the immune system's editing activities, or the host immune system can alter in response to increasing cancer-induced immunosuppression or immune system degeneration. Several mechanisms are responsible for cancer-cell escape, including a loss of antigenicity and/or immunogenicity and developing an immunosuppressive environment. Tumours can increase the secretion of immunosuppressive and tumorigenic factors, including vascular endothelial growth factor (VEGF), IL-10, Galectin-1, and TGFβ, many of which impair CD8+ T cell, NK cell, and macrophage functions.

Furthermore, tumours can recruit large numbers of regulatory immune cells that are the effectors of immunosuppression and reduce effector cytotoxic immune cell functions. These include regulatory T cells myeloid-derived suppressor cells (MDSCs), tumour-associated macrophages, and neutrophils. Tumours instruct regulatory T cells to increase secretion of IL-10 and TGFβ and suppressive-associated CTLA-4 and PD-1 expression is also increased leading to the inactivation of cytotoxic T cells. Similarly, MDSCs produce TGFβ and promote the proliferation of regulatory T cells. Additionally, MDSCs sequester essential amino acids that cytotoxic T cells require, starving T cells of energy. Finally, during an attack, cancer cells upregulate their expression of PD-L1. PD-1 on activated CD8+ T cells binds to the PD-L1 receptor and shuts down T cell cytotoxicity. These mechanisms enable tumours to silence the cytotoxic capabilities of CD8+ T cells and promote a state of T cell anergy and exhaustion.

Cancer immunoediting is an essential framework for understanding the intricate relationships between the immune system and cancer cells. Understanding it is pivotal to developing new and emerging cancer prevention and treatment therapies. As such, the exercise immunologist – and exercise oncologist alike – must acknowledge this intricate relationship to design the most appropriate interventions.

Immuno-oncology: exercise and physical activity

Exercise immunology in the prevention of cancer

Given the pivotal role of the immune system in cancer development, it is clear that the immune system must play a vital role in reducing the risk of cancer in physically fitter and physically active individuals (Emery, Moore, Turner, & Campbell, 2022). However, although higher physical activity levels reduce cancer incidence, it is unlikely to prevent early asymptomatic neoplasms (Frattaroli et al., 2008). Instead, exercise and physical activity likely yield potent immune-enhancing effects that limit the progression of tumours from equilibrium to elimination (Lam, Alexandre, Luben, & Hart, 2018; Moore et al., 2016). The exact mechanisms by which exercise and physical activity have such anti-cancer effects are complex and depend on several host factors, including, but not limited to, the tumour type/mutational burden/severity and the clinical state of the patient (Kathleen A. Ashcraft et al., 2019; Hojman, Gehl, Christensen, & Pedersen, 2018). In a comprehensive review, Emery and colleagues discuss how exercise might modulate the immune system to reduce cancer incidence (Emery et al., 2022).

They first explain how physical activity appears to modulate the cancer immunoediting process and, in doing so, highlight that physical exercise seems less effective in preventing early neoplasia (i.e., fails to eliminate non-immunogenic clones) and is more effective against later stages of cancer (i.e., against immunogenic clones). Thus, it seems that physical activity contains cancers in a 'precancerous' equilibrium state, which – as noted earlier – infers the involvement of the adaptive immune system, particularly T cells – although upon adaptive immune system involvement, NK cell involvement can also be instigated by tumour-antigen-specific immunoglobulin secreted from B cells.

In seeking to understand how physical activity might augment T-cell function to avert cancer outgrowth, they appraise how physical activity affects the determinants of a successful T cell response against immunogenic cancer cells. They use the cancer immunogram (Blank, Haanen, Ribas, & Schumacher, 2016) as a basis for their evaluation, and in doing so, they outline how physical activity affects: (i) general T cell status in blood (e.g., immunosenescence), (ii) T cell infiltration to tissues, (iii) the presence of immune checkpoints associated with T-cell exhaustion and anergy, (iv) the presence of inflammatory inhibitors of T cells, and (v) the presence of metabolic inhibitors of T cells. It is evident that physical activity has the capacity to modulate all of these factors – as noted in other chapters in this book.

Since most cancer cases occur in older adults, immunosenescence likely plays a pivotal role in cancer development (X. Liu, Hoft, & Peng, 2020). Although there is conflicting data on reductions in senescent T cells, being physically active may promote increased production or maintenance of naive CD4+ and CD8+ T cells (Bartlett & Duggal, 2020; Donovan, Bain, Tu, Pyne, & Rao, 2021; Duggal, Pollock, Lazarus, Harridge, & Lord, 2018). This is especially critical given that senescent T cells are antigen-experienced and contribute minimally to eradicating new or developing tumours but can aid in preserving equilibrium. Subsequently, naive T cells, maintained through exercise-responsive increases in IL-7, IL-15, and IL-12, can potently eradicate non-immunogenic tumours (Tasaki, Yamashita, Arai, Nakamura, & Nakao, 2021) – so-called cancer precursors – upon acquiring mutations that promote immunogenicity during equilibrium and escape.

In terms of understanding, if exercise truly impacts the immune system to prevent cancer, it is difficult to test this. Not only will we never know exactly who will develop cancer, but the number of participants (i.e., tens of thousands), inclusion criteria (i.e., specific to each cancer type), and cost (i.e., hundreds of millions) of such a trial limits its likelihood of being completed (Ballard-Barbash et al., 2009).

Two small-sized studies, published in three articles, have assessed immune function in people at a high-risk of developing cancer (Coletta et al., 2021; Deng et al., 2023; Niemiro et al., 2022). In women at a high-risk for breast cancer, 12 weeks of high-intensity interval training (HIIT) was associated with lower changes in naive CD4+, recent thymic CD4+ emigrants (RTE) and CD8+ effector memory cells compared with moderate-intensity continuous training (MICT) (Niemiro et al., 2022). The authors suggest that regardless of the training group, larger increases in aerobic capacity were associated with larger reductions in CD8+ terminally differentiated

CD45RA expressing T cells, cells typically associated with signs of senescence. As the aerobic capacity changes only increased in the HIIT group, HIIT is probably driving this reduction in 'senescent' CD8 T cells. Neither HIIT nor MICT changed NK cell cytotoxicity functions or circulating myokine levels (Coletta et al., 2021). Given that these cells are from circulating blood, it is unclear if these changes will be associated with tumour development. Deng and colleagues investigated the blood and colorectal tissue of patients with Lynch Syndrome following one year of HIIT (Deng et al., 2023). People with Lynch Syndrome have a 60% risk of developing colorectal cancer, while a reduced risk of colorectal cancer is associated with higher levels of physical activity (i.e., > 35 MET-h/week) (Deng et al., 2023). This study showed that HIIT increased aerobic capacity, which was associated with reduced circulating colon tissue levels of inflammatory prostaglandins, which in turn was associated with increased levels of NK and CD8+ T cell infiltration in colonic mucosa. This tissue infiltration appears to be an enhanced immune surveillance mechanism that will aid in eliminating and preventing malignant cell development. Together, these studies suggest that the exercise-mediated reduced risk of cancer is associated with key changes in anti-tumour immune function and increased immune surveillance in tissues where tumour development originates.

Exercise immunology in adults with solid tumours

Exercise interventions and cellular responses during therapy

Many studies in patients with solid tumours assess various physical parameters and patient-reported outcomes. Several studies have assessed circulating soluble factors (see below), but few have evaluated the effects of individual bouts of exercise and/or regular training on immune cells in patients undergoing treatment for solid tumours (Kruijsen-Jaarsma, Revesz, Bierings, Buffart, & Takken, 2013). One reason for this is the complexity of understanding immunological outcomes in these patients. All cancer therapies aim to eradicate the tumour, but almost all have side effects that result in a dysfunctional immune system and altered inflammatory phenotype (Bruni, Angell, & Galon, 2020; Kang et al., 2009).

Highlighting the complexity of assessing immune responses during therapy, Glass and colleagues evaluated 12 weeks of non-linear (60–100% of aerobic capacity) aerobic exercise training ($N = 23$) versus usual care ($N = 21$) in patients receiving standard-of-care cancer therapies (Glass et al., 2015). Although the between-group change in aerobic capacity was a clinically meaningful +4.1 mL/kg/min favouring the exercise training group, there were no between-group differences for changes in numbers of cytotoxic T cells, B cells, NK cells, or monocytes. Trends were observed where the frequency of CD4+, CD8+, and naive CD8+ T cells increased while the usual care group decreased. Given the variability in cancer types tested (i.e., breast, lung, rectum, peritoneal, uterus, abdomen, nasopharynx, and bladder) and therapies received (i.e., half had metastatic disease, ~90% on chemotherapy and the rest radiation therapy), it is difficult to conclude. These trends suggest that exercise may increase the number of immune cells sufficiently to help eradicate the tumour or prevent infectious episodes, but further work is required to determine this.

In a similar study, but this time in men with localised prostate cancer receiving luteinising hormone-releasing hormone agonist (LHRHa) for at least two months – LHRHa is a form of androgen deprivation therapy that reduces tumour growth but also reduces muscle mass and increases fat mass (Galvão et al., 2008) – 10 patients completed 20 weeks of progressive resistance training that caused significant increases in strength and functional fitness (Galvão et al., 2006). Prostate-specific antigen (PSA; a biomarker that can be used to monitor prostate cancer progression) levels did not increase. There were increases in lymphocyte counts at 10 weeks, while leukocyte count was significantly lower at 20 weeks than at 10 weeks and demonstrated a trend towards reduction when compared with baseline. Although results are inconsistent, ADT typically lowers total leukocyte and granulocyte counts, but not lymphocyte counts (Hanson et al., 2023; Swanson, Hammonds, & Jhavar, 2021). Although not discussed in the article, lower neutrophils and monocytes mostly drove these leukocyte reductions, which could suggest an increased risk of infections and reduced cell-mediated inflammation. However, without detailed analyses of cell phenotypes or function or a control group, these changes may be the natural course of ADT (i.e., neutrophils) but could also be exercise-induced enhanced immunosurveillance (i.e., lymphocytes).

The two previous studies assessed the effects of exercise on circulating immune cells while patients were receiving cancer therapies. However, the anti-cancer effect of exercise in solid tumours is likely contingent on the ability of immune cells to infiltrate and destroy tumours. To investigate this, 30 patients (20 exercise and 10 controls) with localised prostate cancer scheduled for radical prostatectomy completed a progressive HIIT programme (Djurhuus et al., 2023). Training was four times weekly for two to eight weeks, consisting of 4–30 sessions depending on scheduled surgery. Of the 30 patients recruited, only 11 exercise and 10 controls completed the study according to the design (per protocol) requirements of ≥75% of sessions from week one to five. For these patients, the per-protocol analysis showed a significant 3.4-fold increase in the number of CD56+ cells (which the authors describe as NK cells, but can also include subsets of T cells, monocytes, and dendritic cells) within the tumours of the exercise group. The prostate tumour control group also increased the number of CD56+ cells, although not significantly, and as such, there were no between-group differences. Critically and in favour of such a short training period, by looking at the number of sessions all participants completed, significant associations existed between the change in tumour-infiltrating lymphocytes and the number of training sessions. However, it is important to note that when all patients were included in the analyses (i.e., an intention to treat), there was no significance for tumour-infiltrating lymphocytes or differences between groups. In addition, the non-exercise healthy tissue displayed a similar degree of change as the exercise prostate cancer group, suggesting other variables could be at play. As such, the role of regular exercise on tumour-infiltrating lymphocytes before surgery requires more investigation. However, if these per-protocol findings hold true, since the prostate tumour is considered a 'cold' tumour with few immune infiltrates, a lymphocyte response like this may promote a cold-to-hot tumour change. If so, these patients may be suitable candidates for immunotherapy before surgery to limit the negative effects of surgery and enhance the effects of prehabilitation. However, this remains hypothetical, with much more work needed to determine whether this occurs.

Interestingly, in similar patients as above, the effects of an individual bout of exercise on tumour-infiltrating lymphocytes the day before radical prostatectomy were assessed (Djurhuus et al., 2022). Twenty patients completed a short session of exercise while 10 patients did not and served as a control group. As with most exercise studies, including in similar cancer patients (Galvão et al., 2008; Hanson et al., 2023; Hanson et al., 2020), there were significant biphasic responses for blood lactate, lymphocyte counts, catecholamines, IL-6, and TNFα. However, there were no between-group differences in the number of NK cells found within the tumours or differences in tumour hypoxia or microvessel density. Clearly, one bout of high-intensity exercise before surgery may not be sufficient to promote tumour infiltration of effector CD56+ cells in prostate cancer. However, as mentioned earlier, T cells are more relevant during equilibrium and escape phases, but the authors did not investigate these. The acute immune mobilisation to exercise appears blunted in patients with prostate cancer (and likely most other cancers), and

Table 10.1 Select soluble inflammatory marker changes with exercise in patients with a solid tumour

Sprod et al., 2010 (Sprod et al., 2010). $N = 38$ patients with breast ($N = 27$) or prostate ($N = 11$) cancer

	Mean (range) Baseline	Post	p Value
IL-6 (pg/mL)			
Usual Care	3.60 (0.00–8.81)	3.75 (0.00–7.76)	0.04
Exercise	1.08 (0.06–2.97)	1.38 (0.29–6.41)	
TNFα (pg/mL)			
Usual Care	9.43 (0.00–28.84)	9.58 (0.00–29.18)	n.s
Exercise	0.57 (0.00–4.18)	2.82 (0.00–35.99)	
sTNF-R (ng/mL)			
Usual Care	0.77 (0.60–0.93)	0.78 (0.60–0.97)	n.s
Exercise	0.76 (0.45–1.48)	0.68 (0.36–1.32)	

Glass et al. 2015 (Glass et al., 2015). $N = 45$ nine solid cancer types

	Median (IQR) Baseline	Post	p Value
HGF (pg/mL)			
Usual Care	226 (132–354)	295 (218–503)	0.008
Exercise	310 (209–499)	236 (130–426)	
IL-4 (pg/mL)			
Usual Care	117 (77–122)	120 (79–138)	0.012
Exercise	115 (77–120)	111 (74–120)	
MIP-1β (pg/mL)			
Usual care	152 (125–199)	159 (136–217)	0.034
Exercise	177 (123–265)	163 (74–120)	
VEGF (pg/mL)			
Usual care	9 (1–18)	11 (3–16)	0.043
Exercise	13 (3–22)	10 (1–20)	
TNFα (pg/mL)			
Usual care	2 (1–3)	3 (1–5)	0.046
Exercise	1 (1–3)	1 (1–3)	

Galvão et al., 2008 (Galvão et al., 2008). $N = 10$ localised prostate cancer receiving ADT

Exercise Only	Mean (SD) Baseline	Post	p Value
IL-6 (pg/mL)	1.8 (0.3)	1.6 (0.5)	n.s.
IL-8 (pg/mL)	8.2 (0.8)	10.6 (1.7)	n.s.
IL-1RA (pg/mL)	286.5 (39.2)	286.2 (41.6)	n.s.
TNFα (pg/mL)	1.8 (0.2)	1.6 (0.2)	n.s.
CRP (pg/mL)	0.91 (0.31)	0.63 (0.25)	n.s.

Galvão et al. 2010 (Galvão et al., 2010). $N = 57$ localised prostate cancer receiving ADT

	Mean (SD) Baseline	Post	p Value
CRP (mg/L)			
Usual care	2.3 (2.6)	4.5 (6.9)	0.008
Exercise	2.7 (3.2)	1.8 (1.1)	

(*Continued*)

Table 10.1 (Continued)

Jones et al., 2009 (Jones et al., 2009). N = 12 Stage I-IIIA NSCLC before curative surgery

Exercise only	Mean (SD) Baseline	Post	p Value
ICAM-1 (ng/mL)	132.2 (34.9)	120.6 (30.7)	0.041
CRP (mg/L)	8.2 (9.3)	6.9 (9.0)	n.s.
MIP-1α (pg/mL)	35.6 (3.9)	34.4 (3.9)	n.s.
IL-6 (pg/mL)	6.5 (5.7)	6.3 (4.2)	n.s.
IL-8 (pg/mL)	22.9 (22.3)	16.5 (12.7)	n.s.
MCP-1 (pg/mL)	214.5 (52.2)	205.2 (36.5)	n.s.
TNFα (pg/mL)	3.7 (4.8)	4.3 (6.0)	n.s.

Harvie et al. (2022) (Harvie et al., 2022). N = 172 early breast cancer during chemotherapy

	Median (IQR) Baseline	Change	p value
CRP (mg/L)			
IER	2.4 (1.6–5.4)	–0.4 (–1.9–0.6)	n.s.
CER	2.9 (1.1–5.4)	–0.1 (–1.9–1.1)	

Schauer et al. (2021) (Schauer et al., 2021). N = 394 breast, prostate, colorectal during curative therapy

Baseline to Post-Treatment	HI: EMD (95% CI)	p value	LMI: EMD (95% CI)	p value
IL-1β (pg/mL)	0.94 (0.72; 1.21)	n.s.	1.02 (0.76; 1.37)	n.s.
IL-6 (pg/mL)	1.39 (1.21; 1.61)	<0.001	1.44 (1.25; 1.67)	<0.001
IL-8 (pg/mL)	1.13 (1; 1.27)	0.037	1.24 (1.1; 1.4)	<0.001
IL-10 (pg/mL)	1.17 (1.07; 1.22)	0.037	1.2 (1.03; 1.39)	0.010
TNFα (pg/mL)	1.14 (1.07; 1.22)	<0.001	1.21 (1.13; 1.29)	<0.001
CRP (mg/L)	1.04 (0.79; 1.36)	n.s.	1.25 (0.94; 1.66)	n.s.
Post-Treatment to Post-Intervention	**HI: EMD (95% CI)**	**p value**	**LMI: EMD (95% CI)**	**p value**
IL-1β (pg/mL)	0.92 (0.69; 1.21)	n.s.	0.99 (0.73; 1.36)	n.s.
IL-6 (pg/mL)	0.73 (0.63; 0.84)	<0.001	0.79 (0.68; 0.92)	<0.001
IL-8 (pg/mL)	0.96 (0.85; 1.08)	n.s.	0.95 (0.84; 1.07)	n.s.
IL-10 (pg/mL)	0.80 (0.69; 0.93)	<0.001	0.86 (0.73; 1.00)	0.043
TNFα (pg/mL)	0.91 (0.85; 0.98)	0.002	0.90 (0.84; 0.97)	<0.001
CRP (mg/L)	0.79 (0.59; 1.06)	n.s.	0.73 (0.55; 0.98)	0.027

IL (interleukin); TNFα (tumour necrosis factor alpha); sTNF-R (soluble TNF receptor); HGF (human growth factor); MIP-1β (macrophage inflammatory protein-1 beta); VEGF (vascular endothelial growth factor); IL-1RA (IL-1 receptor antagonist); CRP (C-reactive protein); ICAM-1 (intercellular adhesion molecule 1); MIP-1α (macrophage inflammatory protein-1 alpha); MCP-1 (monocyte chemoattractant protein-1); IER (intermittent energy restriction); CER (continuous energy restriction); IQR (interquartile range); SD (standard deviation); EMD (estimated mean difference).

exercise training may rejuvenate this response (Galvão et al., 2008; Hanson et al., 2023; Hanson et al., 2020). However, the exercise response of T cells and NK cell numbers are similar, but levels of functional molecules (i.e., perforin) are lower between healthy men and men with prostate cancer receiving ADT (Hanson et al., 2023; Hanson et al., 2020). Prostate cancer is notoriously non-immunogenic; even potent immunotherapy checkpoint inhibitors fail to work in most prostate cancers; thus, one bout of exercise would not be expected to increase tumour-infiltrating lymphocytes. However, perhaps single bouts of exercise may work in other cancer types, especially with immunogenic tumours.

Although in the early stages of research, exercise training may have negligible effects on circulating immune cell counts and possibly cell functions, and if these translate to clinically meaningful effects on tumour-infiltrating lymphocytes in adults with cancer, this will be a significant finding. Individual bouts of exercise can still mobilise immune cells, although in certain situations, the mobilisation appears blunted (Bartlett et al., 2021; Hanson et al., 2023), and immune cells may have reduced effector molecules. Evidence suggests that exercise training may promote a rejuvenation of the acute exercise response, supporting the findings from Djurhuus and colleagues (Djurhuus et al., 2023). More studies are needed in patients with cancer with an ideal experimental model that assesses bouts of exercise before and after regular training and assesses circulating blood and tumour tissue.

Exercise interventions & soluble responses during therapy

Most studies in adults during solid cancer therapy have assessed the responses in soluble cytokines, chemokines, and growth factors. The field is growing rapidly, so this section will discuss some of the most critical literature (Table 10.1). Unfortunately, but expected, results are confounded by differences in clinical approaches, exercise programmes, study size, and the biomarkers assessed. In terms of inflammatory biomarkers, the difficulty lies with the interpretation of the data. Inflammation can be higher because of the tumour, but also because of the treatment of the tumour and the immune system fighting the tumour. Generally, exercise appears to have some beneficial (Galvão, Taaffe, Spry, Joseph, & Newton, 2010; Glass et al., 2015; Schauer et al., 2021), although not always (Harvie et al., 2022), effects on inflammatory markers if it is conducted for ~3 months or longer. Shorter duration studies appear to have less impact on inflammatory markers (Jones et al., 2009; Sprod et al., 2010). It seems that higher-intensity exercises may provide greater benefits than lower-intensity exercises. Finally, it appears that combined resistance training and aerobic exercise may provide the strongest anti-inflammatory response (Khosravi, Stoner, Farajivafa, & Hanson, 2019).

In one of the largest trials to date, Schauer and colleagues assessed six months of low to moderate intensity ($N = 196$) versus high intensity ($N = 198$) combined endurance and resistance training in patients with breast ($N = 298$), prostate ($N = 79$), and colorectal ($N = 17$) cancer preparing for ~3 months of curative therapy (Schauer et al., 2021) (Figure 10.3). During the treatment phase, circulating IL-6, IL-8, IL-10, and TNFα increased by 13–44%, regardless of exercise intensity. In the period following the completion of therapy, circulating IL-6, IL-10, and TNFα generally declined. Levels of C-reactive protein (CRP) reduced during this phase but only in the low- to moderate-intensity group. Although exercise 'normalises' the therapy-induced inflammatory burden, IL-6, IL-8, and TNFα levels remained marginally above before-therapy levels in the low- to moderate-intensity group. Interestingly, patients with breast cancer undergoing chemotherapy appeared to benefit more from high-intensity exercise, as evident with lesser increases in CRP and TNFα. Although the differences in levels of inflammatory markers were marginal, the authors suggest that a higher-intensity exercise approach may be better for reducing some of the inflammatory burden, but not CRP, of these specific patients with cancer.

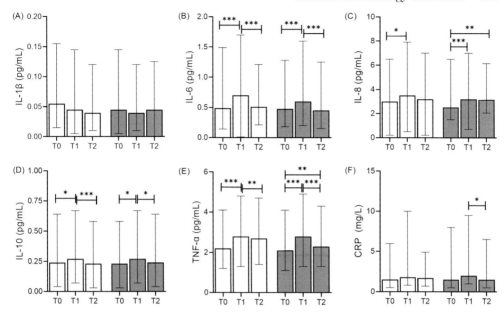

Figure 10.3 Cytokine and acute-phase protein levels before (T0), immediately after adjuvant cancer treatment (T1), and after six months of either high-intensity (white bars) or low- to moderate-
intensity (grey bars) exercise training (T3).

Legend: Treatment was associated with increased levels of IL-6 (B), IL-8 (C), IL-10 (D), TNF-α (E), but not IL-1β (A)
or CRP (F). However, exercise training, regardless of intensity, reduced levels of IL-6, IL-8, IL-10 and TNF-α while
low to moderate intensity exercise reduced CRP. Data are Median with 25th & 75th percentiles. ***P<0.001; **P<0.01;
*P<0.05 difference between selected time points. Adapted from Schauer et al. (Schauer et al., 2021).

In agreement with this proposition, Galvão and colleagues added aerobic exercises to resistance training for 12 weeks in localised prostate cancer receiving ADT (Galvão et al., 2010).
Unlike their earlier pilot study, adding twice-weekly moderate–vigorous-intensity (65–80%
maximum heart rate) aerobic exercises resulted in a 33% reduction in CRP. This significantly
differed from the 95% increase in CRP observed in the control group, suggesting that CRP may
be modifiable by any exercise intensity depending on the cancer population.

Similarly, in the trial by Glass and colleagues described earlier, exercise versus usual care
was associated with significant changes in five cytokine and angiogenic factors (Glass et al.,
2015). Over 30 different cytokines, chemokines, and growth factors were assessed by multiplex
luminometry. TNFα, IL-4, and MIP-1β increased between 2.5% and 50% in the usual care while
remaining stable (TNFα) or decreasing (IL-4 and MIP-1β) in the exercise group. Additionally,
the angiogenic growth factors that aid tumour development, HGF and VEGF, increased in the
usual care while decreasing in the exercise group. Like Schauer and colleagues, this exercise
protocol consisted of a non-linear approach and comprised various exercise intensities, including several high- and maximal-intensity exercise sessions.

In further agreement with non-linear and higher-exercise intensities, Jones and colleagues
(Jones et al., 2009) evaluated the effects of four to six weeks of presurgical exercise training
above 60% VO_2 peak in patients ($N = 12$) with suspected operable lung cancer who did not
receive neoadjuvant therapy. Following exercise training, only ICAM-1 was reduced (~9%).
CRP, IL-6, and IL-8 levels were considerably higher at baseline than typical values observed in
healthy populations, and both CRP (19%) and IL-8 (38%) reduced, albeit non-significantly. The
reductions in ICAM-1 are not fully understood. ICAM-1 is a cell-surface glycoprotein essential

to cell migration, cancer-cell invasion, and metastasis (Bui, Wiesolek, & Sumagin, 2020). Further, ICAM-1 is a possible biomarker of poor prognosis in lung cancer, suggesting that exercise may favourably alter the tumour status before surgery.

The above studies suggest exercise can reduce some inflammatory markers during cancer treatment using a moderate–vigorous–high-intensity approach. In studies that show no effects of exercise training, intensities are generally much lower (Sprod et al., 2010). In patients with breast ($N = 27$) and prostate ($N = 11$) cancer beginning radiation therapy, a four-week home-based exercise programme aimed to increase daily step counts by 5–20% and daily resistance training compared with no exercise controls ($N = 19$). Although exercise increased quality of life and cardiorespiratory fitness (Mustian et al., 2009), there were no changes or differences between groups in circulating IL-6, TNFα, or TNF-R levels. Although this approach did not reduce inflammatory cytokines, it did not increase their concentrations. This suggests that four weeks of increasing daily step counts during radiation therapy does not negatively impact patients' inflammatory balance.

Similarly, Harvie and colleagues assessed the effects of 4.5–6 months of two energy-restriction diets alongside completing physical activity guidelines during adjuvant/neoadjuvant chemotherapy for early breast cancer (Harvie et al., 2022). The focus of this study was weight loss, not exercise prescription. There were no observed changes for CRP. Both intervention groups had minimal changes in body fat and weight. It is unknown whether the participants achieved moderate-intensity exercises; home-based prescriptions typically fall below the prescribed intensities. These data suggest that energy restriction and moderate-intensity physical activity may provide beneficial (i.e., not increasing) CRP responses during breast cancer therapy.

Rather than assessing the soluble responses to exercise training, others evaluated the growth and viability of the tumour cell line by incubating cell lines with patient sera. Dethlefsen and colleagues showed a ~9% reduction in breast cancer cell-line viability when incubated with sera before and after 2 hours of exercise in breast cancer survivors (Dethlefsen et al., 2016). The acute exercise increased levels of lactate, catecholamines, IL-6, IL-8, and TNFα. At the outset, these results may seem counterintuitive that higher lactate and inflammation would cause reduced cell viability. However, other factors released during bouts of exercise, such as SPARC and Oncostatin M, may have counteracted the effects of lactate and inflammation. In a similar study, Devin and colleagues showed that a single bout of high-intensity interval exercise could reduce the growth of colorectal cancer lines in the presence of serum from colorectal cancer survivors (Devin et al., 2019). Again, the acute exercise increased IL-6, IL-8, TNFα, and insulin levels. Unlike the Dethlefsen study, serum was also assayed 2 hours after exercise completion. At 2 hours post-exercise, each of the above markers had returned to baseline levels while cell growth remained lower than baseline, albeit not significantly. This suggests that other factors than the selected cytokines and insulin promote reduced cellular growth. Limitations exist with such cell culture models, however, including a lack of other cells in the culture, especially immune cells, which would otherwise be responsible for preventing outgrowth.

In summary, the effects of exercise training on soluble inflammatory markers are inconsistent. Inflammation appears to either reduce or not increase with exercise training, which has positive clinical relevance for cancer progression and prognosis. Reductions in inflammation appear to favour higher-intensity exercise than lower-intensity exercise. High amounts of physical activity and exercise reduce the risk of cancer, which drives an anti-inflammatory phenotype. Therefore, this physiological response likely crosses over into adults with cancer, and higher intensities are required to promote anti-inflammatory responses. However, it should be noted that although exercise may promote an anti-inflammatory response in healthy adults, short-term elevations of inflammatory cytokines during active cancer therapy may be critical for cancer elimination (Grivennikov, Greten, & Karin, 2010; Zitvogel, Tesniere, Apetoh, Ghiringhelli, & Kroemer, 2008).

Exercise immunology in adults with haematologic malignancies

Unlike solid cancers, there are less exercise-immunology-focused studies in adults with hae-matologic malignancies that have been reviewed elsewhere (Kruijsen-Jaarsma et al., 2013; Sitlinger, Brander, & Bartlett, 2020). This is surprising given that most haematopoietic and lym-phoid neoplasms arise in immune cells, and exercise has the potential to alter immune outcomes in numerous ways. Most haematologic malignancies are associated with increased risks for secondary malignancies, autoimmune diseases, and bacterial and viral infections. Haematologic malignancies include a broad group of haematopoietic and lymphoid neoplasms. It is estimated that there are ~14 new cases of leukaemia (Dobos et al., 2015), ~20 new cases of non-Hodgkin lymphoma (NHL), ~3 new cases of Hodgkin lymphoma, and ~7 new cases of multiple myeloma per 100,000 people per year. While increasing knowledge of cytogenetics and molecular proper-ties greatly adds to prognostication within a particular disease group, considerable variability exists, and these cytogenetic markers are generally not modifiable (Amin et al., 2016).

Understanding modifiable lifestyle factors, including physical activity and exercise levels, is essential to improving outcomes in haematologic malignancies (Knips et al., 2019; Sitlinger et al., 2020). Defining the known effects of exercise immunology in haematological malignan-cies is particularly relevant, as immune dysfunction is associated with significant morbidity and mortality. Additionally, increasing therapeutic options are directed at the immune system, but clinical success is variable, potentially from failure to fully restore immune function (Dholaria, Bachmeier, & Locke, 2018; Pulluri, Kumar, Shaheen, Jeter, & Sundararajan, 2017). Ultimately, exercise could serve as an essential adjunct to haematological cancer care.

Exercise interventions and cellular responses in adults with haematologic malignancies

Chronic lymphocytic leukaemia (CLL) is a disease characterised by malignant B cells where pa-tients have poor immune function. Patients with CLL are generally older at diagnosis but >85% of patients have a five-year survival. As such, they exhibit typical age- and cancer-associated immune dysfunction. To determine whether exercise can improve immunity, Sitlinger and col-leagues (Sitlinger et al., 2021) assessed immune differences between 20 treatment-naive CLL who were grouped based on aerobic capacity levels (VO_2 peak: 34.2 ± 3.3 v 24.9 ± 3.2 mL/kg/min). Although the two groups had similar absolute leukocyte counts, lymphocyte counts, and CD19+/CD5+ CLL cell counts, the group with higher VO_2 peak had marked differences in neu-trophil and NK-cell counts. Higher fitness was associated with a lower NK cell frequency but a greater proportion of mature NKG2A[neg]/KIR+ NK cells. Compared with other studies in healthy adults, newly diagnosed and treatment-naive CLL is associated with higher absolute numbers of CD4+ and CD8+ T cells and NK cells (Forconi & Moss, 2015; Huergo-Zapico et al., 2014). These increases are correlated with tumour burden and are an attempt to maintain equilibrium. However, Sitlinger et al. suggest that higher VO_2 peak is associated with NK and CD8+ T cell frequencies, thus resembling observations in healthy adults. Although it is unclear whether these increases correspond to disease activity (i.e., progression to escape), both high and lower VO_2 peak groups did not have noticeable progression indices. As such, immune function in the higher VO_2 peak group was assumed to be better equipped to manage disease. Further, the higher VO_2 peak group had a lower frequency of inhibitory positive/activatory negative NK cells (i.e. NKG2A[+]/KIR[neg]) and a higher frequency of fully competent mature inhibitory negative/

activatory positive NK cells (i.e. NKG2A^{neg}/KIR^{+}) that are better capable of recognising and killing cancer cells (Béziat, Descours, Parizot, Debré, & Vieillard, 2010). Although cell function was not determined, the higher VO_2 peak group also had lower levels of the composite inflammatory marker GlycA. GlycA is a robust measure of chronic inflammation, immune function, and disease risk and is modifiable by exercise training (Barber et al., 2018; Ritchie et al., 2015). Together, these findings suggest that physical fitness, driven through exercise, may promote better immune function in CLL.

In an attempt to confirm the relationships between aerobic fitness and immune function in CLL, MacDonald and colleagues assessed immune function following 12 weeks of exercise training ($N = 10$) or controls ($N = 6$) in treatment of naive CLL (MacDonald et al., 2021). Using a thrice weekly supervised HIIT combined with two sessions of muscular endurance resistance training, exercise training increased aerobic capacity and muscular strength. Although absolute lymphocyte counts did not change, there was a trend for a smaller increase in CD19+/CD5+ CLL cells. This suggests that exercise training may contribute to slowing CLL progression. In support of this, CD56^{dim} NK cell frequency increased by ~50% with exercise, with a 20% increase in NK cell perforin and granzyme B. Together, this suggests exercise may have increased the ability to identify and kill a cancer cell line of a common secondary cancer found in CLL (i.e., K562), a CLL cell line (i.e., OSU-CLL) and their own CD19+/CD5+ CLL cells. Whether these findings translate to a reduced risk of secondary or recurrent cancer in CLL remains unknown.

The heterogeneity of CLL means some patients have an indolent disease while some are in the equilibrium phase and some are in the escape phase. Nevertheless, in the above study, exercise training appears to have affected the susceptibility of patients' CLL cells, allowing them to be identified by NK cells. One possible explanation for this is that the exercise group increased their levels of circulating β2-microglobulin. β2-microglobulin is a component of MHC-I which might be shed from cells during the shear stress of exercise. Lowering β2-microglobulin on CLL cells would likely promote them as an NK-sensitive target cell, like the K562 cell line, which also lacks MHC-I. Finally, MacDonald and colleagues observed significant associations between changes in muscle strength and immune function, suggesting that physiological fitness is associated with immune function, and when one changes, the other also does (Figure 10.4).

Fewer studies have assessed T cells in CLL. Perry and colleagues compared absolute numbers of regulatory T cells in stable CLL to patients who had increased their absolute lymphocyte counts (ALC) by around 30% over 12 months (Perry et al., 2012). Stable CLL had a lower percentage of pro-tumorigenic regulatory T cells suggesting a relationship between regulatory T cells and CLL progression. To determine whether exercise could affect T cell populations and negate this relationship, they compared four CLL patients to four healthy adults responding to an individual bout (45–60 minutes for CLL and 60–120 minutes for athletes) of treadmill exercise at 70% of heart rate reserve. Although ALC increased following exercise, the numbers of CD4+ T cells were unchanged in CLL. However, the composition of CD4+ cells was altered, with regulatory T cells reduced (−30%) and Th17 cells increased (+49%) 1 hour after completion of the exercise. IL-6 and TGFβ both increased immediately after exercise, while IL-2 was unchanged. Although this study had a small sample size, the absence of a CLL control group and a non-standardised acute exercise protocol, it does suggest that exercise has the potential to alter the frequencies of pro-tumorigenic regulatory T cells and Th17 cells. However, the role of these cells is controversial and not well understood in CLL (Ye, Livergood, & Peng, 2013).

The advantage of these studies in treatment-naive CLL is that patients did not receive cancer therapies. Therefore, we can observe cancer-specific immune disturbances in real-time without

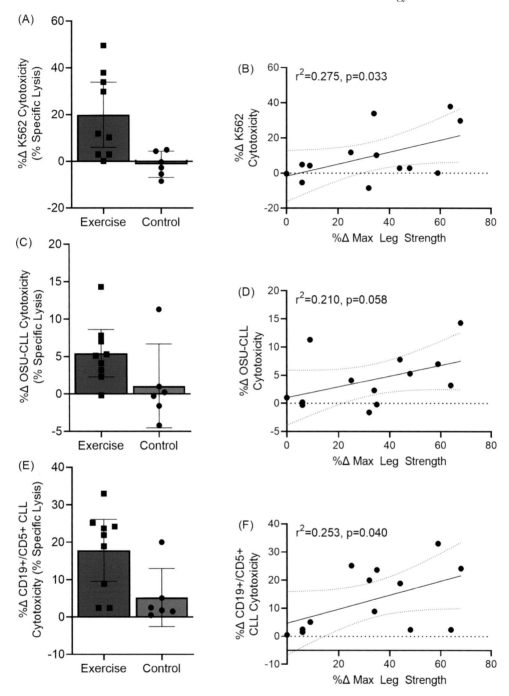

Figure 10.4 Exercise training increases in NK-cell-mediated cancer-cell cytotoxicity and relationships with changes in leg strength following 12 weeks of combined high-intensity interval walking and muscular endurance-based resistance training patients with CLL NK cells improved their ability to kill K562 cell lines (A), a CLL-like cell line (C) and patients own isolated CLL cells (D). These improvements were associated with changes in muscular strength (B, D, F). Panels A, C & E are mean (95% CI). Figures are adapted from MacDonald et al. (MacDonald et al., 2021).

the confounding factors of therapies. However, most patients with haematologic malignancies will receive therapy. What effect exercise has on treatment-initiated CLL remains unknown. However, in other haematologic malignancies, the effects of exercise during treatment have been assessed.

One such study in ten adult patients (aged 18–50) with acute myeloid leukaemia (AML) undergoing inpatient treatments assessed 2 × 30-minute exercise sessions daily (aerobic, strength, and core) at submaximal intensities thrice weekly for three to five weeks (Battaglini et al., 2009). AML is characterised by malignancy in myeloblasts that affect the production and function of neutrophils, monocytes, eosinophils, and basophils. Although lean body mass was reduced during the intervention (a consequence of therapy), significant improvements were observed for cardiorespiratory fitness, fatigue, and depression. There were marginal reductions in IL-6 and non-significant increases in IL-10, suggesting that the pro- to anti-inflammatory balance was being altered. However, IFNγ remained stable, suggesting no improvements in IFNγ-dependent immune function. However, as AML therapy is associated with immune dysfunction, the lack of change for IFNγ likely represents a positive effect and maintains immunity to some degree. Despite being a small study, it highlights that cardiorespiratory fitness can be increased rapidly, as with patients with solid cancer, adaptations to systemic inflammation may require longer or may not occur.

A common treatment for haematologic malignancies is haematopoietic stem-cell transplantation (HSCT) or bone marrow transplantation. Neoadjuvant therapies precede transplant and are associated with worsening physical fitness and immunological conditions. A goal of HSCT is for the patient's lymphocyte counts to recover to normal levels shortly after transplant. To determine whether exercise could counter these effects, Kim and colleagues recruited 35 inpatients during HSCT to complete 30 minutes of bed exercises ($N = 18$) or no exercises ($N = 17$) every day for six weeks (Kim & Kim, 2006). The exercises were very low intensity and involved mostly relaxation techniques and stretching. Lymphocyte and T cell subset counts were assessed before and after HSCT. After six weeks, the exercise group had a lymphocyte count similar to before the transplant, while the control group was ~45% lower than before. Both groups had similar changes in T cells, with CD4+ T cells reducing by ~60% and CD8+ T cells increasing by ~32% following the intervention. The exercise-induced difference in lymphocytes between the groups was therefore driven by non-T cell populations, such as NK cells, B cells, or dendritic cells.

Although the long-term benefits of this observation were not assessed, the exercise-driven immune changes likely contribute to improved clinical outcomes. At the time of the above publication, bed rest was commonly prescribed following HSCT. However, now, inpatients are encouraged to get out of bed and exercise in the ward. In a similar study, but with a longer follow-up time, Hayes and colleagues assessed three months of thrice weekly stretching sessions (Controls: $N = 6$) versus thrice weekly progressive moderate-intensity aerobic and twice-weekly resistance exercises (Exercise: $N = 6$) in patients undergoing HSCT (Hayes, Rowbottom, Davies, Parker, & Bashford, 2003). A combination of small sample size and both groups exercising likely caused no group differences for changes in total leucocytes, lymphocytes, CD3+, CD4+, CD8+ T cells or mitogen-induced T cell proliferation. Compared with healthy data, patients were immunocompromised before and during HSCT. Although total white blood cell and CD8+ T cell counts returned to normal within three months of transplant, it is unclear if the exercise programme facilitated the quicker recovery of the immune cells. However, as in the Kim study (Kim & Kim, 2006), exercise did not delay the recovery. Together, these studies in HSCT highlight the need for appropriate control data to ensure the interpretation of exercise-mediated effects is beneficial and improves immune function in patients.

Exercise interventions and soluble responses in adults with haematologic malignancies

To determine if individual exercise bouts were mediating some of the chronic exercise changes, Zimmer and colleagues enrolled 30 patients with NHL after recently completing chemoimmunotherapy to either a control group (N = 12) or individual bout of exercise (N = 14) (Zimmer et al., 2014). Ten healthy, slightly younger adults were also randomised to a control group (N = 5) or exercise intervention (N = 5). All patients had completed first-line therapies of either bendamustine/rituximab or rituximab and cyclophosphamide, doxorubicin, vincristine, and prednisolone (R-CHOP) protocol 4–24 weeks before. The intervention consisted of 30 minutes of moderate-intensity cycle ergometry. Blood was assessed before the intervention (T1) and 30 minutes post-exercise (T2), unfortunately missing the immediate post-exercise effects. Compared with healthy controls, NHL patients had higher concentrations of macrophage inhibitory factor (MIF) and IL-6, while NK-cell H3K9 histone acetylation was lower, suggesting a greater pro-inflammatory environment and reduced transcriptional activity of lymphocytes in cancer. The response to an individual bout of exercise was similar between NHL controls and NHL who undertook exercise. At T2, IL-6 was higher, and acetylation of the CD8+ T-cell histone H4K45 increased in response to the NHL exercise. There were negative correlations between changes in MIF and H4K45 acetylation in the NHL patients, suggesting an inflammatory-mediated epigenetic regulatory pathway that can be modified by exercise bout. An individual bout of exercise appeared to beneficially alter the epigenetic pattern of NK- and CD8+ T cells, at least in the short term. If these effects translate to long-term exercise training-induced epigenetic modifications, this will be an important step in understanding the biological mechanism of exercise training in NHL.

Sitlinger and colleagues assessed the growth of a CLL-like cell line in the presence of serum obtained from patients with CLL with different aerobic fitness levels (Sitlinger et al., 2021). The higher VO_2 peak group observed significantly slower growth rates than the lower VO_2 peak group. The plasma from the higher VO_2 peak group had lower inflammatory mediators, triglycerides, and higher HDL cholesterol levels but not glucose or LDL. CLL cells are unlike glucose/glutamine-dependent solid tumours and prefer fatty acids and LDL as their primary fuel source (Rozovski et al., 2015; Rozovski, Hazan-Halevy, Barzilai, Keating, & Estrov, 2016). As such, the lower VO_2 peak group plasma likely contained more fuels that the CLL cell line preferred. Other differential circulating factors included extracellular vesicles (EVs); small lipid bilayer particles released from all cells. Sitlinger and colleagues discovered miRNA differences in the smallest EVs, exosomes, between high and low VO_2 peak patients with CLL (Sitlinger et al., 2021). Exosomes are constitutively secreted by CLL cells in response to B-cell receptor activation and stress (Nisticò et al., 2020). The pathophysiological role of these exosomal miRNAs includes the promotion of CLL cell survival by (a) altering transcription in the CLL cell by removing tumour-suppressor miRNAs, (b) altering functions of cells of the tumour microenvironment (TME) to promote tumour progression, and (c) reducing effector functions of immune cells such as CD8+ T cells and NK cells to promote immune evasion and immune suppression (Bobrie, Colombo, Raposo, & Théry, 2011; Whiteside, 2016; Yeh et al., 2015). This study showed that the higher VO_2 peak group had five exosomal miRNAs with higher expression and nine with lower expression. Critically, several of these miRNAs were associated with tumour-suppressor and immune-suppressor actions. Where the exosomes that contained beneficial miRNAs came from remains unknown. However, exercising muscle also secretes large quantities of exosomes, which may alter the balance between tumour- and muscle-derived exosomes.

In summary, exercise and physical activity appear to improve some markers of immunity and inflammation and, at worst, maintain immunity in haematologic malignancies. However, the optimal dose of exercise remains unknown but during HSCT, it is likely that even a little exercise promotes enhanced immune recovery alongside quality of life and wellbeing.

Exercise immunology in adults after completion of cancer therapies

Until now, we have discussed the effects of exercise on immune cells in patients with cancer and during cancer therapies. All treatments are associated with significant side effects that impact patients and limit their abilities to complete day-to-day activities such as exercise. Consequently, it is challenging to complete exercise studies. As such, the majority of exercise oncology and the majority of immunology-focused exercise oncology has been conducted in those who have completed cancer therapies, are in remission and are often considered survivors. Several excellent review articles and meta-analyses have been completed, and we direct the reader to these for insight (Elizabeth S Evans, Hanson, & Battaglini, 2020; Khosravi et al., 2019; Valenzuela et al., 2022).

Exercise interventions and cellular responses in adults after completion of cancer therapies

In a meta-analysis of immune and inflammatory responses to exercise in cancer survivors, Khosravi and colleagues found that exercise training did not alter the cellular immune system (SMD: 0.183, 95% CI: –0.212, –0.529, p = .364). However, fewer studies were explicitly included for immune indices and only NK cell markers were assessed. One of the studies on breast cancer survivors found that being sedentary for eight weeks was associated with non-significant enhanced NK cell cytotoxicity functions compared with thrice weekly 60 minutes of combined training (Nieman et al., 1995). However, the sample size was small (N = 6 per group), and the differences in physical fitness were minor, suggesting that the control group did not maintain a sedentary behaviour. As such, if this study is removed from the meta-analyses, the beneficial effects of exercise on NK cells trend towards significance (SMD: 0.303, 95% CI: −0.021, –0.628, p = .067). However, a more recent meta-analysis supports the findings from Khosravi and suggests that NK cell numbers (MD: 1.47, 95% CI: –0.35, 3.29, p = .113) and function (MD: –0.02, 95% CI: –0.17, 0.14, p = .834) are not altered by exercise training in cancer survivors (Valenzuela et al., 2022). It is unclear what this means. The patients no longer have cancer, may not require as much NK-cell function, and may require better T cell responses. However, breast cancer survivors have a blunted NK cell response to an exercise bout compared with healthy women, suggesting lasting damage that impairs immune responses (Evans et al., 2015). Regardless, exercise during survivorship does not impair NK cell counts and functions, implying that exercise is not deleterious. Cancer survivors are at increased risk of developing secondary cancers, which require optimal NK-cell functions to eliminate. As such, more work is needed to understand why some exercise does and does not improve NK cell functions in cancer survivors.

Although exercise may not alter NK cells during survivorship, this does not mean that other immune cells (such as neutrophils, monocytes, T cells, and unconventional T cells) are not affected. The effects of exercise training on these cells have been assessed in breast cancer survivors within three months (Bartlett et al., 2021; Hanson et al., 2021; Khosravi et al., 2021) and two years (Arana Echarri, Struszczak, Beresford, Campbell, Jones et al., 2023; Arana Echarri, Struszczak, Beresford, Campbell, Thompson et al., 2023) of completing primary treatment. Prior work suggested that there might not be improvements in resting levels of lymphocyte functions (Fairey et al., 2005). As bouts of exercise cause biphasic responses integral to these resting levels, these studies questioned whether the acute effects of exercise would change before and after 16 weeks of progressive supervised combined exercise training and compared the responses to women with no history of cancer (Figure 10.5). In this unique design, survivors' response to exercise bouts was blunted before training (Figure 10.5A). The pre- to post-acute exercise mucosal-associated invariant T-cell (MAIT) mobilisation and the neutrophil phagocytosis of bacteria were blunted in survivors before training (Figure 10.5A and C). After 16 weeks of exercise training, neutrophil functional responses (Figure 10.5D) and mucosal-associated invariant T (MAIT) cell mobilisation were partially restored (Bartlett et al., 2021; Hanson et al., 2021). With this, MAIT cell mobilisation and neutrophil phagocytosis changes pre- to post-exercise were similar to that in healthy women, who represented the 'optimal' immune response. Although it is unclear precisely what this means, it is plausible that exercise has improved the

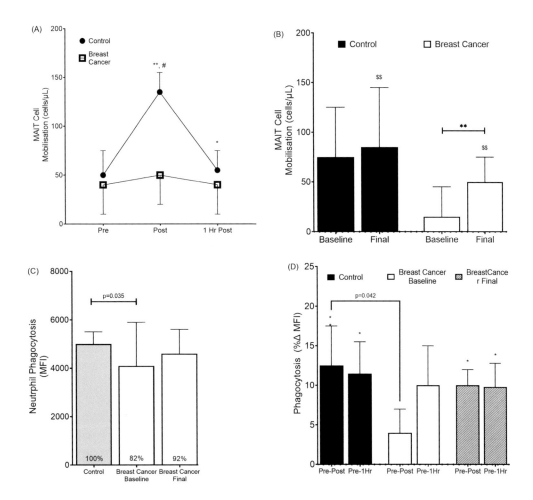

Figure 10.5 Mucosal-associated T-cell (MAIT) and neutrophil responses to acute exercise before and after 16 weeks of exercise training MAIT cell mobilisation to acute exercise is impaired in breast cancer survivors before training (A) and following 16 weeks of exercise training, the mobilisation improves to levels similar to healthy women (B). Neutrophil's ability to phagocytose E. coli is impaired at baseline and recovers to similar levels as healthy women following exercise training (C). Similar to MAIT cells, neutrophils phagocytic capacity is impaired in response to acute exercise but recovers to similar levels in healthy women after training (D). All data are means and SD. In panels A & B, *p<0.05, **p<0.1 versus controls, #p<0.05 from Pre and $$p<0.01 from Baseline. In panel D, *p<0.05, **p<0.01 for groups change score. Figures adapted from Hanson et al. and Bartlett et al. (Bartlett et al., 2021; Hanson et al., 2021).

ability of these immune cells to respond to a stressor. Such stressors include infections and tumour challenges that potentially cause hospitalisation and relapse. MAIT cell numbers, for instance, are resistant to chemotherapy-induced reductions (Dusseaux et al., 2011) and produce more mitogen-stimulated TNFα after exercise (Hanson et al., 2019). If the response to stressors, as suggested by acute exercise following therapy, is dysfunctional, perhaps this contributes to the increased infections observed in survivors after therapy completion. Further, longitudinal studies are required to determine if the exercise-induced partial improvements to a stressor translate to these important clinical outcomes.

Exercise interventions and soluble responses in adults after completion of cancer therapies

In terms of systemic inflammation, Khosravi and colleagues' meta-analysis suggests exercise robustly reduces pro-inflammatory levels (SMD: −0.2, 95% CI: −0.4, −0.1, $p < .001$) (Khosravi et al., 2019). From a total of 27 studies ($N = 1,206$ participants) that met the inclusion criteria, the strongest evidence was for combined aerobic and resistance training (SMD: −0.3, 95% CI: −0.5, −1.9, $p < .001$) and for prostate (SMD: −0.5, 95% CI: −0.8, 0.1, $p = .004$) and breast (SMD: −0.2, 95% CI: −0.3, −0.1, $p = .001$) cancers (Khosravi et al., 2019). Both TNFα (SMD: −0.3, 95% CI: −0.5, −0.06, $p = .004$) and CRP (SMD: −0.5, 95% CI: −0.9, −0.06, $p = .025$), two of the most damaging inflammatory molecules, were most sensitive to change by exercise training. These barely scratch the surface of the cytokine network, and many other important immune-enhancing cytokines (e.g., IL-7 and IL-15 secreted by muscle during exercise) were not included due to a lack of studies. Given that cancer survivors are at an increased risk of cardiovascular disease and cardiovascular disease is driven by elevated levels of inflammation (Fiordelisi, Iaccarino, Morisco, Coscioni, & Sorriento, 2019), these results suggest that exercise may protect survivors' health.

The physical activity guidelines for cancer survivors suggest that patient benefits include improved quality of life, less fatigue, reduced anxiety and depression, and better physical function (Campbell et al., 2019). The immune and inflammatory benefits of exercise during survivorship likely contribute to these benefits and may contribute to reduced infections or severity of infections, but more work is required to confirm this.

Future research, knowledge gaps, and potential mechanisms

The ability of exercise immunologists and physiologists to individually prescribe exercise to target each patient's immune system is limited. Although there are understandable variances across all the studies that have assessed the effects of exercise on the immune system in adults with cancer, there are some indications of the possible mechanisms. In humans, these mechanisms include the role of EVs, crosstalk between exercising muscle-derived cytokines and the immune system, circulating factors that promote cancer cell death, metabolic plasticity, and others. In mice, these mechanisms include a complex exercise-dependent regulation of the TME that is still to be shown in humans (Pedersen et al., 2016). Several excellent review articles are available for a more comprehensive insight into all the potential mechanisms (Ashcraft et al. [2016]; Ashcraft et al. [2019]; Buss & Dachs [2020]; Koelwyn, Quail, Zhang, White, & Jones [2017]; Koelwyn, Wennerberg, Demaria, & Jones [2015]; Koelwyn, Zhuang, Tammela, Schietinger, & Jones, 2020; Xu & Rogers [2020]). In this final section, we will outline some of the pertinent possible mechanisms by which exercise affects the immune system in patients with cancer (Figure 10.6).

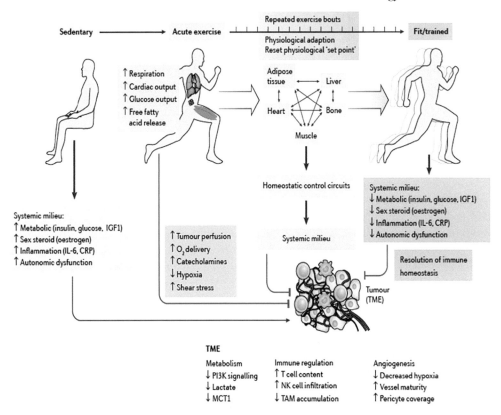

Figure 10.6 Likely exercise-dependent regulation of the tumour microenvironment. Prolonged exposure to physical inactivity is associated with elevated circulating concentrations of numerous growth factors and hormones — a pro-tumorigenic milieu (blue boxes). By contrast, host exposure to acute bouts of exercise stimulates inter-organ signalling achieved by the secretion of hormones, cytokines, and growth factors into the host systemic milieu from various tissues and organs (for example, skeletal muscle, heart, bone, liver and adipose tissue), which can subsequently regulate multiple highly integrated homeostatic control circuits that operate at the cellular, tissue and whole-organismal levels (purple boxes). Over time, chronic exercise-induced perturbation of inter-organ signalling promotes physiological adaptation across homeostatic control circuits (establishment of a higher homeostatic 'set point') that, in concert, stimulates the reprogramming of the systemic milieu, potentially characterized by alterations in the availability (reservoir), mobilization, recruitment, retention and function of specific cell types and/or molecules (green boxes), potentially altering their availability and composition in 'distant' tumour microenvironments (TMEs). Exercise-induced alterations in the systemic milieu influence key regulatory mechanisms in the TME, such as angiogenesis, immune regulation, and metabolism, thus having a cumulative antitumorigenic effect (ochre box). In addition to activating the secretion of numerous factors from skeletal muscle, during acute exercise, blood flow is redirected to the metabolically active skeletal muscle that paradoxically occurs in conjunction with increased tumour blood perfusion and reduced tumour hypoxia. As such, this represents an alternative mechanism of exercise regulation of the TME (red box). CRP, C-reactive protein; IGF1, insulin-like growth factor 1; IL-6, interleukin 6; MCT1, monocarboxylate transporter 1; NK cell, natural killer cell; TAM, tumour-associated macrophage. Figure is taken unedited with permission from Koelwyn et al. (Koelwyn et al., 2017).

Exercise-induced regulation of the TME and circulating systemic factors

It is well established that tumours require and prefer specific metabolic inputs. Although genetic mutations drive cancer, tumours can be phenotyped by dysmetabolism that promotes growth and immune evasion (Bose, Allen, & Locasale, 2020; Hirschey et al., 2015). Tumours alter their metabolic requirements to fulfil uncontrolled proliferation's bioenergetic and signalling demands. The nutrients used by tumours are highly variable, with some cancers dependent on glucose metabolism while others rely on amino acids or lipids, amongst others. In support of this, growth factors, such as insulin and insulin-like growth factors (IGFs) that signal through their respective receptors, activate signalling through PI3K – AKT, mTOR, MAPK, and MYC pathways. In cancer cells, these metabolic pathways are often constitutively activated, allowing for the uptake of nutrients and nucleotides for their survival and hyperproliferation. In preclinical rodent models, exercise has been shown to inhibit activated AKT and mTOR while promoting the activation of AMPK. AMPK regulates energy metabolism by reducing anabolic activities and promoting lipid oxidation, potentially reducing tumour growth/progression.

This dysmetabolism ultimately results in altered fuel availability and dysmetabolism in other tissues, including immune cells. Glucose and glutamine metabolism in the TME results in lactate accumulation. Lactate promotes immunosuppression by recruiting MDSCs into the tumour (Pavlova & Thompson, 2016). Individual bouts of exercise and regular exercise promote glucose uptake by working muscle and activated T cells. This likely reduces glucose availability for cancer cells and the production of lactate. In a preclinical model of breast cancer, exercise training reduced tumour lactate concentrations and reductions of the lactate transporters MCT1 and MCT4 alongside inhibited tumour growth (Aveseh, Nikooie, & Aminaie, 2015). The decrease in lactate transporters suggests that the tumour has begun to shift its preferred choice of glucose to alternative fuel sources that produce fewer lactate by-products. This was likely through glucose uptake by other tissues, but increased blood flow caused by angiogenesis would facilitate faster lactate clearance. Indeed, rodent studies have shown that exercise increases the tumour vasculature's density and maturity, allowing for increased oxygenation (Betof et al., 2015). This physiological angiogenesis overloads the tumour-mediated hypoxia-induced angiogenesis with oxygen and impairs the highly conserved cellular responses to hypoxia-inducible factor-1 (HIF1) (Semenza, 2003). Consequently, the exercise-induced alterations in tumour metabolism, perfusion, and oxygen delivery promote an increased immunosurveillance that may lead to increased eradication of tumours.

As mentioned earlier in this section and previous sections, exercise reduces metabolic intermediaries (e.g., glucose, insulin, and IGF-1) and lipids, sex steroids, and inflammation. To determine the effects of these circulating factors, several studies have utilised an *in vitro* tumour proliferation model with autologous patient serum/plasma (Soares et al., 2021). Although some studies have been completed using healthy adult serum/plasma on cancer cell lines, others have used patients with cancer serum/plasma. Overall, individual bouts of exercise promote an intensity-dependent, favouring higher intensities increase in factors that reduce the viability and impair the growth of cancer cell lines (Soares et al., 2021). Some of these factors are metabolic mediators, such as glucose, lipids, and fatty acids, that normal immune cells also require to function, while some are inflammatory mediators which promote the progression of many malignant cells.

In both the Dethlefsen and Devin studies, no changes in cell growth or viability were observed following a period of exercise training that improved aerobic fitness. This contrasts with the Sitlinger et al (2021) study, where VO_2 peak differences are partially driven by exercise and physical activity exposure. One possible explanation is that both Dethlefsen and Devin did not observe changes in glucose or insulin, both primary metabolic fuels for their cell lines. In contrast, Sitlinger's participants had significant differences in the levels of lipids between groups, the primary fuel source for their cell lines. As such, if exercise promotes a milieu that facilitates changing tumour metabolism, it will likely impact overall tumour survival. The composition of this exercise-induced milieu remains unknown and is likely tumour and host-specific.

Summary

With ever-increasing numbers of people diagnosed, being treated, or living with and beyond cancer, the role of exercise and physical activity in cancer care is pivotal to giving people more control over their outcomes. Several of these outcomes are mediated in part by the immune system. Whether to prevent or reduce the severity of infections and cancer outgrowth, repair damaged tissues, or reduce the risk of other comorbidities, exercise and physical activity can modulate immune responses throughout the cancer continuum. Most of the work so far has focused on soluble markers of inflammation and endocrine dysfunction, which are associated with immune function but are not all directly mediated by immune cells. However, a growing body of work has begun to show exercise-induced immune cell changes that likely facilitate beneficial outcomes before diagnosis, during, and after therapy.

Key Points

1 The heterogeneous biology of each cancer and cancer severity, therapies, and treatments complicate exercise immunology in the realm of cancer.
2 Participation in exercise and higher levels of physical activity are associated with reduced cancer risk. Emerging evidence suggests that exercise likely alters T cell and NK cell functions, which may facilitate this reduced risk of cancer development by trafficking immune cells to sites of tumour development.
3 In patients with solid tumours, exercise training does not appear to alter circulating immune cell numbers, and during bouts of exercise, the immune cell mobilisation appears blunted. However, emerging evidence suggests that exercise may facilitate an increased frequency of lymphocytes within tumours, similar to observations found in preclinical models.
4 In patients with haematologic cancers, exercise training appears to promote a rapid rejuvenation of immune cell numbers during haematopoietic stem-cell transplant. In patients not receiving HSCT, exercise improves NK cell and T cell functions. However, it is unclear what these improvements will translate to.
5 More exercise immunology-focused studies are required to understand the beneficial mechanisms of exercise, the long-term outcomes of changes in immunity, and the optimal dose of exercise needed to promote immune enhancements at different stages of the cancer journey. To do this, assessments of acute exercise-induced immune cell mobilisation and function before, during, and after an exercise training period should become standard in trials.

References

Amin, N. A., Seymour, E., Saiya-Cork, K., Parkin, B., Shedden, K., & Malek, S. N. (2016). A quantitative analysis of subclonal and clonal gene mutations before and after therapy in chronic lymphocytic Leukemia. *Clin Cancer Res, 22*(17), 4525–4535.

Arana Echarri, A., Struszczak, L., Beresford, M., Campbell, J. P., Jones, R. H., Thompson, D., & Turner, J. E. (2023). Immune cell status, cardiorespiratory fitness and body composition among breast cancer survivors and healthy women: a cross sectional study. *Front Physiol, 14*, 1107070.

Arana Echarri, A., Struszczak, L., Beresford, M., Campbell, J. P., Thompson, D., & Turner, J. E. (2023). The effects of exercise training for eight weeks on immune cell characteristics among breast cancer survivors. *Front Sports Act Living, 5*, 1163182.

Ashcraft, K. A., Peace, R. M., Betof, A. S., Dewhirst, M. W., & Jones, L. W. (2016). Efficacy and mechanisms of aerobic exercise on cancer initiation, progression, and metastasis: A critical systematic review of in vivo preclinical data. *Cancer Res, 76*(14), 4032–4050.

Ashcraft, K. A., Warner, A. B., Jones, L. W., & Dewhirst, M. W. (2019). Exercise as adjunct therapy in cancer. *Seminars in Radiation Oncology, 29*(1), 16–24.

Aveseh, M., Nikooie, R., & Aminaie, M. (2015). Exercise-induced changes in tumour LDH-B and MCT1 expression are modulated by oestrogen-related receptor alpha in breast cancer-bearing BALB/c mice. *J Physiol, 593*(12), 2635–2648.

Ballard-Barbash, R., Hunsberger, S., Alciati, M. H., Blair, S. N., Goodwin, P. J., McTiernan, A., … Schatzkin, A. (2009). Physical activity, weight control, and breast cancer risk and survival: Clinical trial rationale and design considerations. *J Natl Cancer Inst, 101*(9), 630–643.

Barber, J. L., Kraus, W. E., Church, T. S., Hagberg, J. M., Thompson, P. D., Bartlett, D. B., … Sarzynski, M. A. (2018). Effects of regular endurance exercise on GlycA: Combined analysis of 14 exercise interventions. *Atherosclerosis, 277*, 1–6.

Bartlett, D. B., & Duggal, N. A. (2020). Moderate physical activity associated with a higher naive/memory T-cell ratio in healthy old individuals: Potential role of IL15. *Age Ageing, 49*(3), 368–373.

Bartlett, D. B., Hanson, E. D., Lee, J. T., Wagoner, C. W., Harrell, E. P., Sullivan, S. A., … Battaglini, C. L. (2021). The effects of 16 weeks of exercise training on neutrophil functions in breast cancer survivors. *Front Immunol, 12*, 733101.

Battaglini, C. L., Hackney, A. C., Garcia, R., Groff, D., Evans, E., & Shea, T. (2009). The effects of an exercise program in Leukemia patients. *Integr Cancer Ther, 8*(2), 130–138.

Betof, A. S., Dewhirst, M. W., & Jones, L. W. (2013). Effects and potential mechanisms of exercise training on cancer progression: A translational perspective. *Brain Behav Immun, 30 Suppl*(0), S75–87.

Betof, A. S., Lascola, C. D., Weitzel, D., Landon, C., Scarbrough, P. M., Devi, G. R., … Dewhirst, M. W. (2015). Modulation of murine breast tumor vascularity, hypoxia and chemotherapeutic response by exercise. *J Natl Cancer Inst, 107*(5).

Béziat, V., Descours, B., Parizot, C., Debré, P., & Vieillard, V. (2010). NK cell terminal differentiation: Correlated stepwise decrease of NKG2A and acquisition of KIRs. *PLoS ONE, 5*(8), e11966.

Blank, C. U., Haanen, J. B., Ribas, A., & Schumacher, T. N. (2016). Cancer immunology. The "Cancer Immunogram". *Science, 352*(6286), 658–660.

Bobrie, A., Colombo, M., Raposo, G., & Théry, C. (2011). Exosome secretion: Molecular mechanisms and roles in immune responses. *Traffic, 12*(12), 1659–1668.

Bose, S., Allen, A. E., & Locasale, J. W. (2020). The molecular link from diet to cancer cell metabolism. *Mol Cell, 78*(6), 1034–1044.

Brewer, J. D., Habermann, T. M., & Shanafelt, T. D. (2014). Lymphoma-associated skin cancer: Incidence, natural history, and clinical management. *Int J Dermatol, 53*(3), 267–274.

Bruni, D., Angell, H. K., & Galon, J. (2020). The immune contexture and Immunoscore in cancer prognosis and therapeutic efficacy. *Nat Rev Cancer, 20*(11), 662–680.

Bui, T. M., Wiesolek, H. L., & Sumagin, R. (2020). ICAM-1: A master regulator of cellular responses in inflammation, injury resolution, and tumorigenesis. *J Leukoc Biol, 108*(3), 787–799.

Bull, F. C., Al-Ansari, S. S., Biddle, S., Borodulin, K., Buman, M. P., Cardon, G., … Willumsen, J. F. (2020). World Health Organization 2020 guidelines on physical activity and sedentary behaviour. *Br J Sports Med, 54*(24), 1451–1462.

Buss, L. A., & Dachs, G. U. (2020). Effects of exercise on the tumour microenvironment. *Adv Exp Med Biol, 1225*, 31–51.

Campbell, K. L., Winters-Stone, K. M., Wiskemann, J., May, A. M., Schwartz, A. L., Courneya, K. S., … Schmitz, K. H. (2019). Exercise guidelines for cancer survivors: Consensus statement from international multidisciplinary roundtable. *Med Sci Sports Exerc, 51*(11), 2375–2390.

Coletta, A. M., Agha, N. H., Baker, F. L., Niemiro, G. M., Mylabathula, P. L., Brewster, A. M., ... Simpson, R. J. (2021). The impact of high-intensity interval exercise training on NK-cell function and circulating myokines for breast cancer prevention among women at high risk for breast cancer. *Breast Cancer Res Treat, 187*(2), 407–416.

Deng, N., Reyes-Uribe, L., Fahrmann, J. F., Thoman, W. S., Munsell, M. F., Dennison, J. B., ... Vilar, E. (2023). Exercise training reduces the inflammatory response and promotes intestinal mucosa-associated immunity in Lynch Syndrome. *Clin Cancer Res*, 29(21), 4361–4372.

Dethlefsen, C., Lillelund, C., Midtgaard, J., Andersen, C., Pedersen, B. K., Christensen, J. F., & Hojman, P. (2016). Exercise regulates breast cancer cell viability: Systemic training adaptations versus acute exercise responses. *Breast Cancer Res Treat, 159*(3), 469–479.

Devin, J. L., Hill, M. M., Mourtzakis, M., Quadrilatero, J., Jenkins, D. G., & Skinner, T. L. (2019). Acute high intensity interval exercise reduces colon cancer cell growth. *J Physiol, 597*(8), 2177–2184.

Dholaria, B. R., Bachmeier, C. A., & Locke, F. (2018). Mechanisms and management of chimeric antigen receptor T-cell therapy-related toxicities. *BioDrugs, 33*(1), 45–60.

Djurhuus, S. S., Schauer, T., Simonsen, C., Toft, B. G., Jensen, A. R. D., Erler, J. T., ... Christensen, J. F. (2022). Effects of acute exercise training on tumor outcomes in men with localized prostate cancer: A randomized controlled trial. *Physiol Rep, 10*(19), e15408.

Djurhuus, S. S., Simonsen, C., Toft, B. G., Thomsen, S. N., Wielsøe, S., Røder, M. A., ... Christensen, J. F. (2023). Exercise training to increase tumour natural killer-cell infiltration in men with localised prostate cancer: A randomised controlled trial. *BJU Int, 131*(1), 116–124.

Dobos, G., Overhamm, T., Bussing, A., Ostermann, T., Langhorst, J., Kummel, S., ... Cramer, H. (2015). Integrating mindfulness in supportive cancer care: A cohort study on a mindfulness-based day care clinic for cancer survivors. *Support Care Cancer, 23*(10), 2945–2955.

Donovan, T., Bain, A. L., Tu, W., Pyne, D. B., & Rao, S. (2021). Influence of exercise on exhausted and senescent T cells: A systematic review. *Front Physiol, 12*, 668327.

Duggal, N. A., Niemiro, G., Harridge, S. D. R., Simpson, R. J., & Lord, J. M. (2019). Can physical activity ameliorate immunosenescence and thereby reduce age-related multi-morbidity? *Nat Rev Immunol, 19*(9), 563–572.

Duggal, N. A., Pollock, R. D., Lazarus, N. R., Harridge, S., & Lord, J. M. (2018). Major features of immunesenescence, including reduced thymic output, are ameliorated by high levels of physical activity in adulthood. *Aging Cell, 17*(2), e12750.

Dusseaux, M., Martin, E., Serriari, N., Péguillet, I., Premel, V., Louis, D., ... Lantz, O. (2011). Human MAIT cells are xenobiotic-resistant, tissue-targeted, CD161hi IL-17-secreting T cells. *Blood, 117*(4), 1250–1259.

Emery, A., Moore, S., Turner, J. E., & Campbell, J. P. (2022). Reframing How physical activity reduces the incidence of clinically-diagnosed cancers: Appraising exercise-induced immuno-modulation as an integral mechanism. *Front Oncol, 12*, 788113.

Evans, E. S., Hackney, A. C., McMurray, R. G., Randell, S. H., Muss, H. B., Deal, A. M., & Battaglini, C. L. (2015). Impact of acute intermittent exercise on natural killer cells in breast Cancer survivors. *Integr Cancer Ther, 14*(5), 436–445.

Evans, E. S., Hanson, E. D., & Battaglini, C. L. (2020). Immune, endocrine, and soluble factor interactions during aerobic exercise in Cancer survivors. In A. Hackney & N. Constantini (Eds.), *Endocrinology of Physical Activity and Sport* (pp. 441–458). New York: Humana Cham.

Fairey, A. S., Courneya, K. S., Field, C. J., Bell, G. J., Jones, L. W., & Mackey, J. R. (2005). Randomized controlled trial of exercise and blood immune function in postmenopausal breast cancer survivors. *J Appl Physiol (1985), 98*(4), 1534–1540.

Fiordelisi, A., Iaccarino, G., Morisco, C., Coscioni, E., & Sorriento, D. (2019). NFkappaB is a key player in the crosstalk between inflammation and Cardiovascular diseases. *Int J Mol Sci, 20*(7), 1599.

Fiuza-Luces, C., Valenzuela, P. L., Gálvez, B. G., Ramírez, M., López-Soto, A., Simpson, R. J., & Lucia, A. (2023). The effect of physical exercise on anticancer immunity. *Nat Rev Immunol, 24*, 282–293.

Flynn, J. M., Andritsos, L., Lucas, D., & Byrd, J. C. (2010). Second malignancies in B-cell chronic lymphocytic leukaemia: Possible association with human papilloma virus. *Br J Haematol, 149*(3), 388–390.

Forconi, F., & Moss, P. (2015). Perturbation of the normal immune system in patients with CLL. *Blood, 126*(5), 573–581.

Frattaroli, J., Weidner, G., Dnistrian, A. M., Kemp, C., Daubenmier, J. J., Marlin, R. O., … Ornish, D. (2008). Clinical events in prostate cancer lifestyle trial: Results from two years of follow-up. *Urology, 72*(6), 1319–1323. Doi:10.1016/j.urology.2008.04.050

Friedenreich, C. M., Neilson, H. K., Farris, M. S., & Courneya, K. S. (2016). Physical activity and cancer outcomes: A precision medicine approach. *Clin Cancer Res, 22*(19), 4766–4775.

Galvao, D. A., Nosaka, K., Taaffe, D. R., Spry, N., Kristjanson, L. J., McGuigan, M. R., … Newton, R. U. (2006). Resistance training and reduction of treatment side effects in prostate cancer patients. *Med Sci Sports Exerc, 38*(12), 2045–2052.

Galvao, D. A., Nosaka, K., Taaffe, D. R., Peake, J., Spry, N., Suzuki, K., … Newton, R. U. (2008). Endocrine and immune responses to resistance training in prostate cancer patients. *Prostate Cancer Prostatic Dis, 11*(2), 160–165.

Galvao, D. A., Taaffe, D. R., Spry, N., Joseph, D., & Newton, R. U. (2010). Combined resistance and aerobic exercise program reverses muscle loss in men undergoing androgen suppression therapy for prostate cancer without bone metastases: A randomized controlled trial. *J Clin Oncol, 28*(2), 340–347.

Glass, O. K., Inman, B. A., Broadwater, G., Courneya, K. S., Mackey, J. R., Goruk, S., … Jones, L. W. (2015). Effect of aerobic training on the host systemic milieu in patients with solid tumours: An exploratory correlative study. *Br J Cancer, 112*(5), 825–831.

Grivennikov, S. I., Greten, F. R., & Karin, M. (2010). Immunity, inflammation, and cancer. *Cell, 140*(6), 883–899.

Gubin, M. M., & Vesely, M. D. (2022). Cancer immunoediting in the era of immuno-oncology. *Clin Cancer Res, 28*(18), 3917–3928.

Hanson, E. D., Bates, L. C., Harrell, E. P., Bartlett, D. B., Lee, J. T., Wagoner, C. W., … Battaglini, C. L. (2021). Exercise training partially rescues impaired mucosal associated invariant t-cell mobilization in breast cancer survivors compared to healthy older women. *Exp Gerontol, 152*, 111454.

Hanson, E. D., Danson, E., Evans, W. S., Wood, W. A., Battaglini, C. L., & Sakkal, S. (2019). Exercise increases mucosal-associated invariant T cell cytokine expression but not activation or homing markers. *Med Sci Sports Exerc, 51*(2), 379–388.

Hanson, E. D., Sakkal, S., Que, S., Cho, E., Spielmann, G., Kadife, E., … Hayes, A. (2020). Natural killer cell mobilization and egress following acute exercise in men with prostate cancer. *Exp Physiol, 105*(9), 1524–1539.

Hanson, E. D., Sakkal, S., Bates-Fraser, L. C., Que, S., Cho, E., Spielmann, G., … Hayes, A. (2023). Acute exercise induces distinct quantitative and phenotypical T cell profiles in men with prostate cancer. *Front Sports Act Living, 5*, 1173377.

Harvie, M., Pegington, M., Howell, S. J., Bundred, N., Foden, P., Adams, J., … Howell, A. (2022). Randomised controlled trial of intermittent vs continuous energy restriction during chemotherapy for early breast cancer. *Br J Cancer, 126*(8), 1157–1167.

Hayes, S. C., Rowbottom, D., Davies, P. S., Parker, T. W., & Bashford, J. (2003). Immunological changes after cancer treatment and participation in an exercise program. *Med Sci Sports Exerc, 35*(1), 2–9.

Hirschey, M. D., DeBerardinis, R. J., Diehl, A. M., Drew, J. E., Frezza, C., Green, M. F., … Wellen, K. E. (2015). Dysregulated metabolism contributes to oncogenesis. *Semin Cancer Biol, 35*(Suppl), S129–150.

Hojman, P., Gehl, J., Christensen, J. F., & Pedersen, B. K. (2018). Molecular mechanisms linking exercise to cancer prevention and treatment. *Cell Metab, 27*(1), 10–21.

Huergo-Zapico, L., Acebes-Huerta, A., Gonzalez-Rodriguez, A. P., Contesti, J., Gonzalez-Garcia, E., Payer, A. R., … Gonzalez, S. (2014). Expansion of NK cells and reduction of NKG2D expression in chronic lymphocytic leukemia. Correlation with progressive disease. *PLoS ONE, 9*(10), e108326.

Jones, L. W., Eves, N. D., Peddle, C. J., Courneya, K. S., Haykowsky, M., Kumar, V., … Reiman, T. (2009). Effects of presurgical exercise training on systemic inflammatory markers among patients with malignant lung lesions. *Appl Physiol Nutr Metab, 34*(2), 197–202.

Kaltenboeck, A., & Bach, P. B. (2018). Value-based pricing for drugs: Theme and variations. *JAMA, 319*(21), 2165–2166.

Kang, D. H., Weaver, M. T., Park, N. J., Smith, B., McArdle, T., & Carpenter, J. (2009). Significant impairment in immune recovery after cancer treatment. *Nurs Res, 58*(2), 105–114.

Khosravi, N., Hanson, E. D., Farajivafa, V., Evans, W. S., Lee, J. T., Danson, E., … Battaglini, C. L. (2021). Exercise-induced modulation of monocytes in breast cancer survivors. *Brain Behav Immun Health, 14*, 100216.

Khosravi, N., Stoner, L., Farajivafa, V., & Hanson, E. D. (2019). Exercise training, circulating cytokine levels and immune function in cancer survivors: A meta-analysis. *Brain Behav Immun, 81*, 92–104.

Kim, S. D., & Kim, H. S. (2006). A series of bed exercises to improve lymphocyte count in allogeneic bone marrow transplantation patients. *Eur J Cancer Care (Engl), 15*(5), 453–457.

Knips, L., Bergenthal, N., Streckmann, F., Monsef, I., Elter, T., & Skoetz, N. (2019). Aerobic physical exercise for adult patients with haematological malignancies. *Cochrane Database Syst Rev, 1*(1), Cd009075.

Koelwyn, G. J., Quail, D. F., Zhang, X., White, R. M., & Jones, L. W. (2017). Exercise-dependent regulation of the tumour microenvironment. *Nat Rev Cancer, 17*(10), 620–632.

Koelwyn, G. J., Wennerberg, E., Demaria, S., & Jones, L. W. (2015). Exercise in regulation of inflammation-immune axis function in cancer initiation and progression. *Oncology (Williston Park), 29*(12), 908–922.

Koelwyn, G. J., Zhuang, X., Tammela, T., Schietinger, A., & Jones, L. W. (2020). Exercise and immunometabolic regulation in cancer. *Nat Metab, 2*(9), 849–857.

Kruijsen-Jaarsma, M., Revesz, D., Bierings, M. B., Buffart, L. M., & Takken, T. (2013). Effects of exercise on immune function in patients with cancer: A systematic review. *Exerc Immunol Rev, 19*, 120–143.

Lam, S., Alexandre, L., Luben, R., & Hart, A. R. (2018). The association between physical activity and the risk of symptomatic Barrett's oesophagus: A UK prospective cohort study. *Eur J Gastroenterol Hepatol, 30*(1), 71–75.

Ligibel, J. A., Bohlke, K., May, A. M., Clinton, S. K., Demark-Wahnefried, W., Gilchrist, S. C., … Alfano, C. M. (2022). Exercise, diet, and weight management during cancer treatment: ASCO guideline. *J Clin Oncol, 40*(22), 2491–2507.

Liu, S., Sun, Q., & Ren, X. (2023). Novel strategies for cancer immunotherapy: counter-immunoediting therapy. *J Hematol Oncol, 16*(1), 38.

Liu, X., Hoft, D. F., & Peng, G. (2020). Senescent T cells within suppressive tumor microenvironments: Emerging target for tumor immunotherapy. *J Clin Invest, 130*(3), 1073–1083.

MacDonald, G., Sitlinger, A., Deal, M. A., Hanson, E. D., Ferraro, S., Pieper, C. F., … Bartlett, D. B. (2021). A pilot study of high-intensity interval training in older adults with treatment naïve chronic lymphocytic leukemia. *Sci Rep, 11*(1), 23137.

Matthews, C. E., Moore, S. C., Arem, H., Cook, M. B., Trabert, B., Håkansson, N., … Lee, I. M. (2020). Amount and intensity of leisure-time physical activity and lower cancer risk. *J Clin Oncol, 38*(7), 686–697.

Moore, S. C., Lee, I., Weiderpass, E., et al. (2016). Association of leisure-time physical activity with risk of 26 types of cancer in 1.44 million adults. *JAMA Internal Medicine, 176*(6), 816–825.

Mustian, K. M., Peppone, L., Darling, T. V., Palesh, O., Heckler, C. E., & Morrow, G. R. (2009). A 4-week home-based aerobic and resistance exercise program during radiation therapy: a pilot randomized clinical trial. *J Support Oncol, 7*(5), 158–167.

Nieman, D. C., Cook, V. D., Henson, D. A., Suttles, J., Rejeski, W. J., Ribisl, P. M., … Nehlsen-Cannarella, S. L. (1995). Moderate exercise training and natural killer cell cytotoxic activity in breast cancer patients. *Int J Sports Med, 16*(5), 334–337.

Niemiro, G. M., Coletta, A. M., Agha, N. H., Mylabathula, P. L., Baker, F. L., Brewster, A. M., … Simpson, R. J. (2022). Salutary effects of moderate but not high intensity aerobic exercise training on the frequency of peripheral T-cells associated with immunosenescence in older women at high risk of breast cancer: A randomized controlled trial. *Immun Ageing, 19*(1), 17.

Nisticò, N., Maisano, D., Iaccino, E., Vecchio, E., Fiume, G., Rotundo, S., … Mimmi, S. (2020). Role of chronic lymphocytic leukemia (CLL)-Derived exosomes in tumor progression and survival. *Pharmaceuticals (Basel), 13*(9).

Pavlova, N. N., & Thompson, C. B. (2016). The emerging hallmarks of cancer metabolism. *Cell Metab, 23*(1), 27–47.

Pedersen, L., Idorn, M., Olofsson, Gitte H., Lauenborg, B., Nookaew, I., Hansen, Rasmus H., … Hojman, P. (2016). Voluntary running suppresses tumor growth through epinephrine- and IL-6-Dependent NK cell mobilization and redistribution. *Cell Metabolism, 23*(3), 554–562.

Perry, C., Herishanu, Y., Hazan-Halevy, I., Kay, S., Bdolach, N., Naparstek, E., … Grisaru, D. (2012). Reciprocal changes in regulatory T cells and Th17 helper cells induced by exercise in patients with chronic lymphocytic leukemia. *Leuk Lymphoma, 53*(9), 1807–1810.

Pulluri, B., Kumar, A., Shaheen, M., Jeter, J., & Sundararajan, S. (2017). Tumor microenvironment changes leading to resistance of immune checkpoint inhibitors in metastatic melanoma and strategies to overcome resistance. *Pharmacol Res, 123*, 95–102.

Ritchie, S. C., Würtz, P., Nath, A. P., Abraham, G., Havulinna, A. S., Kangas, A. J., … Inouye, M. (2015). The Biomarker GlycA Is Associated with Chronic Inflammation and Predicts Long-Term Risk of Severe Infection. *Cell Systems, 1*(4), 293–301.

Rozovski, U., Grgurevic, S., Bueso-Ramos, C., Harris, D. M., Li, P., Liu, Z., … Estrov, Z. (2015). Aberrant LPL expression, driven by STAT3, mediates free fatty acid metabolism in CLL cells. *Mol Cancer Res, 13*(5), 944–953.

Rozovski, U., Hazan-Halevy, I., Barzilai, M., Keating, M. J., & Estrov, Z. (2016). Metabolism pathways in chronic lymphocytic leukemia. *Leuk Lymphoma, 57*(4), 758–765.

Schauer, T., Mazzoni, A. S., Henriksson, A., Demmelmaier, I., Berntsen, S., Raastad, T., … Christensen, J. F. (2021). Exercise intensity and markers of inflammation during and after (neo-) adjuvant cancer treatment. *Endocr Relat Cancer, 28*(3), 191–201.

Schreiber, R. D., Old, L. J., & Smyth, M. J. (2011). Cancer immunoediting: Integrating immunity's roles in cancer suppression and promotion. *Science, 331*(6024), 1565–1570.

Semenza, G. L. (2003). Targeting HIF-1 for cancer therapy. *Nat Rev Cancer, 3*(10), 721–732.

Shankaran, V., Ikeda, H., Bruce, A. T., White, J. M., Swanson, P. E., Old, L. J., & Schreiber, R. D. (2001). IFNgamma and lymphocytes prevent primary tumour development and shape tumour immunogenicity. *Nature, 410*(6832), 1107–1111.

Sitlinger, A., Brander, D. M., & Bartlett, D. B. (2020). Impact of exercise on the immune system and outcomes in hematologic malignancies. *Blood Adv, 4*(8), 1801–1811.

Sitlinger, A., Deal, M. A., Garcia, E., Thompson, D. K., Stewart, T., MacDonald, G. A., … Bartlett, D. B. (2021). Physiological fitness and the pathophysiology of Chronic Lymphocytic Leukemia (CLL). *Cells, 10*(5).

Soares, C. M., Teixeira, A. M., Sarmento, H., Silva, F. M., Rusenhack, M. C., Furmann, M., … Ferreira, J. P. (2021). Effect of exercise-conditioned human serum on the viability of cancer cell cultures: A systematic review and meta-analysis. *Exerc Immunol Rev, 27*, 24–41.

Sprod, L. K., Palesh, O. G., Janelsins, M. C., Peppone, L. J., Heckler, C. E., Adams, M. J., … Mustian, K. M. (2010). Exercise, sleep quality, and mediators of sleep in breast and prostate cancer patients receiving radiation therapy. *Community Oncol, 7*(10), 463–471.

Sung, H., Ferlay, J., Siegel, R. L., Laversanne, M., Soerjomataram, I., Jemal, A., & Bray, F. (2021). Global Cancer Statistics 2020: GLOBOCAN estimates of incidence and mortality worldwide for 36 cancers in 185 countries. *CA Cancer J Clin, 71*(3), 209–249.

Swanson, G. P., Hammonds, K., & Jhavar, S. (2021). Short-term haematogical effects of androgen deprivation and radiotherapy in prostate cancer patients. *Open Journal of Urology, 11*(04), 103.

Tasaki, M., Yamashita, M., Arai, Y., Nakamura, T., & Nakao, S. (2021). IL-7 coupled with IL-12 increases intratumoral T cell clonality, leading to complete regression of non-immunogenic tumors. *Cancer Immunol Immunother, 70*(12), 3557–3571.

Valenzuela, P. L., Saco-Ledo, G., Santos-Lozano, A., Morales, J. S., Castillo-García, A., Simpson, R. J., … Fiuza-Luces, C. (2022). Exercise training and natural killer cells in cancer survivors: Current evidence and research gaps based on a systematic review and meta-analysis. *Sports Med Open, 8*(1), 36.

Vesely, M. D., Kershaw, M. H., Schreiber, R. D., & Smyth, M. J. (2011). Natural innate and adaptive immunity to cancer. *Annu Rev Immunol, 29*, 235–271.

Whiteside, T. L. (2016). Tumor-derived exosomes and their role in cancer progression. *Adv Clin Chem, 74*, 103–141.

Xu, Y., & Rogers, C. J. (2020). Impact of physical activity and energy restriction on immune regulation of cancer. *Transl Cancer Res, 9*(9), 5700–5731.

Ye, J., Livergood, R. S., & Peng, G. (2013). The role and regulation of human Th17 cells in tumor immunity. *Am J Pathol, 182*(1), 10–20.

Yeh, Y. Y., Ozer, H. G., Lehman, A. M., Maddocks, K., Yu, L., Johnson, A. J., & Byrd, J. C. (2015). Characterization of CLL exosomes reveals a distinct microRNA signature and enhanced secretion by activation of BCR signaling. *Blood, 125*(21), 3297–3305.

Zimmer, P., Baumann, F. T., Bloch, W., Schenk, A., Koliamitra, C., Jensen, P., … Elter, T. (2014). Impact of exercise on pro inflammatory cytokine levels and epigenetic modulations of tumor-competitive lymphocytes in Non-Hodgkin-Lymphoma patients-randomized controlled trial. *Eur J Haematol, 93*(6), 527–532.

Zitvogel, L., Tesniere, A., Apetoh, L., Ghiringhelli, F., & Kroemer, G. (2008). Immunological aspects of anticancer chemotherapy. *Bull Acad Natl Med, 192*(7), 1469–1487; discussion 1487–1469.

11 Exercise immunology and autoimmune diseases

Brian J. Andonian, Ana J. Pinto, and Bruno Gualano

Introduction

Autoimmune disorders (diseases) are increasing in prevalence and are estimated to affect at least 4% of the worldwide population. In genetically predisposed individuals, autoimmune disorders occur in association with toxic environmental exposures (e.g., tobacco smoke, pesticides, infectious agents) leading to a breakdown of immune self-tolerance. Via immune recognition of self-antigens and tissue-specific damage, autoimmune disorders result in chronic organ dysfunction and inflammation. Though varied in clinical presentation, diagnosis, and management strategies, autoimmune disorders share common features, such as pain, fatigue, depression, low physical fitness, insufficient physical activity, and poor quality of life (Alexanderson & Lundberg, 2012; Aytekin et al., 2012; Cooney et al., 2011; Dendrou, Fugger, & Friese, 2015; DiMeglio, Evans-Molina, & Oram, 2018; Gualano, Ugrinowitsch, et al., 2010; Strombeck & Jacobsson, 2007).

Currently prescribed therapies for patients with autoimmune disorders have many adverse effects. For instance, glucocorticoids are the cornerstone of the treatment for autoimmune diseases, but chronic use of these drugs has been associated with loss of bone and muscle mass and cardiometabolic dysfunction (Poetker & Reh, 2010). Further, many of the drugs used to treat autoimmune disorders (e.g., glucocorticoids, disease-modifying antirheumatic drugs, and biologics) suppress immune function and increase patients' risk for infection; thus, there is an ongoing need for improved therapies for patients with these diseases.

Most symptoms that characterise autoimmune disorders are driven by and predisposed to sustained inflammation (Fukuda et al., 2010; Machado et al., 2010). Regular exercise training can lead to anti-inflammatory effects in chronic diseases characterised by a low-grade chronic systemic inflammation, such as type 2 diabetes mellitus and congestive heart failure (Petersen & Pedersen, 2005). More recently, this notion has been extended to autoimmune disorders (Perandini et al., 2012).

Exercise training has emerged as a clinical tool aimed at improving symptoms, general health, and quality of life in patients with autoimmune disorders (Perandini et al., 2012). Pathways that exercise training may benefit patients with autoimmune disorders include counteracting inflammation and drug-related adverse effects as well as enhancing anti-inflammatory pharmacological treatment. These putative therapeutic actions are summarised in Figure 11.1 and provide the rationale for prescribing exercise as part of the treatment of patients with a variety of autoimmune disorders.

This chapter highlights the potential role of exercise training in alleviating the inflammatory process, treating symptoms, modifying the disease's natural course, and improving general health in selected autoimmune disorders.

DOI: 10.4324/9781003256991-11

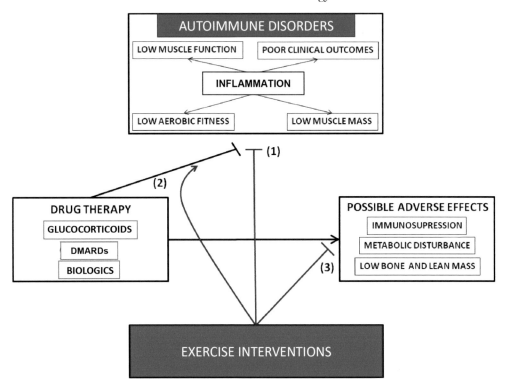

Figure 11.1 Rationale for prescribing exercise for patients with autoimmune disorders.

Legend: 1) Exercise could offset chronic inflammation, low muscle mass, physical capacity and functionality and improve clinical outcomes. 2) Also, exercise could, by mechanisms still poorly understood, combine with anti-inflammatory drugs to improve general symptoms and, perhaps, the disease's natural course. 3) Finally, exercise could partially prevent some drug-related adverse outcomes, such as muscle and bone wasting and immunosuppression. Adapted with permissions from Perandini et al., 2012.

Exercise and type I diabetes mellitus

Introduction

Type I Diabetes (T1D) is a chronic autoimmune disease that results from cell-mediated autoimmune destruction of pancreatic β-cells. Although traditionally characterised as occurring via a single autoimmune event, T1D is now recognised to result from a complex interplay between immune system, metabolism, genome, microbiome, and environmental factors. T1D comprises about 5–10% of diabetes mellitus cases, has an incidence of about 23 cases per 100,000 people, and can occur at any age, with up to 50% of cases being diagnosed in adulthood (DiMeglio et al., 2018).

T1D is characterised by insulin deficiency and resulting hyperglycaemia. Patients with T1D may also present with life-threatening complications (e.g., hypoglycemia and ketoacidosis), and microvascular (e.g., retinopathy, neuropathy, and nephropathy) and macrovascular (e.g., myocardial infarction, cerebrovascular disease) complications. Notably, patients with T1D have a ten times higher risk for cardiovascular events as compared to age-matched healthy adults and cardiovascular diseases are the main cause of mortality among this population (DiMeglio et al., 2018).

Insulin (via injection or pump) is the mainstay treatment for T1D. Other insulin analogues and non-insulin medications for glucose control have been less commonly used as part of drug treatment (DiMeglio et al., 2018). Aside from insulin therapy, healthy eating along with physical activity and exercise have been widely recommended as part of the management of glycaemic control and general health (Colberg et al., 2016; Riddell et al., 2017).

Immune dysfunction in type I diabetes

T1D results from a complex interaction between pancreatic β-cells and innate and adaptive immune systems (Atkinson, Eisenbarth, & Michels, 2014). T1D is thought to be initiated by the presentation of β-cell peptides by antigen-presenting cells. Antigen-presenting cells bearing autoantigens migrate to the pancreatic lymph nodes where they interact with autoreactive CD4+ T cells that activate autoreactive CD8+ T cells. These activated CD8+ T cells return to the islet and lyse β cells expressing immunogenic self-antigens on major histocompatibility complex class I surface molecules. β-cell destruction is further exacerbated by the release of proinflammatory cytokines and reactive oxygen species from macrophages, natural killer cells, and neutrophils. This process is amplified by defects in regulatory T cells sometimes abbreviated to which do not effectively suppress autoimmunity. Finally, activated T cells within pancreatic lymph nodes also stimulate B cells to produce autoantibodies against β-cell proteins (Atkinson et al., 2014).

Physical activity and sedentary behaviour in patients with type I diabetes

Patients with T1D are highly inactive, with only 33% meeting the current physical activity guidelines (McCarthy, Funk, & Grey, 2016). A cross-sectional multicentre study of 18,028 adult patients with T1D also suggested that most patients (63%) do not engage in any recreational exercise (Bohn et al., 2015). Unfortunately, children and adolescents with T1D are similarly inactive (Riddell et al., 2017). Compared with inactive patients, patients with T1D who perform exercise more than 2 days/week had lower odds of diabetes-related complications (i.e., developing severe hypoglycemia, ketoacidosis, retinopathy, and microalbuminuria), hypertension, and dyslipidemia, and higher odds of achieving target glycosylated hemoglobin (HbA1c) and healthy body mass index (BMI) (Bohn et al., 2015). Altogether, these findings in T1D highlight (1) the need to address high levels of physical inactivity and (2) the potential benefits of promoting physical activity and exercise training among T1D patients.

Effects of individual exercise bouts on overall health in patients with type I diabetes

During individual bouts of aerobic exercise, blood glucose concentrations will fall in most patients with T1D on insulin therapy; this phenomenon occurs because exogenous insulin concentrations remain elevated in the systemic circulation, perhaps due to impaired clearance resulting from increased blood flow to subcutaneous adipose tissue (Riddell et al., 2017). Reductions in glucose concentrations during bouts of aerobic exercise are independent of age, occurring in children, adolescents, and adults with T1D (Tonoli et al., 2012). Hypoglycemia develops within about 45 minutes of starting aerobic exercise unless patients ingest carbohydrates prior to and/ or during the session. Of note, greater reductions in blood glucose concentrations during aerobic exercise are observed in trained individuals with T1D who perform higher overall work rates. As such, patients with T1D, independent of aerobic conditioning, typically require either or both increased carbohydrate intake and insulin dose reduction before starting an aerobic exercise session (Riddell et al., 2017).

Compared with moderate-intensity aerobic exercise, resistance exercise is associated with better glucose stability. In fact, resistance exercise could even cause a modest rise in blood glucose in some patients (Riddell et al., 2017). A cross-over study comparing the effects of a 45-minute aerobic exercise session with a 45-minute resistance exercise session demonstrated that resistance exercise resulted in less initial decline in glucose concentrations compared with aerobic exercise (Yardley et al., 2013). A systematic review with meta-analysis corroborated this differential effect of aerobic and resistance exercise on blood glucose responses. Interestingly, the rate of changes in blood glucose during exercise and recovery were not associated with exercise intensity as evidenced by a meta-regression (Garcia-Garcia, Kumareswaran, Hovorka, & Hernando, 2015), indicating that it might be the type of activity and not intensity that underlies differences in glucose responses during exercise. Other training strategies that can attenuate the decrease in blood glucose and minimises the risk of hypoglycemia include (1) performing high-intensity interval sprint training, and (2) performing resistance exercise before starting aerobic exercise (Riddell et al., 2017).

Glucose uptake in skeletal muscle decreases immediately after aerobic exercise cessation, but glucose disposal remains elevated for several hours to replenish glycogen stores. Thus, there is an elevated risk of hypoglycemia in the 24 hours following aerobic exercise, with the greatest risk of nocturnal hypoglycemia occurring when exercise is performed in the afternoon period (Riddell et al., 2017). Interestingly, a cross-over study demonstrated that resistance exercise was associated with more prolonged reductions in glucose concentrations during recovery compared with aerobic exercise (Yardley et al., 2013). These greater effects of resistance training on post-exercise glycemia may explain why reductions in HbA1c concentrations are primarily observed with resistance but not aerobic exercise training (Yardley et al., 2013). However, further studies are required to better elucidate the role of activity type and duration on blood glucose concentrations in T1D during exercise and post-exercise periods.

Effects of exercise training on overall health in patients with type I diabetes

Based on available evidence, regular exercise training can improve glycaemic control (i.e., HbA1c) in patients with T1D. However, due to the low number of studies and variations between training protocol and duration, findings of exercise training effect on T1D outcomes are inconsistent across age groups and between different types of activity (Tonoli et al., 2012).

A systematic review with meta-analysis summarised the effects of exercise training on glycaemic control and other clinical outcomes (Ostman, Jewiss, King, & Smart, 2018). In adults, exercise training increased cardiorespiratory fitness (VO_2 peak; +4.1 ml/kg/min) and reduced body weight (–2.2 kg), BMI (–0.39 kg/m^2), and low-density lipoprotein cholesterol (LDL; –0.21 mmol/L). In children and adolescents, exercise reduced insulin dose (–0.23 IU/kg), waist circumference (–5.4 cm), LDL (-0.31 mmol/L), and triglycerides (–0.21 mmol/L). However, there were no improvements on HbA1c, fasting blood glucose, blood pressure, and high-density lipoprotein cholesterol (HDL) (Ostman et al., 2018). Another meta-analysis combining data from both children/adolescents and adults with T1D demonstrated that exercise training significantly increased cardiorespiratory fitness and reduced HbA1c, daily insulin dosage, and total cholesterol, but did not improve BMI, blood pressure, triglycerides, HDL, and LDL. Sensitivity analysis indicated that improvements were more consistent among children/adolescents, for high frequency of exercise sessions (>3 times/week) and for longer duration of exercise programme (>12 weeks); however, improvements were inconsistent for type of activity (Wu et al., 2019).

Interestingly, though a 12-week combined (aerobic + resistance) exercise training intervention improved cardiorespiratory fitness and muscle strength in patients with T1D, these improvements were blunted in T1D as compared with improvements in healthy controls (Minnock et al., 2022). Thus, T1D may impair adaptations to exercise training. Further studies are needed to test strategies that may potentiate adaptations to exercise in T1D.

In sum, comparing and pooling the effects of exercise training on T1D is limited by both participant and methodological heterogeneity of current studies. Therefore, the most effective exercise prescription is unclear for the management of glycaemic control, cardiometabolic risk factors, and general health in T1D.

Effects of exercise on immune function in patients with type I diabetes

Individual bouts of exercise

Fewer studies have reported the effects of exercise on T1D immune function. One study showed that a 30-minute bout of aerobic exercise resulted in altered kinetics of inflammatory markers in children with T1D as compared with healthy controls. In patients with T1D, exercise-induced interleukin (IL)-6 peak occurred earlier and in greater magnitude than in controls. Other inflammatory markers, including IL-1a, IL-4, IL-8, IL-12p70, IL-17, monocyte chemoattractant protein-1, and macrophage inflammatory protein-1a, have also displayed an early peak and pronounced response to exercise (Rosa et al., 2008). This amplified inflammatory response observed in T1D compared with control occurred despite similar baseline concentrations of most inflammatory markers in children with T1D.

In a study of adults with T1D compared with healthy controls, a 30-minute bout of vigorous exercise differentially affected innate immunity (NK cells) (Curran, Campbell, Powell, Chikhlia, & Narendran, 2020) and acquired immunity (T cells) cell subsets (Curran, Campbell, Drayson, Andrews, & Narendran, 2019). Exercise-induced mobilization of NK cells and myeloid cells were observed in both T1D and controls. However, mature NK cells ($CD56^{dim}CD16^{bright}$) mobilization was blunted in T1D, particularly in early mature KIR^+ (a family of highly polymorphic receptors that serve as key regulators of human NK cell function) NK cells (Curran et al., 2020). This mobilization of innate immune cells highlights the potential of exercise to improve surveillance for infection and modulate autoimmune responses in patients with T1D. Regarding T cells, exercise-induced mobilization of CD4+ and effector memory CD8+ CD45RA+ (EMRA) subsets were blunted in T1D (Curran et al., 2019). This blunted mobilization of differentiated CD8+ EMRA cells might be due to differences in blood glucose, adrenaline receptor density, and sequestration of T cells in the pancreas of patients with T1D.

Exercise training

In humans, a 12-week combined (aerobic and resistance) exercise training intervention increased mitochondrial pyruvate carrier (MPC)-1 mRNA levels in T1D skeletal muscle but did not alter muscle inflammatory markers. Of note, patients with T1D had higher baseline muscle tumor necrosis factor (TNF)- α mRNA levels and phosphorylated p38 mitogen-activated protein kinase as compared with controls. These altered skeletal muscle inflammatory markers were accompanied by blunted adaptations to exercise training (Minnock et al., 2022).

Using a spontaneous mouse model of T1D (non-obese diabetic mouse), Oharomari et al. (2017) investigated the effects of exercise training combined or not with curcumin on T1D progression. Incidence of T1D was similar between trained and untrained mice. Moderate-intensity exercise training (running, 60% max speed, 60 min/day, 5 days/week, 20 weeks) protected pancreatic β cells. Trained groups presented lower insulin concentrations and 50% less immune cell infiltration in pancreatic islets (i.e., insulitis) than untrained groups, but no differences were observed for

IL-6 and TNF-α levels (Oharomari, de Moraes, & Navarro, 2017). These findings provide initial evidence that exercise can mitigate the development of T1D in genetically susceptible mice. Using a model to induce diabetes in rats, Crespilho et al. (2010) demonstrated that alloxan-induced diabetes increased blood neutrophils, which was prevented by a four-week swimming training intervention (5% of body weight, 1 hour/day, 5 days/week). Neutrophils in the diabetic trained rats were comparable to non-diabetic sedentary and trained rats. Finally, no differences were observed for haematocrit, total leucocytes, eosinophils, lymphocytes, and monocytes based on diabetes and training status (Crespilho, de Almeida Leme, de Mello, & Luciano, 2010). These findings suggest exercise training can counteract diabetes-related changes in blood neutrophils.

Based on available evidence, exercise training is a potential immunoregulatory tool for T1D, but further research is warranted regarding the effects of exercise on T1D immune dysfunction, particularly in the longer term.

Exercise prescription for patients with type I diabetes

Children and adults with T1D should follow general public health physical activity (Bull et al., 2020) and exercise (Colberg et al., 2016; Garber et al., 2011) recommendations (Table 11.1). Patients with T1D are also encouraged to increase incidental physical activity and reduce time spent in sedentary behaviour to gain additional health benefits (Bull et al., 2020; Colberg et al., 2016).

Due to the risk of exercise-induced hypoglycaemia, patients with T1D should: (1) check their blood glucose concentration prior to exercise and target an individually tailored range of 90–250 mg/dL [5.0–13.9 mmol/L]; (2) ingest carbohydrate and/or adjust insulin injection or delivery prior to exercise to prevent exercise-induced hypoglycaemia or late onset hypoglycaemia, as directed by a healthcare professional (Figure 11.2); (3) check blood glucose during and after exercise so carbohydrate and insulin adjustment strategies can be implemented if/when needed to maintain stable and safe blood glucose levels; and (4) postpone any exercise if blood glucose <90mg/dL (<5.0 mmol/L), >350 mg/dL (>19.4 mmol/L), or if presenting ketonuria, or intense exercise if >250 mg/dL (>13.9 mmol/L) until these alterations are corrected via carbohydrate intake or insulin adjustments (Colberg et al., 2016; Riddell et al., 2017).

Prior to exercise prescription for T1D, strategies to minimise the risk of exercise-induced hypoglycemia should include consideration of patient goals, current physical activity, and other clinical factors (e.g., comorbidities) (Riddell et al., 2017). As glucose responses to physical activity in T1D are highly variable due to several factors (e.g., insulin regimen, blood glucose concentration, timing of exercise, type, intensity, and duration of activity), healthcare professionals should educate patients on the health-risks of hypoglycemia, the influence of participation in exercise on blood glucose, and how to minimise the risk of hypoglycemia episodes during and after exercise (Colberg et al., 2016; Riddell et al., 2017).

Exercise training is relatively contraindicated for those with a recent severe hypoglycemia episode (i.e., less than 24 hours); however, these individuals may still benefit from low-intensity activities. Overall, the health benefits of being physically active outweighs the risks of being inactive and sedentary in patients with T1D.

Exercise and rheumatoid arthritis

Introduction

Rheumatoid arthritis (RA) is an autoimmune disease characterised by chronic, symmetric inflammatory arthritis of predominantly small joints (i.e., hand/wrist, foot/ankle). RA affects females greater than males at a ratio of approximately 3:1 and has a prevalence of approximately

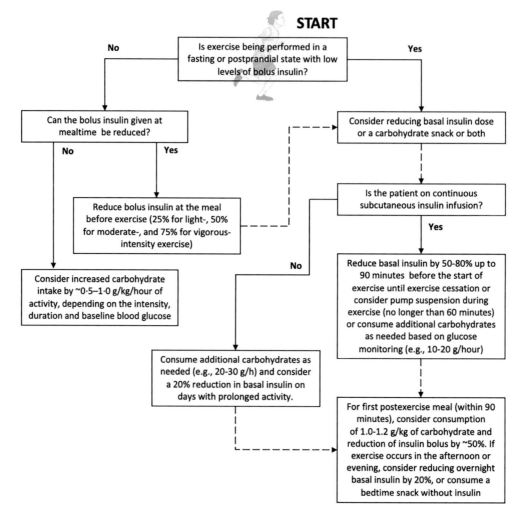

Figure 11.2 Decision tree for aerobic exercise or combined aerobic and anaerobic exercise sessions lasting more than 30 minutes in patients with type 1 diabetes mellitus. Adapted with permissions from Riddell et al. (2017).

1%, making RA one of the most common autoimmune diseases worldwide. Though various biologic pharmacotherapies (e.g., monoclonal antibodies targeting TNF-α and IL-6) have revolutionized RA care since the early 2000s, patients with RA remain at increased risk of early physical disability, age-related cardiometabolic comorbidities (e.g., type 2 diabetes mellitus, coronary artery disease), and death (DeMizio & Geraldino-Pardilla, 2020). Thus, given the ongoing need to improve patient outcomes, more attention has recently been placed on studying interventions in RA to reduce sedentary behaviour and increase physical activity.

Immune dysfunction in RA

The immunopathogenesis of RA is incompletely understood with both environmental (e.g., cigarette smoke) and genetic factors (e.g., HLA-DRB1) contributing to a loss of immunological

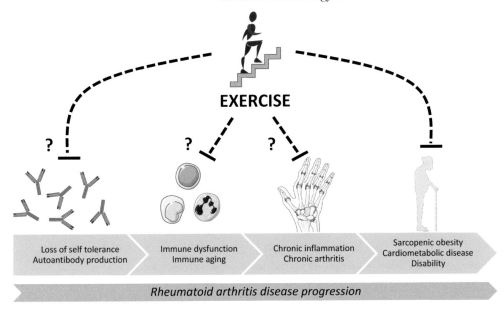

Figure 11.3 Rheumatoid arthritis disease progresses through multiple stages of immune pathology.

Legend: 1) loss of immune self-tolerance and production of rheumatoid factor and anti-citrullinated protein au-toantibodies; 2) immune dysfunction and aging hallmarked by defective DNA repair and metabolic reprogramming; 3) chronic tissue and systemic inflammation; and 4) accumulation of chronic inflammation-associated comorbidities. Exercise and physical activity interventions have potential (?) to impact rheumatoid arthritis disease progression at each stage, however, further study is needed. Figure generated with images from BioIcons and Servier Medical Art, provided by Servier, licensed under a Creative Commons Attribution 3.0 unported license.

self-tolerance and the production of autoantibodies such as rheumatoid factor and anti-cyclic citrullinated protein (CCP) antibodies. Figure 11.3 outlines the stages of immune pathology in RA and potential mechanisms exercise might prevent or improve RA immune dysfunction at each stage. The first stage begins many years before the onset of arthritis symptoms with the loss of self-tolerance and immune recognition of modified proteins (e.g., citrullinated antigens) (Weyand & Goronzy, 2021). Second, following the initial recognition of self-peptides in RA, the transition from asymptomatic autoimmunity (i.e., presence of rheumatoid factor and anti-CCP antibodies) to tissue inflammation occurs in concert with impaired T cell function and immune ageing. This RA immune ageing phenotype is hallmarked by autoantibody production, low peripheral naive CD4+ T cell counts, a contracted T-cell receptor repertoire, telomere fragility, impaired DNA repair, and metabolic reprogramming (Weyand & Goronzy, 2021). Third, in the chronic inflammatory arthritis phase, persistently active proinflammatory macrophages in the RA joint promote synovial fibroblast invasiveness and subsequent joint damage (Weyand & Goronzy, 2021). Finally, chronic tissue and systemic inflammation contribute to cardiometabolic disease and physical disability in RA.

Physical activity and sedentary behaviour in patients with RA

Patients with RA are highly sedentary and inactive and have a very low aerobic capacity (Pinto et al., 2017). Importantly, RA incidence is reduced in persons with high levels of physical activity (Liu et al., 2019). On the contrary, inactive/sedentary RA patients have more severe disease activity on presentation and greater overall CVD risk (Pinto et al., 2017).

Reducing sedentary behaviour is generally regarded as important in the management of RA, but this concept is understudied. Two studies have assessed the effects of interventions designed to reduce sedentary time in persons with RA. First, in a randomised cross-over design, RA participants who completed 3-minute bouts of light-intensity walking (i.e., sitting breaks) every 30 minutes attenuated post-prandial glucose and insulin responses (Pinto et al., 2021). Further, the sitting breaks group also reduced serum concentrations of TNF-α and IL-1β, suggesting a possible anti-inflammatory effect of sitting breaks in RA. Second, in a randomised trial design, RA patients who completed a four-month counseling and text message intervention to reduce daily sitting time decreased sedentary behaviour and improved quality of life, physical function, pain, fatigue, and serum cholesterol levels compared with a control group at 4 months and at 18 months postintervention follow-up (Thomsen et al., 2020). Combined, these studies highlight the potential benefits of reducing sedentary time for improving RA cardiometabolic risk. Further research is needed to assess the impact of modifying sedentary behaviour on inflammation and immune function in RA and other autoimmune disorders.

Effects of individual bouts of exercise on overall health in patients with RA

The effects of individual bouts of exercise in RA are incompletely understood. Available evidence suggests that exercise is associated with impaired autonomic responses in RA. In a study of graded maximal aerobic exercise, RA patients displayed an attenuated heart rate response and lower heart recovery compared with healthy control participants (Pecanha et al., 2018). Further, during isometric knee extension exercise compared with controls, RA participants showed greater increases in mean arterial pressure and muscle sympathetic nerve activity in association with changes in both pro- and anti-inflammatory cytokine profiles (Pecanha et al., 2021). How these altered responses to exercise bouts affect exercise prescription and response to exercise training in RA is not known and a ripe area for future research.

Effects of exercise training on overall health in patients with RA

Exercise training for RA patients was traditionally discouraged due to concerns that excessive exercise would worsen joint disease. On the contrary, research over the past quarter century has consistently showed that exercise training is safe in RA and overall does not increase radiographic damage from arthritis. Current consensus guidelines now recommend exercise training and increasing physical activity for the management of arthritis, including RA (Rausch Osthoff et al., 2018). The list of the benefits of exercise training for RA is ever growing. Similar to other populations, exercise training improves RA muscle mass and strength, cardiorespiratory fitness, cardiometabolic disease risk, pain, fatigue, sleep disturbance, depression, and functional disability (Katz, Andonian, & Huffman, 2020). However, despite research establishing the safety and benefits of exercise training for RA, there remains ongoing questions regarding the impact of exercise on RA inflammatory disease activity.

Animal studies show mixed effects of exercise training on RA joint disease and inflammation depending on the experimental model used. In collagen-induced arthritis rats, the most commonly used model of RA, exercise training decreased synovial hyperplasia, and bone destruction (Shimomura et al., 2018). In SKG/Jcl transgenic inflammatory arthritis mice, exercise training delayed arthritis onset and severity and increased articular cartilage thickness in association with increased systemic expression of IL-10 and IL-15 and inhibition of TNF-α (Kito et al., 2016). In a K/B×N mouse model of inflammatory arthritis, exercise training improved joint inflammation in association with decreased systemic IL-6 and improvements in cardiac and skeletal muscle dysfunction (Huffman et al., 2021). Conversely, in adjuvant-induced arthritis rats, exercise training exacerbated synovial inflammatory infiltrate and destruction (Gonzalez-Chavez, Quinonez-Flores, Espino-Solis, Vazquez-Contreras, & Pacheco-Tena, 2019). Additionally, treadmill running in collagen antibody induced arthritis promoted arthritis chronicity via mechanostrain-induced complement activation and regulatory T cell impairment (Cambre et al., 2019). These findings highlight the need for better understanding in RA the balance between the positive benefits of exercise training and potential detrimental effects of excessive biomechanical stress—an adverse effect that has yet to be shown in the human RA condition.

Several key human studies highlight the potential benefits of aerobic exercise, resistance exercise, and combined exercise training to improve disease activity and systemic inflammation in RA. In a cohort study of traditional aerobic training exercise three times per week for three months, RA participants significantly improved RA inflammatory disease activity as measured by the Disease Activity Score in 28 joints (DAS28) (Wadley et al., 2014). Further, resistance exercise twice weekly for six months significantly improved DAS28 and the inflammatory marker erythrocyte sedimentation rate (ESR) in patients with recent onset RA (Hakkinen, Sokka, Kotaniemi, & Hannonen, 2001). Not surprisingly, combined aerobic and resistance exercise training programmes also improved RA overall cardiovascular risk profiles and disease activity, potentially to a greater extent than either modality alone (Stavropoulos-Kalinoglou et al., 2013). In addition to traditional aerobic exercise programmes, high-intensity interval training (HIIT)—where participants alternated approximately 1-minute bouts of near maximal intensity treadmill walking with bouts of lower intensity—significantly improved DAS28 and ESR (Bartlett et al., 2018). While various exercise training modes for RA have all shown benefits, aggregate data from meta-analyses have yet to conclusively show that exercise training significantly improves RA disease activity or inflammatory markers (Gwinnutt et al., 2022). Thus, further high-quality research is overall needed to better understand the impact of exercise training on RA inflammation.

Effect of exercise on immune function in patients with RA

Individual bouts of exercise

Two recent studies shed light on the potential effects of individual bouts of exercise on RA immune function. After an individual bout of upper and lower extremity resistance exercise, female RA participants significantly decreased systemic concentrations of IL-1b and increased concentrations of IL-1 receptor antagonist, IL-6, and IL-10 (Pereira Nunes Pinto, Natour, de Moura Castro, Eloi, & Lombardi Junior, 2017). Following moderate-to-vigorous treadmill walking for 30 minutes, RA participants decreased serum IL-1β and TNF-α concentrations (Ercan et al., 2022). These post-exercise anti-inflammatory profiles were similar to healthy control cohorts of similar age, sex, and BMI (Ercan et al., 2022; Pereira Nunes Pinto et al., 2017), suggesting that the immunomodulatory immune effects of exercise—which are better studied in non-RA populations—may be translated to RA patients. Further detailed study in RA is needed to better understand both impairments in exercise responses and the potential beneficial effects of individual bouts of exercise on systemic immune function.

Exercise training

RA, like other autoimmune diseases, is hallmarked by loss of self-tolerance and immune dysfunction. Chronic immune activation and inflammation in RA lead to joint and organ damage.

Chronic immune suppression induced by currently available pharmacologic therapies increases RA patients' risk for infection. Unfortunately, no panacea for RA currently exists to control immune activation without immunosuppression; discovery of better immunoregulatory therapeutics is thus a top priority in the fields of rheumatology and immunology.

Though exercise may improve immune function and modify infection risk in otherwise healthy populations, the potential for exercise as a balanced immunomodulatory therapy for RA is not yet well studied. In a study of HIIT for patients with established RA, 10 weeks of training significantly improved innate immune function as evidenced by increased peripheral neutrophil chemotaxis and increased macrophage phagocytosis of *Escherichia coli* (Bartlett et al., 2018). These RA participants also significantly improved inflammatory disease from low–moderate disease activity to remission. Interestingly, despite improvements in inflammatory disease activity, HIIT did not induce changes in RA-circulating inflammatory cytokines (e.g., IL-1β and TNF-α). Similar null responses in RA plasma IL-1β and IL-6 were previously reported in patients with RA and moderate disease activity who completed an eight-week steady-state lower-intensity cycle training program (Baslund et al., 1993); these findings together suggest that the potential beneficial effects of exercise training on RA immune function are not related to changes in circulating inflammatory cytokines. In the same RA HIIT cohort following the intervention, RA CD4+ T-cell oxidative metabolic function also changed in association with significant improvements in cardiorespiratory fitness (i.e., VO_2 peak) (Andonian et al., 2022). In a different intervention of 20 weeks of combined moderate- to high-intensity aerobic and resistance exercise, older (age ≥ 65) patients with RA remained in low-disease activity or disease remission and exhibited decreases in peripheral CD4+Foxp3+CD25+CD127− regulatory T cells and CD19+CD24hiCD38hi regulatory B cells (Andersson et al., 2020). These works highlight the potential for exercise to improve RA immune function without substantially increasing infection risk. Still, a more detailed study is needed regarding the effects of exercise on both RA innate and adaptive immune responses.

The purported pathways that connect exercise to improved immune function in healthy populations may also apply to autoimmune diseases such as RA. Adaptations to biomechanical and metabolic stress during exercise involve signals from multiple intermediate organs, including neuroendocrine, adipose, and skeletal muscle systems. In RA, many of the organ systems key to orchestrating the systemic effects of exercise are impaired at baseline. For example, the biomechanics of exercise are impaired in RA due to chronic joint pain, inflammation, and damage. Further, systemic metabolic function in RA is altered as evidenced by the high prevalence of sarcopenic obesity (i.e., high fat mass, low muscle mass) in RA patients with associated disability and cardiometabolic disease risk (Challal, Minichiello, Boissier, & Semerano, 2016). Whether or not exercise-induced adaptations (e.g., improvements in the biomechanics of physical activity or body composition) alter RA immune function remains to be fully explored.

The overall impact of exercise on chronic RA immune dysfunction is poorly understood and an area for future exercise immunology research (Figure 11.3). Since high levels of physical activity are associated with a lower likelihood of incident RA (Liu et al., 2019), routine physical activity may confer benefits in terms of maintaining immune tolerance in persons at risk for developing RA. Given lifelong exercise can impact immunosenescent pathways (Tylutka, Morawin, Gramacki, & Zembron-Lacny, 2021) (see Chapter 8), exercise training may impart similar anti-ageing effects on RA immune function. It is currently unclear if exercise training can help reverse RA joint pathology and inflammation. However, based on available evidence, exercise training appears to reduce RA chronic inflammation-related comorbidities. Further research is warranted regarding the direct effects of exercise on RA immune dysfunction.

Exercise prescription for patients with RA

Current recommendations for physical activity and exercise for persons with RA agree with public health guidelines for adults (Table 11.1) (Rausch Osthoff et al., 2018). However, there are several special considerations when designing an exercise programme for patients with RA (Table 11.1). Exercise prescription and planning for RA is a collaborative effort between the patient, exercise specialist, and rheumatologist or healthcare provider. During a flare of joint pain, stiffness, and swelling, the patient should discuss with their rheumatologist or healthcare provider regarding medications to reduce RA inflammation while continuing to perform gentle joint movements and light exercise as tolerated. Other considerations involve the timing of exercise for RA during the day. In the morning, when joint stiffness in generally highest, physical activities should focus on light-intensity movements and mobility- and balance-training exercises. Later in the day, higher-intensity aerobic and resistance-training exercises are generally better tolerated. Additionally, since RA commonly affects the small joints of the hands, wrists, feet, and ankles, adjustments may be required for patients to adequately perform certain exercises. These exercise adjustments include modifying aerobic activities to reduce loads to the legs and feet (e.g., cycling, elliptical trainer, and warm water pool exercise) and resistance-training activities to limit stress through the hands and wrists (e.g., resistance bands and cable machines with or without ankle/wrist strap attachments). Furthermore, inclusion of hand- and wrist-specific exercise training may be beneficial to improve RA hand function and pain and assist patients in performing other forms of exercise (Rodriguez Sanchez-Laulhe et al., 2022). Finally, RA patients should be instructed on completing movements within pain-free ranges of motion. If worsening of arthritis pain symptoms with exercise, the RA patient is recommended to consult with both their rheumatologist or healthcare provider and exercise specialist to modify their exercise programme.

Exercise and systemic lupus erythematosus

Introduction

Systemic lupus erythematosus (SLE) is an autoimmune rheumatic disease characterised by autoantibody production (e.g., anti-nuclear antibodies), immune complex and complement deposition, pancytopenia, and a variable presentation of multiorgan inflammation (e.g., dermatitis, nephritis, cerebritis, and arthritis). SLE incidence ranges between 0.3 and 31.5 cases per 100,000 individuals per year with a female predominance of roughly ten women for every man affected by the disease (Fanouriakis, Tziolos, Bertsias, & Boumpas, 2021).

Immune dysfunction in SLE

Immune dysfunction in SLE is evidenced by elevated systemic levels of interferon (IFN)-α, IL-6, TNF-α, IL-10, and soluble TNF receptors (sTNFRs) (Fanouriakis et al., 2021). Chronic inflammation in SLE has also been implicated in a cluster of cardiometabolic risk factors (e.g., dyslipidemia, dysautonomia, atherosclerosis, and insulin resistance), resulting in increased morbidity and mortality (Fanouriakis et al., 2021). SLE immunopathogenesis is a consequence of a complex interplay of genetic, epigenetic, immunoregulatory, ethnic, hormonal, and environmental factors; several mechanisms of these multifactorial connections remain uncertain (Zucchi et al., 2022). Cumulative evidence shows that the T helper 17 (Th17) pathway (e.g., IL-17, IL-21) is involved in several aspects of SLE pathogenesis (Peng et al., 2021; Sippl et al., 2021), whereas IL-18 plays a role in SLE disease progression and activity (Ma, Lam, Lau, & Chan, 2021; Xiang et al., 2021). The reason why SLE is a female-predominant disease is not fully understood, but recent evidence suggests that this sex-bias is related to epigenetically induced modifications in X-linked immune gene expression, namely in B cells-driven autoimmunity (Pyfrom et al., 2021; Yu et al., 2021).

Physical activity and sedentary behaviour in patients with SLE

Patients with SLE are highly inactive (~70% do not achieve physical activity guidelines) and sedentary (~9 hours/day of sedentary time) (Pinto et al., 2017). In this population group, time spent in moderate-to-vigorous physical activity is associated with better physical function, and lower fatigue and pain (Pinto et al., 2017). Thus, promotion of an active lifestyle has potential to improve disease symptoms and general health in SLE.

Effects of individual exercise bouts on overall health in patients with SLE

To our knowledge, there is no study assessing the effects of individual exercise bouts on health-related parameters in SLE, with exception of inflammatory response, which is described in Section 4.6.1.

Effects of exercise training on overall health in patients with SLE

Available evidence suggests that exercise is a safe and effective tool in SLE to improve several clinical outcomes, such as fatigue, depression, physical deconditioning, autonomic and endothelial dysfunction, metabolic disturbances, and impaired quality of life (Sharif et al., 2018). For example, an aerobic exercise programme (three times per week at 70–80% of HR_{max}) led to improvements in fatigue, aerobic capacity, and quality of life; interestingly, continuation of a home-based exercise program following the supervised training period yielded similar benefits (Ramsey-Goldman et al., 2000).

In a randomised controlled exercise training trial, a 12-week moderate-intensity aerobic exercise intervention improved insulin sensitivity in SLE patients with mild/inactive disease. Compared with the control group, the SLE aerobic training group also increased stimulated skeletal muscle AMPK phosphorylation—an important energy sensor involved in muscle glucose uptake and thus a possible mediator of the beneficial metabolic effects of exercise for SLE (Benatti et al., 2018).

Exercise training is also a potential therapeutic tool for children and adolescents with SLE. A 12-week supervised aerobic training programme was safe and effective in improving aerobic conditioning and physical function in a 15-year-old boy with juvenile SLE and antiphospholipid syndrome (Prado et al., 2011). These data were subsequently confirmed in randomised controlled trial of a 12-week aerobic exercise training programme for juvenile SLE, extending the potential therapeutic role of exercise training to paediatric patients suffering from autoimmune rheumatic diseases (Prado et al., 2013).

Of note, some exercise training protocols failed to improve aerobic capacity, lipid profile, and quality of life in patients with SLE, as was expected. Possible contributors to this variability in study results include study differences in exercise prescription, outcomes, and clinical factors (Sharif et al., 2018; Tench, McCarthy, McCurdie, White, & D'Cruz, 2003). Still, overall assessment of the current literature suggests that exercise training has a therapeutic value in the management of SLE. Nonetheless, further well-powered, longer duration, randomised controlled exercise training trials for SLE are necessary.

Effects of exercise on immune function in patients with SLE

Individual bouts of exercise

The effects of individual exercise bouts on immune function and inflammation in SLE have been reported in a series of studies (Perandini et al., 2016; Perandini et al., 2015; Perandini et al., 2014). First, single bouts of either moderate (~50% of VO_2 peak) or vigorous (~70% of VO_2 peak) exercise-induced mild and transient changes in cytokine levels among SLE patients with active and inactive disease, suggesting that individual bouts of aerobic exercise do not trigger inflammation in this population. Additionally, a single bout of moderate-intensity exercise led to gene expression downregulation of innate and adaptive immune-signalling following exercise (e.g., TLR3, IFNG, GATA3, FOXP3, STAT4) with a subsequent up-regulation of these targets occurring upon recovery. Though bouts of exercise regulated gene expression of immune-signalling in blood leucocytes of both SLE patients and healthy controls, SLE patients exhibited fewer modulated genes and less densely connected networks. This latter finding suggests that SLE patients may be partially deficient in triggering a normal exercise-induced immune response.

Exercise training

Fewer studies have reported to the effects of exercise training on SLE immune function. Regular exercise training at a moderate intensity attenuated the inflammatory milieu (e.g., reductions in IL-6, IL-10, and TNF-α levels) in women with SLE, revealing a novel homeostatic immunomodulatory role of exercise in this autoimmune rheumatic disease (Perandini et al., 2014).

Exercise prescription for patients with SLE

Similar to other autoimmune disorders, children and adults with SLE are recommended to follow general public health guidelines for physical activity (Bull et al., 2020) and exercise (Garber et al., 2011) (Table 11.1). SLE patients are also encouraged to increase incidental physical activity and reduce time spent in sedentary behaviour to gain additional health benefits (Bull et al., 2020). Patients with SLE should avoid high-intensity exercise during flares and high disease activity; instead, they should be encouraged to continue lower-intensity physical activity as tolerated and adjust the training program according to the patient's symptoms and comorbidities under the guidance of their rheumatologist or healthcare professional and exercise specialist.

Exercise and multiple sclerosis

Introduction

Multiple sclerosis (MS) is a chronic autoimmune, demyelinating neuroinflammatory disease that affects the brain and spinal cord. The disease affects approximately 2.5 million people worldwide and is a common cause of serious physical disability in young adults, particularly females. Clinical manifestations are variable among patients and include sensory and visual disturbances, motor impairments, fatigue, pain, and cognitive deficits that ultimately compromise quality of life and participation in daily living activities. MS is typically treated with immunosuppressing disease-modifying drugs, which substantially control the inflammatory activity and reduce relapses but not the neurodegenerative processes with resulting physical disability (Dendrou et al., 2015).

Immune dysfunction in MS

Both innate and adaptive immune systems are known to play a role in the pathogenesis of MS. Inflammation only affects the central nervous system in MS, suggesting that T cells and B cells are recruited by specific target antigens that are only expressed in the central nervous system (Thompson, Baranzini, Geurts, Hemmer, & Ciccarelli, 2018). Several mechanisms might initiate autoimmune responses in MS, which will result in a detrimental circle of events: tissue damage leads to the release of antigens to the periphery, which primes new immune responses in the lymphoid tissue, followed by the invasion of lymphocytes into the central nervous system (Thompson et al., 2018). The innate immune system also has an important role in MS progression. Macrophages promote the proinflammatory response of T cells and B cells, which results in tissue damage. Early microglial activation might be one of the initial events in the development of MS lesions. When activated, microglial cells could result in the secretion of proinflammatory cytokines, chemokines, and free radicals (Thompson et al., 2018). During the progressive phase of MS, the contribution of the peripheral immune system decreases and immune responses are thought to be confined to the central nervous system (Thompson et al., 2018).

Physical activity and sedentary behaviour in patients with MS

Physical activity and exercise training can be beneficial for patients with MS by improving general health and disease symptoms (Motl et al., 2017; Sharif et al., 2018). However, this population group tends to be physically inactive, consistently presenting with lower step count and levels of objectively measured moderate-to-vigorous physical activity when compared with the general population (step count: ~5,800 vs. ~9,600 steps/day; moderate-to-vigorous physical activity: 18.4 vs. 27.3 minutes/day). Additionally, this population group also presents with high participation in sedentary behaviour (9.9 hours/day) and low participation in light-intensity physical activity (3.5 hours/day) (Casey, Coote, Galvin, & Donnelly, 2018).

Effects of individual exercise bouts on overall health in patients with MS

Though exercise training confers multiple health benefits for patients with MS, some data exists regarding the impact of individual bouts of exercise on this disease population. To date, a few studies have demonstrated increases in brain-derived neurotrophic factor (BDNF) concentrations following a single 30-minute bout of moderate exercise in adults with MS (Briken et al., 2016; Gold et al., 2003), but no changes were observed in nerve growth factor (Gold et al., 2003), irisin, and IL-6 concentrations (Briken et al., 2016). Increased knowledge of the effects that individual bouts of exercise have and the mechanisms of exercise-induced benefit holds promise to enhance exercise prescription for patients with MS.

Effects of exercise training on overall health in patients with MS

Exercise training has been shown to improve MS lower extremity muscle strength and cardiorespiratory fitness following resistance and aerobic training protocols, respectively. Exercise training also improved MS physical functioning outcomes, including walking speed and balance, fatigue, depression symptoms, sleep quality, and cardiometabolic health (Motl et al., 2017; Sharif et al., 2018). However, there is still insufficient and inconsistent evidence regarding the effects of exercise training on MS cognitive outcomes and quality of life; of note, no study showed worsening of these outcomes following exercise programs. Importantly, exercise training has been shown to be safe for patients with MS and associated with reduced incidence of disease relapses and slowed disability progression (Motl et al., 2017; Sharif et al., 2018).

Effects of exercise on immune function in MS

Using a mouse model of MS (experimental autoimmune encephalomyelitis), Bernardes et al. (2013, 2016) demonstrated that a six-week exercise training program (swimming, 7% of body weight, 30 min/day, 5 days/week) significantly reduced the number of B cells and CD4+ and CD8+ T cells in the spinal cord (Bernardes et al., 2016), IL-6 levels in the spinal cord, and TNF, IL-10, and IL-1β in the brain of trained vs untrained mice (Bernardes et al., 2013). These findings suggest exercise training can reduce leukocyte infiltration and inflammation in the central nervous system.

Exercise prescription in patients with MS

Patients with MS should follow physical activity (Bull et al., 2020) and exercise (Garber et al., 2011) recommendations for the general population, which include performing aerobic, resistance, flexibility, and neuromuscular exercise training (Table 11.1). Patients with MS should consult their healthcare professional for advice on the types and amounts of activity appropriate for their individual needs, clinical manifestations, and treatment programs (Bull et al., 2020; Motl et al., 2017). Exercise adaptations might be necessary depending on mobility and motor impairments and other disease symptoms. Exercise programs should start at low intensity and volume, particularly for those who are inactive and sedentary. Training progression should be slow and gradual, respecting patients' tolerability and clinical manifestations (Motl et al., 2017). Giving the potential benefits associated with participation in physical activity and exercise training programs for MS, it is fundamental to focus on better equipping both healthcare professionals and patients with strategies and resources for the adoption of physical activity/exercise behaviour early on and consistently across the disease course.

Exercise and spondyloarthritis

Introduction

Spondyloarthritis (SpA) is an umbrella term for a group of chronic autoimmune diseases that share similar genetic risk and inflammatory profiles but different clinical phenotypes of inflammatory arthritis. Diseases under the SpA categorization include psoriatic disease, axial SpA/ankylosing spondylitis, inflammatory bowel disease associated arthropathy, and reactive arthritis. Psoriatic disease includes both psoriasis and psoriatic arthritis. Psoriasis is an inflammatory skin condition of erythematous scaling plaques prominently affecting extensor surfaces of extremities, trunk, and the scalp. A subset of patients with psoriatic disease develops psoriatic arthritis hallmarked by polyarticular inflammatory-type arthritis and dactylitis (i.e., diffuse finger or toe swelling). Axial SpA comprises a spectrum of disorders, including ankylosing spondylitis, defined by inflammatory arthritis of the sacroiliac joints and spine. Patients with axial SpA distinctly develop enthesitis or pain and inflammation at entheses (i.e., attachment sites of tendon and ligament to bone such as the Achilles tendon, patellar tendon, and plantar fascia). Perhaps more so than the other autoimmune diseases discussed in this chapter, management guidelines for SpA commonly include strong recommendations for exercise and physical activity to improve pain, function, and disease activity (Gwinnutt et al., 2022).

Immune dysfunction in spondyloarthritis

SpA pathogenesis likely begins with the dysregulation of innate immune responses at key anatomic sites (e.g., skin, gut, and enthesis) (Mauro, Simone, Bucci, & Ciccia, 2021). In SpA, innate and innate-like immune cells—such as innate lymphoid cells, mucosal-associated invariant T cells, and γδ T cells—are gaining recognition for their roles in SpA disease initiation. Further, the strong association between human leukocyte antigen (HLA)-B27 (a common major histocompatibility complex/MHC class 1 allele expressed on all nucleated cells) with ankylosing spondylitis suggests that SpA pathogenesis may be driven by impaired MHC class 1 peptide processing and interactions with CD8+ T cells. Subsequently, adaptive immune dysfunction appears to perpetuate chronic SpA inflammatory disease. SpA is hallmarked by multiple genetic polymorphisms involving the IL-23/IL-17 axis; notably, current treatment paradigms prominently involve targeting IL-23, IL-17, and TNF-α (McGonagle, McInnes, Kirkham, Sherlock, & Moots, 2019). IL-17 production is increased across innate and adaptive immune cells in SpA, though T helper 17 (Th17) cells are thought to be key drivers of inflammatory disease at sites of SpA disease (i.e., skin in psoriasis, synovial fluid in psoriatic arthritis, and enthesis in axial SpA). Questions currently persist regarding the relative roles of innate and adaptive immune responses and the network of cellular interactions that drive inflammatory disease in SpA.

Physical activity and sedentary behaviour in patients with spondyloarthritis

Physical activity and reducing sedentary behaviour play a key role in both preventing and managing disease in SpA. In a mixed cohort of patients with inflammatory arthritis (including RA and SpA), only 29% of participants met current physical activity guidelines of achieving at least 150 minutes/week of moderate-intensity aerobic exercise (Bell, Hendry, & Steultjens, 2022). In psoriatic arthritis, high adiposity and central obesity are associated with increased incidence of disease, whereas high levels of physical activity are associated with a decreased risk of disease (Thomsen et al., 2021). In patients with axial SpA, increased daily step counts and reduced sedentary behaviour are associated with better physical function and quality of life (Coulter, McDonald, Cameron, Siebert, & Paul, 2020). Further study is needed to identify the role of reducing sedentary behaviour on other aspects of SpA clinical management (e.g., disease activity, immune function) across the spectrum of SpA diseases.

Effects of individual exercise bouts on overall health in patients with spondyloarthritis

Though exercise training confers multiple health benefits for patients with SpA, sparse data exists regarding the impact of individual bouts of exercise on this disease population. Increased knowledge of the effect that individual bouts of exercise have and mechanisms of exercise-induced benefit holds promise to enhance exercise prescription for patients with SpA.

Effects of exercise training on overall health in patients with spondyloarthritis

Patients with SpA are at increased risk for cardiovascular disease and associated morbidity and mortality, thus lifestyle interventions are accepted as a key aspect of SpA patient care (Ogdie et al., 2015; Toussirot, 2021). In psoriatic disease, lifestyle factors such as physical inactivity and adiposity contribute to disease risk—perhaps to a greater extent than genetic risk factors (Shen et al., 2022). Adiposity and obesity in psoriatic arthritis are also associated with higher disease activity and worse clinical outcomes (Kumthekar & Ogdie, 2020); lifestyle interventions are thus considered important for patient care. For example, dietary interventions for psoriatic arthritis, with and without weight loss, have positive effects on disease activity (Klingberg et al., 2019; Leite et al., 2022). Additionally, physical activity interventions improve psoriatic arthritis disease activity; combined aerobic and resistance training programs have a higher level of evidence compared with either exercise training modality alone (Gwinnutt et al., 2022). Axial SpA disease may be perpetuated by increased biomechanical stress and microinjury at entheses, in part due to the importance of the enthesis in transmitting biomechanical forces during physical activity (Perrotta, Lories, & Lubrano, 2021). Indeed, in a DBA/1 mouse model of autoimmune inflammatory arthritis, reducing biomechanical stress via hindlimb unloading significantly decreased signs of Achilles tendon enthesitis (Jacques et al., 2014). Thus, in axial SpA, there may be an optimal exercise dose needed to produce the desired benefits of improving disease activity, inflammation, and physical function. If either inadequate or excessive physical activity, altered biomechanical stress may then contribute to SpA inflammatory disease progression (Perrotta et al., 2021). Subsequent study in human SpA is needed to better understand the impact of biomechanical stress on disease pathogenesis and the immune pathways involved.

Effects of exercise on immune function in spondyloarthritis

Despite this wealth of data showing the benefits of lifestyle changes on SpA, little is known regarding the direct effects of these interventions on SpA immune function. Thus, exercise immunology in SpA is a key area for future study.

Exercise prescription for patients with spondyloarthritis

Current recommendations for physical activity and exercise for persons with SpA agree with public health guidelines for adults (Table 11.1) (Rausch Osthoff et al., 2018). Patients should be instructed on completing movements within pain-free ranges of motion. Finally, excessive biomechanical stress from high-intensity/high-impact activities should be avoided, as it may contribute to disease progression. If worsening of arthritis pain symptoms with exercise, the patient is recommended to consult with both their rheumatologist/healthcare provider and exercise specialist to modify their exercise program.

Exercise and myositis

Introduction

Idiopathic inflammatory myopathies (IIM or myositis) are a heterogeneous group of diseases that present with muscular weakness and inflammation of skeletal muscles. IIM clinical classification comprises polymyositis (PM), dermatomyositis (DM), juvenile DM, neoplasm-associated myositis, antisynthetase syndrome, necrotising myopathy, and inclusion body myositis (IBM), with distinct histological, pathological, and therapeutic characteristics. Of note, there is significant clinical overlap between DM and PM as both present with progressive proximal muscle (i.e., hip and shoulder flexors) weakness. Nonetheless, each of the IIM subtypes has a distinct spectrum of target tissues. For instance, necrotising myopathy and IBM predominantly affect skeletal muscle, whereas DM and antisynthetase syndrome are multiorgan diseases, which may also affect the skin, lungs, and/or joints (Lundberg et al., 2021). In IIM, complete recovery of muscle function with pharmacological treatment does not always occur, thus physical disability is a significant concern for these patients.

Immune dysfunction in myositis

IIM phenotypes are thought to develop as a consequence of interactions between genetic and environmental risk factors, in the relative absence of protective factors (Lundberg et al., 2021; Miller, Lamb, Schmidt, & Nagaraju, 2018). It appears that both adaptive and innate immune mechanisms and non-immune mechanisms are implicated in the different types of IIM (Lundberg et al., 2021). Recent data suggest that various risk factors result in each of the major clinical and myositis autoantibody-defined phenotypes, which also differ in many clinical features, responses to therapy, and outcomes; indeed, the pathogenic pathways appear to differ across the several phenotypes (Lundberg et al., 2021). While DM is a microangiopathy mediated by humoral immunity, classically defined PM is characterised by T-cell-induced muscle inflammation and degeneration. The pathogenesis of IBM—which generally presents with more distal muscle weakness compared with DM and PM—is still unknown, but some studies have revealed histological evidence of inflammation in the affected skeletal muscle (Pignone, Fiori, Del Rosso, Generini, & Matucci-Cerinic, 2002). Necrotising myopathy and IBM muscle biopsy samples have characteristic histopathological features that enable distinction from DM or antisynthetase syndrome samples. Furthermore, although DM and antisynthetase syndrome muscle samples share histological similarities, such as prominent perifascicular involvement, transcriptomic analyses have revealed that each has a unique gene expression profile (Lundberg et al., 2021). For instance, the expression of type 1 interferon-inducible genes is substantially higher in DM than in antisynthetase syndrome (Pinal-Fernandez et al., 2019; Rigolet et al., 2019). Of relevance, recent evidence suggests that transcriptomic data can be used to determine a muscle biopsy sample from a patient with DM, antisynthetase syndrome, necrotising myopathy or IBM with >90% accuracy (Pinal-Fernandez et al., 2020).

Physical activity and sedentary behaviour in patients with myositis

There is limited evidence on physical activity levels in myositis. A review of eight studies involving 181 IIM patients (7 out of 8 studies with JDM) showed that accelerometer-derived physical activity is in general reduced in the patients vs. healthy controls. Also, higher levels of physical activity were associated with shorter JDM disease duration, current glucocorticoid use, and lower serum creatine kinase (Oldroyd, Little, Dixon, & Chinoy, 2019).

Effects of individual exercise bouts on overall health in patients with myositis

To our knowledge, there is no study assessing the effects of individual exercise bouts on global markers of health in IIM patients.

Effects of exercise training on overall health in patients with myositis

Anecdotal beliefs remain that exercise could cause a flare, worsen disease activity, and increase inflammation in IIM patients, but studies have shown otherwise. In fact, exercise training seems to be of therapeutic utility in IIM, as it has been shown to improve aerobic capacity (Wiesinger, Quittan, Aringer et al., 1998; Wiesinger, Quittan, Graninger et al., 1998), muscle strength (Gualano, Neves et al., 2010), fatigue (Varju, Petho, Kutas, & Czirjak, 2003), and quality of life (Alexanderson, Stenstrom, Jenner, & Lundberg, 2000) in this disease population.

In a small cohort study, IBM patients ($n = 5$) underwent a three times per week progressive strength training program for 12 weeks. The intervention resulted in improvements in both IBM isometric and dynamic strength, but fatigue and inflammation markers (e.g., IL-2 and natural killer cells) remained unchanged (Spector et al., 1997). Additionally, a low-intensity resistance training (50% of one-repetition maximum (1RM)) program with vascular occlusion resulted in increased thigh cross-sectional area, muscle strength and function, and improved quality of life in a single patient with IBM who was unresponsive to conventional therapy, including "traditional" physical exercises (Gualano, Neves et al., 2010; Gualano, Ugrinowitsch et al., 2010). Of relevance, no evidence of disease flare, muscle damage, or exacerbated inflammation was noted.

In patients with DM and PM, a 12-week, twice per week low-intensity (30% of 1RM resistance exercise training program combined with partial blood flow restriction resulted in improvements in muscle strength and function, muscle mass, and quality of life (Mattar et al., 2014). Additionally, six weeks of moderate aerobic training (60% of HRmax) was shown to improve VO_2 peak, isometric peak torque, and exercise tolerance (Wiesinger, Quittan, Aringer et al., 1998; Wiesinger, Quittan, Graninger et al., 1998). In these studies, training did not affect muscle enzymes, suggesting that this invention is safe for IIM patients (Wiesinger, Quittan, Aringer et al., 1998; Wiesinger, Quittan, Graninger et al., 1998).

Home-based exercise is also a promising and safe strategy to improve muscle health and function in IIM (Alexanderson et al., 2000; Alexanderson, Stenstrom, & Lundberg, 1999). However, null findings have been observed following home-based exercise programs, which could be at least partially attributed to a lower intensity in training regimen (Arnardottir, Alexanderson, Lundberg, & Borg, 2003). Therefore, particular attention should be given to exercise intensity in such programs for IIM.

Exercise training may also be a potential therapeutic tool for children and adolescents with IIM. A 12-week, exercise training program comprising aerobic (70% of VO_2 peak) and resistance (8–12RM) training improved muscle strength and function, aerobic fitness, bone mineral density, and health-related quality of life, without exacerbating disease activity or inflammation in ten children with juvenile DM (Omori et al., 2012). These findings extend the applicability and benefits of exercise to juvenile IIM patients.

Taken together, the still scarce literature suggests that exercise training may have a therapeutic role in IIM. Of relevance, none of the existing studies have shown any serious adverse event associated with exercise, irrespective of the training's characteristics (e.g., strength or aerobic, more or less intensive, home-based, or supervised) or the patients' characteristics (e.g., child or adult, IBM, DM or PM, chronic or active disease).

Effect of exercise on immune function in patients with myositis

Studies on the exercise-induced immune function modulation in myositis are scarce. Svec et al. showed a significant decrease on interleukin IL-7, IL-9, regulated on activation/normal T cell expressed and secreted (RANTES) and TNF-α in IIM patients who underwent a 24-week exercise program (comprising a supervised training of activities of daily living, resistance, and stability, and home-based exercises). In addition, changes in Heat shock protein 90 (Hsp90), an intracellular chaperone protein that mediates immune responses to exercise, were positively associated with changes in IL-7, IL-8, and TNF-α (Svec et al., 2022). Whether exercise modulates inflammatory milieu differently in IIM patients and healthy controls is not determined.

Exercise prescription for patients with myositis

Children and adults with IIM should follow general public health physical activity (Bull et al., 2020) and exercise (Garber et al., 2011) recommendations (Table 11.1). Patients are also

Table 11.1 Exercise prescription and physical activity recommendations and considerations for patients with autoimmune diseases

Condition	Exercise prescription and physical activity recommendations and considerations
All autoimmune diseases	- Aerobic training: light to moderate intensity for short sessions (i.e., 5–10 min) initially for previously sedentary adults and older adults, progressing to 150 min/week of moderate intensity or 75 min/week of vigorous intensity exercise. Children and adolescents should perform 60 min/day of moderate-to-vigorous intensity physical activity. - Resistance training: each major muscle group 2–3 days/week. - Flexibility/mobility training: 2–3 or more days per week. - Neuromuscular training: 2–3 or more days per week, including balance, agility, coordination, and gait-training activities, as well as multifaceted activities such as yoga and tai chi. - Sedentary behaviour: limit the amount of time spent being sedentary by replacing sedentary time with physical activity of any intensity. - Incidental physical activity: when not able to meet physical activity recommendations, patients should aim to engage in physical activity of any intensity according to their abilities.
Type 1 diabetes mellitus	- Check blood glucose concentration prior to exercise. Target range is 90–250 mg/dL (5.0–13.9 mmol/L). - Ingest carbohydrate and/or adjust insulin injection or delivery prior to or at the start of exercise to prevent exercise-induced hypoglycemia or late onset hypoglycemia. - Check blood glucose during and after exercise so carbohydrate and insulin adjustment strategies can be implemented if/when needed to maintain stable and safe blood glucose levels. - Postpone any exercise if blood glucose <90mg/dL (<5.0 mmol/L), >350 mg/dL (>19.4 mmol/L), or presenting ketonuria, or intense exercise if >250 mg/dL (>13.9 mmol/L) until these alterations are corrected.
Rheumatoid arthritis	- Perform hand- and wrist-specific strength and mobility exercise training five days or more per week in patients with prominent hand and wrist arthritis. - During arthritis flare, continue light exercise as tolerated, consult with rheumatologist or health care provider regarding medication changes. - Consider light activities and mobility and balance exercises in the morning during time of peak joint stiffness and higher-intensity aerobic and resistance training in afternoon/evening. - Modify activities to perform movements in pain-free ranges of motion.
Systemic lupus erythematosus	- Patients with high disease activity should avoid high-intensity exercise; they should continue to perform lower-intensity physical activity as tolerated and consult with their rheumatologist or health care provider regarding medication changes. - Adjust the training program according to patients' symptoms, tissue damage, and/or comorbidities in consultation with rheumatologist or health care provider.
Multiple sclerosis	- Modify activities as needed based on mobility and motor impairments, and other disease symptoms. - Start with low-intensity and volume activities, particularly for those who are inactive and sedentary at baseline. - Exercise progression should be slow and gradual, respecting patient's tolerability and clinical manifestations.
Spondyloarthritis	- Excessive biomechanical stress from high-intensity/high-impact activities may contribute to disease progression. - Modify activities to perform movements in pain-free ranges of motion.
Myositis	- Resistance training is mandatory for patients to restore physical function, but aerobic training may also help to prevent cardiometabolic comorbidities. - Exercise progression should be gradual and constant. - Even patients with high disease activity and refractory myositis may benefit from exercise training and increasing physical activity.

encouraged to increase incidental physical activity and reduce time spent in sedentary behaviour to gain additional health benefits (Bull et al., 2020). Resistance training is mandatory for those patients to restore physical function, but aerobic training may also help to prevent cardiometabolic comorbidities. Exercise progression should be gradual and constant. Importantly, even patients with active disease and refractory myositis may benefit from exercise.

Key points

- Autoimmune disorders (diseases) occur due to a loss of immune self-tolerance resulting in tissue-specific immune-mediated damage and chronic inflammation. Common autoimmune disorders include type 1 diabetes, RA, SLE, MS, spondyloarthritis, and myositis.
- Exercise training benefits general health and disease-specific features of autoimmune disorders. The shared benefits of exercise training for autoimmune disorders include improvements in physical fitness and function, quality of life, pain, and cardiometabolic disease risk. Thus, exercise training is recommended as a primary therapy in the management of patients with autoimmune disorders.
- In type 1 diabetes mellitus, exercise training can improve glycaemic control and cardiometabolic risk. Exercise-induced hypoglycemia and late onset hypoglycemia are concerns for these patients and adjustments to dietary carbohydrate intake and insulin therapy before, during, and after exercise are necessary.
- In RA, exercise training is safe and does not contribute to the progression of arthritis symptoms or related damage. On the contrary, exercise training may be a disease-modifying therapy to improve RA inflammatory disease activity.
- In SLE, exercise training improves physical function and cardiometabolic risk. The effects of exercise training on lupus disease activity and immune function are less certain.
- In MS, exercise training improves multiple aspects of general health and disease activity. It is currently unclear if exercise training improves the cognitive dysfunction common in this patient population.
- In spondyloarthritis, exercise training improves multiple aspects of health, including disease activity. However, excessive biomechanical stress may contribute to spondyloarthritis disease onset and progression, and thus high-impact/high-intensity exercise should be performed with caution in this patient population.
- In myositis, resistance training is safe and a cornerstone of therapy to restore physical function. Aerobic training also improves cardiometabolic risk profiles.
- The effects of individual bouts of exercise on autoimmune disorders are incompletely understood. Whether or not individuals with autoimmune disorders have similar physiologic responses to exercise is an area for future research.
- The effects of exercise and physical activity on immune function in autoimmune disorders are largely unknown. Given that exercise has a clear impact on modulating immune function in non-autoimmune disorder populations, there is potential for exercise to improve dysfunctional immune responses in these patient populations. Further study of exercise–immune interactions in autoimmune disorders is a new frontier for the growing field of exercise immunology.

References

Alexanderson, H., & Lundberg, I. E. (2012). Exercise as a therapeutic modality in patients with idiopathic inflammatory myopathies. *Curr Opin Rheumatol, 24*(2), 201–207.

Alexanderson, H., Stenstrom, C. H., & Lundberg, I. (1999). Safety of a home exercise programme in patients with polymyositis and dermatomyositis: A pilot study. *Rheumatology (Oxford), 38*(7), 608–611.

Alexanderson, H., Stenstrom, C. H., Jenner, G., & Lundberg, I. (2000). The safety of a resistive home exercise program in patients with recent onset active polymyositis or dermatomyositis. *Scand J Rheumatol, 29*(5), 295–301.

Andersson, S. E. M., Lange, E., Kucharski, D., Svedlund, S., Onnheim, K., Bergquist, M.,... Gjertsson, I. (2020). Moderate- to high-intensity aerobic and resistance exercise reduces peripheral blood regulatory cell populations in older adults with rheumatoid arthritis. *Immun Ageing, 17*, 12.

Andonian, B. J., Koss, A., Koves, T. R., Hauser, E. R., Hubal, M. J., Pober, D. M.,... Huffman, K. M. (2022). Rheumatoid arthritis T cell and muscle oxidative metabolism associate with exercise-induced changes in cardiorespiratory fitness. *Sci Rep, 12*(1), 7450.

Arnardottir, S., Alexanderson, H., Lundberg, I. E., & Borg, K. (2003). Sporadic inclusion body myositis: Pilot study on the effects of a home exercise program on muscle function, histopathology and inflammatory reaction. *J Rehabil Med, 35*(1), 31–35.

Atkinson, M. A., Eisenbarth, G. S., & Michels, A. W. (2014). Type 1 diabetes. *Lancet, 383*(9911), 69–82.

Aytekin, E., Caglar, N. S., Ozgonenel, L., Tutun, S., Demiryontar, D. Y., & Demir, S. E. (2012). Home-based exercise therapy in patients with ankylosing spondylitis: Effects on pain, mobility, disease activity, quality of life, and respiratory functions. *Clin Rheumatol, 31*(1), 91–97.

Bartlett, D. B., Willis, L. H., Slentz, C. A., Hoselton, A., Kelly, L., Huebner, J. L.,... Huffman, K. M. (2018). Ten weeks of high-intensity interval walk training is associated with reduced disease activity and improved innate immune function in older adults with rheumatoid arthritis: A pilot study. *Arthritis Res Ther, 20*(1), 127.

Baslund, B., Lyngberg, K., Andersen, V., Halkjaer Kristensen, J., Hansen, M., Klokker, M., & Pedersen, B. K. (1993). Effect of 8 wk of bicycle training on the immune system of patients with rheumatoid arthritis. *J Appl Physiol (1985), 75*(4), 1691–1695.

Bell, K., Hendry, G., & Steultjens, M. (2022). Physical activity and sedentary behavior in people with inflammatory joint disease: A cross-sectional study. *Arthritis Care Res (Hoboken), 74*(3), 493–500.

Benatti, F. B., Miyake, C. N. H., Dantas, W. S., Zambelli, V. O., Shinjo, S. K., Pereira, R. M. R.,... Gualano, B. (2018). Exercise increases insulin sensitivity and skeletal muscle AMPK expression in systemic lupus erythematosus: A randomized controlled trial. *Front Immunol, 9*, 906.

Bernardes, D., Brambilla, R., Bracchi-Ricard, V., Karmally, S., Dellarole, A., Carvalho-Tavares, J., & Bethea, J. R. (2016). Prior regular exercise improves clinical outcome and reduces demyelination and axonal injury in experimental autoimmune encephalomyelitis. *J Neurochem, 136(*Suppl 1), 63–73.

Bernardes, D., Oliveira-Lima, O. C., Silva, T. V., Faraco, C. C., Leite, H. R., Juliano, M. A.,... Carvalho-Tavares, J. (2013). Differential brain and spinal cord cytokine and BDNF levels in experimental autoimmune encephalomyelitis are modulated by prior and regular exercise. *J Neuroimmunol, 264*(1–2), 24–34.

Bohn, B., Herbst, A., Pfeifer, M., Krakow, D., Zimny, S., Kopp, F.,... Initiative, D. P. V. (2015). Impact of physical activity on glycemic control and prevalence of cardiovascular risk factors in adults with type 1 diabetes: A cross-sectional multicenter study of 18,028 patients. *Diabetes Care, 38*(8), 1536–1543.

Briken, S., Rosenkranz, S. C., Keminer, O., Patra, S., Ketels, G., Heesen, C.,... Gold, S. M. (2016). Effects of exercise on Irisin, BDNF and IL-6 serum levels in patients with progressive multiple sclerosis. *J Neuroimmunol, 299*, 53–58.

Bull, F. C., Al-Ansari, S. S., Biddle, S., Borodulin, K., Buman, M. P., Cardon, G.,... Willumsen, J. F. (2020). World Health Organization 2020 guidelines on physical activity and sedentary behaviour. *Br J Sports Med, 54*(24), 1451–1462.

Cambre, I., Gaublomme, D., Schryvers, N., Lambrecht, S., Lories, R., Venken, K., & Elewaut, D. (2019). Running promotes chronicity of arthritis by local modulation of complement activators and Impairing T regulatory feedback loops. *Ann Rheum Dis, 78*(6), 787–795.

Casey, B., Coote, S., Galvin, R., & Donnelly, A. (2018). Objective physical activity levels in people with multiple sclerosis: Meta-analysis. *Scand J Med Sci Sports, 28*(9), 1960–1969.

Challal, S., Minichiello, E., Boissier, M. C., & Semerano, L. (2016). Cachexia and adiposity in rheumatoid arthritis. Relevance for disease management and clinical outcomes. *Joint Bone Spine, 83*(2), 127–133.

Colberg, S. R., Sigal, R. J., Yardley, J. E., Riddell, M. C., Dunstan, D. W., Dempsey, P. C.,... Tate, D. F. (2016). Physical activity/exercise and diabetes: A position statement of the american diabetes association. *Diabetes Care, 39*(11), 2065–2079.

Cooney, J. K., Law, R. J., Matschke, V., Lemmey, A. B., Moore, J. P., Ahmad, Y.,... Thom, J. M. (2011). Benefits of exercise in rheumatoid arthritis. *J Aging Res, 2011*, 681640.

Coulter, E. H., McDonald, M. T., Cameron, S., Siebert, S., & Paul, L. (2020). Physical activity and sedentary behaviour and their associations with clinical measures in axial spondyloarthritis. *Rheumatol Int, 40*(3), 375–381.

Crespilho, D. M., de Almeida Leme, J. A., de Mello, M. A., & Luciano, E. (2010). Effects of physical training on the immune system in diabetic rats. *Int J Diabetes Dev Ctries, 30*(1), 33–37.

Curran, M., Campbell, J., Drayson, M., Andrews, R., & Narendran, P. (2019). Type 1 diabetes impairs the mobilisation of highly-differentiated CD8+T cells during a single bout of acute exercise. *Exerc Immunol Rev, 25*, 64–82.

Curran, M., Campbell, J. P., Powell, E., Chikhlia, A., & Narendran, P. (2020). The mobilisation of early mature CD56dim-CD16bright NK cells is blunted following a single bout of vigorous intensity exercise in Type 1 Diabetes. *Exerc Immunol Rev, 26*, 116–131.

DeMizio, D. J., & Geraldino-Pardilla, L. B. (2020). Autoimmunity and inflammation link to cardiovascular disease risk in rheumatoid arthritis. *Rheumatol Ther, 7*(1), 19–33.

Dendrou, C. A., Fugger, L., & Friese, M. A. (2015). Immunopathology of multiple sclerosis. *Nat Rev Immunol, 15*(9), 545–558.

DiMeglio, L. A., Evans-Molina, C., & Oram, R. A. (2018). Type 1 diabetes. *Lancet, 391*(10138), 2449–2462.

Ercan, Z., Deniz, G., Yentur, S. B., Arikan, F. B., Karatas, A., Alkan, G., & Koca, S. S. (2022). Effects of acute aerobic exercise on cytokines, klotho, irisin, and vascular endothelial growth factor responses in rheumatoid arthritis patients. *Ir J Med Sci, 192*, 491–497. https://link.springer.com/article/10.1007/s11845-022-02970-7

Fanouriakis, A., Tziolos, N., Bertsias, G., & Boumpas, D. T. (2021). Update on the diagnosis and management of systemic lupus erythematosus. *Ann Rheum Dis, 80*(1), 14–25.

Fukuda, W., Omoto, A., Oku, S., Tanaka, T., Tsubouchi, Y., Kohno, M., & Kawahito, Y. (2010). Contribution of rheumatoid arthritis disease activity and disability to rheumatoid cachexia. *Mod Rheumatol, 20*(5), 439–443.

Garber, C. E., Blissmer, B., Deschenes, M. R., Franklin, B. A., Lamonte, M. J., Lee, I. M.,... American College of Sports, M. (2011). American college of sports medicine position stand. Quantity and quality of exercise for developing and maintaining cardiorespiratory, musculoskeletal, and neuromotor fitness in apparently healthy adults: guidance for prescribing exercise. *Med Sci Sports Exerc, 43*(7), 1334–1359.

Garcia-Garcia, F., Kumareswaran, K., Hovorka, R., & Hernando, M. E. (2015). Quantifying the acute changes in glucose with exercise in type 1 diabetes: A systematic review and meta-analysis. *Sports Med, 45*(4), 587–599.

Gold, S. M., Schulz, K. H., Hartmann, S., Mladek, M., Lang, U. E., Hellweg, R.,... Heesen, C. (2003). Basal serum levels and reactivity of nerve growth factor and brain-derived neurotrophic factor to standardized acute exercise in multiple sclerosis and controls. *J Neuroimmunol, 138*(1–2), 99–105.

Gonzalez-Chavez, S. A., Quinonez-Flores, C. M., Espino-Solis, G. P., Vazquez-Contreras, J. A., & Pacheco-Tena, C. (2019). Exercise exacerbates the transcriptional profile of hypoxia, oxidative stress and inflammation in rats with adjuvant-induced arthritis. *Cells, 8*(12), 1–30. https://www.mdpi.com/2073-4409/8/12/1493

Gualano, B., Neves, M., Jr., Lima, F. R., Pinto, A. L., Laurentino, G., Borges, C.,... Ugrinowitsch, C. (2010). Resistance training with vascular occlusion in inclusion body myositis: A case study. *Med Sci Sports Exerc, 42*(2), 250–254.

Gualano, B., Ugrinowitsch, C., Neves, M., Jr., Lima, F. R., Pinto, A. L., Laurentino, G.,... Roschel, H. (2010). Vascular occlusion training for inclusion body myositis: a novel therapeutic approach. *J Vis Exp, 40*, 1894.

Gwinnutt, J. M., Wieczorek, M., Cavalli, G., Balanescu, A., Bischoff-Ferrari, H. A., Boonen, A.,... Verstappen, S. M. M. (2022). Effects of physical exercise and body weight on disease-specific outcomes of people with rheumatic and musculoskeletal diseases (RMDs): Systematic reviews and meta-analyses informing the 2021 EULAR recommendations for lifestyle improvements in people with RMDs. *RMD Open, 8*(1), e002168. https://pubmed.ncbi.nlm.nih.gov/35361692/

Hakkinen, A., Sokka, T., Kotaniemi, A., & Hannonen, P. (2001). A randomized two-year study of the effects of dynamic strength training on muscle strength, disease activity, functional capacity, and bone mineral density in early rheumatoid arthritis. *Arthritis Rheum, 44*(3), 515–522.

Huffman, K. M., Andonian, B. J., Abraham, D. M., Bareja, A., Lee, D. E., Katz, L. H.,... White, J. P. (2021). Exercise protects against cardiac and skeletal muscle dysfunction in a mouse model of inflammatory arthritis. *J Appl Physiol (1985), 130*(3), 853–864.

Jacques, P., Lambrecht, S., Verheugen, E., Pauwels, E., Kollias, G., Armaka, M.,... Elewaut, D. (2014). Proof of concept: enthesitis and new bone formation in spondyloarthritis are driven by mechanical strain and stromal cells. *Ann Rheum Dis, 73*(2), 437–445.

Katz, P., Andonian, B. J., & Huffman, K. M. (2020). Benefits and promotion of physical activity in rheumatoid arthritis. *Curr Opin Rheumatol, 32*(3), 307–314.

Kito T, T. T., Nishii, K., Sakai, K., Matsubara, M., & Yamada, K. (2016). Effectiveness of exercise-induced cytokines in alleviating arthritis symptoms in arthritis model mice. *Okajimas Folia Anat Jpn, 93*(3), 81–88.

Klingberg, E., Bilberg, A., Bjorkman, S., Hedberg, M., Jacobsson, L., Forsblad-d'Elia, H.,... Larsson, I. (2019). Weight loss improves disease activity in patients with psoriatic arthritis and obesity: An interventional study. *Arthritis Res Ther, 21*(1), 17.

Kumthekar, A., & Ogdie, A. (2020). Obesity and psoriatic arthritis: A narrative review. *Rheumatol Ther, 7*(3), 447–456.

Leite, B. F., Morimoto, M. A., Gomes, C. M. F., Klemz, B. N. C., Genaro, P. S., Shivappa, N.,... Pinheiro, M. M. (2022). Dietetic intervention in psoriatic arthritis: The DIETA trial. *Adv Rheumatol, 62*(1), 12.

Liu, X., Tedeschi, S. K., Lu, B., Zaccardelli, A., Speyer, C. B., Costenbader, K. H.,... Sparks, J. A. (2019). Long-term physical activity and subsequent risk for rheumatoid arthritis among women: A prospective Cohort study. *Arthritis Rheumatol, 71*(9), 1460–1471.

Lundberg, I. E., Fujimoto, M., Vencovsky, J., Aggarwal, R., Holmqvist, M., Christopher-Stine, L.,... Miller, F. W. (2021). Idiopathic inflammatory myopathies. *Nat Rev Dis Primers, 7*(1), 86.

Ma, J., Lam, I. K. Y., Lau, C. S., & Chan, V. S. F. (2021). Elevated Interleukin-18 receptor accessory protein Mediates enhancement in reactive oxygen species production in neutrophils of systemic lupus erythematosus patients. *Cells, 10*(5). https://www.mdpi.com/2073-4409/10/5/964

Machado, P., Landewe, R., Braun, J., Hermann, K. G., Baker, D., & van der Heijde, D. (2010). Both structural damage and inflammation of the spine contribute to impairment of spinal mobility in patients with ankylosing spondylitis. *Ann Rheum Dis, 69*(8), 1465–1470.

Mattar, M. A., Gualano, B., Perandini, L. A., Shinjo, S. K., Lima, F. R., Sa-Pinto, A. L., & Roschel, H. (2014). Safety and possible effects of low-intensity resistance training associated with partial blood flow restriction in polymyositis and dermatomyositis. *Arthritis Res Ther, 16*(5), 473.

Mauro, D., Simone, D., Bucci, L., & Ciccia, F. (2021). Novel immune cell phenotypes in spondyloarthritis pathogenesis. *Semin Immunopathol, 43*(2), 265–277.

McCarthy, M. M., Funk, M., & Grey, M. (2016). Cardiovascular health in adults with type 1 diabetes. *Prev Med, 91*, 138–143.

McGonagle, D. G., McInnes, I. B., Kirkham, B. W., Sherlock, J., & Moots, R. (2019). The role of IL-17A in axial spondyloarthritis and psoriatic arthritis: Recent advances and controversies. *Ann Rheum Dis, 78*(9), 1167–1178.

Miller, F. W., Lamb, J. A., Schmidt, J., & Nagaraju, K. (2018). Risk factors and disease mechanisms in myositis. *Nat Rev Rheumatol, 14*(5), 255–268.

Minnock, D., Annibalini, G., Valli, G., Saltarelli, R., Krause, M., Barbieri, E., & De Vito, G. (2022). Altered muscle mitochondrial, inflammatory and trophic markers, and reduced exercise training adaptations in type 1 diabetes. *J Physiol, 600*(6), 1405–1418.

Motl, R. W., Sandroff, B. M., Kwakkel, G., Dalgas, U., Feinstein, A., Heesen, C.,... Thompson, A. J. (2017). Exercise in patients with multiple sclerosis. *Lancet Neurol, 16*(10), 848–856.

Ogdie, A., Yu, Y., Haynes, K., Love, T. J., Maliha, S., Jiang, Y.,... Gelfand, J. M. (2015). Risk of major cardiovascular events in patients with psoriatic arthritis, psoriasis and rheumatoid arthritis: A population-based cohort study. *Ann Rheum Dis, 74*(2), 326–332.

Oharomari, L. K., de Moraes, C., & Navarro, A. M. (2017). Exercise training but not curcumin supplementation decreases immune cell infiltration in the pancreatic islets of a genetically susceptible model of type 1 diabetes. *Sports Med Open, 3*(1), 15.

Oldroyd, A., Little, M. A., Dixon, W., & Chinoy, H. (2019). A review of accelerometer-derived physical activity in the idiopathic inflammatory myopathies. *BMC Rheumatol, 3*, 41.

Omori, C. H., Silva, C. A., Sallum, A. M., Rodrigues Pereira, R. M., Luciade Sa Pinto, A., Roschel, H., & Gualano, B. (2012). Exercise training in juvenile dermatomyositis. *Arthritis Care Res (Hoboken), 64*(8), 1186–1194.

Ostman, C., Jewiss, D., King, N., & Smart, N. A. (2018). Clinical outcomes to exercise training in type 1 diabetes: A systematic review and meta-analysis. *Diabetes Res Clin Pract, 139*, 380–391.

Pecanha, T., Meireles, K., Pinto, A. J., Rezende, D. A. N., Iraha, A. Y., Mazzolani, B. C.,... Roschel, H. (2021). Increased sympathetic and haemodynamic responses to exercise and muscle metaboreflex activation in post-menopausal women with rheumatoid arthritis. *J Physiol, 599*(3), 927–941.

Pecanha, T., Rodrigues, R., Pinto, A. J., Sa-Pinto, A. L., Guedes, L., Bonfiglioli, K.,... Roschel, H. (2018). Chronotropic incompetence and reduced heart rate recovery in rheumatoid arthritis. *J Clin Rheumatol, 24*(7), 375–380.

Peng, X., Lu, Y., Wei, J., Lin, T., Lu, Q., Liu, Q., & Ting, W. J. (2021). A cohort study of T helper 17 cell-related cytokine levels in tear samples of systemic lupus erythematosus and Sjogren's syndrome patients with dry eye disease. *Clin Exp Rheumatol, 39 Suppl 133*(6), 159–165.

Perandini, L. A., de Sa-Pinto, A. L., Roschel, H., Benatti, F. B., Lima, F. R., Bonfa, E., & Gualano, B. (2012). Exercise as a therapeutic tool to counteract inflammation and clinical symptoms in autoimmune rheumatic diseases. *Autoimmun Rev, 12*(2), 218–224.

Perandini, L. A., Sales-de-Oliveira, D., Mello, S. B., Camara, N. O., Benatti, F. B., Lima, F. R.,... Gualano, B. (2014). Exercise training can attenuate the inflammatory milieu in women with systemic lupus erythematosus. *J Appl Physiol (1985), 117*(6), 639–647.

Perandini, L. A., Sales-de-Oliveira, D., Mello, S., Camara, N. O., Benatti, F. B., Lima, F. R.,... Gualano, B. (2015). Inflammatory cytokine kinetics to single bouts of acute moderate and intense aerobic exercise in women with active and inactive systemic lupus erythematosus. *Exerc Immunol Rev, 21*, 174–185.

Perandini, L. A., Sales-de-Oliveira, D., Almeida, D. C., Azevedo, H., Moreira-Filho, C. A., Cenedeze, M. A.,... Gualano, B. (2016). Effects of acute aerobic exercise on leukocyte inflammatory gene expression in systemic lupus erythematosus. *Exerc Immunol Rev, 22*, 64–81.

Pereira Nunes Pinto, A. C., Natour, J., de Moura Castro, C. H., Eloi, M., & Lombardi Junior, I. (2017). Acute effect of a resistance exercise session on markers of cartilage breakdown and inflammation in women with rheumatoid arthritis. *Int J Rheum Dis, 20*(11), 1704–1713.

Perrotta, F. M., Lories, R., & Lubrano, E. (2021). To move or not to move: The paradoxical effect of physical exercise in axial spondyloarthritis. *RMD Open, 7*(1), e001480. https://pubmed.ncbi.nlm.nih.gov/33547227/

Petersen, A. M., & Pedersen, B. K. (2005). The anti-inflammatory effect of exercise. *J Appl Physiol (1985), 98*(4), 1154–1162.

Pignone, A., Fiori, G., Del Rosso, A., Generini, S., & Matucci-Cerinic, M. (2002). The pathogenesis of inflammatory muscle diseases: On the cutting edge among the environment, the genetic background, the immune response and the dysregulation of apoptosis. *Autoimmun Rev, 1*(4), 226–232.

Pinal-Fernandez, I., Casal-Dominguez, M., Derfoul, A., Pak, K., Plotz, P., Miller, F. W.,... Mammen, A. L. (2019). Identification of distinctive interferon gene signatures in different types of myositis. *Neurology, 93*(12), e1193–e1204.

Pinal-Fernandez, I., Casal-Dominguez, M., Derfoul, A., Pak, K., Miller, F. W., Milisenda, J. C.,... Mammen, A. L. (2020). Machine learning algorithms reveal unique gene expression profiles in muscle biopsies from patients with different types of myositis. *Ann Rheum Dis, 79*(9), 1234–1242.

Pinto, A. J., Meireles, K., Pecanha, T., Mazzolani, B. C., Smaira, F. I., Rezende, D.,... Gualano, B. (2021). Acute cardiometabolic effects of brief active breaks in sitting for patients with rheumatoid arthritis. *Am J Physiol Endocrinol Metab, 321*(6), E782–E794.

Pinto, A. J., Roschel, H., de Sa Pinto, A. L., Lima, F. R., Pereira, R. M. R., Silva, C. A.,... Gualano, B. (2017). Physical inactivity and sedentary behavior: Overlooked risk factors in autoimmune rheumatic diseases? *Autoimmun Rev, 16*(7), 667–674.

Poetker, D. M., & Reh, D. D. (2010). A comprehensive review of the adverse effects of systemic corticosteroids. *Otolaryngol Clin North Am, 43*(4), 753–768.

Prado, D. M., Benatti, F. B., de Sa-Pinto, A. L., Hayashi, A. P., Gualano, B., Pereira, R. M.,... Roschel, H. (2013). Exercise training in childhood-onset systemic lupus erythematosus: a controlled randomized trial. *Arthritis Res Ther, 15*(2), R46.

Prado, D. M., Gualano, B., Pinto, A. L., Sallum, A. M., Perondi, M. B., Roschel, H., & Silva, C. A. (2011). Exercise in a child with systemic lupus erythematosus and antiphospholipid syndrome. *Med Sci Sports Exerc, 43*(12), 2221–2223.

Pyfrom, S., Paneru, B., Knox, J. J., Cancro, M. P., Posso, S., Buckner, J. H., & Anguera, M. C. (2021). The dynamic epigenetic regulation of the inactive X chromosome in healthy human B cells is dysregulated in lupus patients. *Proc Natl Acad Sci U S A, 118*(24), e2024624118. https://pubmed.ncbi.nlm.nih.gov/34103397/

Ramsey-Goldman, R., Schilling, E. M., Dunlop, D., Langman, C., Greenland, P., Thomas, R. J., & Chang, R. W. (2000). A pilot study on the effects of exercise in patients with systemic lupus erythematosus. *Arthritis Care Res, 13*(5), 262–269.

Rausch Osthoff, A. K., Niedermann, K., Braun, J., Adams, J., Brodin, N., Dagfinrud, H.,... Vliet Vlieland, T. P. M. (2018). 2018 EULAR recommendations for physical activity in people with inflammatory arthritis and osteoarthritis. *Ann Rheum Dis, 77*(9), 1251–1260.

Riddell, M. C., Gallen, I. W., Smart, C. E., Taplin, C. E., Adolfsson, P., Lumb, A. N.,... Laffel, L. M. (2017). Exercise management in type 1 diabetes: a consensus statement. *Lancet Diabetes Endocrinol, 5*(5), 377–390.

Rigolet, M., Hou, C., Baba Amer, Y., Aouizerate, J., Periou, B., Gherardi, R. K.,... Authier, F. J. (2019). Distinct interferon signatures stratify inflammatory and dysimmune myopathies. *RMD Open, 5*(1), e000811.

Rodriguez Sanchez-Laulhe, P., Luque-Romero, L. G., Barrero-Garcia, F. J., Biscarri-Carbonero, A., Blanquero, J., Suero-Pineda, A., & Heredia-Rizo, A. M. (2022). An exercise and educational and self-management program delivered with a smartphone App (CareHand) in Adults With rheumatoid arthritis of the hands: Randomized controlled trial. *JMIR Mhealth Uhealth, 10*(4), e35462.

Rosa, J. S., Oliver, S. R., Mitsuhashi, M., Flores, R. L., Pontello, A. M., Zaldivar, F. P., & Galassetti, P. R. (2008). Altered kinetics of interleukin-6 and other inflammatory mediators during exercise in children with type 1 diabetes. *J Investig Med, 56*(4), 701–713.

Sharif, K., Watad, A., Bragazzi, N. L., Lichtbroun, M., Amital, H., & Shoenfeld, Y. (2018). Physical activity and autoimmune diseases: Get moving and manage the disease. *Autoimmun Rev, 17*(1), 53–72.

Shen, M., Xiao, Y., Jing, D., Zhang, G., Su, J., Lin, S.,... Liu, H. (2022). Associations of combined lifestyle and genetic risks with incident psoriasis: A prospective cohort study among UK Biobank participants of European Ancestry. *J Am Acad Dermatol, 87*(2):343–350. https://pubmed.ncbi.nlm.nih.gov/35427684/

Shimomura, S., Inoue, H., Arai, Y., Nakagawa, S., Fujii, Y., Kishida, T.,... Kubo, T. (2018). Treadmill running ameliorates destruction of articular cartilage and subchondral bone, not only synovitis, in a rheumatoid arthritis rat model. *Int J Mol Sci, 19*(6), 1653. https://www.ncbi.nlm.nih.gov/pmc/articles/PMC6032207/

Sippl, N., Faustini, F., Ronnelid, J., Turcinov, S., Chemin, K., Gunnarsson, I., & Malmstrom, V. (2021). Arthritis in systemic lupus erythematosus is characterized by local IL-17A and IL-6 expression in synovial fluid. *Clin Exp Immunol, 205*(1), 44–52.

Spector, S. A., Lemmer, J. T., Koffman, B. M., Fleisher, T. A., Feuerstein, I. M., Hurley, B. F., & Dalakas, M. C. (1997). Safety and efficacy of strength training in patients with sporadic inclusion body myositis. *Muscle Nerve, 20*(10), 1242–1248.

Stavropoulos-Kalinoglou, A., Metsios, G. S., Veldhuijzen van Zanten, J. J., Nightingale, P., Kitas, G. D., & Koutedakis, Y. (2013). Individualised aerobic and resistance exercise training improves cardiorespiratory fitness and reduces cardiovascular risk in patients with rheumatoid arthritis. *Ann Rheum Dis, 72*(11), 1819–1825.

Strombeck, B., & Jacobsson, L. T. (2007). The role of exercise in the rehabilitation of patients with systemic lupus erythematosus and patients with primary Sjogren's syndrome. *Curr Opin Rheumatol, 19*(2), 197–203.

Svec, X., Storkanova, H., Spiritovic, M., Slaby, K., Oreska, S., Pekacova, A.,... Tomcik, M. (2022). Hsp90 as a myokine: Its association with systemic inflammation after exercise interventions in patients with myositis and healthy subjects. *Int J Mol Sci, 23*(19), 11451. https://pubmed.ncbi.nlm.nih.gov/36232755/

Tench, C. M., McCarthy, J., McCurdie, I., White, P. D., & D'Cruz, D. P. (2003). Fatigue in systemic lupus erythematosus: A randomized controlled trial of exercise. *Rheumatology (Oxford), 42*(9), 1050–1054.

Thompson, A. J., Baranzini, S. E., Geurts, J., Hemmer, B., & Ciccarelli, O. (2018). Multiple sclerosis. *The Lancet, 391*(10130), 1622–1636.

Thomsen, R. S., Nilsen, T. I. L., Haugeberg, G., Gulati, A. M., Kavanaugh, A., & Hoff, M. (2021). Adiposity and physical activity as risk factors for developing psoriatic arthritis: Longitudinal data from a population-based study in Norway. *Arthritis Care Res (Hoboken), 73*(3), 432–441.

Thomsen, T., Aadahl, M., Beyer, N., Hetland, M. L., Loppenthin, K. B., Midtgaard, J.,... Esbensen, B. A. (2020). Sustained long-term efficacy of motivational counseling and text message reminders on daily sitting time in patients with rheumatoid arthritis: Long-term follow-up of a randomized, parallel-group trial. *Arthritis Care Res (Hoboken), 72*(11), 1560–1570.

Tonoli, C., Heyman, E., Roelands, B., Buyse, L., Cheung, S. S., Berthoin, S., & Meeusen, R. (2012). Effects of different types of acute and chronic (training) exercise on glycaemic control in type 1 diabetes mellitus: A meta-analysis. *Sports Med, 42*(12), 1059–1080.

Toussirot, E. (2021). The risk of cardiovascular diseases in axial spondyloarthritis. Current insights. *Front Med (Lausanne), 8*, 782150.

Tylutka, A., Morawin, B., Gramacki, A., & Zembron-Lacny, A. (2021). Lifestyle exercise attenuates immunosenescence; flow cytometry analysis. *BMC Geriatr, 21*(1), 200.

Varju, C., Petho, E., Kutas, R., & Czirjak, L. (2003). The effect of physical exercise following acute disease exacerbation in patients with dermato/polymyositis. *Clin Rehabil, 17*(1), 83–87.

Wadley, A. J., Veldhuijzen van Zanten, J. J., Stavropoulos-Kalinoglou, A., Metsios, G. S., Smith, J. P., Kitas, G. D., & Aldred, S. (2014). Three months of moderate-intensity exercise reduced plasma 3-nitrotyrosine in rheumatoid arthritis patients. *Eur J Appl Physiol, 114*(7), 1483–1492.

Weyand, C. M., & Goronzy, J. J. (2021). The immunology of rheumatoid arthritis. *Nat Immunol, 22*(1), 10–18.

Wiesinger, G. F., Quittan, M., Aringer, M., Seeber, A., Volc-Platzer, B., Smolen, J., & Graninger, W. (1998). Improvement of physical fitness and muscle strength in polymyositis/dermatomyositis patients by a training programme. *Br J Rheumatol, 37*(2), 196–200.

Wiesinger, G. F., Quittan, M., Graninger, M., Seeber, A., Ebenbichler, G., Sturm, B.,... Graninger, W. (1998). Benefit of 6 months long-term physical training in polymyositis/dermatomyositis patients. *Br J Rheumatol, 37*(12), 1338–1342.

Wu, N., Bredin, S. S. D., Guan, Y., Dickinson, K., Kim, D. D., Chua, Z.,... Warburton, D. E. R. (2019). Cardiovascular health benefits of exercise training in persons living with type 1 diabetes: A systematic review and Meta-Analysis. *J Clin Med, 8*(2), 253. https://pubmed.ncbi.nlm.nih.gov/30781593/

Xiang, M., Feng, Y., Wang, Y., Wang, J., Zhang, Z., Liang, J., & Xu, J. (2021). Correlation between circulating interleukin-18 level and systemic lupus erythematosus: A meta-analysis. *Sci Rep, 11*(1), 4707.

Yardley, J. E., Kenny, G. P., Perkins, B. A., Riddell, M. C., Balaa, N., Malcolm, J.,... Sigal, R. J. (2013). Resistance versus aerobic exercise: acute effects on glycemia in type 1 diabetes. *Diabetes Care, 36*(3), 537–542.

Yu, B., Qi, Y., Li, R., Shi, Q., Satpathy, A. T., & Chang, H. Y. (2021). B cell-specific XIST complex enforces X-inactivation and restrains atypical B cells. *Cell, 184*(7), 1790–1803 e1717.

Zucchi, D., Elefante, E., Schiliro, D., Signorini, V., Trentin, F., Bortoluzzi, A., & Tani, C. (2022). One year in review 2022: Systemic lupus erythematosus. *Clin Exp Rheumatol, 40*(1), 4–14.

12 Exercise immunology and cardiometabolic diseases

Mark Ross, Graeme Koelwyn, and Alex Wadley

The immune landscape of cardiometabolic disease

The immune system is not only tasked with fighting infections and protecting the body from pathogens but also plays a crucial role in tissue maintenance and repair. The term "immune response" is used to describe a variety of functions which includes inflammation, so it is often difficult to separate an "immune response" or "immune function" with "inflammation". Indeed, inflammation is a physiological mechanism to enable the restoration of damaged/infected tissue. However, there is a fine balance between inflammation which is physiological in nature and that which is pathophysiological/pathological. Atherosclerotic cardiovascular disease (ASCVD), type 2 diabetes mellitus (T2DM), obesity, and the metabolic syndrome (MetS) are all characterised by unresolved inflammation, yielding a chronic pro-inflammatory phenotype, evident by high circulating pro-inflammatory cytokines, such as interleukin-6 (IL-6), tumour necrosis factor-α (TNF-α), and C-reactive protein (CRP). The tissue origin of these inflammatory factors includes skeletal muscle, adipose tissue, and liver but also cells of the innate and adaptive immune system, implicating these cells in the aetiology of the disease(s).

As highlighted above, the immune system is intricately involved in tissue maintenance and repair. For example, we know that neutrophils, monocytes, and certain subsets of lymphocytes can exert pro-angiogenic effects, either contributing to blood vessel growth or repair in the context of tissue damage, or adaptation to exercise. Some lymphocytes also regulate/dampen the immune response to infection and tissue damage to modulate hyperinflammatory responses. For example, in response to a myocardial infarction or stroke, many immune cells are recruited to the infarct, which can instigate repair of the tissue leading to functional and perfusion recovery of the myocardium). But a balance is required, as when an inflammatory response is exaggerated or prolonged, this can lead to impaired recovery of the tissue, and therefore regulating this process is crucial. The following sections will discuss the involvement of both the innate and adaptive immune systems in the onset and progression of cardiometabolic diseases.

Role of innate immunity in cardiometabolic disease

Cells that comprise the innate immune system, namely monocytes, macrophages, and neutrophils, are immune sentinels that have an essential role in systemic and tissue homeostasis. It is therefore not surprising that these cells are also integral in maintaining cardiometabolic health (Nahrendorf, 2018). However, numerous cardiometabolic risk factors are associated with the overproduction of innate immune cells with an altered inflammatory phenotype that can lead to cardiometabolic disease, including hyperlipidaemia, hyperglycaemia, and obesity. For example, in individuals with these risk factors, circulating monocytes (and to some degree,

DOI: 10.4324/9781003256991-12

neutrophils) possess dysregulated functions, most notably a hyperinflammatory phenotype, increasing cytokine production at rest and following immune challenge – a phenomenon known as innate immune training – driven by cell-specific metabolic and epigenetic adaptations. Within tissues (e.g., adipose, atherosclerotic plaques), macrophages also display a dominance towards inflammatory phenotypes, or M1-like skewing, versus M2-like anti-inflammatory phenotypes. Together, these shifts drive systemic and tissue-specific chronic inflammation and subsequent onset of cardiometabolic disease, with high circulating numbers of monocytes and neutrophils, and an elevated neutrophil-to-lymphocyte ratio all associated with increased risk of cardiometabolic disease.

In established ASCVD, within the arterial wall, sterile inflammation (i.e., inflammation in the absence of a microorganism) is the consequence of intramural retention of cholesterol-rich lipoproteins, which is caused by an imbalance in cholesterol homeostasis at the systemic and cell-specific level, promoting arterial wall deposition. These lipoproteins subsequently exhibit various pro-inflammatory modifications such as oxidation, nitration, and aggregation, resulting in increased monocyte recruitment, which differentiate into macrophages. Prior to entering the plaque and differentiation into macrophages, circulating monocytes have a pro-inflammatory phenotype in these patients (elevated cytokine and chemokine production at rest and upon stimulation, driven, for example, by innate immune training), which also extends to their progenitors in the bone marrow. Initially, monocyte-derived macrophages in the arterial wall clear the inflammatory lipoproteins, but over time, this process breaks down, leading to foam cell accumulation (macrophages engorged in lipid), promoting plaque formation. Plaque formation is also due in part to defective phagocytosis and a failure of macrophage efferocytosis or clearing of apoptotic cells from the arterial wall. As this process progresses, macrophages also display a reduced migratory capacity and maintained hyperinflammatory phenotype, attracting additional monocytes, T cells and neutrophils, establishing a vicious, non-resolving cycle. Over time, this cycle results in continued macrophage content expansion, increasing plaque size and development of a necrotic core, and overall plaque instability. This manifests in eventual plaque rupture, thrombosis formation, arterial blockage, and ischaemia, leading to major cardiovascular events, such as myocardial infarction.

Following an acute event such as a myocardial infarction, inflammatory monocytes and neutrophils orchestrate the initial inflammatory response in the heart to danger signals released during ischaemia-reperfusion injury. During this inflammatory phase, monocytes differentiate into macrophages with an inflammatory phenotype, which subsequently shift towards a reparative phenotype during infarct repair. Myocardial infarction also triggers numerous local and systemic effects (sympathetic nervous signalling, alarmin release from dying cardiac tissue, increased circulating inflammatory cytokines such as IL-1β), which cumulatively signal to haematopoietic progenitors in the bone marrow and spleen, resulting in the activation of myelopoiesis and downstream monocyte and neutrophil production (Nahrendorf, 2018). These factors also contribute to sustained increased monocyte and neutrophil levels post infarction, as well as cardiac macrophage levels, which can contribute to progression to heart failure.

Finally, specific to MetS and T2DM, chronic systemic and tissue-specific inflammation are also known to interfere with insulin signalling, typified by elevated circulating numbers of monocytes and neutrophils that propagate insulin resistance, and adipose tissue dysfunction and disease onset. Higher circulating monocyte and neutrophil numbers in people with obesity and T2DM also results from increased bone marrow myelopoiesis (Nagareddy et al., 2013). These cells typically have a higher propensity to migrate into metabolic active tissues, such as adipose tissue, creating a perpetual cycle that results in macrophage accumulation (crown-like structures) and defective insulin signalling. Animal data indicates that this may be driven by inflammatory cytokines, damage-associated molecular patterns, and hyperglycaemia. Further, these cells display similar activated and inflammatory phenotypes, alongside defective phagocytosis, as seen with ASCVD (also increasing future CVD risk). Intriguingly, reduced *in vitro* phagocytosis of *Escherichia coli* (*E. coli*) Bioparticles® in T2DM is inversely associated with fasting glucose and HbA1c (Lecube, Pachon, Petriz, Hernandez, & Simo, 2011). This significantly improved upon a five-day intensive medical intervention (combined metformin, insulin, and/or diet optimisation) to treat high blood glucose levels in those diabetics with a HbA1c of >8%, suggesting that improvements in cardiometabolic profile may subsequently positively alter innate immune cell function.

Role of adaptive immunity in cardiometabolic disease

The cellular adaptive immune system comprises lymphocytes, including both T and B cells. The principal immunological role of these cells is to coordinate a specific attack on foreign pathogens upon failure of the innate immune system to eliminate these risks. T and B cells reside in various tissues, such as the spleen, gut, lungs, muscle, lymph, and in blood, and are trafficked to specific sites of inflammation as part of immune surveillance and priming, via complex communication pathways. Circulating lymphocyte profiles are altered in individuals with cardiometabolic diseases such as ASCVD, T2DM, and in those with MetS. People with ASCVD and T2DM typically exhibit a high circulating number of T and B cells compared with people without these diseases and the frequency of these cells is positively associated with CVD mortality risk (Phillips, Carroll, Gale, Drayson, & Batty, 2011). It is likely that these cells are influenced by the risk factors in cardiometabolic disease, with lymphocyte numbers associated with components of the MetS, such as circulating triglycerides, postprandial glucose levels, and HDL cholesterol levels.

The relationship between high lymphocyte and T cell number with high cardiometabolic risk is driven by the phenotype of the T cell, with effector T cells (such as effector memory T cells) typically elevated in individuals with cardiometabolic conditions, and concomitantly, naive T cells being lower (see Chapter 8). This is also evident in established cardiometabolic disease, with the frequency of peripheral blood effector T cells being associated with the severity of atherosclerotic disease in the coronary and carotid arteries (Rattik et al., 2019). T cells from people with T2DM or ASCVD also exhibit increased pro-inflammatory function, as expected given the effector phenotype of the expanded T cell pool. These cells are characterised by increased production of interferon (IFN)-γ and TNF-α (Bansal et al., 2017), both being cytokines that contribute to low-grade inflammation and exert potent damaging effects on the vasculature.

T cell senescence (see Chapter 8) is also associated with poor cardiometabolic health. Senescent T cells (identified as CD28null T cells) which are avid producers of pro-inflammatory factors, are elevated in patients with established CVD, hypertension, and T2DM (Dworacka et al., 2007). This is further demonstrated in *Cytomegalovirus* (CMV) seropositive individuals, who typically have a higher number and proportion of senescent T cells (Di Benedetto et al., 2015). These individuals are at higher risk of CVD and often exhibit vascular dysfunction, as well as impaired response to vaccines and virus infections. The CMV-dependent elevated levels of senescent T cells are a result of lifelong CMV re-activation, which persists without elimination, and this re-activation results in an expansion of CMV-specific cytotoxic T cells (Kim, Kim, & Shin, 2015). Over time, these cells undergo replicative senescence and exert negative effects on peripheral tissues due to their pro-inflammatory phenotype (see Chapter 8).

Circulating regulatory T cells are typically associated with anti-inflammatory processes through suppressing effector T cell functions and are often studied in T2DM due to the dysregulation of the auto-immune response in this disease. In T2DM, regulatory T cells decrease with disease severity, with higher number associated with reduced risk of developing ASCVD (Wang et al., 2018). The direct involvement of regulatory T cells in disease has been demonstrated in animal studies, with the transfer of regulatory T cells from normotensive mice into Angiotensin-II-induced hypertensive mice, resulting in a marked reduction in arterial blood pressure, and improved vascular function, implicating these cells in modulating inflammation-induced hypertension (Matrougui et al., 2011), possibly through IL-10 mediated-suppression of Th17 cell differentiation. An often-overlooked subpopulation of T cells, γδ T cells (carriers of the T cell receptor [TCR] γ and δ chains), may also be implicated in cardiometabolic disease. These cells make up 1–10% of T cells in peripheral blood, but are more common in mucosal tissues, including the skin, lung, and intestine. Immunologically, these cells are more like innate immune cells than adaptive immune cells, often acting within the first line of defence. Their involvement in cardiometabolic disease is poorly understood compared with their CD4+ and CD8+ counterparts. However, γδ T cells are often present within atherosclerotic plaques, and the lack and/or depletion of these cells results in reduced atherosclerotic disease in mice (Vu et al., 2014). In the Multi-Ethnic Study of Atherosclerosis cohort study, a high proportion of γδ T cells were associated with poor cardiac function (Sinha et al., 2021), which could be due to the role these cells play in promoting the infiltration of innate immune cells, namely neutrophils and macrophages, into cardiac tissue. B cells, another subset of lymphocytes (distinct from T cells) which have a primary role to produce and release antibodies in response to infection, as well as to present antigens to T cells, are also affected by disease status. B cells can be recruited to the vasculature in response to arterial injury and have been implicated in T2DM, hypertension, and atherosclerotic disease.

A primary focus of research when investigating lymphocytes in cardiometabolic diseases is often the impact of lymphocytes on disease onset and progression due to inflammatory processes. However, what is often overlooked is the impact of the condition on adaptive immune cell-mediated immunity. Observations demonstrate those with ASCVD and other metabolic conditions have increased infection risk compared with those without disease, and severity of infection. Indeed, in people with T2DM, lymphocyte proliferative responses are low (Chang & Shaio, 1995), which may be a result of the phenotypic switch observed in patients. As previously highlighted, CD4+ naive T cell counts are lower in patients with advanced ASCVD (Gaddis et al., 2021; Rattik et al., 2019), and these cells also have impaired proliferative capacities (Gaddis et al., 2021), which may relate to increased infection risk (and increased severity of consequences to infection) in ASCVD or metabolic conditions.

Assessment of circulating lymphocytes may provide insight into relationships between disease status and lymphocyte number and function; however, investigations into tissue-resident/infiltrating cells provide additional context for the role of the adaptive immune system in cardiometabolic disease pathology. For example, atherosclerotic plaques contain a significant number of inflammatory cells, including lymphocytes, the majority of which are T activated or senescent T cells (Fernandez et al., 2019). These cells potentiate inflammation within the arterial wall and promoting further infiltration of macrophages. B cells found in the artery of those with established atherosclerotic disease, like T cells, exhibit a pro-inflammatory phenotype, expressing high levels of TNF-α, IL-6, and granulocyte macrophage colony-stimulating factor (GM-CSF) (Hamze et al., 2013). These cells may also produce self-reactive IgG antibodies that can cross-react with the bacterial wall of Gram-negative bacteria. This interestingly implicates infection as a key player in atherosclerotic plaque development.

T cells may not solely act to promote inflammation but can also be protective in peripheral tissues. For example, regulatory T cells can reduce monocyte infiltration into vascular tissue, and their subsequent differentiation into macrophages and foam cells. This also has the knock-on effect of reducing TNF-α release which may contribute to the observed improvement in arteriolar endothelium-dependent relaxation. In high fat-fed mice, the level of regulatory T cells residing in adipose tissue was lower than in controls, and these were inversely associated with inflammatory factors in visceral fat, such as TNFα, IL-6, and RANTES (Feuerer et al., 2009), which could further exacerbate monocyte infiltration into adipose tissue. Adaptive immune cells also infiltrate adipose tissue (Nishimura et al., 2009; Trim et al., 2022) significantly contributing to metabolic health and disease, including contributing to insulin resistance. Indeed CD8+ T-cell infiltration in adipose tissue in a mouse model of obesity appears to precede macrophage accumulation, resulting in diminished insulin sensitivity, and deletion of the CD8+ T cells ameliorated these effects (Nishimura et al., 2009), providing a contributing mechanism and possible target for treating obesity-associated diseases.

Section summary

The immune system participates in the onset and progression of cardiometabolic disease (Figure 12.1). The innate immune system contributes through the pro-inflammatory shift in monocytes and macrophages contributing to plaque formation and altered insulin sensitivity of peripheral tissues. Additionally, myelopoiesis in the bone marrow results in further production of pro-inflammatory innate immune cells, which exacerbates disease. The adaptive immune system contributes through pro-inflammatory lymphocyte phenotypes, resulting in elevated production and release of pro-inflammatory cytokines, and a dysregulated immune response. This promotes cardiometabolic inflammation, which contributes to disease progression through plaque (ASCVD) infiltration, and/or adipose tissue dysfunction (T2DM, MetS). However, it is well established that physical activity and/or structured exercise can favourably modulate cardiometabolic risk factors. These benefits may be partly underpinned by alterations in the immunological mechanisms described above, and thus, will be the focus of the next sections (Figure 12.1).

Box 12.1   Type 1 Diabetes – an auto-immune cardiometabolic disease?

When we think of diabetes as a cardiometabolic disease, type 2 diabetes mellitus (T2DM) typically comes to mind. However, type 1 diabetes mellitus (T1DM), classified as an auto-immune disease (due to self-reactive immune destruction of pancreatic islet cells), often results in perturbations in the cardiovascular system, such as vascular dysfunction, elevated pro-inflammatory factors, resulting in increased risk of ASCVD compared with people without diabetes (2–3-fold in men, 3–5-fold in women) (Schnell et al., 2013). These effects are predominantly due to long-term hyperglycaemia, which negatively affects the micro- and macrovascular beds, causing accelerated atherosclerosis.

In addition to the traditional cardiovascular risk factors (HbA1C, lipid profile), alterations in immune cell number and/or function may also contribute to elevated ASCVD risk in T1DM. The differentiation of T cells towards a more effector state is involved in the pathogenesis of the disease, but also contributes to cardiometabolic disease development, and similar involvement of the innate cellular immune system is known. However, the role of exercise to alleviate cardiometabolic disease risk and/or progression is not well understood, and a key area for future research. See Chapter 11 for more details.

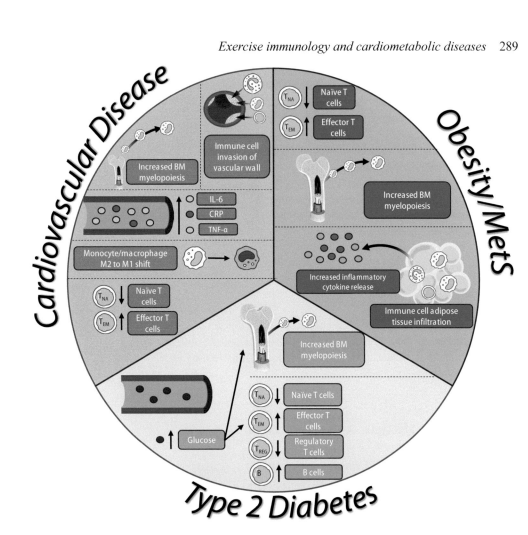

Figure 12.1 Immune landscape of cardiometabolic disease.

Legend: Atherosclerotic Cardiovascular disease is typified by increased bone marrow (BM) myelopoiesis, altered my-eloid and lymphocyte frequencies, which results in increased inflammation (IL-6, CRP, TNF-α) and development of immune cell ladened atherosclerotic plaques. Obesity and factors associated with the metabolic syndrome (MetS) results in adipose immune cell infiltration, increased myeloid cell production by the bone marrow, and an expansion of lymphocyte numbers. Type 2 diabetes results in increased bone marrow production of myeloid cells and altered T cell phenotypes. Parts of the figure were drawn by using pictures from Servier Medical Art. Servier Medical Art by Servier is licensed under a Creative Commons Attribution 3.0 Unported License (https://creativecommons.org/licenses/by/3.0/).

Exercise, physical activity, and cardiometabolic disease

It is well established that exercise and physical activity reduce the risk and progression of ASCVD, MetS, and T2DM. Mechanisms include improving cardiac and vascular functions, and improving the metabolic profile of the individual, including improved fasting glucose and lipid profiles. It is also known that exercise and physical activity strongly affect the immune system, both in response to individual exercise bouts (Peake, Neubauer, Walsh, & Simpson, 2017), and long-term repeated bouts during exercise training (Woods et al., 1999). However, what is relatively less known, is whether exercise and/or physical activity influence cardiometabolic disease risk and progression via modulation of the immune system.

Individual bouts of exercise mobilise immune cells, including neutrophils, monocytes, and lymphocytes, with different mobilisation patterns depending on the immune cell subset. In people with cardiometabolic disease, individual bouts of exercise can mobilise immune cells, but this acute response is different compared with healthy controls (Harbaum et al., 2016). The exact relevance for people with cardiometabolic disease is not known, but it has been proposed that for both the innate and adaptive immune cells, that individual bouts of exercise may promote the movement of immune cells (such as effector and senescent T cells, which have been linked with inflammation in cardiometabolic disease) into the circulation, subsequently promoting their apoptosis (programmed cell death) in peripheral tissues in the hours post-exercise. This may (after regular acute stimuli over time) leave 'vacant space' for the production and maturation of new naive T cells and immature monocytes resulting in a clinically relevant shift in the immune cell pool (Simpson, 2011; Wonner, Wallner, Orso, & Schmitz, 2018). This shift (reduced effector and senescent T cells, M1-like monocytes/macrophages, and increased production of naive T cells and immature monocytes) may reflect reduced inflammation, and subsequent immune cell accumulation in adipose and vascular tissues (Figure 12.2).

The focus for the remainder of this chapter will be determining the influence of exercise and physical activity, as well as cardiorespiratory fitness, on the immune system among patients with cardiometabolic disease. Cross-sectional and intervention studies will be summarised, and in places, studies examining individuals with risk factors for cardiometabolic diseases are included.

Exercise, innate immunity, and mechanisms implicated in the development of cardiometabolic disease

Primary prevention of disease refers to preventing its onset before any ill-health occurs (for example, in healthy people to prevent the onset of ASCVD). Exercise appears to provide a clinically beneficial impact on the population through keeping a check on inflammation mediated by the innate immune system. Evidence shows that long-term exercise training can modulate the production of innate immune cells, which protects against cardiometabolic disease. In mice, exercise training promotes bone marrow haematopoietic stem cell and progenitor (HSPC) quiescence, decreasing basal bone marrow myelopoiesis, resulting in lower circulating monocytes and neutrophils compared with non-exercising controls (Frodermann et al., 2019). This was attributed to exercise-induced diminished leptin signalling to the stromal haematopoietic bone marrow niche. Decreased basal innate immune cell availability following chronic exercise has also been shown in mice exhibiting cardiometabolic risk factors, including obesity, hypercholesterolemia, and hypertension (Carvalho et al., 2023). Despite regular exercise lowering basal haematopoietic progenitor activity, the ability of the bone marrow to rapidly stimulate haematopoiesis and release mature immune cells during models of acute infection is not impacted (Frodermann et al., 2019).

Importantly, long-term exercise training, when appropriately dosed, does not dampen, and can actually enhance, emergency haematopoietic responses related to cardiometabolic disease risk. For example, in mice, six weeks of exercise prior to myocardial infarction does not dampen emergency myelopoiesis that is required for infarct healing (Frodermann et al., 2019) and enhances myeloid cell production in response to the development of acute heart failure (Feng et al., 2022). In the latter study, exercise training for two weeks before and after induction of heart failure induced by isoproterenol resulted in increased circulating myeloid-derived suppressor cells (MDSCs - immature monocytes and neutrophils) in the circulation, bone marrow, spleen, and heart. This occurred alongside enhanced protection from heart failure through preserved cardiac function (e.g., fractional shortening, ejection fraction) and morphological changes compared to sedentary controls. Ablation of MDSCs mitigated these protective effects, suggesting the increased availability and recruitment of these cells with exercise, which can suppress inflammatory processes within cardiac tissue, resulting in heart failure protection. Together, these data suggest that while exercise decreases basal innate immune cell production protecting against cardiometabolic disease, it does not mitigate acute myelopoietic responses, which are also needed for cardiometabolic disease protection.

Cross-sectional studies in patients at risk for cardiometabolic disease support preclinical findings, showing that exercise is associated with decreased innate immune cell production and peripheral blood concentration. In sedentary men with obesity, higher cardiorespiratory fitness (peak oxygen consumption; VO_2 peak) mitigates obesity-associated increases in CD14++CD16+ intermediate and CD14+CD16++ non-classical monocyte populations (Dorneles, da Silva, Boeira et al., 2019). Similar findings have also been shown in women with obesity, whereby circulating total monocyte and neutrophil counts were lower in those with high versus low VO_2 peak (Michishita, Shono, Inoue, Tsuruta, & Node, 2008). A recent study by Wadley et al. (2021) reported that *ex vivo* tethering and migration of monocyte subsets were reduced in people with obesity who were more physically active. This data indicates that being more physically active may reduce the propensity of peripheral blood monocytes to migrate into metabolically active tissues (e.g., adipose tissue and liver), thus reducing ASCVD risk. Reductions to the pro-inflammatory and pro-migratory function of these cells may result from a combination of training-related weight loss (and thus lower inflammatory cues from adipose tissue), and/ or direct phenotypic changes to the innate immune cells.

Interventional studies also show that exercise training reduces circulating levels of innate immune cells, as well as cell functions, albeit findings are mixed. In older sedentary overweight adults, 12 weeks of resistance training decreased CD14+CD16− inflammatory monocytes by ~35%, independent of weight loss (Markofski et al., 2014). In older (55+ years) sedentary individuals with increased cardiovascular risk (e.g., hypertension, body mass index [BMI] > 28kg·m^2), a 16-week intervention aimed to reduce sitting and increase physical activity resulted in a reduction in the percentage of circulating monocytes as well as an increase in the lymphocyte-to-monocyte ratio (Noz et al., 2019). Physical activity also attenuated 24-hour IL-6, IL-8, and IL-10 cytokine production from peripheral blood mononuclear cells (PBMCs) (which include monocytes and lymphocytes, but not neutrophils) in response to toll-like receptor agonists, which activate immune responses. Exercise also attenuated IL-1β, IL-6, and IL-8 production from PBMCs following stimulation. There was also a strong, inverse correlation between walking time and IL-1β, IL-6, IL-8, and IL-10 production following inflammatory stimulation. Similarly, in breast cancer survivors (who are high risk for cardiometabolic disease due to direct and indirect [i.e., sedentary lifestyle] effects of cancer treatment), 16 weeks of combined aerobic and resistance exercise decreased the percentage of IL-1β and IL-6+ CD16+CD14− inflammatory monocytes, as well as the mean fluorescent intensity (representing cytokine levels per cell) of IL-1β in both CD16+CD14− inflammatory and CD16+CD14+ intermediate monocytes (Khosravi et al., 2021). Obesity-associated increases in non-classical monocytes proportions and increased expression of HLA-DR by intermediate monocytes are also reduced by eight weeks of high-intensity interval training (de Matos et al., 2019). Further, in adults with prediabetes (defined as glycated haemoglobin [HbA1c] between 5.7% and 6.4% and/or a Canadian Diabetes Risk score > 21), just two weeks (ten sessions total) of progressive moderate-intensity exercise training or high-intensity interval training decreased TLR4 expression, but not TLR2, on both neutrophils and CD14+ monocytes (Robinson et al., 2015). No differences were shown in circulating cell numbers in response to either intervention or inflammatory cytokine production in response to LPS in whole blood cultures.

Not all studies show that exercise influences the immune system in the context of cardio-metabolic disease, however. For example, among people with obesity (with ~47% meeting MetS criteria), randomisation to 18-month lifestyle intervention (diet, increased exercise, and behavioural change) had no appreciable effect on circulating monocyte subsets compared with the control group (van der Valk et al., 2022). Further, a dose-dependent decrease in circulating neutrophils, but not monocytes, was observed following six months of aerobic exercise in sedentary, overweight/obese postmenopausal women with elevated blood pressure (Johannsen et al., 2012). Specifically, a reduction in neutrophils was shown at a dose equivalent of energy expenditure equal to 12 kcal·kg·week (exercising at ~50% VO_2 peak) compared with both 4 kcal·kg·week and control. Follow-up analyses also suggested that women with the highest baseline total whole blood cell counts had the largest reductions in neutrophil counts.

Together, these data suggest that in general, aerobic and/or resistance exercise training decrease innate immune cell production and availability, as well as the inflammatory and tissue-homing functions of these cells, in people at risk of cardiometabolic disease. However, these effects are likely to be dependent on multiple factors, not limited to patient risk factor status, degree of exercise-associated loss of adipose tissue, exercise dose and scheduling, and cardiometabolic disease subtype.

Exercise, innate immunity, and mechanisms of established cardiometabolic disease

Exercise training also modulates innate immune cell number and function in animal models and patients with established cardiometabolic disease. In mice, sometimes so-called 'knock out' models are used in mechanistic investigations. A 'knockout' mouse is a laboratory mouse in which a gene has been inactivated or 'knocked out' to investigate the influence of that gene or protein on biological or physiological function. In mice, apolipoprotein E knock-out mice (*ApoE-/-*; a model of ASCVD) show that exercise training can, in general, also modify both the number and function of innate immune cells in the circulation and atherosclerotic plaques, which is associated with the deceleration of disease progression. In mice with established atherosclerosis, 10 weeks of voluntary wheel running lowered circulating and aortic plaque monocytes/macrophages and neutrophils, alongside reductions in plaque size compared with control mice (Frodermann et al., 2019). Reductions in plaque size and macrophage content with exercise have also been shown with swimming exercise in a similar mouse model of atherosclerosis (Pellegrin et al., 2009), as well as in mice with diabetic atherosclerosis (diabetes induced via streptozotocin injection) (Kadoglou et al., 2013).

Exercise training-induced reductions in plaque size and innate immune cell infiltration also extend to other mouse models of atherosclerosis. In the *Ldlr−/−* atherosclerotic model (hyper-cholesterolemic mice lacking the LDL receptor [*Ldlr−/−*] fed a high fat diet), eight weeks of treadmill running not only reduced plaque size and macrophage content in eight-week-old male mice but also altered *ex vivo* macrophage function (Rentz et al., 2020). Specifically, bone marrow derived macrophages isolated from exercised vs control mice showed a decreased migratory capacity towards monocyte chemoattractant protein-1 (MCP-1; a monocyte/macrophage chemoattractant) and reduced basal inflammatory cytokine expression. Further, transplantation of bone marrow from *Ldlr−/−* exposed to 16 weeks of exercise into 7-week *Ldlr−/−* mice subsequently fed a high-fat diet for eight weeks resulted in attenuated plaque size and reduced macrophage content in recipient mice compared with transplant from controls (Rentz et al.,

2020). These data suggest that exercise-induced bone marrow (myelopoietic) alterations may be causal in protecting from the progression of atherosclerosis in this model. In sum, studies across mouse models of atherosclerosis suggest that exercise training decreases the progression of established atherosclerosis, and this occurs through the modulation of innate immunity at the systemic (bone marrow, circulation), tissue (plaque) and cell functional-specific levels.

Exercise training also protects from complications following myocardial infarction that co-incide with modulation of the innate immune system. In rats, eight weeks of isometric exercise training following myocardial infarction (consisting of vertical weighted grip exercises) significantly improved cardiac function, such as left ventricular ejection fraction, and reduced infarct size, compared with sedentary control (Zhang et al., 2021). This coincided with greater accumulation of CD68+ myeloid cells and greater cardiac tissue-specific levels of MCP-1, suggesting these cells contribute to improved cardiac remodelling. Further, eight weeks of exercise in rats with myocardial infarction-induced heart failure led to changes in peritoneal macrophage number and function (Batista, Santos, Oliveira, Seelaender, & Costa Rosa, 2007). Specifically, myocardial infarction heart failure induced increases in peritoneal macrophage number were inhibited by exercise. Functionally, macrophages displayed enhanced phagocytic function, inhibition of high fat diet-induced increases in macrophage chemotaxis, and inhibition of TNF-α production in response to lipopolysaccharide (Garcia-Bragado et al., 1979) following exercise exposure.

Exercise training-induced changes in innate immune cell availability and function also coincide with improved outcomes in animal models of T2DM (Belotto et al., 2010; Thakur, Gonzalez, Pennington, Nargis, & Chattopadhyay, 2016). For example, in the KK-Ay mouse model of diabetes, which have specific genetic mutations that lead to the development of obesity, impaired glucose tolerance, resting hyperglycaemia, and insulin resistance, eight weeks of treadmill running in male mice attenuated the diabetes-induced increase in renal expression of MCP-1 and infiltration of macrophages, which occurred alongside improved markers of diabetic nephropathy (Ishikawa et al., 2012). Further, in the *db/db model*, a model of T2DM and obesity, just two weeks of exercise attenuated diabetes-induced increases in cardiac macrophage content compared with sedentary control, which coincided with reductions in TNF-α cardiac tissue gene expression (Botta et al., 2013). Together, these data suggest that exercise-induced modulation of innate immunity improves multiple sequelae of diabetes pathogenesis.

Comparably fewer studies have evaluated exercise-induced changes in innate immunity in humans with established cardiometabolic diseases. In patients with atherosclerosis (carotid stenosis >50%), moderate levels of self-reported physical activity levels (27–75 metabolic equivalent [MET] hrs/wk) were associated with a lower percentage of circulating inflammatory monocytes and a higher percentage of non-classical monocytes compared with non-active (<27 MET hrs/wk) and highly active individuals (>75 MET hrs/wk) (Mura et al., 2022). Physical activity, however, was not associated with macrophage infiltration of plaques in patients with asymptomatic atherosclerosis who were undergoing carotid endarterectomy (Mury et al., 2020). Further, in a small cohort of females diagnosed with T2DM ($n = 4$), 12 weeks of high-intensity interval training resulted in global transcriptomic changes in circulating monocytes (Hamelin Morrissette et al., 2022). Specifically, RNA sequencing of CD14+ monocytes (which includes classical and intermediate monocyte populations) identified 56 differentially expressed genes that were downregulated pre/post intervention. Pathway enrichment of these downregulated genes included immune cell activation, as well as cellular adhesion and migration. Overall, these human studies provide early albeit limited evidence that exercise modulates innate immunity in patients with established cardiometabolic disease (Figure 12.2).

Box 12.2   Individual bouts of exercise and regular exercise training – the influence on haematopoietic stem and progenitor cells

Haemopoietic stem/progenitor cells (HSPCs) are unspecialised cells made in the bone marrow with multilineage potential for producing up to ~10^9 mature lymphoid and my-eloid blood cells in the circulation per day. The build-up of monocytes, neutrophils, and macrophages in vascular, adipose, and liver tissue are central in the development and progression of chronic inflammation and increased cardiometabolic disease risk. A feedforward cycle is driven by the heightened release of catecholamines and other danger signals (e.g., inflammatory cytokines, damage-associated molecular patterns) that enhance hae-matopoiesis and subsequent production bias of myeloid cells from the bone marrow niche (Pittet, Nahrendorf, & Swirski, 2014). This is reflected by higher numbers of monocytes and neutrophils in the peripheral blood of people with cardiometabolic diseases (Idz-kowska et al., 2015), and these cell counts are positively associated with various adverse health outcomes (Garofallo et al., 2019).

Accumulating evidence indicates that regular bouts of exercise can impact the distribution and activity of myeloid cells, thus lowering chronic inflammation. It has been reported that between 8 and 12 weeks of exercise can lower the number of total monocytes and neutrophils (Johannsen et al., 2012) in blood and reduce the composition of CD16+ monocytes (Timmerman, Flynn, Coen, Markofski, & Pence, 2008), which may be a result of changes in HSPC production and release with exercise. In fact, six weeks of regular exercise can regulate haematopoiesis in the bone marrow of mice (Frodermann et al., 2019). This was restricted to lymphoid and myeloid cells, but not erythroid progenitor cells and was reflected by reduced numbers of mature lymphocytes, monocytes, and neutrophils following training. Exercise and physical activity may therefore confer protection against cardiometabolic disease by directly reducing chronic inflammatory leukocyte output from the bone marrow.

Interestingly, the trafficking of HSPCs from the bone marrow to peripheral blood after single bouts of exercise may also be of importance in the maintenance of chronic cardiometabolic health. Individual bouts of exercise elicit an increase in the concentration of HSPCs in peripheral blood in a biphasic manner. A rapid and transient initial mobilisation of HSPCs from marginal pools within the circulation is followed by their mobilisation from the bone marrow approximately 3–24 hours later (Kruger et al., 2015). These dynamic changes accompany the migration of senescent lymphoid and myeloid cells out of peripheral blood in exercise recovery. The appearance of HSPCs in peripheral blood after individual bouts of exercise, when repeated over time, might be a mechanism to rejuvenate the composition of peripheral blood immune cells and lower numbers of inflammatory leucocytes. These changes would in turn lower the risk of cardiometabolic disease.

Exercise, adaptive immunity, and mechanisms of cardiometabolic disease

The cellular adaptive immune system is intricately linked to the development and/or progression of various cardiometabolic diseases, as explained at the beginning of this chapter. The altered frequencies in a variety of tissues (such as blood, adipose tissue, pancreas, and vascular), phenotype, and inflammatory/regulatory functions suggest that therapeutics to target these cells may provide promise for the prevention and/or treatment of such cardiometabolic diseases. Exercise is one such intervention. This section will discuss the current evidence demonstrating how regular exercise can modulate lymphocyte numbers and functions, with specific reference to cardiometabolic disease (Figure 12.2).

A potential mechanism by which regular exercise or physical activity may affect the adaptive immune system is through shifting the circulating and tissue lymphocyte pool from a pro-inflammatory to an anti-inflammatory phenotype. In healthy people, lifelong exercise is characterised by higher naive T cells, and lower effector and senescent T cells in the peripheral blood compartment (Duggal, Pollock, Lazarus, Harridge, & Lord, 2018). Similarly, higher cardiorespiratory fitness (an indicator of physical activity status) is associated with reduced proportions of effector (Dorneles, da Silva, Boeira et al., 2019; Dorneles, da Silva, Peres, & Romao, 2019; Spielmann et al., 2011) and senescent T cells (Spielmann et al., 2011), and increased regulatory T cell populations (Dorneles, da Silva, Boeira et al., 2019; Dorneles, da Silva, Peres et al., 2019) (see Chapter 8). Further associations were evident between cardiorespiratory fitness and cytokines associated with regulatory and effector T cell functions, with IL-10 and IL-33 elevated, and TNF-α lower in lean and obese adults with high vs. low VO_2 (Dorneles, da Silva, Boeira et al., 2019). The IL-33 differences may account for the differences in regulatory T cell frequencies between high and low VO_2 peak groups, with IL-33 being responsible for orchestrating the maintenance of regulatory T cell populations. Activation status of CD4+ cells were also affected by fitness status: individuals (both lean and obese) with low VO_2 peak (50–80% of the mean VO_2 peak in total sample) exhibited higher levels of HLA-DR expression on these cells compared with people with high VO_2 peak (>95% of the mean VO_2 peak) (Dorneles et al., 2019). While activation of immune cells is essential for the immune response to foreign invaders, sustained HLA-DR expression is associated with inflammation and may contribute to elevated pro-inflammatory cytokine release.

Exercise training studies in humans with cardiometabolic disease are rare, and the results are equivocal, so most insight comes from animal studies. In murine models of CVD, such as induced ischaemic cardiomyopathy (Chen et al., 2018) or post-myocardial infarction heart failure (Batista et al., 2008) demonstrate some improvements in indices of lymphocyte-specific inflammation. For example, 12 weeks of exercise training in rats with induced ischaemic cardiomyopathy demonstrated a decrease in the Th17 to regulatory T cell ratio (Chen et al., 2018), indicative of a more regulatory, less pro-inflammatory T cell pool, while 8–10 weeks of regular treadmill exercise (60-min a·day, five days·a week) resulted in a 2.2-fold increase in lymphocyte production of IL-2 (Batista et al., 2008), suggestive of improved immunoregulatory function. However, no changes in lymphocyte proliferation were observed in response to mitogen stimulation in the latter study. Studies in humans with ASCVD demonstrate no effect of regular exercise training on lymphocyte phenotypes or function. In cardiac transplant patients, six weeks of exercise training (three sessions per week, 50% maximal tolerated power for 4 minutes, 1-minute maximal tolerated power for 1 minute × 9) resulted in no changes in total circulating lymphocyte, T or B cell counts; however, these patients were taking immunosuppressive medication which may have influenced results (Zhao et al., 1998). In adults with pre-hypertension (130–139/85–59 mmHg), cycling exercise (three days·a week, 12 weeks) resulted in no changes in CD4+ or CD8+ T cell number, or activation status, as indicated by the expression of CD60, CD25, and CCR5, results which were mirrored by the parallel animal study (Mazur et al., 2018). Similar findings were observed in individuals defined to be at risk of myocardial infarction (based on serum complement or CRP levels), whereby a six-month individualised supervised exercise training programme resulted in no changes in circulating lymphocytes (CD3+, CD4+, CD8+, B cells, HLA-DR+ cells, γδ or NK cells) (Smith, Dykes,

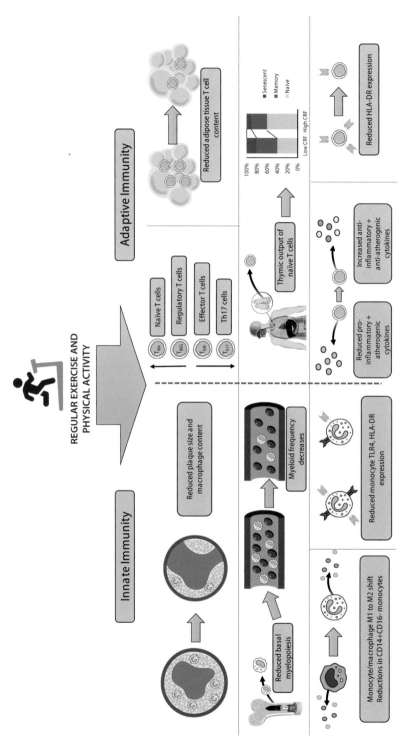

Figure 12.2 High cardiorespiratory fitness, regular exercise, and physical activity can reshape the immune system to a more favourable cardiometabolic profile.

Legend: Exercise and physical activity can reduce atherosclerotic plaque size and macrophage content, possibly through reductions in basal bone marrow myelopoiesis, which also results in reduced myeloid frequency in circulation. Exercise can also shift monocytes/macrophages from an M1-like inflammatory phenotype to an M2-like anti-inflammatory phenotype, characterised by reduced inflammatory cytokine release and activation status. For the adaptive immune system, physical activity and/or exercise can result in altered lymphocyte profile, with increased naïve and anti-inflammatory T cells, and a decrease in effector and senescent T cell pools, through an increase in thymic output. Lymphocyte pro-inflammatory cytokine release and activation status may also be reduced. Parts of the figure were drawn by using pictures from Servier Medical Art. Servier Medical Art by Servier is licensed under a Creative Commons Attribution 3.0 Unported License (https://creativecommons.org/licenses/by/3.0/).

Douglas, Krishnaswamy, & Berk, 1999). However, PBMC production of atherogenic cytokines (TNF-α, IFN-γ, and IL-1α) were lower, and production of anti-atherogenic compounds (IL-10, transforming growth factor-β [TGF-β], IL-4) rose, in response to the training programme.

Studies recruiting people with T2DM and obesity have shown very promising effects on exercise for modifying disease mechanisms. For example, regular Tai Chi exercise for 12 weeks appears to be effective in increasing regulatory T cell frequency and proportion, as well as reducing the proportion of CD8+ cytotoxic T cells in patients with T2DM (Yeh et al., 2007). The same exercise programme in the same patient population also resulted in increased T-bet (immune cell transcription factor involved in Th1 function and tissue migration) expression in CD4+ Th cells, which may suggest improved CD4+ T cell function (Yeh et al., 2009). In people with obesity, even short exercise training interventions can result in significant T cell changes. Just one week of high-intensity interval training (3 bouts of 10×60 s at 85–90% maximum heart rate [HR_{max}] alternated with 75 seconds of recovery at 50% HR_{max}, bouts separated by 48 hours) resulted in an increased frequency of regulatory T cells, which was associated with improvement in VO_2 peak (Dorneles, da Silva, Boeira et al., 2019). In mice fed a high-fat diet, regular exercise training (treadmill exercise, 5 times a week, 16 weeks) resulted in a reduction in the CD8+ T cell accumulation in visceral adipose tissue, which subsequently resulted in less macrophage accumulation (Kawanishi, Mizokami, Yano, & Suzuki, 2013) normally observed in such models of obesity (Nishimura et al., 2009). It has also been examined whether 8 weeks of exercise training can modulate renal lymphocyte numbers (Amaral et al., 2018), due to their role in promoting the development of angiotensin II-induced hypertension (Nava, Quiroz, Vaziri, & Rodriguez-Iturbe, 2003). Renal tissue lymphocytes were lower in exercised rats with T2DM compared with sedentary controls, and this was accompanied by reduced macrophage infiltration and reduction in renal tissue oxidative stress. In non-insulin-dependent T2DM patients, three months of regular endurance exercise (2 times week, 45 minutes cycling) resulted in no significant circulating blood T cell changes, despite improvements in metabolic profile (Wenning et al., 2013). However, the researchers did observe a decline in CD19+ B cells, which may prove beneficial due to the role elevated CD19+ B cells play in auto-immune disease risk (Musette & Bouaziz, 2018). Despite the evidence demonstrating mixed effects, there is potential for exercise to significantly alter the cellular adaptive component in diabetes and/or obesity, but the evidence in individuals with established ASCVD is lacking. As such, while changes in lymphocyte numbers, phenotypes, and/or function may contribute to improved metabolic health in these cohorts, more evidence is needed before providing specific exercise recommendations to patients (Figure 12.2).

Chapter summary

Regular exercise can result in a switch from a pro-inflammatory to an anti-inflammatory state through modulating both the cellular innate and adaptive immune systems. These include altered myelopoiesis (reduced inflammatory myeloid cell production) and shifts in the monocyte/macrophage (from an M1-like to M2-like phenotype) and T cell (phenotype reductions in effector cells and increased naive T cells) pool. Studies report reduced tissue infiltration of pro-inflammatory macrophages and effector lymphocytes which may reduce vascular and adipose tissue inflammation. Most of the work to-date in humans has focused on individuals with risk factors for the development of cardiometabolic disease, rather than in clinical patient cohorts. The evidence in secondary prevention in cardiometabolic disease states primarily comes from animal models, which provides mechanistic insight into the possible benefits of exercise training for patients, but more work is needed in this field to be able to provide evidence-based recommendations.

Key points

- The cellular immune system and the risk for the development and the progression of cardiometabolic disease are inextricably linked. The pro-inflammatory signals from both innate and adaptive immune cells can directly contribute to the development of ASCVD and T2DM.
- Individuals with established disease exhibit increased myelopoiesis (resulting in increased pro-inflammatory monocyte and tissue-resident macrophage number) and altered lymphocyte populations which are characterised by a shift towards a pro-inflammatory phenotype (increased counts and proportion of effector and senescent T cells). These cells infiltrate into peripheral tissues, including adipose and vascular tissues, which exert deleterious effects on local cells, causing cellular dysfunction and cell death.
- Regular exercise can alter both the innate and adaptive cellular immune systems, both in primary and secondary prevention, which include reduced myelopoiesis, shifts in the monocyte/macrophage (from M1-like to M2-like shift) and lymphocyte pools (reduced effector and senescent, and increased naive T cell numbers).
- These alterations in both innate and adaptive cellular immune systems often correlate with reductions in pro-inflammatory markers (such as TNF-α, CRP), and improved metabolic indices/markers, such as improved lipid profile, reduced fasting glucose and HbA1c.
- Further evidence is needed to support the immunoregulatory role of exercise in people with cardiometabolic diseases to help inform specific exercise training recommendations.

References

Amaral, L. S. B., Souza, C. S., Volpini, R. A., Shimizu, M. H. M., de Braganca, A. C., Canale, D., … Soares, T. J. (2018). Previous exercise training reduces markers of renal oxidative stress and inflammation in streptozotocin-Induced Diabetic Female Rats. *J Diabetes Res, 2018*, 6170352.

Bansal, S. S., Ismahil, M. A., Goel, M., Patel, B., Hamid, T., Rokosh, G., & Prabhu, S. D. (2017). Activated T lymphocytes are essential drivers of pathological remodeling in ischemic heart failure. *Circ Heart Fail, 10*(3), e003688.

Batista, M. L., Jr., Santos, R. V., Oliveira, E. M., Seelaender, M. C., & Costa Rosa, L. F. (2007). Endurance training restores peritoneal macrophage function in post-MI congestive heart failure rats. *J Appl Physiol (1985), 102*(5), 2033–2039.

Batista, M. L., Jr., Santos, R. V., Lopes, R. D., Lopes, A. C., Costa Rosa, L. F., & Seelaender, M. C. (2008). Endurance training modulates lymphocyte function in rats with post-MI CHF. *Med Sci Sports Exerc, 40*(3), 549–556.

Belotto, M. F., Magdalon, J., Rodrigues, H. G., Vinolo, M. A., Curi, R., Pithon-Curi, T. C., & Hatanaka, E. (2010). Moderate exercise improves leucocyte function and decreases inflammation in diabetes. *Clin Exp Immunol, 162*(2), 237–243.

Botta, A., Laher, I., Beam, J., Decoffe, D., Brown, K., Halder, S., … Ghosh, S. (2013). Short term exercise induces PGC-1alpha, ameliorates inflammation and increases mitochondrial membrane proteins but fails to increase respiratory enzymes in aging diabetic hearts. *PLOS ONE, 8*(8), e70248.

Carvalho, V. H. C., Wang, Q., Xu, X., Liu, L., Jiang, W., Wang, X., … Qiu, S. (2023). Long-term exercise preserves pancreatic islet structure and beta-cell mass through attenuation of islet inflammation and fibrosis. *FASEB J, 37*(3), e22822.

Chang, F. Y., & Shaio, M. F. (1995). Decreased cell-mediated immunity in patients with non-insulin-dependent diabetes mellitus. *Diabetes Res Clin Pract, 28*(2), 137–146.

Chen, Z., Yan, W., Mao, Y., Ni, Y., Zhou, L., Song, H., … Shen, Y. (2018). Effect of aerobic exercise on treg and Th17 of rats with ischemic cardiomyopathy. *J Cardiovasc Transl Res, 11*(3), 230–235.

de Matos, M. A., Garcia, B. C. C., Vieira, D. V., de Oliveira, M. F. A., Costa, K. B., Aguiar, P. F., … Rocha-Vieira, E. (2019). High-intensity interval training reduces monocyte activation in obese adults. *Brain Behav Immun, 80*, 818–824.

Di Benedetto, S., Derhovanessian, E., Steinhagen-Thiessen, E., Goldeck, D., Muller, L., & Pawelec, G. (2015). Impact of age, sex and CMV-infection on peripheral T cell phenotypes: Results from the Berlin BASE-II Study. *Biogerontology, 16*(5), 631–643.

Dorneles, G. P., da Silva, I., Boeira, M. C., Valentini, D., Fonseca, S. G., Dal Lago, P., … Romao, P. R. T. (2019). Cardiorespiratory fitness modulates the proportions of monocytes and T helper subsets in lean and obese men. *Scand J Med Sci Sports, 29*(11), 1755–1765.

Dorneles, G. P., da Silva, I. M., Peres, A., & Romao, P. R. T. (2019). Physical fitness modulates the expression of CD39 and CD73 on CD4(+) CD25(-) and CD4(+) CD25(+) T cells following high intensity interval exercise. *J Cell Biochem, 120*(6), 10726–10736.

Duggal, N. A., Pollock, R. D., Lazarus, N. R., Harridge, S., & Lord, J. M. (2018). Major features of immunesenescence, including reduced thymic output, are ameliorated by high levels of physical activity in adulthood. *Aging Cell, 17*(2), e12750.

Dworacka, M., Winiarska, H., Borowska, M., Abramczyk, M., Bobkiewicz-Kozlowska, T., & Dworacki, G. (2007). Pro-atherogenic alterations in T-lymphocyte subpopulations related to acute hyperglycaemia in type 2 diabetic patients. *Circ J, 71*(6), 962–967.

Feng, L., Li, G., An, J., Liu, C., Zhu, X., Xu, Y., … Qi, Z. (2022). Exercise training protects against heart failure Via expansion of myeloid-derived suppressor cells through regulating IL-10/STAT3/S100A9 pathway. *Circ Heart Fail, 15*(3), e008550.

Fernandez, D. M., Rahman, A. H., Fernandez, N. F., Chudnovskiy, A., Amir, E. D., Amadori, L., … Giannarelli, C. (2019). Single-cell immune landscape of human atherosclerotic plaques. *Nat Med, 25*(10), 1576–1588.

Feuerer, M., Herrero, L., Cipolletta, D., Naaz, A., Wong, J., Nayer, A., … Mathis, D. (2009). Lean, but not obese, fat is enriched for a unique population of regulatory T cells that affect metabolic parameters. *Nat Med, 15*(8), 930–939.

Frodermann, V., Rohde, D., Courties, G., Severe, N., Schloss, M. J., Amatullah, H., … Nahrendorf, M. (2019). Exercise reduces inflammatory cell production and cardiovascular inflammation via instruction of hematopoietic progenitor cells. *Nat Med, 25*(11), 1761–1771.

Gaddis, D. E., Padgett, L. E., Wu, R., Nguyen, A., McSkimming, C., Dinh, H. Q., … Hedrick, C. C. (2021). Atherosclerosis impairs naive CD4 T-cell responses via disruption of glycolysis. *Arterioscler Thromb Vasc Biol, 41*(9), 2387–2398.

Garcia-Bragado, F., Vilardell, M., Caralps, A., Bosch, J. A., Gordo, P., Magrina, N., … Tornos, J. (1979). Essential mixed IgG-IgM cryoimmunoglobulinemia. Clinical and anatomopathological remission with cyclophosphamide and prednisone treatment. *Rev Clin Esp, 155*(2), 149–151.

Garofallo, S. B., Portal, V. L., Markoski, M. M., Dias, L. D., de Quadrosa, A. S., & Marcadenti, A. (2019). Correlations between traditional and nontraditional indicators of adiposity, inflammation, and monocyte subtypes in patients with stable coronary artery disease. *J Obes, 2019*, 3139278.

Hamelin Morrissette, J., Tremblay, D., Marcotte-Chenard, A., Lizotte, F., Brunet, M. A., Laurent, B., … Geraldes, P. (2022). Transcriptomic modulation in response to high-intensity interval training in monocytes of older women with type 2 diabetes. *Eur J Appl Physiol, 122*(4), 1085–1095.

Hamze, M., Desmetz, C., Berthe, M. L., Roger, P., Boulle, N., Brancherau, P., … Guglielmi, P. (2013). Characterization of resident B cells of vascular walls in human atherosclerotic patients. *J Immunol, 191*(6), 3006–3016.

Harbaum, L., Renk, E., Yousef, S., Glatzel, A., Luneburg, N., Hennigs, J. K., … Klose, H. (2016). Acute effects of exercise on the inflammatory state in patients with idiopathic pulmonary arterial hypertension. *BMC Pulm Med, 16*(1), 145.

Idzkowska, E., Eljaszewicz, A., Miklasz, P., Musial, W. J., Tycinska, A. M., & Moniuszko, M. (2015). The role of different monocyte subsets in the pathogenesis of atherosclerosis and acute coronary syndromes. *Scand J Immunol, 82*(3), 163–173.

Ishikawa, Y., Gohda, T., Tanimoto, M., Omote, K., Furukawa, M., Yamaguchi, S., … Tomino, Y. (2012). Effect of exercise on kidney function, oxidative stress, and inflammation in type 2 diabetic KK-A(y) mice. *Exp Diabetes Res, 2012*, 702948.

Johannsen, N. M., Swift, D. L., Johnson, W. D., Dixit, V. D., Earnest, C. P., Blair, S. N., & Church, T. S. (2012). Effect of different doses of aerobic exercise on total white blood cell (WBC) and WBC subfraction number in postmenopausal women: results from DREW. *PLOS ONE, 7*(2), e31319.

Kadoglou, N. P., Moustardas, P., Kapelouzou, A., Katsimpoulas, M., Giagini, A., Dede, E., … Liapis, C. D. (2013). The anti-inflammatory effects of exercise training promote atherosclerotic plaque stabilization in apolipoprotein E knockout mice with diabetic atherosclerosis. *Eur J Histochem, 57*(1), e3.

Kawanishi, N., Mizokami, T., Yano, H., & Suzuki, K. (2013). Exercise attenuates M1 macrophages and CD8+ T cells in the adipose tissue of obese mice. *Med Sci Sports Exerc, 45*(9), 1684–1693.

Khosravi, N., Hanson, E. D., Farajivafa, V., Evans, W. S., Lee, J. T., Danson, E., … Battaglini, C. L. (2021). Exercise-induced modulation of monocytes in breast cancer survivors. *Brain Behav Immun Health, 14*, 100216.

Kim, J., Kim, A. R., & Shin, E. C. (2015). Cytomegalovirus infection and memory T cell inflation. *Immune Netw, 15*(4), 186–190.

Kruger, K., Pilat, C., Schild, M., Lindner, N., Frech, T., Muders, K., & Mooren, F. C. (2015). Progenitor cell mobilization after exercise is related to systemic levels of G-CSF and muscle damage. *Scand J Med Sci Sports, 25*(3), e283–291.

Lecube, A., Pachon, G., Petriz, J., Hernandez, C., & Simo, R. (2011). Phagocytic activity is impaired in type 2 diabetes mellitus and increases after metabolic improvement. *PLOS ONE, 6*(8), e23366.

Markofski, M. M., Flynn, M. G., Carrillo, A. E., Armstrong, C. L., Campbell, W. W., & Sedlock, D. A. (2014). Resistance exercise training-induced decrease in circulating inflammatory CD14+CD16+ monocyte percentage without weight loss in older adults. *Eur J Appl Physiol, 114*(8), 1737–1748.

Matrougui, K., Abd Elmageed, Z., Kassan, M., Choi, S., Nair, D., Gonzalez-Villalobos, R. A., … Partyka, M. (2011). Natural regulatory T cells control coronary arteriolar endothelial dysfunction in hypertensive mice. *Am J Pathol, 178*(1), 434–441.

Mazur, M., Glodzik, J., Szczepaniak, P., Nosalski, R., Siedlinski, M., Skiba, D., … Mikolajczyk, T. P. (2018). Effects of controlled physical activity on immune cell phenotype in peripheral blood in prehypertension - studies in preclinical model and randomised crossover study. *J Physiol Pharmacol, 69*(6), 875–887.

Michishita, R., Shono, N., Inoue, T., Tsuruta, T., & Node, K. (2008). Associations of monocytes, neutrophil count, and C-reactive protein with maximal oxygen uptake in overweight women. *J Cardiol, 52*(3), 247–253.

Mura, M., Weiss-Gayet, M., Della-Schiava, N., Chirico, E., Lermusiaux, P., Chambion-Diaz, M., … Pialoux, V. (2022). Monocyte phenotypes and physical activity in patients with carotid atherosclerosis. *Antioxidants (Basel), 11*(8).

Mury, P., Mura, M., Della-Schiava, N., Chanon, S., Vieille-Marchiset, A., Nicaise, V., … Pialoux, V. (2020). Association between physical activity and sedentary behaviour on carotid atherosclerotic plaques: An epidemiological and histological study in 90 asymptomatic patients. *Br J Sports Med, 54*(8), 469–474.

Musette, P., & Bouaziz, J. D. (2018). B cell modulation strategies in autoimmune diseases: New concepts. *Front Immunol, 9*, 622.

Nagareddy, P. R., Murphy, A. J., Stirzaker, R. A., Hu, Y., Yu, S., Miller, R. G., … Goldberg, I. J. (2013). Hyperglycemia promotes myelopoiesis and impairs the resolution of atherosclerosis. *Cell Metab, 17*(5), 695–708.

Nahrendorf, M. (2018). Myeloid cell contributions to cardiovascular health and disease. *Nat Med, 24*(6), 711–720.

Nava, M., Quiroz, Y., Vaziri, N., & Rodriguez-Iturbe, B. (2003). Melatonin reduces renal interstitial inflammation and improves hypertension in spontaneously hypertensive rats. *Am J Physiol Renal Physiol, 284*(3), F447–454.

Nishimura, S., Manabe, I., Nagasaki, M., Eto, K., Yamashita, H., Ohsugi, M., … Nagai, R. (2009). CD8+ effector T cells contribute to macrophage recruitment and adipose tissue inflammation in obesity. *Nat Med, 15*(8), 914–920.

Noz, M. P., Hartman, Y. A. W., Hopman, M. T. E., Willems, P., Tack, C. J., Joosten, L. A. B., … Riksen, N. P. (2019). Sixteen-week physical activity intervention in subjects with increased cardiometabolic risk shifts innate immune function towards a less proinflammatory state. *J Am Heart Assoc, 8*(21), e013764.

Peake, J. M., Neubauer, O., Walsh, N. P., & Simpson, R. J. (2017). Recovery of the immune system after exercise. *J Appl Physiol (1985), 122*(5), 1077–1087.

Pellegrin, M., Miguet-Alfonsi, C., Bouzourene, K., Aubert, J. F., Deckert, V., Berthelot, A., … Laurant, P. (2009). Long-term exercise stabilizes atherosclerotic plaque in ApoE knockout mice. *Med Sci Sports Exerc, 41*(12), 2128–2135.

Phillips, A. C., Carroll, D., Gale, C. R., Drayson, M., & Batty, G. D. (2011). Lymphocyte cell counts in middle age are positively associated with subsequent all-cause and cardiovascular mortality. *QJM, 104*(4), 319–324.

Pittet, M. J., Nahrendorf, M., & Swirski, F. K. (2014). The journey from stem cell to macrophage. *Ann N Y Acad Sci, 1319*(1), 1–18.

Rattik, S., Engelbertsen, D., Wigren, M., Ljungcrantz, I., Ostling, G., Persson, M., … Bjorkbacka, H. (2019). Elevated circulating effector memory T cells but similar levels of regulatory T cells in patients with type 2 diabetes mellitus and cardiovascular disease. *Diab Vasc Dis Res, 16*(3), 270–280.

Rentz, T., Wanschel, A., de Carvalho Moi, L., Lorza-Gil, E., de Souza, J. C., Dos Santos, R. R., & Oliveira, H. C. F. (2020). The anti-atherogenic role of exercise is associated with the attenuation of bone marrow-derived macrophage activation and migration in hypercholesterolemic mice. *Front Physiol, 11*, 599379.

Robinson, E., Durrer, C., Simtchouk, S., Jung, M. E., Bourne, J. E., Voth, E., & Little, J. P. (2015). Short-term high-intensity interval and moderate-intensity continuous training reduce leukocyte TLR4 in inactive adults at elevated risk of type 2 diabetes. *J Appl Physiol (1985), 119*(5), 508–516.

Schnell, O., Cappuccio, F., Genovese, S., Standl, E., Valensi, P., & Ceriello, A. (2013). Type 1 diabetes and cardiovascular disease. *Cardiovasc Diabetol, 12*, 156.

Simpson, R. J. (2011). Aging, persistent viral infections, and immunosenescence: Can exercise "make space"? *Exerc Sport Sci Rev, 39*(1), 23–33.

Sinha, A., Rivera, A. S., Doyle, M. F., Sitlani, C., Fohner, A., Huber, S. A., … Psaty, B. M. (2021). Association of immune cell subsets with cardiac mechanics in the multi-ethnic study of atherosclerosis. *JCI Insight, 6*(13), e149193.

Smith, J. K., Dykes, R., Douglas, J. E., Krishnaswamy, G., & Berk, S. (1999). Long-term exercise and atherogenic activity of blood mononuclear cells in persons at risk of developing ischemic heart disease. *JAMA, 281*(18), 1722–1727.

Spielmann, G., McFarlin, B. K., O'Connor, D. P., Smith, P. J., Pircher, H., & Simpson, R. J. (2011). Aerobic fitness is associated with lower proportions of senescent blood T-cells in man. *Brain Behav Immun, 25*(8), 1521–1529.

Thakur, V., Gonzalez, M., Pennington, K., Nargis, S., & Chattopadhyay, M. (2016). Effect of exercise on neurogenic inflammation in spinal cord of Type 1 diabetic rats. *Brain Res, 1642*, 87–94.

Timmerman, K. L., Flynn, M. G., Coen, P. M., Markofski, M. M., & Pence, B. D. (2008). Exercise training-induced lowering of inflammatory (CD14+CD16+) monocytes: A role in the anti-inflammatory influence of exercise? *J Leukoc Biol, 84*(5), 1271–1278.

Trim, W. V., Walhin, J. P., Koumanov, F., Bouloumie, A., Lindsay, M. A., Chen, Y. C., … Thompson, D. (2022). Divergent immunometabolic changes in adipose tissue and skeletal muscle with ageing in healthy humans. *J Physiol, 600*(4), 921–947.

van der Valk, E. S., Mulder, D. S., Kouwenhoven, T., Nagtzaam, N. M. A., van Rossum, E. F. C., Dik, W. A., & Leenen, P. J. M. (2022). Monocyte adaptations in patients with obesity during a 1.5 year lifestyle intervention. *Front Immunol, 13*, 1022361.

Vu, D. M., Tai, A., Tatro, J. B., Karas, R. H., Huber, B. T., & Beasley, D. (2014). Gammadeltat cells are prevalent in the proximal aorta and drive nascent atherosclerotic lesion progression and neutrophilia in hypercholesterolemic mice. *PLOS ONE, 9*(10), e109416.

Wadley, A., Roberts, M., Creighton, J., Thackray, A., Stensel, D., & Bishop, N. (2021). Higher levels of physical activity are associated with reduced tethering and migration of pro-inflammatory monocytes in males with central obesity. *Exerc Immunol Rev, 27*, 54–66.

Wang, M., Chen, F., Wang, J., Zeng, Z., Yang, Q., & Shao, S. (2018). Th17 and Treg lymphocytes in obesity and Type 2 diabetic patients. *Clin Immunol, 197*, 77–85.

Wenning, P., Kreutz, T., Schmidt, A., Opitz, D., Graf, C., Voss, S., … Brixius, K. (2013). Endurance exercise alters cellular immune status and resistin concentrations in men suffering from non-insulin-dependent type 2 diabetes. *Exp Clin Endocrinol Diabetes, 121*(8), 475–482.

Wonner, R., Wallner, S., Orso, E., & Schmitz, G. (2018). Effects of acute exercise on monocyte subpopulations in metabolic syndrome patients. *Cytometry B Clin Cytom, 94*(4), 596–605.

Woods, J. A., Ceddia, M. A., Wolters, B. W., Evans, J. K., Lu, Q., & McAuley, E. (1999). Effects of 6 months of moderate aerobic exercise training on immune function in the elderly. *Mech Ageing Dev, 109*(1), 1–19.

Yeh, S. H., Chuang, H., Lin, L. W., Hsiao, C. Y., Wang, P. W., & Yang, K. D. (2007). Tai chi chuan exercise decreases A1C levels along with increase of regulatory T-cells and decrease of cytotoxic T-cell population in type 2 diabetic patients. *Diabetes Care, 30*(3), 716–718.

Yeh, S. H., Chuang, H., Lin, L. W., Hsiao, C. Y., Wang, P. W., Liu, R. T., & Yang, K. D. (2009). Regular Tai Chi Chuan exercise improves T cell helper function of patients with type 2 diabetes mellitus with an increase in T-bet transcription factor and IL-12 production. *Br J Sports Med, 43*(11), 845–850.

Zhang, X., Zheng, Y., Geng, C., Guan, J., Wang, L., Zhang, X., … Lu, X. (2021). Isometric exercise promotes arteriogenesis in rats after myocardial infarction. *J Biomed Res, 35*(6), 436–447.

Zhao, Q. M., Mettauer, B., Epailly, E., Falkenrodt, A., Lampert, E., Charloux, A., … Lonsdorfer, J. (1998). Effect of exercise training on leukocyte subpopulations and clinical course in cardiac transplant patients. *Transplant Proc, 30*(1), 172–175.

13 Environmental exercise immunology

David B. Pyne, William Trim, and Samuel J. Oliver

Introduction

Individuals participating in sports, exercise, and physical activity, and people working in a variety of occupational settings, often contend with the demands of adverse environmental conditions. These demands may exceed the homeostatic limits of the immune system, resulting in impaired immune function and increased risk of illness and/or infection. The effects of environmental stressors on the immune system depend on several factors including: each individual's immune system, the degree of exposure to environmental stress, exposure to pathogens, physical fitness, age, psychological considerations, neuro-endocrine regulation, and the presence of any underlying medical condition(s).

In sports, exercise, and physical activity, there are a wide range of individuals in terms of age and level of performance, and school-age children competing in junior sports. Popular recreational pursuits, including walking, hiking, and mountaineering can also expose individuals to challenging conditions. Occupational settings that expose individuals to adverse environmental conditions when performing physically active tasks include the military, spaceflight, first responders such as firefighters, police and paramedics, mining, construction, agriculture, and other industry workers. Both exercisers and workers are keen to maintain good health and avoid the negative effects of environmental stress on their activities and performance.

The modern history of research in exercise, extreme environments, and immune function essentially covers the period from the late 1990s through to the present day. Two leading proponents of this research were, initially Roy Shephard in Canada (Shephard, 1998) and subsequently Neil Walsh in the UK. Their work covered athletes, recreational exercisers, and particularly military personnel, and investigated changes in cellular and soluble immune measures under several environmental conditions. Other notable research includes the proposed physiological models of heat stress and immune function by Fabian Lim (Singapore) and Laurel Mackinnon (Australia), Lisa Leon and colleagues (USA) linking changes in muscle, immune measures, and the onset of heat stress pathophysiology, Brian Crucian and colleagues (USA) investigating immune changes and clinical conditions during long-haul space flight, Dominique Gagnon (Finland) and colleagues on exercise, immunity, and cold stress, and Jong-Shyan Wang and colleagues (Taiwan) evaluating the effects of hypoxia on immune cell distribution and function.

The immune system provides host defence against a wide variety of pathogens and pollutants and is centrally involved in the processes of tissue repair and inflammatory control. In healthy individuals, the immune system is robust, regulates within homeostatic limits, and has a high level of counter-redundancy. When the immune system is disturbed, dysregulation of function beyond homeostatic limits occurs, and this dysregulation is implicated in many illnesses, diseases, pathophysiologies, and autoimmune disorders. Models predicting how the immune

DOI: 10.4324/9781003256991-13

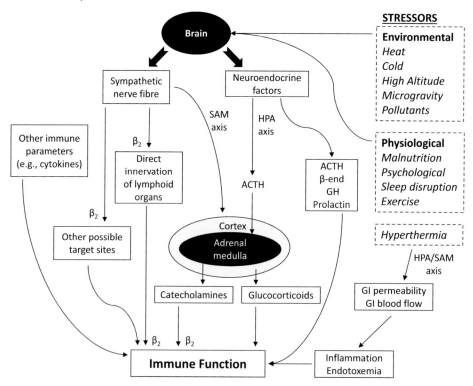

Figure 13.1 Overview of the potential modulators of immune function under stress. Environmetnal stressors such as heat, cold, high altitude, or microgravity may indirectly influence immune function through the initiation of a stress-hormone response involving the hypothalamic-pituitary-adrenal axis and sympatheticoadrenal-medullary axis. Hyperthermia may also have a direct effect on immune function via impacts on the GI tract. ACTH = adrenocorticotropic hormone, GH = growth hormone, GI = gastrointestinal, HPA = hypothalamic-pituitary-adrenal, SAM = sympathetic adrenal medullary.

system responds to exercise in conjunction with environmental stressors have evolved from a neuro-endocrine focused foundation (Figure 13.1) to include other factors, including crosstalk from a range of other body systems. Understanding exercise and immunoregulation has continued to evolve from 1990s neuro-endocrine models (Pedersen & Hoffman-Goetz, 2000) to the involvement of exercise-induced myokines from skeletal muscle and communication between the muscle, liver, and the gut-brain axis (Bay & Pedersen, 2020).

Assessing the impact of exercise alongside other stressors on immune function requires consideration of numerous factors. Research on environmental exercise immunology is a priority to ensure real-world issues are addressed, and researchers obtain appropriate recognition of their study outcomes. Investigators should address the real-world importance of research in the context of tight physiological regulation and homeostasis (Albers et al., 2013). The notion of individual responders to exercise and environmental demands is also an important consideration. Individual risk factors influence how any given individual responds to exercise and physical activity including excessive exercise or training loads, lack of variation in physical load, inadequate unloading and/or of recovery, seasonal effects, and long-haul international travel (Derman et al., 2022).

This chapter identifies the key effects of environmental stressors and exercise on immune function and evaluates the strengths and weaknesses of laboratory, field, and clinical research. Practical recommendations are presented for individuals, practitioners, and researchers, and the chapter considers future directions in real-world and research settings. The environmental conditions discussed include: ambient temperature (both hot and cold conditions), altitude hypoxia, air pollution, and space flight.

Heat stress

The challenges of training and competing in hot environments are well known to most athletes and for those in occupational settings with similar demands. The effects of the season from the extremes of summer and winter weather often dictate the scheduling of sporting practices and competitions, and the timing of events and activities over a given day or week. There is also increasing interest in the effects of climate change on the preparation and conduct of sporting, social and occupational pursuits. The issues for all these individuals centre on limiting impairments to exercise performance, work tasks, and the risk of heat illness.

The principles of body temperature regulation have been well described and form the basis of managing individuals and teams (groups) exercising in heat and humidity. Exercise in the heat is associated with increased core and skin temperatures, higher heart rate, increased concentrations of hormones, and alterations in substrate utilisation. The immune disturbances associated with heat illness and more severe heat stroke include the suppression of immune cells and their functions, suppression of cell-mediated immunity, translocation of lipopolysaccharide (LPS), suppression of anti-LPS antibodies, increased macrophage activity due to muscle tissue damage, and increased concentration of circulating inflammatory and pyrogenic cytokines (Lim & Mackinnon, 2006). Observational data derived from field-based studies show that immune disturbance is implicated in the aetiology of heat stroke, particularly where core temperatures typically exceed 40°C (Walsh & Whitham, 2006). However, simple explanations of a linear relationship between increases in core temperature and alteration in immune function are complicated by data showing that in some settings, immune disturbance is evident in modest environmental conditions and core temperatures of <40°C. Collectively, these data point to a combination of thermoregulatory and immunological factors that influence the maintenance or otherwise of immunological control and resistance to infection.

Thermoregulatory and inflammatory models of exercise-induced heat illness

Over the last 20 years, there has been extensive interest in the mechanisms of heat illness, primarily settling on two pathways: the classic thermoregulatory model where exercise capacity is impaired with an increase in thermal strain (as a function of environmental conditions and exercise demand) and hyperthermia-induced alterations in cardiovascular, nervous system, and skeletal muscle function. The second pathway involves an inflammatory model of heat illness. This latter model involves exercise-induced increases in immune mediators, environmental influences, and metabolic responses to fever of either inflammatory or non-inflammatory origin that culminate in heat illness. Field studies of thermoregulatory changes in parallel with immune measures in response to exercise are challenging given technical constraints.

Models of immune dysregulation and exertional heat illness following exercise have been proposed and evaluated through the 1990s and 2000s, yet despite two further decades of research to understand these processes, the scientific discussions continue. The involvement of immune function disturbance as a contributor to exercise-induced heat complications during or after exercise was first described at the end of the last century (Shephard & Shek, 1999). These investigators highlighted the likely involvement of immune disturbance to both heat exposure and heavy exercise, and an increased risk of heat illness and possibly heat fatalities. After this work, a dual pathway mode of heat stroke was proposed that centred on the proposition that the underlying pathology is triggered by hyperthermia but driven by endotoxemia (Lim & Mackinnon, 2006). Endotoxins are released when bacteria die and remnants cross the gastrointestinal barrier into the bloodstream. At the same time, another model emerged that also challenged the orthodoxy that exercise in the heat poses a greater threat to immune function than thermoneutral conditions (Walsh & Whitham, 2006). In one study, innate immune responses were compared for three representative activities of a marathon, construction work, and walking at 4 km/h in hot and humid conditions (Presbitero et al., 2021). Exposure to moderate heat stress increased core temperatures to 37°C or 38°C, which activated the innate immune system as an adaptive (protective) response in the initial hours after exercise. However, further exercise impaired aspects of the innate immune system compromising subsequent inflammatory control for several hours. Dose response issues of duration of exposure e.g., minutes (walking) to several hours per day (construction work) would likely influence the underlying immune response to heat and other environmental factors.

Heat acclimation training – immune implications

Heat acclimation (artificial heat) and acclimatisation (natural heat *in situ*) training is commonly undertaken to prepare active individuals for exercise in the heat and to limit the risk of heat illnesses. A variety of different methods of heat training have been developed, including self-paced exercise, constant work rate exercise, passive heating, post-exercise heating, controlled hyperthermia, and controlled heart rate. However, the differential effects of these methods of limiting heat illness and/or promoting improved immune tolerance to heat, are not well described. The issue of how this specific form of training raises the issue of complementary and/or shared pathways of thermoregulation and inflammatory control. Molecular regulation is likely involved as gene expression studies show a close temporal alignment between the elevation of interleukin (IL)-1 and the time course of increases in core body temperature (Helwig & Leon, 2011).

There is emerging evidence of the active participation of skeletal muscles in the pathogenesis of exertional heat stroke. Heat stroke has been associated with rhabdomyolysis, which reflects skeletal muscle damage and subsequent synthesis and secretion of intramuscular bioactive molecules including cytokines, chemokines, and acute phase proteins into the circulation (Laitano et al., 2021). The release of these molecules can shift the overall inflammatory status from anti- to pro-inflammatory, affecting other organ systems. The activation of innate immunity can determine whether a victim of exertional heat stroke is ready to return to physical activity or experiences a prolonged convalescence.

Given the possibility of heat-induced disturbance to immune regulation, it is prudent for athletes, coaches, and support staff to evaluate options for the inclusion of heat alleviation strategies in both acute and chronic forms (Gibson et al., 2020). This approach would likely apply to individuals in other occupational settings facing the dual demands of physical activity and heat/humidity. Repeated exposure to exercise and heat facilitates partial thermoregulatory adaptation without attenuating resting immune functions (Keaney et al., 2021). Once or twice-a-day heat prescribed in a non-consecutive training sequence over 12 days can promote heat and exercise tolerance without impairing immune function (Willmott et al., 2018). This heat training programme elicited a modest acute post-exercise increase in the cytokines IL-6, TNF-α, and cortisol concentrations, but there were no cumulative increases observed over the entire heat acclimation training period. Athletes and active individuals can be advised they can undertake prescribed heat acclimation training in confidence without necessarily increasing the underlying risk of infection-related illness. A recent study indicated that a 10-day heat acclimation programme had no negative effects on mucosal immunity, as indicated by measurements of salivary IgA (Alkemade et al., 2022).

Cold stress

Sedentary and active people are exposed to a range of cold stresses. In people with sufficient clothing, cold stress will be generally localised to the face and upper airway. In contrast, in lightly dressed individuals, such as athletes, peripheral cold exposure causes vasoconstriction and shivering thermogenesis that are initiated by autonomic and endocrine systems, which are also potent modulators of the immune system (Gagnon et al., 2014). A popular belief is that catching a chill can lead to infection. Indeed, cold weather has long been linked to respiratory infection with winter epidemics documented by Hippocrates around 400 BC (Moriyama et al., 2020). Cold exposure has also been hypothesised as a potential cause for the increased incidence of respiratory symptoms commonly reported by athletes. One potential reason for these increases in infection is an alteration in immune function. Contrastingly, regular cold-water immersion, showering, and swimming, have been practised in some countries for centuries, and are gaining popularity, because of the perceived mental and physical health benefits, including improved immune health (Massey et al., 2022). This section will carefully unpick the evidence to determine whether cold stress is a threat, or indeed beneficial, to the immune system of sedentary and active people.

Cold weather and infection

Epidemiological research in the general population reveals low ambient temperature is associated with a greater incidence and virulence of some respiratory infections, including influenza and SARS-CoV-2 (Nottmeyer & Sera, 2021). Similar observations have also been made in active populations with the strongest evidence to date from research showing three days of cold weather preceded the onset of a respiratory infection in Finnish soldiers (Mäkinen et al., 2009). Interpretation of these findings requires caution as the greater infection incidence identified with cold weather may also be explained by coinciding low humidity and low solar radiation, including vitamin D-dependent and independent effects (Harrison et al., 2021; Tateo et al., 2022). Another explanation for the increased infection during cold weather is that people spend more time together indoors which consequently increases their exposure to pathogens. To understand the independent effect of cold on the immune system of sedentary and active people, it is necessary to examine the evidence from research that compares immune responses between cold intervention and control groups.

Passive cold stress and immune function

Few studies have examined cold stress *per se* on the immune system and careful consideration of the interactive effect of cold stress type (air *vs* water) and physical activity level (sedentary *vs* active) is needed when interpreting these studies. In sedentary healthy individuals, moderate cold stress that causes hypothermia (core body temperature $\leq 36°C$) has been shown to reduce innate cellular and adaptive soluble immune function, i.e., neutrophil degranulation and saliva

IgA (sIgA) secretion, respectively (Costa et al., 2010). It is also well-established that hypothermia during and after surgical operations increases wound infection and slows wound healing (Rauch et al., 2021). Nevertheless, cold-water immersion, causing a 1°C reduction in core temperature, the day after inoculation with the common cold *rhinovirus* did not lead to a greater incidence of respiratory infection in healthy men compared with those men that were inoculated but not exposed to the cold (Douglas et al., 1968). Although rare in active people exercising in cold air, hypothermia has been observed in those exercising in water, e.g., long-distance swimmers, and therefore exercising in water compared with air may present a greater risk to the immune system.

Immune function responses to exercise in the cold

The intensity of physical activity commonly completed by active populations produces sufficient heat to maintain or increase core temperature and thereby limits direct cold stress to the upper airway and periphery even in sub-freezing cold-air temperatures (Gagnon et al., 2014). As exercise greatly exaggerates upper-airway cooling, due to the significant increase in ventilation, it is important to consider the local effects of cooling on upper-airway immune function in active populations. Indeed, the increased upper-airway infection, previously reported in those training in cold weather (Mäkinen et al., 2009), may be explained by local upper-airway tissue cooling and peripheral vasoconstriction that reduces blood flow and oxygen tension, impairs immune cell trafficking and function, and has been shown to aid viral replication (Foxman et al., 2015). Breathing cold air reduces the temperature of upper-airway mucous membranes and saliva where plasma cells that produce antibodies like sIgA reside. Tomasi and colleagues pioneering exercise immunology research 40 years ago demonstrated that elite cross-country skiers had lower sIgA secretion immediately after competition in freezing conditions (Tomasi et al., 1982). However, later carefully controlled laboratory research demonstrated a similar decline in sIgA after prolonged vigorous cycling exercise (2 hours at 70% VO_2 max) in sub-freezing and thermoneutral conditions (Walsh et al., 2002). Therefore, the decline in mucosal immunity observed in these studies was more likely a consequence of prolonged vigorous exercise rather than cold temperature. This highlights the importance of including a work-matched exercising control group to isolate the additional effect of cold on exercising immune responses.

In contrast to prolonged endurance exercise, completing shorter-duration vigorous endurance exercise in cold air may be beneficial for mucosal immunity. A decline in sIgA secretion rate observed in male athletes after 1 hour of cycling exercise at 21°C was prevented by completing the same exercise at 5°C (Akimoto et al., 2009), and sIgA secretion rate was higher after running for 30 minutes in 1°C than after running in 24°C in female athletes (Mylona et al., 2002). Indeed, compared with exercising in thermoneutral/temperate conditions, completing exercise in the cold reduces thermoregulatory and cardiovascular strain, and blunts the rise in the catecholamine norepinephrine, which is an important effector in the immune system (Figure 13.1) (Gagnon et al., 2014). This reduced catecholamine response may also explain the blunted leucocytosis and reduced systemic inflammation, as indicated by pro-inflammatory cytokines (IL-17, IP-10, MIP-1β, MCP-1), which has been observed after moderate exercise in the cold compared with temperate conditions (Gagnon et al., 2014).

A further potentially beneficial effect of cold exposure on clinically important outcomes of immune health is highlighted in one study of more than 3,000 people that demonstrated those people finishing their usual warm shower with 30–90 seconds of cold water reported less sickness absence over 90 days than those who maintained their usual warm showers (Buijze et al., 2016). Nevertheless, the only study to examine the effect of regular cold exposure in an active population, and include a control group, demonstrated that over 13 winter weeks, respiratory infection incidence, duration, and severity were not different between indoor pool and outdoor cold-water swimmers who swam in ≤18°C cold water without a wetsuit at least twice per week (Collier et al., 2021). A particular strength of this research was that it showed the infection was not different between cold-water and indoor swimmers, and their co-habiting non-swimming partners (Collier et al., 2021). As cold water induces significantly greater thermal stress than cold air, it is reasonable to speculate that regular cold-air exposure may not increase respiratory infection risk in healthy active populations. Future research using adequate exercise and co-habiting control groups is sorely needed to confirm this idea.

In summary, the effect of cold stress on immune health is dependent on the severity of cold exposure, and the activity level of the exposed individual. Severe cold exposure leading to hypothermia impairs innate cellular and adaptive soluble aspects of immune function, increases surgical wound infection risk in patients, and may increase upper respiratory tract infection (URTI) risk in healthy persons. However, mild cold exposure appears on balance to pose a limited additional risk, or may even be beneficial, for immune health in active otherwise healthy people. Future research using adequate control groups is required to further explore whether intermittent cold exposure alone or coupled with exercise may improve immune health and reduce the burden of respiratory infection.

Altitude stress

Approximately 500 million people live at an altitude of ≥1,500 m. People are also periodically exposed to the effects of altitude when travelling to high mountains for work and leisure, e.g., athletes, soldiers, miners, skiers, trekkers, and mountaineers (Derman et al., 2022; Oliver et al., 2012). The cardinal stressor faced by those at altitude is hypobaric hypoxia, although cold, low humidity, and increased solar radiation are also commonly experienced. The reduction in barometric air pressure at altitude lowers the partial pressure of oxygen in the air, and at each step in the oxygen cascade, alveoli–artery–tissue–cell, causing hypoxia in all physiological systems and impairing exercise capacity (Mazzeo, 2008). Hypoxia may also affect the immune system as the compensatory physiological responses to improve oxygen delivery and utilisation in a hypoxic environment, including increases in ventilation, cardiac output, systemic vasodilation, and carbohydrate oxidation, are orchestrated by autonomic and endocrine actions which are also potent effectors of the immune system (Mazzeo, 2008). It is increasingly important to understand the combined effects of hypoxia and exercise on immune health as hypoxic training methods, which have been adopted by athletes for more than 50 years, are also now being explored for their potential therapeutic effects in sedentary and clinical populations. In this section, we examine the evidence to determine whether the additional stress posed by altitude and hypoxia alter immune function and the risk of respiratory infection.

Altitude stress and respiratory infection

Respiratory infection symptoms are commonly reported to be increased in athletes travelling to altitude-training camps (Derman et al., 2022), and trekkers and mountaineers at high altitude (Oliver et al., 2012). In contrast, recent research indicates that modest altitude is associated with a similar or lower incidence of SARS-CoV-2 infection rates than observed at low altitude (Stephens et al., 2021). Possible explanations for lower infection at altitude include less exposure to the potentially immunosuppressive effects of air pollution and allergens and reduced pathogen viability due to greater solar radiation. However, perhaps the most obvious explanation for these reported differences in respiratory infection is that athletes, trekkers, and mountaineers, unlike the general population, are exposed to the combined stresses of altitude and arduous exercise, which in combination may lead to alterations to aspects of immune function and a greater risk of respiratory infection. Confidently interpreting whether URTI symptoms are a genuine infection at altitude is challenging as other plausible causes exist for the symptoms, such as upper-airway irritation from breathing cold-dry air and altitude illness. To better understand the relationship between altitude and respiratory infection, future research should confirm the presence of infection using PCR analysis on nasopharyngeal and throat swabs rather than rely solely on symptom questionnaires (see Chapter 3). Moreover, this highlights the importance of examining other immune measures to understand the effect of altitude on immune health. Indeed, as infection also requires the presence of a pathogen, even a profoundly depressed immune system may go unsuspected clinically. The relationship between altitude and immune function is complex and likely depends on the severity of hypoxic exposure and people's physical activity level.

Immune function responses to exercise in moderate and severe hypoxic conditions

Well-controlled laboratory studies show similar effects on markers of innate and adaptive immunity in blood and mucosal tissue when a single bout of exercise is completed at sea level and in moderate hypoxic conditions (>15% fraction of inspired oxygen, equivalent <2,700 m) (Svendsen et al., 2016; Wang & Lin, 2010). The limited research examining exercise training in moderate hypoxic conditions suggests exercise posed no additional effect on immune function in athletes during a three-week live-high train-high altitude-training camp at 2,100 m (Pyne et al., 2000). In contrast to moderate hypoxic conditions, immune function responses are altered compared with sea level when exercise is completed in severe hypoxic conditions (≤15% fraction of inspired oxygen, equivalent ≥2,700 m), with the evidence to date suggesting the direction of the immune response is dependent on the immune function marker assessed and the exercise and hypoxic duration.

In a comprehensive series of studies in sedentary men, Wang and colleagues showed that compared with a single bout of exercise for 30 minutes at sea level, the same exercise completed in severe hypoxia (12% fraction of inspired oxygen, equivalent ~4,500 m) led to increased neutrophil function [i.e., phagocytic activity and oxidative burst (Wang & Chiu, 2009)], increased NK cell number and function [i.e., cytotoxic protein concentrations (Wang & Wu, 2009)], but reduced lymphocyte function [i.e., decreased proliferation and increased apoptosis (Wang & Lin, 2010)]. Wang and colleagues then revealed that repeating moderate exercise for 30 minutes in severe hypoxia five days per week for four weeks, as an intermittent hypoxic exercise training programme, led to several beneficial adaptations in innate and adaptive immune function compared with the exercise training at sea level. These adaptations included increased neutrophil phagocytic activity (Chen et al., 2018), increased NK cell number and activity (Wang & Weng, 2011), decreased lymphocyte apoptosis, and increased lymphocyte activation, and circulating levels of IFN-γ, which collectively suggest enhanced T cell mediated antiviral immunity (Wang et al., 2011). In addition, the beneficial effects of intermittent hypoxic exercise training on innate and adaptive mucosal immunity have also been reported in athletes (Born et al., 2016).

Studies exploring the effects of more prolonged severe hypoxic conditions on immune function in active populations have typically been conducted at altitude and indicate unaffected or enhanced innate (i.e., NK cell activity, neutrophil function), humoral and soluble adaptive immunity (Morabito et al., 2016) but impaired T cell-mediated immune function [i.e., decreased *in vitro* T cell activation and proliferation, including a reduction in the pro-inflammatory cytokine IFN-γ (Ermolao et al., 2009)]. Altitude exposure has also been shown to impair *in vivo* immune function

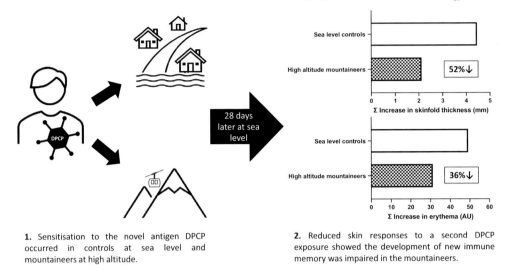

1. Sensitisation to the novel antigen DPCP occurred in controls at sea level and mountaineers at high altitude.

2. Reduced skin responses to a second DPCP exposure showed the development of new immune memory was impaired in the mountaineers.

Figure 13.2 In vivo immune function reduced by high altitude exposure. DPCP = diphenylcyclopropenone.

[Figure 13.2 (Oliver et al., 2013)]. The importance of using *in vivo* immune measures has been highlighted because they involve an integrated, multi-cellular response that is clinically relevant (Albers et al., 2013). Oliver and colleagues assessed *in vivo* immune function by the application (sensitisation) of a novel antigen to the lower back of rested healthy adult controls at sea level and rested mountaineers 28 hours after passive ascent by a cable car to 3,777 m. Exactly 28 days after the sensitisation, the strength of the immune memory was determined by re-exposing the sea-level controls and mountaineers to a low-dose series of the same antigen and measuring the skinfold thickness and erythema responses. In comparison with the sea-level controls, the mountaineers' skinfold thickness and erythema responses were reduced indicating impaired development of new immune memory. Blood oxygenation saturation at the time of the initial sensitisation was closely related to the strength of the immune memory 28 days later. That is, mountaineers with the lowest blood oxygenation at sensitisation had the poorest *in vivo* immune function. This outcome suggests that systemic hypoxia caused by hypobaria at altitude is a key determinant of the impaired cell-mediated immunity observed in active populations under hypoxic conditions.

T cells have a high density of adrenergic receptors and so the increased noradrenaline observed at altitude is one possible mechanism for the observed reduction in T-cell function in active populations at altitude (Ermolao et al., 2009). Although cortisol and noradrenaline are typically shown to increase at altitude; only noradrenaline was related to the impairment of *in vitro* and *in vivo* immune function. As T-cell functions are involved in host defences against bacterial, viral, and fungal infections, these impairments, at least in part, may be responsible for the increased respiratory infection observed at altitude. An additional explanation for the infection at altitude is evidenced by data showing a single 60-minute bout of endurance exercise in severe hypoxia caused increased gut barrier permeability (Hill et al., 2020).

In summary, exercising in hypoxia has limited effects on innate, humoral, and soluble adaptive immunity, whereas T cell-mediated immunity is impaired. Encouragingly, eloquent research by Wang and colleagues highlights how carefully controlled repeated 30-minute doses of hypoxia coupled with exercise may be beneficial for immune health. Future research should determine if intermittent hypoxic exercise training can beneficially pre-condition the immune system before travel to altitude.

Spaceflight

Spaceflight represents a uniquely stressful environment for humans (Crucian et al., 2018). Not only do astronauts experience Earth-based physical and psychological stressors, they also experience novel stressors including cosmic radiation and microgravity. The immunological impact of spaceflight was first realised during the Apollo 7 mission, where all crew members contracted head colds, and the Apollo 13 mission where one astronaut developed a severe bacterial-induced fever mid-mission, dramatised in the 1995 film 'Apollo 13'. Following these incidents, NASA recognised the hazards of spaceflight and implemented pre-flight quarantine and research programmes. Thus dawned the age of space immunology research.

Investigations employing a variety of *in vitro* and *in vivo* approaches into the effect of spaceflight on the immune system over the last half-century, have revealed substantial immunological perturbations in this environment. Both arms of the immune system are influenced by the stress of spaceflight, with profound implications for astronaut health and in response to adverse medical events. However, logistical restraints (e.g., obtaining samples mid-flight around mission objectives) and confounding factors (i.e., post-landing stress influencing samples obtained upon return) make this area of research incredibly complex. Factors including latent viral reactivation, radiation, and microgravity exposure have also been implicated in the spaceflight-induced immune dysfunction.

Neutrophils are increased in peripheral blood following short- (<2 weeks) and long-duration (3–12 months or longer) spaceflight (Crucian et al., 2008), though these findings were likely confounded by the impact of stress hormone-induced neutrophilia during landing. Nonetheless, neutrophil chemotaxis, adhesion to endothelial cells, and phagocytic capacity are impaired *in vitro* following only nine days of spaceflight (Kaur et al., 2004). Likewise, monocytes exhibit impaired phagocytic and oxidative burst capacities in astronauts (Rykova et al., 2008), whereas their numbers have not shown consistent responses between studies. Similar to monocytes and neutrophils, NK cells display reduced killing capacity and cytotoxicity during short- and long-duration spaceflight, and their numbers are increased during flight, but decline upon landing (Bigley et al., 2018; Crucian et al., 2008).

The adaptive immune system is also impacted by spaceflight with most research concentrating on T cells. Pre-flight CD8+ T cells are increased in peripheral blood, which continues throughout the mission, and is evident across all lymphocytes during short-, but not long-duration flights (Crucian et al., 2008; Rykova et al., 2008). Strikingly, T-cell activation to mitogen stimulation is dramatically reduced by spaceflight (Cogoli et al., 1984), resulting in impaired cytokine production capacity (Crucian et al., 2015). It has been suggested that T-cell activation impairments during spaceflight are focused not on the primary (mitogen binding/patching/capping) or tertiary (IL-1 production/cell-cell contacts) signals, but at the secondary signal (IL-2/IL-2R signalling) and associated downstream signalling [reviewed in Guéguinou et al. (2009)]. Conversely, B cells are minimally impacted by spaceflight (30–90 days) (Spielmann et al., 2019), suggesting antibody—e.g., induced by vaccination—could be maintained and may offer therapeutic potential against latent viral reactivation.

Latent viral reactivation during spaceflight

Factors including radiation, take-off, and landing stress, and impairments to the immune system collectively contribute to viral reactivation in-flight (Figure 13.3). Over half of all astronauts aboard the International Space Station (ISS) have undergone *herpes simplex virus* reactivation (Crucian et al., 2018). Similarly, *Varicella zoster virus* has been detected in ~60% of astronauts during flight, causing allergy-like symptoms (Crucian et al., 2016). Significant re-activations occur primarily in-flight, with the magnitude and frequency of shedding directly correlated with

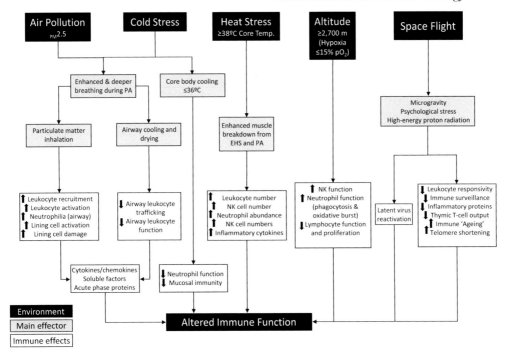

Figure 13.3 Overview of the effects of extreme environmental factors on host immune function during exercise. Air pollutants, thermal stress (both cold and hot), high altitude, and space flight represent some of the most-extreme environmental conditions faced by humans during both recreational (voluntary exercise, sports) and operational (fire fighters, astronauts, soldiers) activities that have differential effects on host immune competency. In brief, larger and deeper breathing can promote both particulate matter inhalation (Pollution) and airway cooling/drying (Cold Stress) that leads to leukocyte recruitment and activation (Pollution) or impaired leukocyte trafficking and function (Cold Stress), ultimately influencing airway inflammation, and increased risk of upper respiratory tract infections. Core body cooling (hypothermia) although rare in active people can occur during exercise in water and may impair circulating and mucosal immune function. Hyperthermia, likewise, can induce myriad immune system alterations, including changes in both the abundance and activity of both innate and adaptive immune cells. Exercise at high altitude is also suggested to potentially influence the capacity of the immune system to defend against invading pathogens/ immunological challenges, which may predispose those exercising at altitude to impaired viral clearance, increased viral replication, and incidence of opportunistic infections. Space flight represents a wholly unique environmental stressor to astronauts—and the public in the near future—that causes immunological ageing, impaired leukocyte responsivity, activity, and the production of new T-lymphocytes, which all promote (and occur in conjunction with) latent viral reactivation, opportunistic infections, and potentially elevated cancer risks in later life. Here, we summarise the main influence these environments can impart on the host immune system with specific reference to during exercise/physical activity, split into layers from the environmental factor (Black Boxes; 'Environment'), the cause of, and physiological response to said environment (Grey Boxes; 'Main effector'), and finally, the immunological responses to these stressors (White Boxes; 'Immune effects'). EHS = exertional heat stress, NK = natural killer, PA = physical activity, pm2.5, = particulate matter <2.5μm, pO2 = partial pressure of oxygen.

the mission duration (Crucian et al., 2016). Indeed, astronauts aboard the ISS for 180 days exhibit latent and lytic Epstein–Barr Virus (EBV) expression resembling patterns seen during infectious mononucleosis (Crucian et al., 2013). These observations suggest diminished immunity as a likely culprit for latent virus reactivation, causing significant adverse health events (e.g., rash, malaise). As CD8+ T cells regulate latency and replication of latent viruses (e.g., EBV), their impairment during spaceflight may facilitate the frequently reported latent viral reactivation in astronauts. It is also common for one or more latent viruses to reactivate simultaneously during flight, with implications for astronaut health and capacity to perform their work, and a potent systemic innate immune response during missions to counter these re-activations.

The result of latent viral reactivation during spaceflight is that viral shedders exhibit higher circulating cytokine concentrations, biased towards a Th_2 shift in T cells. Furthermore, immune surveillance/combative capacities against latent viruses may also be impaired. Thus, spaceflight-associated stressors impair the immune system allowing for latent viral reactivation, which further perturbs astronaut immune dysfunction. With deep space exploration on the horizon, these findings pose a considerable warning for the risk of a wide range of adverse medical events during missions.

Gravity-sensing

The most unique space-associated stressor is the lack of gravity. Our evolution was shaped under Earth's constant gravitational force, so the removal of this stimulus has a profound impact on cellular biology. To assess how the loss of gravity impacts leucocytes, studies have been undertaken in outer orbit, on parabolic flights, and using microgravity simulation machines. Gravity-sensing is important for cell cycle, epigenetic, and chromatin regulation; and motility, apoptosis, cytokine expression, and activation in T cells (Tauber et al., 2015). Early work found T cell activation by Concanavalin A was reduced 97% in microgravity (Cogoli et al., 1984). Even 20 seconds of parabolic flight was sufficient to decrease CD3 and IL-2R surface expression in resting human T cells (Tauber et al., 2015). Similarly, Boonyaratanakornkit et al. (2005) found a ~90% reduction in genes related to T cell activation following mitogen activation in 1G *versus* simulated microgravity (9 vs 99 genes respectively). Specifically, activation-associated signal transduction pathways (JAK/STAT), transcription factors involved in mitogenesis, chemokine induction, and activation markers (*IL2*, *IL2RA*, *SLAMF1*, *IFNG*, *CD69*, TNF-superfamily genes) were selectively impaired in microgravity in stimulated T- and NK cells relative to 1G controls (Boonyaratanakornkit et al., 2005; Chang et al., 2012; Martinez et al., 2015; Spatz et al., 2021). Thymic function is also impacted by spaceflight, with reduced abundance in blood post-flight of T cell receptor excision circle (a marker of naive T cell development) (Benjamin et al., 2016), suggesting impaired T cell development and output with microgravity exposure (Figure 13.3). Collectively, microgravity exposure—both simulated and during actual spaceflight—may elicit system-wide immune activation defects, impaired T cell output, effector/memory T cell generation capabilities, and impaired anti-tumour cytotoxic capabilities in NK cells (Crucian et al., 2018; Mylabathula et al., 2020; Spatz et al., 2021).

Radiation

Radiation profoundly impacts living organisms, driving DNA damage/mutations, most pertinently evidenced in the injuries, sicknesses, and increased mortality of survivors of the atomic bomb attacks on Hiroshima and Nagasaki during World War II (1939–1945). Similarly, the use of radiation to damage tumour cell DNA to limit their growth/replication also highlights its potent effect on our cells. High-energy proton radiation outside of Earth's protective magnetosphere is a considerable health risk to astronauts spending prolonged periods in space [Figure 13.3 (Gueguinou et al., 2009)]. Indeed, astronauts display greater age-specific risks for developing melanoma (Reynolds et al., 2021). Thus, with a three-year mission to Mars expected to deliver a whole body dose of >1 sievert, radiation exposure poses a substantial risk to astronaut health both during—by inducing EBV lytic gene transcription and reactivation (Mehta et al., 2018)—and post-mission (by increasing mortality risk). Several investigations into the impact of radiation on cells in orbit have been conducted, highlighting increased DNA damage, mutations, chromosomal aberrations, and aberrant differentiation (Ohnishi et al., 2009). It remains to be determined how radiation impacts overall immunity in humans during spaceflight, but radiation exposure likely plays an important role.

In summary, spaceflight represents a novel, uniquely stressful environment that causes considerable disturbance to the immune system. It is, therefore, paramount to identify the mechanistic basis for these immune perturbations so appropriate countermeasures can be developed to ensure astronaut safety during exploration-class missions (e.g., Mars/asteroids).

Air pollution

Air pollution, from burning fossil fuels, burdens both industrial and rural areas alike, with its detrimental effects to human health being first documented by Hippocrates in ca. 400 BC. Air pollution comes in several forms, including ozone, carbon monoxide, and nitrogen dioxide, which are associated with reduced life expectancy, cardiopulmonary disease, cognitive decline, and reduced exercise capacity (Bernstein et al., 2004). Even acute exposure in healthy individuals reduces airway function compared with less-polluted regions, which is worsened by exercise due to the intake of greater volumes of polluted air via the mouth rather than the nose, resulting in reduced pollutant filtration before entering the respiratory tract [Figure 13.3 (Carlisle & Sharp, 2001)]. Consequently, pollution increases lung inflammation/irritation and impairs half-marathon performance (Hodgson et al., 2022). Pollutant exposure may also outweigh some of the beneficial cardiopulmonary effects of exercise and promote upper respiratory tract complications. For example, one study found residing near busy roads was associated with increased hospitalisation from asthma and increased prevalence/severity of wheezing and allergenic rhinitis (Diaz-Sanchez et al., 2003). Controlled ozone exposure experiments on healthy humans have directly attributed pollutants to chest discomfort and airway hyperresponsiveness (Koren et al., 1989). Similarly, in the context of exercise, URTI symptoms (e.g., coughing/sore throat) are increased in polluted air (Joad et al., 2007). As URTIs are a common occurrence in athletes that negatively impact performance and health, factors increasing URTI prevalence (i.e., pollutants) require consideration. The COVID-19 pandemic has certainly highlighted the involvement of air-borne pathogens and the potential effects of poor indoor ventilation.

The immune system plays a fundamental role in the pathogenesis of URTIs, and the body's response to pollutants. The first line of defence against invading pollutants are solid barriers including mucous and ciliated membranes in the nasal cavity and upper respiratory tract that contain neutralising agents to further counter foreign bodies. Secondly, leucocytes throughout the entire respiratory tract produce mediators against, or actively phagocytose foreign bodies. These immune responses to invading foreign bodies are essential for airway defence. However, hyperactive/prolonged responses such as those induced by exercising in polluted air are detrimental, leading to excessive inflammation and airway constriction (Giles & Koehle, 2014), possibly driving increased URTI prevalence.

The precise immunological response to inhaled pollutants is multifaceted. Inhaled pollutants both activate and damage epithelial cells, inducing cytokine/chemokine production, thereby driving leukocyte recruitment (Bernstein et al., 2004). One hour post-exposure, IL-6, IL-8, and CXCL-1 are increased in sputum and bronchial biopsy samples, leading to neutrophil and mast cell recruitment into the bronchia (Bernstein et al., 2004). These responses may be further exacerbated when exercising in polluted air, with increased neutrophilia in bronchioalveolar lavage specimens. Chronic pollutant exposure was also shown to impair NK cell cytotoxicity in overweight females who regularly exercise within 150 m of a major road (Williams Lori et al., 2009). Whether these immune responses influence the susceptibility to URTIs is unclear, but the effect of pollution on airway health is striking and likely underpins the detrimental effects of pollution on exercise performance. Furthermore, striking evidence is now linking air pollution (specifically $_{PM}2.5$; particulate matter ~3% the width of a human hair) with increased lung and oropharyngeal cancer risk in never-smokers, not by direct pollutant effects, but instead by increased inflammatory signals (e.g., IL-1β) driving mutations. Therefore, given almost 99% of the global population live within high $_{PM}2.5$ regions, the potential for pollution driving global cancer rates is considerable and warrants attention. Government public health directives may be needed to counter this threat (i.e., increasing awareness of the benefits of pollutant-filtering masks in at-risk regions).

Operational implementation of measures to counter immune stress

Implementation of countermeasures requires planning, resources, and an operational framework. In athletic settings, the usual approach is pre-season medical screening which can include an assessment of immune status, and then individual plans. In military or other occupation settings, pre-operation screening is useful to identify underlying immune issues or risk of heightened susceptibility for immune issues during operations. Overseas deployment into challenging conditions is a key consideration, similar to athletes changing hemispheres for international competitions. Environmental stress can also impact cognitive performance (Martin et al., 2019) and evolving knowledge of gut microbiota and brain function will require further studies of individuals under environmental stress, exercise, and nutritional countermeasures.

A recent development in exercise immunology is the development of diagnostic testing to identify a heightened risk of a compromised immune system and an increased risk of common illnesses or infections. Two promising approaches are an extended exercise test akin to the heat tolerance test (Alele et al., 2021), and a two-bout exercise protocol to assess the effects of repeated exercise on the immune system similar to the Meeusen two-bout exercise test (Vrijkotte et al., 2019). This test protocol involves two incremental graded exercise tests until exhaustion on a cycle ergometer or treadmill with 4 hours of rest in between. Blood samples are typically taken before and after each of the two tests to assess baseline, post-exercise, and recovery concentrations of selected haematological, hormonal, and/or immunological measures. This type of functional testing has yielded useful outcomes in individuals who are underperforming, presenting with fatigue and/or other signs and symptoms of overtraining, and potentially immune system perturbations.

The selection of immune biomarkers to monitor immune responses to challenging environments is a perennial question (Haunhorst et al., 2022). Systematic reviews show the response to an individual bout of heavy resistance exercise yields larger increases in the concentrations of adrenaline, noradrenaline, and growth hormone immediately after the exercise. Although cortisol is widely measured in blood and saliva, its levels likely reflect circadian rhythms and the effects of various exercise interventions. Results on other biomarkers, including brain-derived neurotrophic and adrenocorticotrophic factors were more variable and the utility of these markers is unclear. Biomarker concentrations typically return to their baseline value within 1 hour after the completion of the exercise. Future work will further clarify the utility of immune and biomarker measures in both clinical and research settings.

The continuing evolution of medical technology in the areas of immunology, metabolomics, epigenetics, phenomics, and point-of-care devices is underpinning new approaches in both laboratory and field settings. Research that was previously limited to the laboratory can now be conducted in a wide range of field settings with a variety of specimens, including capillary blood, saliva, and faeces. At present, much of this technology is expensive, requires specialist training and technical support, and is not widely accessible to the sporting, exercise, and industry communities. Multi 'omics' technology has the potential to transform our understanding without demanding huge logistical requirements other than a single blood draw/fingertip bloods. Further developments will reduce costs, simplify processes, and broaden accessibility, to facilitate wider access for clinicians and researchers.

Summary

Many studies show that a combination of exercise and environmental stress can elicit a range of perturbations in cellular and soluble immune measures (Figure 13.3). Researchers have also proposed several regulatory models to explain mechanisms and mediators of the immune response to exercise and environmental stress. A range of countermeasures to preserve immune function in these settings have evolved including acclimation strategies and therapeutic and dietary interventions.

Key points

- Individuals in a wide variety of exercise and occupational settings can be exposed to various environmental stresses, including heat and humidity, cold conditions, altitude hypoxia, weightlessness, and air pollution.
- Studies show a range of perturbations in cellular and soluble immune parameters in response to exercise. The classic thermoregulatory model describes scenarios where exercise capacity is impaired with an increase in thermal strain, and hyperthermia-induced alterations in cardiovascular, nervous system, and skeletal muscle function.
- A second pathway involves an inflammatory model of heat illness where exercise-induced increases in immune mediators, environmental influences, and metabolic response to fever of either inflammatory or non-inflammatory origin can perturb the immune system.
- Severe cold exposure leading to hypothermia impairs innate cellular and adaptive soluble aspects of immune function, increases surgical wound infection risk in patients and may increase URTI risk in healthy persons. In contrast, mild cold exposure appears on balance to pose a limited additional risk, and may even be beneficial, for immune health in active healthy people.
- Exercising in hypoxia has limited effects on innate, humoral, and soluble adaptive immunity, whereas T cell-mediated immunity is impaired by hypoxia and provides one explanation for the increased upper respiratory infection reported by those performing at altitude.
- Evidence to date indicates that repeated short doses of hypoxia coupled with exercise, i.e., intermittent hypoxic exercise training programmes, lead to several beneficial adaptations in innate and adaptive immune function. Consequently, intermittent hypoxic exercise training may be an effective countermeasure to pre-condition the immune system before travel to altitude.
- Spaceflight represents a unique, novel stressor for the immune system. Multiple factors, including radiation, microgravity, and physical workload play a role in space flight-associated immune dysfunction. These changes underpin the increased risk of adverse medical events during missions, highlighting the importance of immunological research into potential countermeasures to defend astronaut health in future long-duration spaceflight missions.

- Air pollution is increasing in many urban settings and induces acute and chronic airway inflammation that may be exacerbated by exercise in polluted regions. These impairments may counter some of the cardiopulmonary health benefits of exercise.
- Practitioners and researchers should consider practical strategies that avoid or limit the harmful effects of environmental stress, and lifestyle choices that promote a healthy immune system. Individuals need to know what exercise and environments to avoid, and which ones to embrace for adaptative and preventative strategies akin to heat acclimation and hypoxia adaptation commonly employed by athletes and other individuals.
- Emerging opportunities in point-of-care diagnostic analysers, phenomics areas, including metabolomics and epigenetics, data science, and personalised medicine, will facilitate a new generation of studies that examine the immune response to various combinations of exercise and environmental stress.

References

Akimoto, T., Kim, K., Yamauchi, R., Izawa, S., Hong, C., Aizawa, K., … Suzuki, K. (2009). Exercise in, and adaptations to a cold environment have no effect on SIgA. *J Sports Med Phys Fitness, 49*(3), 315–319.

Albers, R., Bourdet-Sicard, R., Braun, D., Calder, P. C., Herz, U., Lambert, C., … Sack, U. (2013). Monitoring immune modulation by nutrition in the general population: Identifying and substantiating effects on human health. *Br J Nutr, 110*(Suppl 2), S1–30.

Alele, F. O., Malau-Aduli, B. S., Malau-Aduli, A. E. O., & Crowe, M. J. (2021). Haematological, biochemical and hormonal biomarkers of heat intolerance in military personnel. *Biology (Basel), 10*(10), 1068.

Alkemade, P., Gerrett, N., Daanen, H. A. M., Eijsvogels, T. M. H., Janssen, T. W. J., & Keaney, L. C. (2022). Heat acclimation does not negatively affect salivary immunoglobulin-A and self-reported illness symptoms and wellness in recreational athletes. *Temperature (Austin), 9*(4), 331–343.

Bay, M. L., & Pedersen, B. K. (2020). Muscle-organ crosstalk: Focus on immunometabolism. *Front Physiol, 11*, 567881.

Benjamin, C. L., Stowe, R. P., St. John, L., Sams, C. F., Mehta, S. K., Crucian, B. E., … Komanduri, K. V. (2016). Decreases in thymopoiesis of astronauts returning from space flight. *JCI Insight, 1*(12), e88787.

Bernstein, J. A., Alexis, N., Barnes, C., Bernstein, I. L., Bernstein, J. A., Nel, A., … Williams, P. B. (2004). Health effects of air pollution. *J Allergy Clin Immunol, 114*(5), 1116–1123.

Bigley, A. B., Agha, N. H., Baker, F. L., Spielmann, G., Kunz, H. E., Mylabathula, P. L., … Simpson, R. J. (2018). NK cell function is impaired during long-duration spaceflight. *J Appl Physiol (1985), 126*(4), 842–853.

Boonyaratanakornkit, J. B., Cogoli, A., Li, C. F., Schopper, T., Pippia, P., Galleri, G., … Hughes-Fulford, M. (2005). Key gravity-sensitive signaling pathways drive T cell activation. *Faseb J, 19*(14), 2020–2022.

Born, D. P., Faiss, R., Willis, S. J., Strahler, J., Millet, G. P., Holmberg, H. C., & Sperlich, B. (2016). Circadian variation of salivary immunoglobin A, alpha-amylase activity and mood in response to repeated double-poling sprints in hypoxia. *Eur J Appl Physiol, 116*(1), 1–10.

Buijze, G. A., Sierevelt, I. N., van der Heijden, B. C., Dijkgraaf, M. G., & Frings-Dresen, M. H. (2016). The effect of cold showering on health and work: A randomized controlled trial. *PLoS One, 11*(9), e0161749.

Carlisle, A. J., & Sharp, N. C. (2001). Exercise and outdoor ambient air pollution. *Br J Sports Med, 35*(4), 214–222.

Chang, T. T., Walther, I., Li, C. F., Boonyaratanakornkit, J., Galleri, G., Meloni, M. A., … Hughes-Fulford, M. (2012). The Rel/NF-κB pathway and transcription of immediate early genes in T cell activation are inhibited by microgravity. *J Leukoc Biol, 92*(6), 1133–1145.

Chen, Y. C., Chou, W. Y., Fu, T. C., & Wang, J. S. (2018). Effects of normoxic and hypoxic exercise training on the bactericidal capacity and subsequent apoptosis of neutrophils in sedentary men. *Eur J Appl Physiol, 118*(9), 1985–1995.

Cogoli, A., Tschopp, A., & Fuchs-Bislin, P. (1984). Cell sensitivity to gravity. *Science, 225*(4658), 228–230.

Collier, N., Lomax, M., Harper, M., Tipton, M., & Massey, H. (2021). Habitual cold-water swimming and upper respiratory tract infection. *Rhinology, 59*(5), 485–487.

Costa, R. J., Smith, A. H., Oliver, S. J., Walters, R., Maassen, N., Bilzon, J. L., & Walsh, N. P. (2010). The effects of two nights of sleep deprivation with or without energy restriction on immune indices at rest and in response to cold exposure. *Eur J Appl Physiol, 109*(3), 417–428.

Crucian, B., Babiak-Vazquez, A., Johnston, S., Pierson, D. L., Ott, C. M., & Sams, C. (2016). Incidence of clinical symptoms during long-duration orbital spaceflight. *Inter J Gen Med, 9*, 383–391.

Crucian, B., Stowe, R., Mehta, S., Uchakin, P., Quiriarte, H., Pierson, D., & Sams, C. (2013). Immune system dysregulation occurs during short duration spaceflight on board the space shuttle. *J Clin Immunol, 33*(2), 456–465.

Crucian, B., Stowe, R. P., Mehta, S., Quiriarte, H., Pierson, D., & Sams, C. (2015). Alterations in adaptive immunity persist during long-duration spaceflight. *npj Microgravity, 1*, 15013.

Crucian, B. E., Choukèr, A., Simpson, R. J., Mehta, S., Marshall, G., Smith, S. M., … Sams, C. (2018). Immune System dysregulation during spaceflight: Potential countermeasures for deep space exploration missions. *Front Immunol, 9*.

Crucian, B. E., Chouker, A., Simpson, R. J., Mehta, S., Marshall, G., Smith, S. M., … Sams, C. (2018). Immune system dysregulation during spaceflight: Potential countermeasures for deep space exploration missions. *Front Immunol, 9*, 1437.

Crucian, B. E., Stowe, R. P., Pierson, D. L., & Sams, C. F. (2008). Immune system dysregulation following short- vs long-duration spaceflight. *Aviat Space Environ Med, 79*(9), 835–843.

Derman, W., Badenhorst, M., Eken, M., Gomez-Ezeiza, J., Fitzpatrick, J., Gleeson, M., … Schwellnus, M. (2022). Risk factors associated with acute respiratory illnesses in athletes: A systematic review by a subgroup of the IOC consensus on 'acute respiratory illness in the athlete'. *Br J Sports Med, 56*(11), 639–650.

Diaz-Sanchez, D., Proietti, L., & Polosa, R. (2003). Diesel fumes and the rising prevalence of atopy: An urban legend? *Curr Allergy Asthma Rep, 3*(2), 146–152.

Douglas, R. G., Lindgren, K. M., & Couch, R. B. (1968). Exposure to cold environment and rhinovirus. *New England J Med, 279*(14), 742–747.

Ermolao, A., Travain, G., Facco, M., Zilli, C., Agostini, C., & Zaccaria, M. (2009). Relationship between stress hormones and immune response during high-altitude exposure in women. *J Endocrinol Invest, 32*(11), 889–894.

Foxman, E. F., Storer, J. A., Fitzgerald, M. E., Wasik, B. R., Hou, L., Zhao, H., … Iwasaki, A. (2015). Temperature-dependent innate defense against the common cold virus limits viral replication at warm temperature in mouse airway cells. *Proc Natl Acad Sci U S A, 112*(3), 827–832.

Gagnon, D. D., Gagnon, S. S., Rintamäki, H., Törmäkangas, T., Puukka, K., Herzig, K. H., & Kyröläinen, H. (2014). The effects of cold exposure on leukocytes, hormones and cytokines during acute exercise in humans. *PLoS One, 9*(10), e110774.

Gibson, O. R., James, C. A., Mee, J. A., Willmott, A. G. B., Turner, G., Hayes, M., & Maxwell, N. S. (2020). Heat alleviation strategies for athletic performance: A review and practitioner guidelines. *Temperature (Austin), 7*(1), 3–36.

Giles, L. V., & Koehle, M. S. (2014). The health effects of exercising in air pollution. *Sports Med, 44*(2), 223–249.

Guéguinou, N., Huin-Schohn, C., Bascove, M., Bueb, J. L., Tschirhart, E., Legrand-Frossi, C., & Frippiat, J. P. (2009). Could spaceflight-associated immune system weakening preclude the expansion of human presence beyond Earth's orbit? *J Leukoc Biol, 86*(5), 1027–1038.

Harrison, S. E., Oliver, S. J., Kashi, D. S., Carswell, A. T., Edwards, J. P., Wentz, L. M., … Walsh, N. P. (2021). Influence of vitamin D supplementation by simulated sunlight or oral D3 on respiratory infection during military training. *Med Sci Sports Exerc, 53*(7), 1505–1516.

Haunhorst, S., Bloch, W., Ringleb, M., Fennen, L., Wagner, H., Gabriel, H. H. W., & Puta, C. (2022). Acute effects of heavy resistance exercise on biomarkers of neuroendocrine-immune regulation in healthy adults: A systematic review. *Exerc Immunol Rev, 28*, 36–52.

Helwig, B. G., & Leon, L. R. (2011). Tissue and circulating expression of IL-1 family members following heat stroke. *Physiol Genomics, 43*(19), 1096–1104.

Hill, G. W., Gillum, T. L., Lee, B. J., Romano, P. A., Schall, Z. J., & Kuennen, M. R. (2020). Reduced inflammatory and phagocytotic responses following normobaric hypoxia exercise despite evidence supporting greater immune challenge. *Appl Physiol Nutr Metab, 45*(6), 628–640.

Hodgson, J. R., Chapman, L., & Pope, F. D. (2022). Amateur runners more influenced than elite runners by temperature and air pollution during the UK's Great North Run half marathon. *Sci Total Environ, 842*, 156825.

Joad, J. P., Sekizawa, S., Chen, C. Y., & Bonham, A. C. (2007). Air pollutants and cough. *Pulm Pharmacol Ther, 20*(4), 347–354.

Kaur, I., Simons, E. R., Castro, V. A., Mark Ott, C., & Pierson, D. L. (2004). Changes in neutrophil functions in astronauts. *Brain Behav Immun, 18*(5), 443–450.

Keaney, L. C., Kilding, A. E., Merien, F., Shaw, D. M., & Dulson, D. K. (2021). Upper respiratory tract symptom risk in elite field hockey players during a dry run for the Tokyo Olympics. *Eur J Sport Sci, 22*(12), 1827–1835.

Koren, H. S., Devlin, R. B., Graham, D. E., Mann, R., McGee, M. P., Horstman, D. H., … et al. (1989). Ozone-induced inflammation in the lower airways of human subjects. *Am Rev Respir Dis, 139*(2), 407–415.

Laitano, O., Oki, K., & Leon, L. R. (2021). The role of skeletal muscles in exertional heat stroke pathophysiology. *Int J Sports Med, 42*(8), 673–681.

Lim, C. L., & Mackinnon, L. T. (2006). The roles of exercise-induced immune system disturbances in the pathology of heat stroke: The dual pathway model of heat stroke. *Sports Med, 36*(1), 39–64.

Mäkinen, T. M., Juvonen, R., Jokelainen, J., Harju, T. H., Peitso, A., Bloigu, A., … Hassi, J. (2009). Cold temperature and low humidity are associated with increased occurrence of respiratory tract infections. *Respir Med, 103*(3), 456–462.

Martin, K., McLeod, E., Périard, J., Rattray, B., Keegan, R., & Pyne, D. B. (2019). The impact of environmental stress on cognitive performance: A systematic review. *Hum Factors, 61*(8), 1205–1246.

Martinez, E. M., Yoshida, M. C., Candelario, T. L., & Hughes-Fulford, M. (2015). Spaceflight and simulated microgravity cause a significant reduction of key gene expression in early T-cell activation. *Am J Physiol Regul Integr Comp Physiol, 308*(6), R480–488.

Massey, H., Gorczynski, P., Harper, C. M., Sansom, L., McEwan, K., Yankouskaya, A., & Denton, H. (2022). Perceived impact of outdoor swimming on health: Web-based survey. *Interact J Med Res, 11*(1), e25589.

Mazzeo, R. S. (2008). Physiological responses to exercise at altitude: An update. *Sports Med, 38*(1), 1–8.

Mehta, S. K., Bloom, D. C., Plante, I., Stowe, R., Feiveson, A. H., Renner, A., … Pierson, D. L. (2018). Reactivation of latent epstein-barr virus: A comparison after exposure to gamma, proton, carbon, and iron radiation. *Int J Mol Sci, 19*(10).

Morabito, C., Lanuti, P., Caprara, G. A., Guarnieri, S., Verratti, V., Ricci, G., … Mariggiò, M. A. (2016). Responses of peripheral blood mononuclear cells to moderate exercise and hypoxia. *Scand J Med Sci Sports, 26*(10), 1188–1199.

Moriyama, M., Hugentobler, W. J., & Iwasaki, A. (2020). Seasonality of respiratory viral infections. *Annu Rev Virol, 7*(1), 83–101.

Mylabathula, P. L., Li, L., Bigley, A. B., Markofski, M. M., Crucian, B. E., Mehta, S. K., … Simpson, R. J. (2020). Simulated microgravity disarms human NK-cells and inhibits anti-tumor cytotoxicity in vitro. *Acta Astronautica, 174*, 32–40.

Mylona, E., Fahlman, M. M., Morgan, A. L., Boardley, D., & Tsivitse, S. K. (2002). s-IgA response in females following a single bout of moderate intensity exercise in cold and thermoneutral environments. *Int J Sports Med, 23*(6), 453–456.

Nottmeyer, L. N., & Sera, F. (2021). Influence of temperature, and of relative and absolute humidity on COVID-19 incidence in England - A multi-city time-series study. *Environ Res, 196*, 110977.

Ohnishi, T., Takahashi, A., Nagamatsu, A., Omori, K., Suzuki, H., Shimazu, T., & Ishioka, N. (2009). Detection of space radiation-induced double strand breaks as a track in cell nucleus. *Biochem Biophys Res Commun, 390*(3), 485–488.

Oliver, S. J., Macdonald, J. H., Harper Smith, A. D., Lawley, J. S., Gallagher, C. A., Di Felice, U., & Walsh, N. P. (2013). High altitude impairs in vivo immunity in humans. *High Alt Med Biol, 14*(2), 144–149.

Oliver, S. J., Sanders, S. J., Williams, C. J., Smith, Z. A., Lloyd-Davies, E., Roberts, R., ... Macdonald, J. H. (2012). Physiological and psychological illness symptoms at high altitude and their relationship with acute mountain sickness: A prospective cohort study. *J Travel Med, 19*(4), 210–219.

Pedersen, B. K., & Hoffman-Goetz, L. (2000). Exercise and the immune system: Regulation, integration, and adaptation. *Physiol Rev, 80*(3), 1055–1081.

Presbitero, A., Melnikov, V. R., Krzhizhanovskaya, V. V., & Sloot, P. M. A. (2021). A unifying model to estimate the effect of heat stress in the human innate immunity during physical activities. *Sci Rep, 11*(1), 16688.

Pyne, D. B., McDonald, W. A., Morton, D. S., Swigget, J. P., Foster, M., Sonnenfeld, G., & Smith, J. A. (2000). Inhibition of interferon, cytokine, and lymphocyte proliferative responses in elite swimmers with altitude exposure. *J Interferon Cytokine Res, 20*(4), 411–418.

Rauch, S., Miller, C., Bräuer, A., Wallner, B., Bock, M., & Paal, P. (2021). Perioperative hypothermia-A narrative review. *Int J Environ Res Public Health, 18*(16), 8749.

Reynolds, R., Little, M. P., Day, S., Charvat, J., Blattnig, S., Huff, J., & Patel, Z. S. (2021). Cancer incidence and mortality in the USA Astronaut Corps, 1959–2017. *Occup Environ Med, 78*(12), 869–875.

Rykova, M. P., Antropova, E. N., Larina, I. M., & Morukov, B. V. (2008). Humoral and cellular immunity in cosmonauts after the ISS missions. *Acta Astronautica, 63*, 697–705.

Shephard, R. J. (1998). Immune changes induced by exercise in an adverse environment. *Can J Physiol Pharmacol, 76*(5), 539–546.

Shephard, R. J., & Shek, P. N. (1999). Immune dysfunction as a factor in heat illness. *Crit Rev Immunol, 19*(4), 285–302.

Spatz, J. M., Fulford, M. H., Tsai, A., Gaudilliere, D., Hedou, J., Ganio, E., ... Gaudilliere, B. (2021). Human immune system adaptations to simulated microgravity revealed by single-cell mass cytometry. *Sci Rep, 11*(1), 11872.

Spielmann, G., Agha, N., Kunz, H., Simpson, R. J., Crucian, B., Mehta, S., ... Campbell, J. (2019). B cell homeostasis is maintained during long-duration spaceflight. *J Appl Physiol (1985), 126*(2), 469–476.

Stephens, K. E., Chernyavskiy, P., & Bruns, D. R. (2021). Impact of altitude on COVID-19 infection and death in the United States: A modeling and observational study. *PLoS One, 16*(1), e0245055.

Svendsen, I. S., Hem, E., & Gleeson, M. (2016). Effect of acute exercise and hypoxia on markers of systemic and mucosal immunity. *Eur J Appl Physiol, 116*(6), 1219–1229.

Tateo, F., Fiorino, S., Peruzzo, L., Zippi, M., De Biase, D., Lari, F., & Melucci, D. (2022). Effects of environmental parameters and their interactions on the spreading of SARS-CoV-2 in North Italy under different social restrictions. A new approach based on multivariate analysis. *Environ Res, 210*, 112921.

Tauber, S., Hauschild, S., Paulsen, K., Gutewort, A., Raig, C., Hürlimann, E., ... Ullrich, O. (2015). Signal transduction in primary human T lymphocytes in altered gravity during parabolic flight and clinostat experiments. *Cell Physiol Biochem, 35*(3), 1034–1051.

Tomasi, T. B., Trudeau, F. B., Czerwinski, D., & Erredge, S. (1982). Immune parameters in athletes before and after strenuous exercise. *J Clin Immunol, 2*(3), 173–178.

Vrijkotte, S., Roelands, B., Pattyn, N., & Meeusen, R. (2019). The overtraining syndrome in soldiers: Insights from the sports domain. *Mil Med, 184*(5–6), e192–e200.

Walsh, N. P., Bishop, N. C., Blackwell, J., Wierzbicki, S. G., & Montague, J. C. (2002). Salivary IgA response to prolonged exercise in a cold environment in trained cyclists. *Med Sci Sports Exerc, 34*(10), 1632–1637.

Walsh, N. P., & Whitham, M. (2006). Exercising in environmental extremes: A greater threat to immune function? *Sports Med, 36*(11), 941–976.

Wang, J. S., Chen, W. L., & Weng, T. P. (2011). Hypoxic exercise training reduces senescent T-lymphocyte subsets in blood. *Brain Behav Immun, 25*(2), 270–278.

Wang, J. S., & Chiu, Y. T. (2009). Systemic hypoxia enhances exercise-mediated bactericidal and subsequent apoptotic responses in human neutrophils. *J Appl Physiol (1985), 107*(4), 1213–1222.

Wang, J. S., & Lin, C. T. (2010). Systemic hypoxia promotes lymphocyte apoptosis induced by oxidative stress during moderate exercise. *Eur J Appl Physiol, 108*(2), 371–382.

Wang, J. S., & Weng, T. P. (2011). Hypoxic exercise training promotes antitumour cytotoxicity of natural killer cells in young men. *Clin Sci (Lond), 121*(8), 343–353.

Wang, J. S., & Wu, C. K. (2009). Systemic hypoxia affects exercise-mediated antitumor cytotoxicity of natural killer cells. *J Appl Physiol (1985), 107*(6), 1817–1824.

Williams Lori, A., Ulrich, C. M., Larson, T., Wener, M. H., Wood, B., Campbell, P. T., … De Roos, A. J. (2009). Proximity to traffic, inflammation, and immune function among women in the Seattle, Washington, area. *Environ Health Perspect, 117*(3), 373–378.

Willmott, A. G. B., Hayes, M., James, C. A., Dekerle, J., Gibson, O. R., & Maxwell, N. S. (2018). Once- and twice-daily heat acclimation confer similar heat adaptations, inflammatory responses and exercise tolerance improvements. *Physiol Rep, 6*(24), e13936.

14 Exercise immunology, nutrition, and immunometabolism

Brandt D. Pence

Introduction

Sufficient nutrient intake is critical to maintaining immune function. Activated immune cells mount a variety of energy-consuming processes, including migration, proliferation, and protein production, and even quiescent immune cells may account for ~20% of the basal metabolic rate for maintenance functions (Straub et al., 2010). Infections cause sickness behaviours including lethargy or decreased locomotor activity that are thought to conserve energy for anti-pathogen immune responses (Ganeshan et al., 2019). This reduces the overall metabolic rate, but as discussed Replace with "in this chapter", shifting nutrient metabolism in immune cells also speeds the utilisation of limited carbohydrate energy stores (primarily circulating glucose and liver glycogen). Active infection also causes anorexia (suppressed appetite and eating behaviour) which is thought to be an evolutionary strategy to deny critical nutrients to pathogenic microbes and to decrease energy expenditure needed for food-seeking behaviours that were highly energy-dependent in early humans (Dantzer, 2023). Thus, appropriate intake of dietary macronutrients (carbohydrates, fats, and proteins) is important for maintaining immune responses.

There is a wide body of nutrition research related to sport and exercise performance which is outside the scope of this chapter and is thoroughly reviewed elsewhere (Kerksick et al., 2022a, b). Indeed, many sport, exercise, and nutrition studies, do not examine immune function, and instead focus on optimal training strategies and performance among athletes. The impact of nutrition in the context of exercise immunology has largely been conducted with the idea of countering, what was interpreted at the time, as being short-term exercise-induced immunosuppression. As outlined in Chapters 1 and 9, it is now generally thought that most volumes and intensities of exercise are largely beneficial to immunological health, especially when combined with appropriate nutrient intake. However, the rationale for most nutrition and exercise immunology studies is that inadequate nutrient availability may be detrimental to immune function. Deficiencies or excesses of specific nutrients may alter the immune response by 'direct' and/or 'indirect' mechanisms. A nutritional deficiency is thought to have a 'direct effect' when the nutritional factor has primary activity within the immune system (e.g., as a fuel source) and an 'indirect effect' when the primary activity affects all cellular material or another organ system that acts as an immune regulator. A reduction in the availability of carbohydrate for example (e.g., decreased blood glucose concentration during prolonged exercise) might decrease immune-cell energy metabolism and protein synthesis (e.g., cytokine or antibody production) and is a 'direct effect'. Alternatively, decreased blood glucose availability might have an 'indirect effect' on immune function too through its stimulatory effect on the secretion of stress hormones. Indeed, stress hormones such as cortisol and adrenaline are widely acknowledged to have a strong influence on immune function.

DOI: 10.4324/9781003256991-14

The first part of this chapter will provide an overview of energy intake, dietary macronutrients and micronutrients, and the effect on immune function in general, and in regulating changes caused by exercise. This overview is followed by a brief exploration of dietary strategies which may influence immunity by improving the efficiency of nutrient utilisation. Finally, this chapter will close with a discussion of the links between exercise, nutrition, and immune-cell metabolism (often termed immunometabolism). Immunometabolism is an emerging discipline based on the observed links between immune-cell activation and shifts in the metabolic pathways used by these cells for ATP production.

Energy intake, macronutrients, micronutrients, and exercise immunology

Energy intake

Aspects of cellular and humoral immunity have been shown to be impaired in soldiers surviving for 12 days on ration packs providing only half of their daily energy requirements (around 1,800 kcal) compared with a control group consuming sufficient energy to maintain energy balance (Booth, Coad, Forbes-Ewan, Thomson, & Niro, 2003). Another study has shown that soldiers undertaking an eight-week training program with a modest daily energy deficit (of around 500 kcal) exhibited lower resting lymphocyte and monocyte counts (Diment et al., 2012). A shorter-term study examining more extreme energy deficit has shown that a seven-day fast lowered total T cell numbers and CD4+ helper T cell numbers and impaired interleukin (IL)-2 release in lymphocyte cultures in response to bacterial stimulation (Savendahl & Underwood, 1997). In addition, a 36-hour fast has been shown to decrease both neutrophil chemotaxis and oxidative burst, although this was reversed in just 4 hours with re-feeding (Walrand et al., 2001). While these studies provide evidence for low-energy availability influencing aspects of immune function, it could be argued that prolonged fasting (e.g., for periods of up to seven days) has less relevance to the general population taking part in sport and exercise. However, some athletes adopt very low-energy diets and periods of fasting in sports where leanness or low body weight are thought to confer an advantage (e.g., gymnastics, dance) or to 'make weight' for competition (e.g., boxing, martial arts, rowing). As further rationale for examining low-energy availability and immune function, the subclinical disorder 'anorexia athletica' has been associated with an increased susceptibility to infection (Beals & Manore, 1994).

Opposite to the concept that low-energy availability could impair immune function, another body of work has examined dietary restriction as an intervention that might be beneficial for immunity, usually in the context of ageing. This idea is based upon the concept of hormesis which is a cellular or organismal response to intermittent stressors, usually leading to greater stress resilience (Mattson, 2008). Some dietary compounds, especially phytochemicals, are thought to promote better immune (and general cellular) function through hormetic means (Ademowo et al., 2019). In recent years, significant effort has been made to characterise the hormetic effects of dietary restriction and therefore energy balance on physiology, including immunity, with mixed results. Mild caloric restriction, such that an energy deficit is created without malnutrition, has been shown to increase thymic function during aging in mice (Spadaro et al., 2022), but a similar study found impaired T cell function and reduced survival with West Nile virus infection in old mice undergoing caloric restriction (Goldberg et al., 2015). Likewise, intermittent fasting and time-restricted eating dietary strategies have received considerable research attention in recent years. These dietary interventions may reduce inflammation and promote immune function, although study results have been mixed and are highly dependent on context and population (He et al., 2023). Dietary restriction and mimetic agents such as rapamycin continue to receive significant interest in a variety of fields related to longevity and disease, and the positive effects may be attributable to increased resilience of the immune system.

Carbohydrates

Carbohydrates are organic macromolecules of carbon, hydrogen, and oxygen, of which mono- and disaccharide sugars such as glucose, fructose, and sucrose, are the most common from a nutritional standpoint. Carbohydrates provide approximately four kilocalories of energy per gram consumed, and simple nutritional carbohydrates are primarily metabolised through conversion or breakdown to glucose or fructose and subsequent entry into the glycolysis pathway. Glycolysis provides rapid energy production through the generation of adenosine triphosphate (ATP) and nicotinamide adenine dinucleotide (NADH), at the expense of lesser efficiency compared with other catabolic pathways discussed below. Carbohydrates have been widely examined for their role in sports and performance (Kerksick et al., 2022a).

Carbohydrate is the macronutrient which has received most attention in exercise immunology. Carbohydrates have been investigated for potential nutritional benefit on immune function primarily in the context of supplementation immediately before, during, or after exercise. The rationale for carbohydrate supplementation is to maintain or to restore muscle and liver glycogen stores to avoid hypoglycemia during prolonged exercise. Indeed, glycogen stores are limited – usually less than 500 grams – and the amount of glycogen stored in the body prior to exercise influences the exercise-induced hormone and inflammatory response. Ingesting carbohydrates (e.g., 30–60 grams per hour) during vigorous exercise leads to reductions in post-exercise inflammation, as was first shown in the late 1990s (Nehlsen-Cannarella et al., 1997; Nieman et al., 1997) and consuming a high-carbohydrate diet for several days before exercise has similar effects. This concept was demonstrated by Bishop et al. (2001) where changes in plasma cortisol, interleukin-6 (IL-6) and interleukin-1 receptor antagonist (IL-1ra) were examined after 1 hour of cycling at 60% VO_2 max immediately followed by a 30-minute time trial (work rate around 80% VO_2 max). For the three days prior to exercise, participants consumed either a high-carbohydrate diet (>70% of total dietary energy was from carbohydrate) or a low-carbohydrate diet (< 10% of total dietary energy was from carbohydrate). Compared with a high-carbohydrate diet, plasma cortisol was consistently higher during exercise and for 2 hours later, when consuming a low-carbohydrate diet. In addition, while IL-6 and IL-1ra did not differ due to diet during exercise, their levels were substantially higher in the 2 hours after exercise when eating a low-carbohydrate diet (Bishop, Walsh, Haines, Richards, & Gleeson, 2001). These findings examining the effects of carbohydrate intake and post-exercise inflammation have subsequently been replicated and extended by numerous groups over at least two decades (Nieman & Wentz, 2019).

Carbohydrate supplementation has also been shown to influence aspects of cell-mediated immunity. For example, one study examined T cell migration *ex vivo* towards supernatants from a human rhinovirus-infected bronchial epithelial cell line. In a randomised, cross-over, double-blind design, seven trained males ran for 2 hours at 60% VO_2 peak on two occasions with regular ingestion of either a 6.4% w/v glucose and maltodextrin solution (CHO trial) or placebo solution (PLA trial). Plasma glucose concentration was higher in the CHO trial than PLA trial after exercise. The migration of CD4+ and CD8+ cells toward supernatants from infected cells decreased following exercise but the migration of CD4+ cells was approximately 35% higher in the CHO trial than in the PLA trial at 1-hour post-exercise (Bishop, Walker, Gleeson, Wallace, & Hewitt, 2009). Other studies have reported that carbohydrate supplementation helps maintain other anti-viral T cell functions following vigorous exercise, including cytokine production and proliferation (Lancaster et al., 2005; Bishop et al., 2005) and that carbohydrate also influences innate immune cells, by limiting the fluctuations in neutrophil function in the hours after exercise (Scharhag et al., 2002; Bishop et al., 2003). While many studies have reported that carbohydrate supplementation influences aspects of immune function after vigorous exercise, studies examining moderate or intermittent exercise, reported minimal to no effect on a variety of immune function parameters (Nieman et al., 1999; Davison et al., 2016) – most likely due to the smaller magnitude of exercise-induced immune change with less-demanding exercise. Less information is available about long-term effects of carbohydrate intake on immune function and exercise, and high-carbohydrate diets or short-term carbohydrate supplementation have variable or mildly positive effects on measured immune function and inflammatory markers as summarised elsewhere (Gunzer et al., 2012). Some evidence indicates that insufficient carbohydrate intake dysregulates immune function (Mitchell et al., 1998), possibly due to the preferential use of glucose as a fuel during activation by many immune cells as discussed in the immunometabolism sections below.

Fats

Dietary fats are a broad class of biomolecule which collectively includes mono-, di-, and triglycerides, sterols, and other types of lipids. Stored lipids in the form of fatty acids are energy dense, yielding about nine kilocalories per gram when catabolised. Lipids also are important constituents of various cellular membranes (typically in the form of a phospholipid bilayer), are critical to the function of fat-soluble vitamins A, D, E, and K, and are precursors to a variety of eicosanoids and other molecules which are important for inflammation and immune function. Lipids are stored in relative abundance even in lean individuals, so there is little evidence that insufficient dietary fat intake is a concern with respect to immune function. Essential polyunsaturated fatty acids (PUFAs), either omega-3 (alpha-linoleic acid) or omega-6 (linoleic acid), must be obtained from the diet and are precursors to eicosanoids and other bioactive lipids with important immune-modulating properties.

Dietary deficiencies in essential PUFAs are uncommon, and there is little consensus on the utility of supplementing the diet with these fats for the purpose of maintaining immunity. Indeed, in non-exercise studies, it is unclear whether PUFA supplementation is beneficial among healthy people. For example, it has been shown that supplementation with omega-3 PUFAs (e.g. eicosapentaenoic acid) decreases natural killer (NK) cell cytotoxicity at rest (Thies et al., 2001). However, one study in untrained men reported that the inflammatory response to eccentric exercise-induced muscle damage, evaluated by increases in circulating IL-6 and C-reactive protein, was attenuated by 14 days of daily supplementation with docosahexaenoic acid (an omega-3 PUFA: 800 mg), but this was combined with other supplements with antioxidant and anti-inflammatory actions, including tocopherols (300 mg) and flavonoids (100 mg hesperetin and 200 mg quercetin) (Phillips, Childs, Dreon, Phinney, & Leeuwenburgh, 2003). Another study indicated that supplementation with a combination of 2.4 g/day of fish oils containing omega-3 PUFAs (eicosapentaenoic acid and docosahexaenoic acid) for six weeks had no effect on selected immune responses to prolonged exercise (Toft et al., 2000). Some studies have reported a beneficial effect of omega-3 PUFA supplementation in the context of exercise, showing higher NK-cell cytotoxicity 3 hours after exercise with krill oil supplementation compared with control (Da Boit et al., 2015). However, most other immune function metrics were unaffected, as was exercise performance and systemic measures of inflammation and oxidative stress.

Conversely, substantial evidence suggests high levels of fat intake cause immune system dysfunction. Adipose tissue expansion during overnutrition leads to adipocyte cell death and macrophage-driven inflammation, and in turn to insulin resistance (see Chapter 12) (Lee & Dixit, 2020). Obese adipose tissue itself is also an endocrine organ which regulates immune-cell function both locally and systemically (see Chapter 6). Obesity is associated with increased severity of influenza (Green & Beck, 2017) and COVID-19 (Singh et al., 2022) as well as diminished vaccine responses (Painter et al., 2015; Nasr et al., 2022), so diets with high fat content or with caloric excess in general are likely to be detrimental to immune function. Indeed, a high fat diet may even limit the benefits of regular exercise. For example, NK cell cytotoxicity decreased following a seven-week endurance training program in previously untrained men who consumed a fat-rich diet (62% daily energy intake as fat) whereas NK cell cytotoxicity increased in men who undertook endurance training while consuming a carbohydrate-rich diet (65% daily energy intake as carbohydrate) (Pedersen, Helge, Richter, Rohde, & Kiens, 2000). Furthermore, saturated fatty acids are themselves pro-inflammatory mediators and can activate inflammatory signalling pathways in immune cells (Huang et al., 2012). Little is known about the effect of exercise on immune-modulating properties of saturated fatty acids from a mechanistic perspective, but moderate exercise training is well known to lessen inflammation and improve immune function in obese adipose tissue, as extensively reviewed elsewhere (Winn et al., 2021) and described in Chapter 6.

Proteins

Amino acids are nitrogen-containing organic compounds that form the building blocks of proteins. In addition to their role as contractile elements in skeletal muscle, proteins perform a variety of functions in the body. Proteins are structural constituents of cells, catalyse enzymatic reactions, and transduce cellular signals among other roles. The production of proteins is a major contributor to cellular energy expenditure. In the immune system, many cells produce cytokines and other protein hormones during activation in order to signal to other cells, and cellular energy is also consumed during cellular signal propagation via protein phosphorylation.

Dietary protein deficiencies are relatively rare, so increasing dietary protein intake is unlikely to have a substantial effect on immune function during exercise or competitive sport for most individuals. The chief risk of protein insufficiency is in people who are fasting, attempting to cut weight via undernutrition, or who consume diets without significant animal protein intake (e.g., vegetarian and vegan diets). Insufficient intake of protein is associated with impaired immune function and increased risk of infections (Li et al., 2007). The effects of dietary protein deficiency on the immune system include atrophy of lymphoid tissue, decreased T cell numbers and T cell proliferation response to mitogens, decreased CD4+ to CD8+ T cell ratio, and decreased macrophage phagocytic activity and IL-1 production (Daly, Reynolds, Sigal, Shou, & Liberman, 1990).

Some evidence also suggests that individual amino acids regulate aspects of immune function, and in the 1990s, glutamine received a lot of attention. Glutamine is an important energy source for activated immune cells by providing an alternative to glucose metabolism for fueling the tricarboxylic acid (TCA) cycle (Kelly & Pearce, 2020), as covered more thoroughly later in this chapter. Indeed, in the 1990s, the so-called "glutamine hypothesis" was formed to explain the observation that lymphocyte function measurements (e.g., cytokine production, proliferation, cytotoxicity) made in blood in the hours after vigorous exercise were less than pre-exercise values. It is now known that these observations were driven by the populations of cells present in blood at the time of sampling, when most effector cells (e.g., NK cells, memory T cells) have left the blood post-exercise, migrating to tissues. Thus, with fewer effector cells present in the samples collected post-exercise – even when accounting for total cell numbers overall in the assay – it is no surprise that functional measurements were lower than pre-exercise values. This concept has been reviewed elsewhere (Campbell & Turner, 2018) and is covered in Chapters 1 and 4 (see Box 4.1). However, the "glutamine hypothesis" gained traction because of consistent reports that exercise bouts decreased plasma glutamine availability (Parry-Billings et al., 1992). Around this time, many studies investigated the effect of oral glutamine supplementation during and after exercise on various indices of immune function but failed to find any effect. For example, one study provided a glutamine solution (0.1 g/kg body mass) during exercise, immediately post-exercise and 30 minutes after prolonged bouts of cycling at 75% VO_2 max, which prevented the fall in the plasma glutamine concentration, but did not prevent the fall in mitogen-induced lymphocyte proliferation and NK cell cytotoxicity (Rohde, MacLean, & Pedersen, 1998). In addition, studies examining the influence of glutamine supplementation on respiratory tract infections were inconsistent, with some studies reporting no effect (Castell, Poortmans, & Newsholme, 1996) and others reporting a reduced incidence of infections (Castell & Newsholme, 1997). However, a variety of other studies found no support for the "glutamine hypothesis" and glutamine supplementation was eventually discounted as having effects on immunity in an exercise context or among athletes (Hiscock & Pedersen, 2002).

Other studies examined supplementation with the branched-chain amino acids (BCAAs) leucine, isoleucine, and valine, because they are precursors to glutamine, and provide acetyl-CoA and other substrates to the TCA cycle (Kelly & Pearce, 2020). For example, one study reported that supplementation (6 g/day for 15 days) with BCAAs before a triathlon or a 30 km run prevented the fall in plasma glutamine and the decline in mitogen-stimulated lymphocyte proliferation observed in the placebo control group after exercise (Bassit et al., 2002). Many other specific amino acids support various aspects of immune function and influence other aspects of physiology in the context of exercise as comprehensively reviewed elsewhere (Kelly & Pearce, 2020). For example, L-arginine supplementation is likely beneficial for increasing nitric oxide-mediated blood flow and fitness, although possibly only in the context of disease and not among healthy people (Cruzat et al., 2014) and L-citrulline supplementation has also been shown to preserve oxidative burst capacity of neutrophils after exercise (Sureda et al., 2009). However, information about the immune-modulating effects of these amino acids with exercise is sparse, and a small-scale meta-analysis has found little basis for their effects on inflammation or oxidative stress in the context of exercise (Porto et al., 2023).

Micronutrients and other supplements

In addition to the macronutrients above, micronutrients have been studied for their immune-regulating roles, including in the context of exercise. Nutritional deficiencies in micronutrients such as vitamins and minerals are relatively unusual, so basic mechanistic evidence for the roles of many of these substances in regulating immune-cell function are usually not applicable to exercise and immuno-nutrition. Several minerals such as iron, magnesium, and zinc may be lost to sweat during high-intensity exercise, which may contribute to deficiencies in these micronutrients in athletes. However, there is little to no evidence that supplementing these minerals during exercise has appreciable effects on immune function (Bermon et al., 2017). The influence of many other supplements on either exercise-induced immune responses or susceptibility to upper respiratory tract infections has been examined, including bovine colostrum, β-glucans, polyphenols, caffeine, probiotics, selenium; herbal supplements, such as ginseng, echinacea, and garlic; and also water- and fat-soluble vitamins. It is beyond the scope of this chapter to review these supplements in detail, and readers are directed to comprehensive reviews (Bermon et al., 2017;

Walsh, 2019). The remainder of this section will focus on two micronutrients that have some of the strongest evidence for their influence on immune function: vitamin D and vitamin C.

Vitamin D may have a beneficial impact on the immune system, especially during winter months where synthesis from UV radiation (sun exposure) is limited. Vitamin D deficiency has been associated with increased infections, for example, in one study of 756 military conscripts, those with serum 25(OH)D less than 40 nmol/l (N = 24, 3.6%) experienced significantly more days of absence from duty due to respiratory infections than those with serum 25(OH)D greater than 40 nmol/l (Laaksi et al., 2007). A follow-up study of 164 military conscripts, showed that supplementation with 10 µg/day vitamin D for six months prevented the typical wintertime fall in serum 25(OH)D (data for N = 58) but did not alter the number of days absent due to respiratory tract infections. However, the proportion of men remaining healthy throughout the six-month period was greater in the intervention group (51%) than in the placebo group (36%; [Laaksi et al., 2010]). Similar results have been found among athletes with lower vitamin D status who exhibited higher respiratory illness rates (Halliday et al., 2011) and winter vitamin D supplementation increased salivary antimicrobial protein production (He et al., 2013). Vitamin D promotes expression of a wide variety of genes important in the functions of most immune cells (Bermon et al., 2017), so supplementation may be beneficial for immunity in athletes and exercisers who are deficient in vitamin D.

Vitamin C also has a substantial evidence base for influencing immune function. In a double-blind placebo-controlled study, daily supplementation of 1,500 mg of vitamin C (or placebo) within a carbohydrate beverage for seven days before an ultra-marathon race, showed no effect on oxidative stress, cytokine, or immune function measures during and after the race (Nieman et al., 2002). In contrast, it has been reported that four weeks of combined supplementation with vitamin C (500 mg/day) and vitamin E (400 iu/day equivalent to around 270 mg which is around 27 times the RDA of 10 mg) before a 3-hour knee-extension exercise protocol reduced muscle IL-6 release and reduced the systemic rise in circulating IL-6 and cortisol (Fischer et al., 2004). Overall, meta-analyses have shown that daily doses of more than 200 mg vitamin C were more effective in preventing or treating the common cold than placebo (Douglas, Hemila, Chalker, & Treacy, 2007). However, athletes should also consider that high-dose vitamin C supplementation can blunt some training adaptations (Ristow et al., 2009).

Summary – energy intake, macronutrients, micronutrients, and exercise immunology

Many studies in exercise immunology have examined the effects of macronutrients and micronutrients on the risk of infections in athletic populations, or on the isolated aspects of immune function. Although it is beyond the scope of this chapter to review these studies in detail, readers are directed to several reviews that cover this topic in comprehensive detail, including making practical recommendations for athletes (Bermon et al., 2017; Walsh, 2019). Most importantly, when interpreting research examining whether nutrient intake or supplementation influences aspects of immunity, readers are strongly encouraged to reflect on methodological considerations and interpretations of the immunological measurements made, as outlined in detail elsewhere (Albers et al., 2013). We will return to the effects of some of these nutritional interventions on immunometabolism later in this chapter.

Immunometabolism

In recent decades, there has been an explosion of interest in the regulation of immune function by metabolism. This is not a completely new concept, as it was shown that glucose metabolism increases in activated phagocytes at least as far back as the 1960s (Oren et al., 1963). Several developments in the 1990s and early 2000s, however, substantially accelerated interest in the bidirectional links between metabolism and immunity. First, local increases in tissue inflammation were shown to regulate metabolism, when the pro-inflammatory cytokine tumor necrosis factor (TNF)-α was shown to directly cause insulin resistance in adipocytes (Hotamisligil et al., 1993). Second, changes in cellular metabolism were shown to reprogram immune-cell function and phenotype. The following sections cover several key metabolic pathways regulating immune-cell function.

Glycolysis

Much of the earliest work in this latter area showed that immune cells, especially T cells and macrophages, relied on glycolysis for energy production when activated by pro-inflammatory stimuli (O'Neill et al., 2016). From a bioenergetics perspective, this is sensible and can be thought of as analogous to high-intensity exercise. Under acute stress conditions (such as infection or inflammation), immune cells have a substantial need for rapid adenosine triphosphate (ATP) production to meet demands for cytokine biosynthesis, cytoskeletal rearrangement, chemotaxis, cell division, and a variety of other cellular processes integral to immune function. While glycolytic energy production is less efficient than other metabolic pathways, intermediate enzymes can be activated far more quickly and can produce ATP at an overall faster rate while substrate supply is adequate. Glycolysis also provides important metabolic intermediates which support cell division and growth, including substrates for the pentose phosphate pathway (PPP), acetyl-CoA, and amino acid precursors (O'Neill et al., 2016). Moreover, this enhanced glycolysis occurs even in the presence of substantial oxygen. This increase in aerobic glycolysis was long known to occur in cancer cells and is commonly termed the Warburg effect. Krawczyk et al. identified Warburg metabolism in dendritic cells (DCs) activated by lipopolysaccharide (LPS) (Krawczyk et al., 2010), and macrophages, NK cells, T cells, and B cells all undergo this process upon activation (O'Neill et al., 2016). Additionally, interfering with glycolytic metabolism can transition immune cells between phenotypes, as was probably first shown by reprogramming pro-inflammatory T helper 17 (T_H17) cells to regulatory T cells via suppressing glycolysis (Shi et al., 2011). Activation of glycolysis during immune activation occurs in both hypoxia-inducible factor (HIF)-1α-dependent and -independent manners, and a variety of glycolytic enzymes are induced or suppressed during activation to shunt glycolytic intermediates into the TCA cycle (from pyruvate), PPP (from glucose-6-phosphate), or to lactate production as necessary (O'Neill et al., 2016).

TCA cycle

The TCA (Krebs) cycle is also a central regulator of immunometabolism. Generally, anti-inflammatory or longer-lived immune cells (such as CD4+ T_{regs}, M2 macrophages, and CD8+ memory T cells) maintain an intact TCA cycle in the mitochondria, while pro-inflammatory or proliferative reprogramming of these cells (e.g., to CD4+ Th17 cells, M1 macrophages, or effector CD8+ T cells) involves TCA cycle breaks that shunt metabolite intermediates to other pathways. These TCA cycle breaks primarily cause accumulation of two intermediates with immunopotentiating functions: citrate and succinate (Zasłona & O'Neill, 2020). Excess citrate is produced via downregulation of the isocitrate lyase enzyme and is used for fatty acid biosynthesis and the production of prostaglandins during immune-cell activation, after being transported outside the mitochondria by citrate transporters (O'Neill et al., 2016). Likewise, accumulation of succinate, via decreases in succinate dehydrogenase activity, activates HIF-1α to promote pro-inflammatory cytokine production, especially interleukin (IL)-1β (Tannahill et al., 2013). Interestingly, citrate and succinate also have immunoregulatory functions. Citrate is a substrate for the production of itaconate, which acts in an anti-inflammatory fashion in activated immune cells by suppressing succinate dehydrogenase (SDH)-dependent reactive oxygen species (ROS) production (Lampropoulou et al., 2016) and promoting nuclear translocation of the anti-inflammatory and antioxidant protein Nrf2 (Mills et al., 2018). Likewise, accumulated succinate, when released to the extracellular environment, acts as an endocrine inhibitor of inflammation and promotes immune-cell reprogramming through the extracellular succinate receptor (SUCNR1) (Keiran et al., 2019). These anti-inflammatory effects of accumulating TCA cycle intermediates may be integral to resolving the inflammatory response and avoiding tissue pathology resulting from chronic inflammation.

Fatty acid oxidation

A transformative point in the study of immunometabolism was the identification that mitochondrial metabolism, and especially fatty acid oxidation (FAO), is integral to reprogramming of immune cells to an anti-inflammatory state. Vats et al. showed that anti-inflammatory M2 macrophages depend on FAO regulated by signal transducer and activator of transcription 6 (STAT6) and PPARγ-co-activator 1β (PGC-1β) (Vats et al., 2006), and that suppressing this reprogrammed macrophages to glycolytic metabolism and a pro-inflammatory M1 phenotype. Fatty acid oxidation also promotes CD4+ regulatory T cells and suppresses effector Th1, Th2, and Th17 polarisation, and FAO is integral to memory T cell and memory B cell persistence as well (O'Neill et al., 2016). Conversely, pro-inflammatory and activated immune cells generally increase fatty acid synthesis (FAS) and downregulate FAO, as shown for monocytes (Ecker et al., 2010) as well as other immune-cell types. FAS is thought to support cellular proliferation in activated lymphocytes and biosynthesis of various signalling molecules under pro-inflammatory conditions (O'Neill et al., 2016), and is thus integral to immune-cell activation. The regulation of immune-cell function by fatty acid metabolism may also extend beyond the balance between FAO and FAS, as the type of fatty acid (e.g., saturated vs. polyunsaturated) involved seems to also play a role.

Other metabolic pathways

While glycolysis and mitochondrial metabolic pathways have been the most widely studied, several other pathways are known to support immune-cell functioning or phenotypic reprogramming. The PPP incorporates G6P from glycolysis for the production of pentose sugars (important precursors to nucleotides) and NADPH, an important precursor to FAS. The PPP is elevated typically in pro-inflammatory cells, including M1 macrophages and activated dendritic cells (DCs) (O'Neill et al., 2016). NADPH from the PPP is also a precursor to superoxide, which is used by immune cells (including neutrophils and monocytes) during respiratory burst to kill bacteria. Likewise, amino acid metabolic pathways support various immune-cell functions, including but not limited to protein biosynthesis. For example, the anaplerotic metabolism of glutamine allows the synthesis of the TCA cycle intermediate α-ketoglutarate from glutamate, which has been shown to support anti-inflammatory M2 polarisation in macrophages (Jha et al., 2015). Conversely, arginine serves as a precursor to nitric oxide (NO) through conversion to citrulline and processing by inducible nitric oxide synthase (iNOS), a pathway that has been shown to be important in maintaining pro-inflammatory polarisation of M1 macrophages

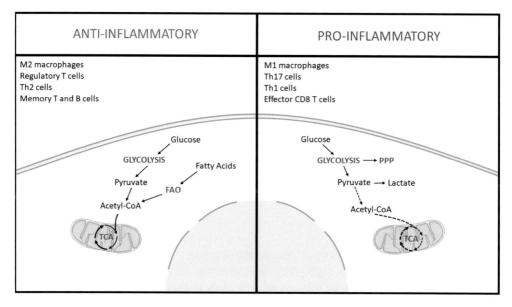

Figure 14.1 Metabolic pathways activated during immune-cell polarisation.

Legend: "Anti-inflammatory cells, such as those listed on the left, utilize glucose and fatty acids, an intact TCA cycle, and aerobic metabolism to produce ATP primarily in the mitochondria. Pro-inflammatory cells, such as those listed on the right, have breaks in the TCA cycle, increased lactate production from glycolysis, and activation of the pentose phosphate pathway (PPP)."

(MacMicking et al., 1997). A schematic of major metabolic pathways induced in anti-inflammatory vs. pro-inflammatory immune cells is included in Figure 14.1.

Dysregulation of immunometabolism

The immune system is tightly involved in the maintenance of homeostasis, especially through its role in the inflammatory process. Dysregulated immunity is therefore a contributing factor to many disease states and degenerative conditions. Investigations into the link between immuno-metabolic reprogramming and disease are still in very early stages, but it is clear that a variety of diseases and conditions are correlated to changes in immune-cell metabolism.

Cardiovascular disease

Atherosclerosis is the most common cardiovascular disease and is characterised by lipid re-tention in the arterial wall, leading to plaque development. As these plaques progress, they are at risk of destabilization and rupture, which can block blood flow and lead to myocardial infarction or stroke. Arterial plaques contain a variety of infiltrating immune cells, especially monocytes/macrophages but also including T and B cells and others (Ketelhuth et al., 2019). Macrophages in the plaque take up modified lipids to become pro-inflammatory foam cells, and excessive inflammation leads to plaque growth and eventual necrosis and rupture. Macrophages in the plaque have been shown to increase glucose uptake (Folco et al., 2011), which reflects pro-inflammatory polarisation. This may be due directly to the exposure to modified lipids such as oxidized low-density lipoprotein, which increases glucose uptake, ROS production, and in-flammatory cytokine production in monocytes (Bekkering et al., 2014). This has led to some interest in targeting immunometabolism to treat atherosclerosis, although to date these studies are in very early stages.

Obesity

Obesity is a condition of excess body mass which is correlated with type 2 diabetes and other metabolic disorders and increases the risk of a variety of chronic diseases including cardiovascular disease and cancer. Obesity also increases severity of infection (Khwatenge et al., 2021), and inflammation is a causative factor in insulin resistance in adipose tissue as described above. These observations suggest dysregulated immune-cell function during obesity. Indeed, obese adipose tissue promotes the infiltration and pro-inflammatory polarisation of macrophages, as well as B and T cell infiltration (Khwatenge et al., 2021). Pro-inflammatory repolarisation of macrophages and lymphocytes suggest changes in immunometabolism, especially glycolytic reprogramming. The core of obese adipose tissue is hypoxic, and it is thought that glycolytic reprogramming may confer a survival advantage to adipose-infiltrating immune cells under these oxygen-limited conditions (Ghesquière et al., 2014).

Cancer

Like obesity, cancer is a chronic condition of dysregulated metabolism. Many tumors are highly glycolytic and compete with infiltrating immune cells for glucose, potentially suppressing pro-inflammatory repolarisation of the latter and thereby suppressing anti-tumor immunity (Roy et al., 2020). While many immune cells have some degree of metabolic flexibility and may utilise alternate metabolic programs (such as FAO in place of glycolysis) when preferred nutrients are unavailable (Roy et al., 2020), resulting immune responses may be less efficient or effective. Likewise, metabolites produced in the tumor microenvironment can also reprogram immune cells to suppress anti-tumor immunity. A prominent example of this is lactate, which is produced in high levels by cancer cells and can downregulate the expression of glycolytic enzymes in macrophages, T cells, and NK cells (Roy et al., 2020). Lactate has also recently been shown to cause epigenetic changes which suppress pro-inflammatory cytokine production (Zhang et al., 2019). A variety of other metabolic intermediates, such as acetyl-CoA and succinate, may also regulate epigenetics in a similar fashion, suggesting another mode by which metabolism affects immune function.

Infection

In addition to chronic diseases, various infectious diseases also reprogram immunometabolism. Dendritic cells have been shown to undergo glycolytic reprogramming during influenza infection, and ROS production and gene expression consistent with glycolytic reprogramming have been demonstrated in influenza-infected macrophages (Bahadoran et al., 2020). Interestingly, mitochondrial respiration also increases in influenza-infected DCs, suggesting a general metabolic program for increasing ATP production. This is in contrast to many experimental inflammation models such as LPS stimulation, in which glycolysis increases are concomitant with the suppression of mitochondrial respiration (O'Neill et al., 2016). The recent pandemic severe acute respiratory syndrome coronavirus-2 (SARS-CoV-2) also reprograms immune cells, including by increasing glycolysis and promoting lipid droplet storage in directly infected monocytes (Pence, 2021). SARS-CoV-2 may alter metabolism in immune cells in part through direct binding of virion proteins such as spike (Cory et al., 2021) or envelope to cellular receptors. Immunometabolism is thus a promising emerging target for therapies against infectious diseases.

Ageing

Finally, while not a disease in the strict sense, ageing is a condition which increases risk and/ or severity of nearly all chronic and infectious diseases. Ageing increases systemic inflammation and suppresses immune responses, and these are often linked to shifts in phenotypes of circulating and tissue immune cells (Nikolich-Žugich, 2018). Ageing increases the proportion of memory T cells at the expense of naive T cell subsets, and this may be driven by increased reliance on mitochondrial metabolism in aged T cells (Bharath et al., 2020). Oxidative metabolism is also increased in aged platelets, which may in part explain increased reactivity in these cells during ageing (Van Avondt et al., 2023). Conversely, maximal mitochondrial oxygen consumption decreases in aged monocytes (Pence & Yarbro, 2018), which may drive shifts observed to more pro-inflammatory phenotypes and also suppress anti-pathogen responses (Yarbro et al., 2020). Neutrophils also undergo functional changes leading to deficient responses during ageing (Van Avondt et al., 2023). However, the degree to which this is determined by metabolic changes is questionable, as there appears to be minimal effect of ageing on levels of metabolic intermediates in neutrophils despite significant changes in gene expression (Lu et al., 2021). Metabolism of NAD^+ is also dysregulated during ageing, especially in macrophages, and this is linked to cellular senescence, mitochondrial dysfunction, and increased pro-inflammatory cytokine production (Yarbro et al., 2020). NAD^+ is a coenzyme which is critical to almost all major metabolic pathways including glycolysis, the TCA cycle, and FAO. As such, alterations in immunometabolism, precipitated by ageing, may underlie the development of age-related diseases such as cancer and cardiovascular disease. However, little is known about this important area, and most studies have provided observations of metabolic dysfunction during ageing without fully elucidating the underlying mechanism(s).

Nutrition and immunometabolism

Host metabolism is tightly regulated by a network of nutrient-sensing pathways, which serve to reprogram cellular metabolism to maintain ATP production to meet energy demands under diverse physiological conditions. As such, interventions which target nutrient sensors and/or transporters broadly are likely to regulate immunometabolism, in addition to metabolism in other cells. This has naturally led to some interest in studying nutritional strategies to target immune-cell metabolism, especially in the context of suppressing inflammation and inflammatory diseases.

Nutrient sensors and transporters

Cells must control intracellular levels of a variety of lipids, amino acids, carbohydrates, and micronutrients, only some of which can be endogenously synthesized. Cells therefore rely on intracellular and extracellular receptors and enzymes which sense and react to changes in the levels of these nutrients in order to maintain homeostasis. These sensors surveil nutrient levels either by direct binding or by recognition of proxies, such as the cellular ratio of ATP to adenosine monophosphate (AMP) (Efeyan et al., 2015). Additionally, nutrient transporters are expressed by immune cells in order to take up extracellular nutrients for metabolism.

Many nutrient sensors have been implicated in the regulation of immunometabolism in some form, and several have been studied relatively extensively. The mechanistic target of rapamycin (mTOR) is a central regulator of protein and lipid synthesis and so controls many anabolic processes (Efeyan et al., 2015) in response to various stimuli, including exercise and nutrient intake. In the immune system, mTOR activation also participates in glycolysis-associated HIF-1α activation (Cheng et al., 2014) and consequent inflammatory cytokine production, and is therefore a critical component of immune-cell activation. The energy-sensing molecule AMP-activated protein kinase (AMPK) is activated when the ATP:AMP ratio in the cell decreases, signalling low ATP availability. AMPK activation increases catabolic metabolism (including FAO) and suppresses FAS and lipid synthesis (Richter & Ruderman, 2009), and counteracts the activation of mTOR. AMPK activation is consequently a mechanism for anti-inflammatory activation of immune cells, and small molecular AMPK activators are of interest as anti-inflammatory therapies (Garcia & Shaw, 2017).

A variety of specific transporters for carbohydrates, amino acids, fatty acids, and metabolites are expressed by immune cells. Glucose transport is primarily mediated by GLUT1 and is increased during pro-inflammatory activation of macrophages and T cells, and these cells also upregulate amino acid transport to provide substrate for protein production during activation (Weiss & Angiari, 2020). Conversely, anti-inflammatory cells such as M2 macrophages rely on fatty acid uptake through FATP1 and other transporters to support energy production via FAO (Johnson et al., 2016). Immune cells additionally express a variety of metabolite transporters, including monocarboxylate transporters critical to regulating cellular levels of lactate, pyruvate, succinate, etc. While activated immune cells produce and excrete large amounts of these metabolites (as covered in the following exerkines and metabolites section), whether uptake of these metabolites can support ATP production in immune cells is something of an open question.

Caloric restriction

Caloric restriction (CR) is a common nutritional intervention aimed at weight loss or promotion of longevity. Such dietary plans require a deficit of caloric intake compared with *ad libitum* feeding, usually around 15% in humans but up to 40% or more in rodent models, while maintaining necessary intake of vitamins and minerals and without restriction on particular foods or macronutrients. As a result of reduced nutrient intake and therefore nutrient availability, nutrient sensors in immune (and other) cells may have reduced activity. CR activates AMPK and suppresses mTOR and HIF-1α (Blagosklonny, 2010), promoting an anti-inflammatory phenotype overall. CR has been shown to suppress Th17 differentiation through AMPK activation and resulting suppression of ACC1, which in turn reduces FAS necessary to support Th17 differentiation (Wang et al., 2019). Likewise, CR increases intracellular NAD^+ levels, which may suppress M1 macrophage differentiation through obviating the need for NAD^+ salvage pathway activation. To date, this has not been directly demonstrated, although CR does promote M2 macrophage polarisation and suppress M1 polarisation as shown in a mouse-ageing model (Ma et al., 2020). Substantial work (Madeo et al., 2019) has been published with potential CR mimetics including mTOR inhibitors (e.g., rapamycin), AMPK activators, and NAD precursors (e.g., nicotinamide riboside) which may yield insight into potential anti-inflammatory mechanisms of CR, even if direct evidence for a CR effect is currently lacking.

Nutrient timing

A variety of feeding pattern diets have become popular in recent years, based mostly on short-term fasting. Intermittent fasting (IF) is a limited fasting pattern in which normal eating patterns are (typically) maintained four to five days a week, with no food intake on the other two to three days. In contrast, time-restricted feeding is a fasting pattern in which food is eaten only over a certain period of time each day, often 6–8 hours, with no food intake during the remainder of the day. These practices do not necessarily (but can) produce a caloric deficit, and are touted as having metabolic benefits including increased insulin sensitivity (Wang & Wu, 2022). These fasting patterns are thought to promote FAO, reduce blood glucose, and increase ketone bodies such as β-hydroxybutyrate (β-HB), conceivably promoting an anti-inflammatory state. While support for these particular pathways is currently limited in the literature, IF has been demonstrated to suppress Th17 and increase Treg cells in the gastrointestinal tract (Cignarella et al., 2018). Short-term fasting may also increase autophagy in immune cells, which is broadly considered to be anti-inflammatory (Okawa et al., 2021).

Nutrient restriction

Dietary restriction of specific amino acids may also serve to modulate immunometabolism, principally based on the resulting regulation of nutrient-sensing pathways. Restrictions of leucine, tryptophan, and methionine have individually been shown to be anti-inflammatory, probably based on the inhibition of mTOR signalling, activation of AMPK, or modulation of the gut microbiome (Okawa et al., 2021). Likewise, carbohydrate restriction and ketogenic diets (KD) modulate immune function, at least in part through anti-inflammatory actions of the ketone body β-HB. β-HB inhibits the NLRP3 inflammasome and suppresses pro-inflammatory cytokine production in myeloid cells (Youm et al., 2015), linking ketogenesis to an anti-inflammatory program. KD has also been demonstrated to reduce the Th17/Treg ratio in epileptic patients through inhibition of mTOR (Ni et al., 2016), providing a potential explanation for the known efficacy of KD in treating epilepsy.

Supplements and nutraceuticals

Various dietary supplements and natural products are also known to modulate immunometabolism, although in many cases the best evidence of this is from *in vitro* studies. For example, a variety of plant-derived polyphenols have immunomodulatory properties (Man et al., 2020). Diets rich in polyphenols modulate the gut microbiome, potentially promoting the production of microbial-derived metabolites such as short chain fatty acids, BCAAs, bile acids, and others which have known anti-inflammatory or immunopotentiating effects (Man et al., 2020). Polyphenolic compounds such as curcumin (Campbell et al., 2019) and epigallocatechin gallate (Cai et al., 2021) also directly modulate immunometabolism by inhibiting glycolytic reprogramming in myeloid cells such as dendritic cells and macrophages.

Mimicking KD, direct administration of β-HB suppresses cytokine production *in vitro* in macrophages and *in vivo* in mouse models of various inflammatory diseases (Youm et al., 2015). Conversely, BCAAs appear critical to mTOR-dependent pro-inflammatory functions, with a BCAA-limited diet suppressing Treg numbers in mice (Ikeda et al., 2017). These studies and others suggest that specific dietary components may be efficacious in modulating immunometabolism to treat disease. However, almost all such studies have used highly purified and concentrated forms, and there is so far little evidence that sufficient intake of these compounds (for the purposes of targeting immunometabolism) can be achieved in a whole foods diet alone.

Several other common dietary constituents may affect immunometabolism, although little direct evidence is available in some cases. Caffeine intake has diverse effects on the immune system, and particularly appears to suppress the functions of many immune cells, including macrophages and lymphocytes (Horrigan et al., 2006), and so caffeine may be efficacious in suppressing aberrant inflammation found in many chronic diseases. However, evidence for the direct effects of caffeine (i.e., as opposed to indirect effects based on caffeine intake) on immune

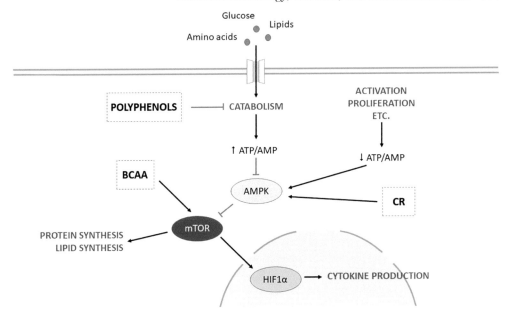

Figure 14.2 Selected mechanisms governing nutritional control of immunometabolism.

Legend: "Key nutrient sensors including AMPK and mTOR are regulated by nutritional strategies. Activation of AMPK suppresses mTOR and inflammation, while activation of mTOR activates HIF-1α and pro-inflammatory cytokine production. Caloric restriction (CR) is thought to activate AMPK to stimulate an anti-inflammatory program, while branched chain amino acids (BCAA) activate mTOR to promote inflammation. Polyphenolic compounds such as curcumin or epigallocatechin gallate are thought to directly target catabolic metabolism, especially glycolysis, thereby reducing ATP production."

function is limited. In contrast, omega-3 fatty acids are well-known regulators of immunometabolism, both as precursors to anti-inflammatory eicosanoids (Saini & Keum, 2018) and through promotion of FAO and M2 polarisation in macrophages (Rombaldova et al., 2017). Selected proposed mechanisms for nutritional control of immunometabolism are shown in Figure 14.2.

Exercise and immunometabolism

While some evidence links nutrition to changes in immunometabolism, to date, there has been very little research on exercise and immunometabolism despite the well-known effects of physical activity on metabolism. Frequent aerobic exercise suppresses inflammation (at least in individuals with existing inflammatory conditions) and promotes immune function in a general sense, although the mechanisms by which this occurs are still obscure or contested (Nieman & Pence, 2020). Given what is known about regulation of immunometabolism as outlined above, it is easily conceivable that exercise may regulate immunity by altering immune-cell metabolism. While evidence is limited, some support for this does exist.

Direct evidence

To date, only a few studies have directly examined immunometabolism changes due to exercise. High-intensity interval training for 10 weeks did not alter CD4+ T cell preference for oxidative metabolism on average in individuals with rheumatoid arthritis, although individual subject increases in T cell oxidative metabolism were correlated to increases in circulating proportions of naive CD4+ T cells (Andonian et al., 2022), supporting a possible link between metabolic reprogramming and cellular phenotypes with exercise. A two-week high-intensity interval training program also failed to increase immune-cell respiration (Hedges et al., 2019), despite increasing mitochondrial respiration in muscle cells, suggesting that immune-cell and muscle metabolism are not necessarily linked. Likewise, an acute exercise bout did not alter mitochondrial respiration in active adults (Theall et al., 2021). However, this was measured in peripheral blood mononuclear cell (PBMC, essentially a purified sample of lymphocytes and monocytes), and so results may have been biased by the known effect of exercise in transiently redistributing immune-cell populations (Gustafson et al., 2021). A similar lack of an acute exercise effect on mitochondrial respiration in PBMCs was found in collegiate swimmers (Stampley et al., 2023). Media conditioned with exercise-derived serum also had no effect on oxygen consumption of CD4+ T cells (Palmowski et al., 2021), although exercise may have reduced glucose consumption in these cells after extended culture. Conversely, Janssen et al. found that PBMCs had increased mitochondrial basal and maximal respiration in high-fit compared with low-fit women (Janssen et al., 2022), while glycolytic parameters were unchanged. A 12-week exercise training program also increased PBMC respiration in HIV-positive individuals (Kocher et al., 2017). PBMCs have been shown to have gene transcription patterns reflective of increased mitochondrial function in exercise-trained individuals (Dias et al., 2015; Liu et al., 2017). Exercise training also reverses lymphocyte mitochondrial dysfunction induced by hypoxia (Tsai et al., 2016). These conflicting results suggest that the participants examined, the exercise paradigm, training duration, and other factors are likely determinants of the efficacy of exercise in modulating immunometabolism.

Exerkines and metabolites

Further indirect support for a potential role for exercise on modulating immunometabolism comes from well-established links between exercise-associated signalling molecules and immunometabolic reprogramming. Exercise increases cytokine production, including IL-6, IL-10, and growth differentiation factor-15, and these have been shown to promote pathways, leading to increased mitochondrial metabolism in macrophages (Mauer et al., 2014; Ip et al., 2017; Jung et al., 2018). Extracellular vesicles are also released during exercise (Nederveen et al., 2021) and may carry protein, metabolite, or other cargo which has immunometabolic effects (Akbar et al., 2021). Likewise, metabolic intermediates may be released into the extracellular space during exercise to impact immune function. The effect of lactate on immunometabolism has been extensively studied in the context of cancer (de la Cruz-López et al., 2019), so exercise-produced lactate is a plausible mechanism for anti-inflammatory effects of high-intensity exercise. Succinate also accumulates in muscle during exercise and is released into the extracellular space when intracellular pH decreases (Reddy et al., 2020), and extracellular succinate promotes M2 macrophage polarization through binding to SUCNR1 (Keiran et al., 2019). Therefore, although direct evidence is lacking at present, exercise increases concentrations of a number of molecules in circulation with anti-inflammatory or immunomodulatory properties, further supporting a potential effect of exercise on immunometabolism.

Conclusions

Immunometabolism has become an intense area of study in the past decade, driven by the recognition that metabolic reprogramming drives immune-cell function and is thus a promising target for therapeutics. Immunometabolism is dysregulated in a number of chronic and infectious disease states, so behavioural therapies (such as nutrition and exercise) known to modulate cellular metabolism are of particular interest due to their safety profiles and inexpensive nature compared with pharmaceutical therapies. While some evidence exists that various dietary and exercise interventions can promote metabolic reprogramming in immune cells (especially by promoting anti-inflammatory phenotypes), much of this evidence is indirect, and further study is needed both to describe mechanisms of action and to define the most effective interventions for use on an individualized basis.

References

Ademowo OS, Dias HKI, Pararasa C, Griffiths HR. (2019). Nutritional hormesis in a modern environment. In Rattan S.I.S. & Kyriazis M. (Eds.) *The Science of Hormesis in Health and Longevity* (pp 75-86). .

Akbar N, Paget D, Choudhury RP. (2021). Extracellular vesicles in innate immune cell programming. *Biomed* 9(7):713.

Albers, R., Bourdet-Sicard, R., Braun, D., Calder, P. C., Herz, U., Lambert, C., ... Sack, U. (2013). Monitoring immune modulation by nutrition in the general population: Identifying and substantiating effects on human health. *Br J Nutr* 110(Suppl 2):S1–30.

Andonian BJ, Koss A, Koves TR, Hauser ER, Hubal MJ, Pober DM, ... MacIver NJ. (2022). Rheumatoid arthritis T cell and muscle oxidative metabolism associate with exercise-induced changes in cardiorespiratory fitness. *Sci Reports* 12(1):1–13.

Bahadoran A, Bezavada L, Smallwood HS. (2020). Fueling influenza and the immune response: Implications for metabolic reprogramming during influenza infection and immunometabolism. *Immunol Rev* 295(1):140–166.

Bassit RA, Sawada LA, Bacurau RF, Navarro F, Martins E, Jr., Santos R V, ... Costa Rosa LF. (2002). Branched-chain amino acid supplementation and the immune response of long-distance athletes. *Nutrition* 18(5):376–379.

Beals KA, Manore MM. (1994). The prevalence and consequences of subclinical eating disorders in female athletes. *Int J Sport Nutr* 4(2):175–195.

Bekkering S, Quintin J, Joosten LAB, Van Der Meer JWM, Netea MG, Riksen NP. (2014). Oxidized low-density lipoprotein induces long-term proinflammatory cytokine production and foam cell formation via epigenetic reprogramming of monocytes. *Arterioscler Thromb Vasc Biol* 34(8):1731–1738.

Bermon, S., Castell, L. M., Calder, P. C., Bishop, N. C., Blomstrand, E., Mooren, F. C., ... Nagatomi, R. (2017). Consensus statement immunonutrition and exercise. *Exerc Immunol Rev* 23:8–50.

Bharath LP, Agrawal M, McCambridge G, Nicholas DA, Hasturk H, Liu J, ... Liu R. (2020). Metformin enhances autophagy and normalizes mitochondrial function to alleviate aging-associated inflammation. *Cell Metab* 32(1):44–55.e6.

Bishop NC, Walker GJ, Bowley LA, Evans KF, Molyneux K, Wallace FA, Smith AC. (2005). Lymphocyte responses to influenza and tetanus toxoid in vitro following intensive exercise and carbohydrate ingestion on consecutive days. *J Appl Physiol* 99(4):1327–1335.

Bishop NC, Walker GJ, Gleeson M, Wallace FA, Hewitt CR. (2009). Human T lymphocyte migration towards the supernatants of human rhinovirus infected airway epithelial cells: Influence of exercise and carbohydrate intake. *Exerc Immunol Rev* 15:127–144.

Bishop NC, Walsh NP, Haines DL, Richards EE, Gleeson M. (2001). Pre-exercise carbohydrate status and immune responses to prolonged cycling: II. Effect on plasma cytokine concentration. *Int J Sport Nutr Exerc Metab* 11(4):503–512.

Bishop NC, Walsh NP, Scanlon GA. (2003) Effect of prolonged exercise and carbohydrate on total neutrophil elastase content. *Med Sci Sports Exerc* 35(8):1326–1332.

Blagosklonny MV. (2010). Calorie restriction: Decelerating mTOR-driven aging from cells to organisms (including humans). *Cell Cycle* 9(4):683–688.

Booth CK, Coad RA, Forbes-Ewan CH, Thomson GF, Niro PJ. (2003). The physiological and psychological effects of combat ration feeding during a 12-day training exercise in the tropics. *Mil Med* 168(1):63–70.

Cai F, Liu S, Lei Y, Jin S, Guo Z, Zhu D,… Zhao H. (2021). Epigallocatechin-3 gallate regulates macrophage subtypes and immunometabolism to ameliorate experimental autoimmune encephalomyelitis. *Cell Immunol* 368: 104421.

Campbell JP, Turner JE. (2018). Debunking the myth of exercise-induced immune suppression: Redefining the impact of exercise on immunological health across the lifespan. *Front Immunol* 9:648.

Campbell NK, Fitzgerald HK, Fletcher JM, Dunne A. (2019). Plant-derived polyphenols modulate human dendritic cell metabolism and immune function via AMPK-dependent induction of heme oxygenase-1. *Front Immunol* 10(March):345.

Castell LM, Newsholme EA. (1997). The effects of oral glutamine supplementation on athletes after prolonged, exhaustive exercise. *Nutrition* 13(7–8):738–742.

Castell LM, Poortmans JR, Newsholme EA. (1996). Does glutamine have a role in reducing infections in athletes? *Eur J Appl Physiol Occup Physiol* 73(5):488–490. Doi:10.1007/BF00334429

Cheng SC, Quintin J, Cramer RA, Shepardson KM, Saeed S, Kumar V, … Martens JH (2014). mTOR- and HIF-1alpha-mediated aerobic glycolysis as metabolic basis for trained immunity. *Science* 345(6204):1250684.

Cignarella F, Cantoni C, Ghezzi L, Salter A, Dorsett Y, Chen L,… Weinstock GM. (2018). Intermittent fasting confers protection in CNS autoimmunity by altering the gut microbiota. *Cell Metab* 27(6): 1222–1235.e6.

Cory TJ, Emmons RS, Yarbro JR, Davis KL, Pence BD. (2021). Metformin suppresses monocyte immunometabolic activation by SARS-CoV-2 spike protein Subunit 1. *Front Immunol* 12: 733921.

Cruzat VF, Krause M, Newsholme P. (2014). Amino acid supplementation and impact on immune function in the context of exercise. *J Int Soc Sport Nutr* 11(1):1–13.

Da Boit M, Mastalurova I, Brazaite G, McGovern N, Thompson K, Gray SR. (2015). The effect of krill oil supplementation on exercise performance and markers of immune function. *PLoS One* 10(9): e0139174.

Daly JM, Reynolds J, Sigal RK, Shou J, Liberman MD. (1990). Effect of dietary protein and amino acids on immune function. *Crit Care Med* 18(2 Suppl):S86–93.

Dantzer R. (2023). Evolutionary aspects of infections: Inflammation and sickness behaviors. *Curr Top Behav Neurosci*, 61: 1–14.

Davison G, Kehaya C, Diment BC, Walsh NP. (2016) Carbohydrate supplementation does not blunt the prolonged exercise-induced reduction of in vivo immunity. *Eur J Nutr* 55(4):1583–1593.

de la Cruz-López KG, Castro-Muñoz LJ, Reyes-Hernández DO, García-Carrancá A, Manzo-Merino J. (2019) Lactate in the regulation of tumor microenvironment and therapeutic approaches. *Front Oncol* 9: 1143.

Dias RG, Silva MS, Duarte NE, Bolani W, Alves CR, Junior JR, … de Oliveira PA (2015) PBMCs express a transcriptome signature predictor of oxygen uptake responsiveness to endurance exercise training in men. *Physiol Genomics* 47(2):13–23.

Diment BC, Fortes MB, Greeves JP, Casey A, Costa RJ, Walters R, Walsh NP. (2012). Effect of daily mixed nutritional supplementation on immune indices in soldiers undertaking an 8-week arduous training programme. *Eur J Appl Physiol* 112(4):1411–1418.

Douglas RM, Hemila H, Chalker E, Treacy B. (2007). Vitamin C for preventing and treating the common cold. *Cochrane Database Syst Rev* (3):CD000980.

Ecker J, Liebisch G, Englmaier M, Grandl M, Robenek H, Schmitz G. (2010). Induction of fatty acid synthesis is a key requirement for phagocytic differentiation of human monocytes. *Proc Natl Acad Sci U S A* 107(17):7817–7822.

Efeyan A, Comb WC, Sabatini DM. (2015) Nutrient-sensing mechanisms and pathways. *Nature* 517(7534):302–310.

Fischer CP, Hiscock NJ, Penkowa M, Basu S, Vessby B, Kallner A, … Pedersen BK. (2004). Supplementation with vitamins C and E inhibits the release of interleukin-6 from contracting human skeletal muscle. *J Physiol* 558(Pt 2):633–645.

Folco EJ, Sheikine Y, Rocha VZ, Christen T, Shvartz E, Sukhova GK, Di Carli MF, Libby P. (2011). Hypoxia but not inflammation augments glucose uptake in human macrophages: Implications for imaging atherosclerosis with 18fluorine-labeled 2-deoxy-D-glucose positron emission tomography. *J Am Coll Cardiol* 58(6):603–614.

Ganeshan K, Nikkanen J, Man K, Leong YA, Sogawa Y, Maschek JA, …, Chagwedera DN. (2019) Energetic Trade-Offs and Hypometabolic States Promote Disease Tolerance. *Cell* 177(2):399–413.e12.

Garcia D, Shaw RJ. AMPK. (2017). Mechanisms of cellular energy sensing and restoration of metabolic balance. *Mol Cell* 66(6):789–800.

Ghesquière B, Wong BW, Kuchnio A, Carmeliet P. (2014). Metabolism of stromal and immune cells in health and disease. *Nature* 511(7508):167–176.

Goldberg EL, Romero-Aleshire MJ, Renkema KR, Ventevogel MS, Chew WM, Uhrlaub JL, … Limesand K. (2015). Lifespan-extending caloric restriction or mTOR inhibition impair adaptive immunity of old mice by distinct mechanisms. *Aging Cell* 14(1):130–138.

Green WD, Beck MA. (2017) Obesity impairs the adaptive immune response to influenza virus. *Ann Am Thorac Soc* 14(Suppl 5):S406–S409.

Gunzer W, Konrad M, Pail E. (2012) Exercise-induced immunodepression in endurance athletes and nutritional intervention with carbohydrate, protein and fat — What is possible, what is not? *Nutr* 4(9):1187–1212.

Gustafson MP, Wheatley-Guy CM, Rosenthal AC, Gastineau DA, Katsanis E, Johnson BD, Simpson RJ. (2021) Exercise and the immune system: Taking steps to improve responses to cancer immunotherapy. *J Immunother Cancer* 9(7):e001872.

Halliday TM, Peterson NJ, Thomas JJ, Kleppinger K, Hollis BW, Larson-Meyer DE. (2021). Vitamin D status relative to diet, lifestyle, injury, and illness in college athletes. *Med Sci Sports Exerc* 43(2):335–343.

He C, Handzlik M, Fraser WD, Muhamad A, Preston H, Richardson A, Gleeson M. (2013). Influence of vitamin D status on respiratory infection incidence and immune function during 4 months of winter training in endurance sport athletes. *Exerc Immunol Rev*:19: 86–101.

He Z, Xu H, Li C, Yang H, Mao Y. (2023). Intermittent fasting and immunomodulatory effects: A systematic review. *Front Nutr* 10:1048230.

Hedges CP, Woodhead JST, Wang HW, Mitchell CJ, Cameron-Smith D, Hickey AJR, Merry TL. (2019). Peripheral blood mononuclear cells do not reflect skeletal muscle mitochondrial function or adaptation to high-intensity interval training in healthy young men. *J Appl Physiol* 126(2):454–461.

Hiscock N, Pedersen BK. (2002) Exercise-induced immunodepression–plasma glutamine is not the link. *JAP* 93(3):813–822.

Horrigan LA, Kelly JP, Connor TJ. (2006) Immunomodulatory effects of caffeine: Friend or foe? *Pharmacol Ther* 111(3):877–892.

Hotamisligil GS, Shargill NS, Spiegelman BM. (1993). Adipose expression of tumor necrosis Factor-α: Direct role in obesity-linked insulin resistance. *Science* 259(5091):87–91.

Huang S, Rutkowsky JM, Snodgrass RG, Ono-Moore KD, Schneider DA, Newman JW, …, Hwang DH. (2012). Saturated fatty acids activate TLR-mediated proinflammatory signaling pathways. *J Lipid Res* 53(9):2002–2013.

Ikeda K, Kinoshita M, Kayama H, Nagamori S, Kongpracha P, Umemoto E, …, Kurakawa T. (2017). Slc3a2 Mediates branched-chain amino-acid-dependent maintenance of regulatory T cells. *Cell Rep* 21(7):1824–1838.

Ip WKE, Hoshi N, Shouval DS, Snapper S, Medzhitov R. (2017). Anti-inflammatory effect of IL-10 mediated by metabolic reprogramming of macrophages. *Science* 356(6337):513–519.

Janssen JJE, Lagerwaard B, Porbahaie M, Nieuwenhuizen AG, Savelkoul HFJ, Van Neerven RJJ, …, De Boer VCJ. (2022). Extracellular flux analyses reveal differences in mitochondrial PBMC metabolism between high-fit and low-fit females. *Am J Physiol Endocrinol Metab* 322(2): E141-E153.

Jha AK, Huang SCC, Sergushichev A, Lampropoulou V, Ivanova Y, Loginicheva E, …, Stewart KM. (2015). Network integration of parallel metabolic and transcriptional data reveals metabolic modules that regulate macrophage polarization. *Immunity* 42(3):419–430.

Johnson AR, Qin Y, Cozzo AJ, Freemerman AJ, Huang MJ, Zhao L, …, Milner JJ. (2016) Metabolic reprogramming through fatty acid transport protein 1 (FATP1) regulates macrophage inflammatory potential and adipose inflammation. *Mol Metab* 5(7):506–526.

Jung SB, Choi MJ, Ryu D, Yi HS, Lee SE, Chang JY, …, Kim YK. (2018). Reduced oxidative capacity in macrophages results in systemic insulin resistance. *Nat Commun* 9(1): 1551.

Keiran N, Ceperuelo-Mallafré V, Calvo E, Hernández-Alvarez MI, Ejarque M, Núñez-Roa C, …, Maymó-Masip E, (2019). SUCNR1 controls an anti-inflammatory program in macrophages to regulate the metabolic response to obesity. *Nat Immunol* 20(5):581–592.

Kelly B, Pearce EL. (2020). Amino assets: How amino acids support immunity. *Cell Metab* 32(2):154–175.

Kerksick CM, Arent S, Schoenfeld BJ, Stout JR, Campbell B, Wilborn CD, …, Kalman D (2022a). International society of sports nutrition position stand. *Nutrient Timing* 14(33).

Kerksick CM, Wilborn CD, Roberts MD, Smith-Ryan A, Kleiner SM, Jäger R, …, Cooke M. (2022b) ISSN exercise & sports nutrition review update: Research & recommendations 15(1).

Ketelhuth DFJ, Lutgens E, Bäck M, Binder CJ, Van Den Bossche J, Daniel C, …, Hoefer I. (2019). Immunometabolism and atherosclerosis: Perspectives and clinical significance: A position paper from the working group on atherosclerosis and vascular biology of the European society of cardiology. *Cardiovasc Res* 115(9):1385–1392.

Khwatenge CN, Pate M, Miller LC, Sang Y. (2021). Immunometabolic dysregulation at the intersection of obesity and COVID-19. *Front Immunol* 12: 124302.

Kocher M, McDermott M, Lindsey R, Shikuma CM, Gerschenson M, Chow DC, …, Hetzler RK. (2017). Short communication: HIV patient systemic mitochondrial respiration improves with exercise. *AIDS Res Hum Retroviruses* 33(10):1035–1037.

Krawczyk CM, Holowka T, Sun J, Blagih J, Amiel E, DeBerardinis RJ, …, Jung E. (2010). Toll-like receptor-induced changes in glycolytic metabolism regulate dendritic cell activation. *Blood* 115(23):4742–4749.

Laaksi I, Ruohola JP, Tuohimaa P, Auvinen A, Haataja R, Pihlajamaki H, Ylikomi T. (2007). An association of serum vitamin D concentrations < 40 nmol/L with acute respiratory tract infection in young Finnish men. *Am J Clin Nutr* 86(3):714–717.

Laaksi I, Ruohola JP, Mattila V, Auvinen A, Ylikomi T, Pihlajamaki H. (2010). Vitamin D supplementation for the prevention of acute respiratory tract infection: a randomized, double-blinded trial among young Finnish men. *J Infect Dis* 202(5):809–814.

Lampropoulou V, Sergushichev A, Bambouskova M, Nair S, Vincent EE, Loginicheva E, …, Ma X. (2016). Itaconate links inhibition of succinate dehydrogenase with macrophage metabolic remodeling and regulation of inflammation. *Cell Metab* 24(1):158–166.

Lancaster GI, Khan Q, Drysdale PT, Wallace F, Jeukendrup AE, Drayson MT, Gleeson M. (2005). Effect of prolonged exercise and carbohydrate ingestion on type 1 and type 2 T lymphocyte distribution and intracellular cytokine production in humans. *J Appl Physiol* 98(2):565–571.

Lee AH, Dixit VD. (2020) Dietary regulation of immunity. *Immunity* 53(3):510–523.

Li P, Yin YL, Li D, Kim WS, Wu G. (2007). Amino acids and immune function. *Br J Nutr* 98(2):237–252.

Liu D, Wang R, Grant AR, Zhang J, Gordon PM, Wei Y, Chen P. (2017). Immune adaptation to chronic intense exercise training: new microarray evidence. *BMC Genomics* 18(1):29.

Lu RJ, Taylor S, Contrepois K, Kim M, Bravo JI, Ellenberger M, Sampathkumar NK, Benayoun BA. (2021). Multi-omic profiling of primary mouse neutrophils predicts a pattern of sex- and age- related functional regulation. *Nat Aging* 1(8):715–733.

Ma S, Sun S, Geng L, Song M, Wang W, Ye Y, …, Zou Z. (2020). Caloric restriction reprograms the single-cell transcriptional landscape of rattus norvegicus aging. *Cell* 180(5):984–1001.e22.

MacMicking J, Xie QW, Nathan C. (1997). Nitric oxide and macrophage function. *Annu Rev Immunol*:15323–15: 323-350.

Madeo F, Carmona-Gutierrez D, Hofer SJ, Kroemer G. (2019). Caloric restriction mimetics against age-associated disease: Targets, mechanisms, and therapeutic potential. *Cell Metab* 29(3):592–610.

Man AWC, Zhou Y, Xia N, Li H. (2020). Involvement of gut microbiota, microbial metabolites and interaction with polyphenol in host immunometabolism. *Nutrients* 12(10):1–29.

Mattson MP. (2008). Hormesis defined. *Ageing Res Rev* 7(1):1.

Mauer J, Chaurasia B, Goldau J, Vogt MC, Ruud J, Nguyen KD, …, Hausen AC. (2014). Signaling by IL-6 promotes alternative activation of macrophages to limit endotoxemia and obesity-associated resistance to insulin. *Nat Immunol* 15(5):423–430.

Mills EL, Ryan DG, Prag HA, Dikovskaya D, Menon D, Zaslona Z, …, Costa ASH. (2018). Itaconate is an anti-inflammatory metabolite that activates Nrf2 via alkylation of KEAP1. *Nature* 556(7699):113–117.

Mitchell JB, Pizza FX, Paquet A, Davis BJ, Forrest MB, Braun WA. (1998). Influence of carbohydrate status on immune responses before and after endurance exercise. *J Appl Physiol* 84(6):1917–1925.

Nasr MJC, Geerling E, Pinto AK. (2022). Impact of obesity on vaccination to SARS-CoV-2. *Front Endocrinol* 13: 898810.

Nederveen JP, Warnier G, Di Carlo A, Nilsson MI, Tarnopolsky MA. (2021). Extracellular vesicles and exosomes: Insights from exercise science. *Front Physiol* 11: 604274.

Nehlsen-Cannarella SL, Fagoaga OR, Nieman DC, Henson DA, Butterworth DE, Schmitt RL,…, Warren BJ (1997). Carbohydrate and the cytokine response to 2.5 h of running. *J Appl Physiol* 82(5):1662–1667.

Ni FF, Li CR, Liao JX, Wang GB, Lin SF, Xia Y, Wen JL. (2016). The effects of ketogenic diet on the Th17/Treg cells imbalance in patients with intractable childhood epilepsy. *Seizure:*38: 17–22.

Nieman DC, Fagoaga OR, Butterworth DE, Warren BJ, Utter A, Davis JM, Henson DA, Nehlsen- Cannarella SL. (1997). Carbohydrate supplementation affects blood granulocyte and monocyte trafficking but not function after 2.5 h or running. *Am J Clin Nutr* 66(1):153–159.

Nieman DC, Henson DA, McAnulty SR, McAnulty L, Swick NS, Utter AC, … Morrow JD. (2002). Influence of vitamin C supplementation on oxidative and immune changes after an ultramarathon. *J Appl Physiol (1985)* 92(5):1970–1977.

Nieman DC, Nehlsen-Cannarella SL, Fagoaga OR, Henson DA, Shannon M, Davis JM, …, Hisey CL. (1999). Immune response to two hours of rowing in elite female rowers. *Int J Sports Med* 20(7):476–481.

Nieman DC, Pence BD. (2020). Exercise immunology: Future directions. *J Sport Heal Sci* 9(5):432–445.

Nieman DC, Wentz LM. (2019). The compelling link between physical activity and the body's defense system. *J Sport Heal Sci* 8(3):201–217.

Nikolich-Žugich J. (2018). The twilight of immunity: Emerging concepts in aging of the immune system. *Nat Immunol* 19(1):10–19.

Okawa T, Nagai M, Hase K. (2021). Dietary intervention impacts immune cell functions and dynamics by inducing metabolic rewiring. *Front Immunol* 11:623989 .

O'Neill LAJ, Kishton RJ, Rathmell J. (2016). A guide to immunometabolism for immunologists. *Nat Rev Immunol* 16(9):553–565.

Oren R, Farnham AE, Saito K, Milofsky E, Karnovsky ML. (1963). Metabolic patterns in three types of phagocytizing cells. *J Cell Biol* 17(3):487–501.

Painter SD, Ovsyannikova IG, Poland GA. (2015). The weight of obesity on the human immune response to vaccination. *Vaccine* 33(36):4422.

Palmowski J, Gebhardt K, Reichel T, Frech T, Ringseis R, Eder K,…, Krüger K. (2021). The impact of exercise serum on selected parameters of CD4+ T cell metabolism. *Immuno* 1(3):119–131.

Parry-Billings M, Budgett R, Koutedakis Y, Blomstrand E, Brooks S, Williams C, … Newsholme EA. (1992). Plasma amino acid concentrations in the overtraining syndrome: possible effects on the immune system. *Med Sci Sports Exerc* 24(12):1353–1358.

Pedersen BK, Helge JW, Richter EA, Rohde T, Kiens B. (2000). Training and natural immunity: Effects of diets rich in fat or carbohydrate. *Eur J Appl Physiol* 82(1–2):98–102.

Pence BD. (2021). Aging and monocyte immunometabolism in COVID-19. *Aging* 13(7):9154–9155.

Pence BD, Yarbro JR. (2018). Aging impairs mitochondrial respiratory capacity in classical monocytes. *Exp Gerontol*:108112–108117.

Phillips T, Childs AC, Dreon DM, Phinney S, Leeuwenburgh C. (2003). A dietary supplement attenuates IL-6 and CRP after eccentric exercise in untrained males. *Med Sci Sports Exerc* 35(12):2032–2037.

Porto AA, Gonzaga LA, Benjamim CJR, Valenti VE. (2023). Absence of effects of L-Arginine and L- Citrulline on inflammatory biomarkers and oxidative stress in response to physical exercise: A systematic review with meta-analysis. *Nutrients* 15(8):1995.

Reddy A, Bozi LHM, Yaghi OK, Mills EL, Xiao H, Nicholson HE, …, Paulo JA. (2020). pH-Gated succinate secretion regulates muscle remodeling in response to exercise. *Cell* 183(1):62–75.e17.

Richter EA, Ruderman NB. (2009). AMPK and the biochemistry of exercise: Implications for human health and disease. *Biochem J* 418(2):261–275.

Ristow M, Zarse K, Oberbach A, Kloting N, Birringer M, Kiehntopf M, … Bluher M. (2009). Antioxidants prevent health-promoting effects of physical exercise in humans. *Proc Natl Acad Sci U S A* 106(21):8665–8670.

Rohde T, MacLean DA, Pedersen BK. (1998). Effect of glutamine supplementation on changes in the immune system induced by repeated exercise. *Med Sci Sports Exerc* 30(6):856–862.

Rombaldova M, Janovska P, Kopecky J, Kuda O. (2017). Omega-3 fatty acids promote fatty acid utilization and production of pro-resolving lipid mediators in alternatively activated adipose tissue macrophages. *Biochem Biophys Res Commun* 490(3):1080–1085.

Roy DG, Kaymak I, Williams KS, Ma EH, Jones RG. (2020). Immunometabolism in the tumor microenvironment. *Annu Rev Cancer Biol* 5: 137–159.

Saini RK, Keum YS. (2018). Omega-3 and omega-6 polyunsaturated fatty acids: Dietary sources, metabolism, and significance — A review. *Life Sci* 203:255–267.

Savendahl L, Underwood LE. (1997). Decreased interleukin-2 production from cultured peripheral blood mononuclear cells in human acute starvation. *J Clin Endocrinol Metab* 82(4):1177–1180.

Scharhag J, Meyer T, Gabriel HHW, Auracher M, Kindermann W. (2002). Mobilization and oxidative burst of neutrophils are influenced by carbohydrate supplementation during prolonged cycling in humans. *Eur J Appl Physiol* 87(6):584–587.

Shi LZ, Wang R, Huang G, Vogel P, Neale G, Green DR, Chi H. (2011). HIF1α-dependent glycolytic pathway orchestrates a metabolic checkpoint for the differentiation of TH17 and Treg cells. *J Exp Med* 208(7):1367–1376.

Singh R, Rathore SS, Khan H, Karale S, Chawla Y, Iqbal K, …, Tekin A. (2022). Association of obesity with COVID-19 severity and mortality: An updated systemic review, meta-analysis, and meta-regression. *Front Endocrinol* 13:780872.

Spadaro O, Youm Y, Shchukina I, Ryu S, Sidorov S, Ravussin A, …, Aladyeva E. (2022). Caloric restriction in humans reveals immunometabolic regulators of health span. *Science* 375(6581):671–677.

Stampley JE, Cho E, Wang H, Theall B, Johannsen NM, Spielmann G, Irving BA. (2023). Impact of maximal exercise on immune cell mobilization and bioenergetics. *Physiol Rep* 11(11):e15753.

Straub RH, Cutolo M, Buttgereit F, Pongratz G. (2010). Energy regulation and neuroendocrine–immune control in chronic inflammatory diseases. *J Intern Med* 267(6):543–560.

Sureda A, Córdova A, Ferrer MD, Tauler P, Pérez G, Tur JA, Pons A. (2009). Effects of L-citrulline oral supplementation on polymorphonuclear neutrophils oxidative burst and nitric oxide production after exercise. *Free Radic Res* 43(9):828–835.

Tannahill GM, Curtis AM, Adamik J, Palsson-Mcdermott EM, McGettrick AF, Goel G, …, Bernard NJ. (2013). Succinate is an inflammatory signal that induces IL-1β through HIF-1α. *Nature* 496(7444):238–242.

Theall B, Stampley J, Cho E, Granger J, Johannsen NM, Irving BA, Spielmann G. (2021). Impact of acute exercise on peripheral blood mononuclear cells nutrient sensing and mitochondrial oxidative capacity in healthy young adults. *Physiol Rep* 9(23 e15147).

Thies F, Nebe-von-Caron G, Powell JR, Yaqoob P, Newsholme EA, Calder PC. (2001). Dietary supplementation with eicosapentaenoic acid, but not with other long-chain n-3 or n-6 polyunsaturated fatty acids, decreases natural killer cell activity in healthy subjects aged >55 y. *Am J Clin Nutr* 73(3):539–548.

Toft AD, Thorn M, Ostrowski K, Asp S, Moller K, Iversen S, … Pedersen BK. (2000). N-3 polyunsaturated fatty acids do not affect cytokine response to strenuous exercise. *J Appl Physiol (1985)* 89(6):2401–2406.

Tsai HH, Chang SC, Chou CH, Weng TP, Hsu CC, Wang JS. (2016). Exercise training alleviates hypoxia-induced mitochondrial dysfunction in the lymphocytes of sedentary males. *Sci Rep* 6: 35170.

Van Avondt K, Strecker JK, Tulotta C, Minnerup J, Schulz C, Soehnlein O. (2023). Neutrophils in aging and aging-related pathologies. *Immunol Rev* 314(1):357–375.

Vats D, Mukundan L, Odegaard JI, Zhang L, Smith KL, Morel CR, …, Murray PJ. (2006).Oxidative metabolism and PGC-1β attenuate macrophage-mediated inflammation. *Cell Metab* 4(1):13–24.

Walrand S, Moreau K, Caldefie F, Tridon A, Chassagne J, Portefaix G, … Boirie Y. (2001). Specific and nonspecific immune responses to fasting and refeeding differ in healthy young adult and elderly persons. *Am J Clin Nutr* 74(5):670–678.

Walsh NP. (2019). Nutrition and athlete immune health: New perspectives on an old paradigm. *Sports Med* 49(Suppl 2):153–168.

Wang X, Zhou Y, Tang D, Zhu Z, Li Y, Huang T, …, Yu W. (2019). ACC1 (Acetyl Coenzyme A Carboxylase 1) is a potential immune modulatory target of cerebral ischemic stroke. *Stroke* 50(7):1869–1878.

Wang Y, Wu R. (2022). The effect of fasting on human metabolism and psychological health. *Dis Markers*. 2022:5653739.

Weiss HJ, Angiari S. (2020). Metabolite transporters as regulators of immunity. *Metabolites* 10(10):1–18.

Winn NC, Cottam MA, Wasserman DH, Hasty AH. (2021). Exercise and adipose tissue immunity: Outrunning inflammation. *Obesity* 29(5):790–801.

Yarbro JR, Emmons RS, Pence BD. (2020). Macrophage immunometabolism and inflammaging: Roles of mitochondrial dysfunction, cellular senescence, CD38, and NAD. *Immunometabolism* 2(3):e200026.

Youm YH, Nguyen KY, Grant RW, Goldberg EL, Bodogai M, Kim D, …, Planavsky N. (2015). The ketone metabolite β-hydroxybutyrate blocks NLRP3 inflammasome–mediated inflammatory disease. *Nat Med* 21(3):263–269.

Zasłona Z, O'Neill LAJ. (2020). Cytokine-like roles for metabolites in immunity. *Mol Cell* 78(5):814–823.

Zhang D, Tang Z, Huang H, Zhou G, Cui C, Weng Y, …, Kim S. (2019). Metabolic regulation of gene expression by histone lactylation. *Nature* 574:575–580.

15 Exercise immunology and interactions with psychological stress

Courtney A. Bouchet and Monika Fleshner

Introduction

Immune responses protect against infections, eliminate senescent and neoplastic cells, contribute to brain development, muscle repair, and wound healing, sustain tolerance to foreign proteins and commensal bacteria, remove toxins, and facilitate liver function. Much is known about how immune cells distinguish self from non-self, damaged cells from undamaged cells, infected cells from uninfected cells, and the nature of receptor-ligand activation signals. While immune activation can arise from a variety of sources, tight regulation is required to ensure rapid and targeted responses against these threats. In particular, the sympathetic nervous system (SNS) has a key role in activating innate and adaptive immune cells via the release of catecholamines, while the Hypothalamic-Pituitary–Adrenal (HPA) axis is involved in immune regulation and control, facilitating the return to homeostasis via the release of glucocorticoids. When exposed to acute short-term psychological stress, the SNS instigates a "fight or flight" response which increases immune surveillance and activates various components of the innate immune system (including neutrophils and macrophages) and adaptive immune system (including T cells and NK cells). In parallel, HPA axis activation ensures that the immune response mounted against the potential pathogenic threat remains controlled, measured, and transient. When exposure to the psychological stressor is repeated over extended periods of time, immune-cell sensitivity to glucocorticoids is reduced (Cohen et al., 2012), which ultimately leads to chronic immune activation, low-grade inflammation, and increased risk of infection. More recently, immunologists have recognised the critical need to understand the complex regulation of these responses (Afonina, Zhong, Karin, & Beyaert, 2017). Many immune-related pathologies, for example, are due to excessive or inappropriate immune responses. While there is good evidence that exercise is generally beneficial for immune function, it is possible that these effects are brought about by protecting the immune system from so-called "immune-disruptive challenges", including long-term stress.

Immune-disruptive challenges can lead to immune system regulation failure that is associated with autoimmunity, cancer, neurodegeneration, inflammatory bowel disease, and latent viral reactivation (Bennett et al., 2012; Houston, 2023; Klopack, 2023; Santos, Ostler, Harrison, & Jones, 2023). Many factors can disrupt immune responses. Some cannot be avoided, such as aging, genetics, and sex, whereas others are difficult to avoid – but not impossible to manage – such as stress (Segerstrom & Miller, 2004). The term 'stress' is widely used in many fields, from physics and materials science to engineering and biology. The consensus among scientists is that in 1926, Hans Selye first introduced stress in a biological context. He offered the following definition: "the non-specific response of the body to any demand placed upon it." This vague definition has resulted in the use of stress as a noun, a verb, and an adjective. When preparing his book, "The Stress of Life," Seyle acknowledged that "stress, in addition to

DOI: 10.4324/9781003256991-15

being itself, is also the cause of itself, and the result of itself" (Seyle, 1978). Since then, stress biologists describe stress or the stress response as the consequence of exposure to a stressor and a stressor as an aversive event that the organism would choose to avoid if it could.

In 2018, the National Institutes of Mental Health, the government agency that provides major funding for basic and clinical research in the United States, encouraged researchers to appreciate several key factors when studying stress biology. First, the stress response results in a spectrum of changes ranging from adaptive to maladaptive. In other words, stress is not always bad. Second, the stress response impacts multiple physiological systems with unique chronologies and impacted systems may in turn influence each other. For example, acute traumatic and chronic stressors may disturb sleep before impacting the immune system and disturbed sleep contributes to the development of immune dysregulation. Third, stress affects males and females differently. The research on sex differences on the immune response to stress is relatively recent since women have historically been underrepresented and understudied in biological studies. To counter this dearth of data, the National Institutes of Health in the United States enacted the Revitalization Act of 1993, which led to an uptake in the inclusion of males and females in research globally. This mandate has resulted in a plethora of studies reporting sex differences in the nature and health consequences of stressor exposure (Galovski, Blain, Chappuis, & Fletcher, 2013; Schoech, Allie, Salvador, Martinez, & Rivas, 2021; Strewe et al., 2019; Weinstock, Razin, Schorer-Apelbaum, Men, & McCarty, 1998; Zavala, Fernandez, & Gosselink, 2011).

Adverse experiences and stressors can modulate the central and peripheral nervous systems, the endocrine system, the immune system, and more recently, the gut microbiota (Mika & Fleshner, 2016; Mika, Rumian, Loughridge, & Fleshner, 2016; Mika et al., 2015; Thaiss, 2023; Wang et al., 2023). Systems impacted by stressors display adaptation strategies at the genetic, protein, cellular, and systems level; however, adaptation can fail which is sometimes described as the 'tipping point'. Exceeding the tipping point for the immune system, for example, can result in inappropriate inflammatory responses and impaired aspects of immunity (Fleshner, 2013; M. Fleshner, Bellgrau, Laudenslager, Watkins, L. R. & Maier, 1995; Fleshner, Nguyen, Cotter, Watkins, & Maier, 1998). It is possible that leading a physically active lifestyle increases this 'tipping point' (see Figure 15.1). Indeed, there are compelling human studies showing that

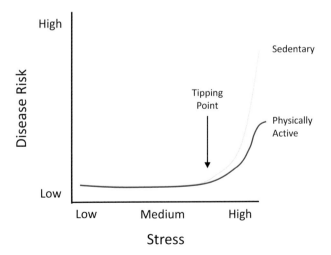

Figure 15.1 Regular exercise and physical activity impacts immunity and disease risk when the immune system is disrupted, for example, by exposure to acute traumatic, repeated, or chronic stressors.

excessive short-term acute and long-term chronic psychological stressors can exceed organisms' capacity to adapt, resulting in adaptation failure and exhaustion of stress-responsive systems. For example, meta-analysis based on 293 independent studies found that chronic psychological stressors, such as dementia caregiving, living with a disability, and unemployment, have negative effects on multiple measures of immune system functioning and contribute to immune dysregulation (Segerstrom & Miller, 2004). These findings indicate that exposure to psychological stressors can exceed the 'tipping point' and contribute to dysregulation of immune function.

Immune-disruptive challenges

Stress

The impact of psychological stress on poor health outcomes is well documented, including links with alterations to aspects of immune system functioning (Morey, Boggero, Scott, & Segerstrom, 2015). This has been demonstrated across multiple facets of the immune system, ranging from increased circulation of pro-inflammatory cytokines in adults exposed to chronic psychological stressors, to impaired *in vivo* immune responses such as wound healing following marital disputes (Kiecolt-Glaser et al., 2005). However, the dichotomic impact of psychological stress on immunity mainly depends on the duration and intensity of the stressor. For example, the Trier Social Stress Test, a public speaking task that includes a surprise mental arithmetic test (Allen et al., 2017), induces the rapid release of IL-1β, IL-6, and TNF-α (Marsland, Walsh, Lockwood, & John-Henderson, 2017), which return to baseline 2 hours after test completion. However, when adults are exposed to sustained and intense psychological stress (e.g., as marked by low socioeconomic status), the same pro-inflammatory cytokines are elevated in circulation chronically (Knight et al., 2016), which in turn can lead to the development of inflammatory diseases such as Rheumatoid Arthritis (Straub, 2014). Further, a recent study identified impairments in mitogen-stimulated proliferation of lymphocytes from caregivers of patients with Alzheimer's disease and after spouse loss (Wilson et al., 2020). Interestingly however, following bereavement, lymphocyte proliferation capabilities improved over time, especially in those with a robust social-support network. Taken together, it is clear that sustained exposure to intense psychological stressors has deleterious impact on immunity.

Exercise

While repeated short-duration bouts of moderate-to-vigorous intensity exercise has many beneficial effects on health and immune system functioning, extreme volumes and intensities of exercise, especially if undertaken over weeks and months can lead to overexertion and can disrupt aspects of immune system functioning, potentially increasing risk of some diseases (Knechtle & Nikolaidis, 2018; Nakayasu et al., 2023). For example, an ultramarathon is classified as any race longer than a traditional marathon of 42.195 km (which typically takes between 3 and 4 hours to complete, depending on training status). Ultra-marathons however, consist of exercising for large chunks of a day (e.g., 6–12 hours), whole days (i.e., 24 hours) or even multiple days. These events and the training for them place substantial strain on physiology and many studies have reported prolonged muscle damage, inflammation, and alterations to aspects of immunity that can last for at least a month (Turner, Bennett, Bosch, Griffiths, & Aldred, 2014). Indeed, ultra-endurance events and overexertion might be associated with impaired immune surveillance (as discussed in Chapter 9) and thus, extreme forms of exercise could also be an immune-disruptive challenge, as with chronic long-term psychological stress. However, most studies in exercise immunology have examined less extreme forms of exercise, such as regular marathon running. Despite the shorter duration compared to ultra-endurance events, one of the early prominent studies in exercise immunology published in 1983 showed that a marathon run increased the risk of developing infections and was one of the first studies indicating that exercise could be an immune-disruptive challenge. For example, in 1983, Peters and Bateman studied 150 ultramarathon runners and reported that 33% of the athletes' experienced symptoms of upper respiratory tract infection compared to controls (15%). In addition, runners with the faster race times had the highest incidence and those with slow race times had no greater incidence of upper respiratory tract infection than non-running controls (Peters & Bateman, 1983). Furthermore, infections were positively correlated with the number of kilometers ran per week during training, with athletes who ran less than 65 km per week exhibiting significantly fewer symptomatic cases of upper respiratory infections compared to athletes who ran more than 65 km per week during training. While it is now generally accepted that many of these older studies in exercise immunology did not fully consider confounding factors (e.g., exposure to infections by crowds) and that bouts of exercise probably do not impair immune function (Simpson et al., 2020) (see Chapter 9) it is worth emphasising that most studies do not consider interaction between exercise and stress when examining the immune system.

Interaction between stress and exercise

The previous section reviewed findings from animal and human studies that psychological or physical stressors contribute to immune dysregulation. However, the "*stress of life*" often involves complex stressors that cross a wide range of psychological and physical challenges. For example, the Spielmann laboratory conducted a series of studies that revealed the greatest negative immune and health impacts of an individual bout of exercise to exhaustion and excessive training is seen when combined with high levels of psychological stress associated with competition and academic study (Theall et al., 2020; H. Wang et al., 2023). Indeed, the overwhelming body of exercise immunology literature presented in this textbook is mostly focused on immunological changes in response to individual bouts of exercise or exercise training across the lifespan and healthspan, but few studies have attempted to identify the role played by extraneous factors such as the allostatic stress load, or cumulative burden of chronic stressors, for participants. In these two studies, Theall et al. (2020) and Wang et al. (2023) found that elite level collegiate swimmers (n = 15, age 19.9 ± 0.7 years old) exhibited poor sleep quality, low levels of energy and elevated symptoms of stress and anxiety as measured by the validated Daily analysis of Life Demands of Athletes questionnaire when peak competitive season and finals coincided (Theall et al., 2020; Wang et al., 2023). This heightened level of stress was not only associated with impaired performance during a maximal swimming test, it was also associated with impaired immune-cell trafficking in response to this individual bout of maximal exercise (Theall et al., 2020), suggesting that the immune response to physical stressor is influenced by psychological stress load.

The confounding effect of psychological stress on the immune changes to individual bouts of exercise are not limited to athletic populations. Edwards et al. (2018) using a similar experimental approach to the Spielmann group but in healthy recreationally active men (n = 64), reported that exposure to an exercise treadmill challenge impacted immune responses to a lesser extent if participants reported moderate levels of anxiety prior to the exercise challenge compared to those that reported lower levels. In this elegant study, Edwards assessed the confounding effects of anxiety and overall psychological well-being and stress on contact hypersensitivity to a novel antigen (Diphenylcyclopropenone – DPCP), an *in vivo* measure of immune function (as described in Chapter 3), before and after 120 minutes of moderate-intensity running at 60% VO2$_{max}$. Interestingly, participants who self-reported moderate levels of anxiety exhibited a greater increase in skinfold thickening in response to the DPCP challenge than those that reported not to be anxious. Furthermore, they reported a positive correlation between pre-exercise anxiety levels and post-exercise *in vivo* immune response ($r = 0.39$). While these findings appear to suggest that a moderate level of stress may be beneficial for the immune response in healthy adults, it is important to highlight that participants with high levels of stress and anxiety were not included in this research. Considering that sustained exposure to psychological and physiological stressors has been associated with increased the risk of infections in adults of all ages (Segerstrom, 2004), these findings highlight the clear synergies between psychological and physical stressors, further advocating for exercise immunologists to account for participants' allostatic stress load when interpreting results.

It is worth highlighting that the source of stress which might interact with exercise could come from a variety of experiences, such as academic studying, pressure in the setting of employment or elite sporting performance, personal relationship or family issues including illness, bereavement, caregiving, or general complications linked to socioeconomic status. Many of these experiences can lead to stress, anxiety, depression, and can also impact other aspects of life, such as health behaviour choices (e.g., diet, alcohol, drug use) and sleep. While it is beyond the scope of this chapter to review how all of these stressors can influence aspects of immunity and the immune response to exercise, this concept is becoming increasingly more appreciated, and readers are directed to the review by Walsh (2018) which specifically highlights the role of stress in elite sport settings (Walsh, 2018). However, a key area that has received attention in exercise immunology is poor sleep quantity and quality – which is well known to be impacted by stress – and also has detrimental impacts on various aspects of immunity, including resting immune competency and the immune response to exercise. While total sleep deprivation, such as 64 hours of total sleep restriction, has been identified to negatively impact T cell counts in healthy adults almost 30 years ago (Dinges et al., 1994), such extreme exposure to altered sleep patterns is uncommon in the general population. However, recent studies have shown that even disruption over 8 hours of nighttime sleep is sufficient to alter the exercise-induced pattern of T cell mobilization in healthy cyclists following a 40km time trial (Ingram, Simpson, Malone, & Florida-James, 2015). In a similar study design, it has been shown that sleep fragmentation increases the magnitude of lymphocyte mobilization in response to exercise – which might be indicative of exaggerated inflammatory responses to exercise – but had no effect on neutrophil function (Ellis, Lee, & Turner, 2021). Deleterious effects of altered sleep quality and quantity on immune outcomes have also been replicated in so-called "tactical athletes," such as warfighters. Indeed, a recent study showed that restricting sleep for 2 hours of night over 72 hours in military personnel was associated with significant reductions in wound healing time (using a blister model as described in Chapter 3), (Smith et al., 2018) mediated by altered immune-cell infiltration and cytokine release within the blisters (Smith et al., 2022). Considering the prevalence of improper sleep quality and quantity in the modern era, the impact of sleep on psychological stress and immune function is of clear clinical relevance – and the current evidence to date implies that there is interaction between stress and exercise when considering these factors as potential immune-disruptive challenges.

Exercise and physical activity reduce stress-associated increases in illness

In 1988, Brown and Siegel conducted an important study that is less well known to the exercise immunology community but yielded important results. Experimental participants were 364 females, mean age of 13 years and 10 months who attended a private secondary school. Using previously validated surveys, stress levels, hours spent weekly on physically active behaviours, and illnesses were measured at two time points. The results showed that high levels of stress increased the number of illnesses for girls who were less physically active but not highly physically active girls. During times of low stress, disease incidence was not impacted by differences between physically active and sedentary lifestyles. This is one of the first reports supporting the supposition that physical activity promotes health by buffering the negative impact of stress on incidences of disease (Brown & Siegel, 1988). However, a limitation of the findings by Brown and Siegel is how disease data were collected (i.e., health questionnaires were used to assess disease symptoms, which is different from disease risk). Nonetheless, the findings reported by Brown and Siegel have been replicated in both human and animal studies and extended using improved measures of disease risk. These findings indicate that adaptations produced by regular exercise and physical activity promote resistance and resilience to stress, empowering an organism to cope with and endure severe and chronic stressors before reaching the tipping point and experiencing stress-associated increases in disease risk. The following cited publications establish that regular exercise and physical activity reduce the impact of stress-associated increases in disease risk, severity, and recovery. These include evidence-based research on the role of adequate nutritional intake and various micronutrients, such as phenolic acids on disease risk (Chen 1995, Afnan, Saleem et al. 2022), overall reduction in risks of developing cardiovascular and respiratory diseases (Aich, Potter et al., 2009, Poirier, Gendron et al., 2023, Slusher & Acevedo, 2023), cancer (Adraskela, Veisaki et al., 2017), and poor metabolic and cognitive health (Pruimboom, Raison et al., 2015, Joisten, Rademacher et al., 2019). In addition, studies linking poor psychological health, chronic exposure to stress, impaired immune function, and overall disease risks are plentiful (Malecki, Nikodemova et al., 2021, Simpson, Bosslau et al., 2021, Park, Prochnow et al., 2023, Warreman, Nooteboom et al., 2023, Yazgan, Bartlett et al., 2023).

Exercise and physical activity reduce stress-associated immune dysregulation

The previous section introduced evidence that exercise and physical activity reduce stress-associated increases in disease, disease risk, severity, and recovery. The following section describes evidence that regular, moderate-intensity exercise and physical activity reduce stress-associated changes in immunity and suggests how such changes in immunity could impact disease occurrence and severity.

Inappropriate inflammation

Physical and psychological stressors trigger inappropriate inflammation and exacerbate inflammatory diseases. There are many examples in the human literature establishing that a physically active lifestyle reduces a cadre of disease with inflammatory etiologies, including cardiovascular disease (Ding & Xu, 2023; Slusher & Acevedo, 2023), multiple sclerosis (Silveira, Jeng, Cutter, & Motl, 2023), inflammatory bowel disease (Popa, Pirlog, Alexandru, & Gheonea, 2022; Schneider et al., 2023), and neuroinflammatory diseases (Gajewski et al., 2022). Given that regular exercise and physical activity reduce the impact of stress on inflammatory-disease exacerbation (Ding & Xu, 2023; Kucuk, 2022), it is feasible that the constraint of stress-induced inflammation plays a role (Yang, Gao, Du, Yang, & Jiang, 2017). In addition to inflammatory mechanisms, exercise and physical activity also change leukocyte migration patterns across the body (Kunz et al., 2018), delay immunosenescence (Niemiro et al., 2022; Padilha, Von Ah Morano, Kruger, Rosa-Neto, & Lira, 2022; Rosa-Neto et al., 2022), and modulate the gut microbiota (Mika et al., 2016; Mika et al., 2015; Noce et al., 2023). All of these changes could also contribute to how exercise and physical activity reduce stress-associated disease exacerbation and increased disease risk. Experiments testing the stress-buffering impact of regular exercise and physical activity on disease-specific immunity, however, are rare.

Viral reactivation

Stressor exposure can dysregulate and suppress many aspects of acquired immunity including viral surveillance and antibody responses. Nearly 90% of adults carry one or more latent viruses (Chen, Deng, & Pan, 2022). There is compelling evidence from both human and animal studies that traumatic acute and chronic stressors allow the reactivation of latent viruses such as Epstein–Barr Virus, Varicella Zoster Virus, Herpes Simplex Virus, and Hepatitis B. The mechanisms responsible for stress-induced viral reactivation are not fully understood. Potential mechanisms reported in the literature are stress hormones (glucocorticoids [cortisol in humans] and epinephrine), stress-induced transcription factors, inappropriate inflammation, altered regulatory T cell function, reduced levels of anti-viral antibody (Ives & Bertke, 2017; Klopack, 2023; Lin et al., 2017; Matundan, Wang, Jaggi, Yu, & Ghiasi, 2021; Santos et al., 2023; Yu et al., 2018). In fact, many of these mechanisms may work together to reactivate viral replication. Santos et al. (2023), for example, recently reported that activation of the glucocorticoid receptor by stress hormones and a stress-induced transcription factor called slug, work together to wake-up or reactivate replication of Herpes Simplex Virus 1 (HSV-1). HSV-1 is the virus that causes cold sores. It infects trigeminal ganglionic neurons and can evade the immune system. Exposure to immune-disruptive stressors is a well-known trigger of HSV-1 reactivation and a flare-up of cold sores. Lacking, however, are studies testing if regular physical activity reduces or prevents stress-associated viral reactivation. The literature highlighting the effect of exposure to stressors on the reactivation of latent viruses by suppressing immune surveillance and increasing viral replication is plentiful. Indeed, a wide range of latent viruses have been shown to reactivate in response to acute and chronic psychological stress, such as HSV-1 (Bonneau, 1996; Ashcraft & Bonneau, 2008; Ortiz, Sheridan et al., 2003; Freeman, Sheridan et al., 2007; Uchakin, Parish et al., 2011; Yu, Geng et al., 2018), HSV-2 (Ives & Bertke, 2017), latent Hepatitis B Virus (Tashiro, Ogawa et al., 2017), Cytomegalovirus (Santos Rocha, Hirao et al., 2018). This is especially relevant for tactical athletes exposed to sustained psychological and physical stressors as part of their professions, including astronauts exposed to isolation and microgravity (Mehta, Cohrs et al., 2004; Pierson, Stowe et al., 2005; Mehta, Bloom et al., 2018; Mehta, Laudenslager et al., 2014; Masini, Bonetto et al., 2022) and warfighters exposed to operational stressors (Hatch-McChesney, Radcliffe et al., 2023).

In vivo antibody responses to vaccination

There is evidence that regular exercise and physical activity prevent stress- and age-associated declines in the *in vivo* antibody responses to novel antigens (Fleshner, 2000; Jasnow, Drazen, Huhman, Nelson, & Demas, 2001; Moraska & Fleshner, 2001; Sun, Pence, Wang, & Woods, 2019) (also see Chapter 7). For example, rats injected with keyhole limpet hemocyanin (KLH) and exposed to an acute, traumatic stressor (90 min, intermittent unpredictable and inescapable electrical shocks to their tail) compared with controls that remain in their home cages have lower anti-KLH antibody in blood. This finding has been replicated many times, and much is understood about the hormonal and cellular changes mediating this effect (Fleshner, Brennan, Nguyen, Watkins, & Maier, 1996; Fleshner, Brohm, Laudenslager, Watkins, & Maier,

1993; Fleshner, Deak, Nguyen, Watkins, & Maier, 2001; M. Fleshner, Hermann, J., Lockwood, Watkins, L.R., Laudenslager, M.L. & Maier, 1995; Fleshner, Laudenslager, Simons, & Maier, 1989; Fleshner et al., 1998; Gazda, Smith, Watkins, Maier, & Fleshner, 2003; Laudenslager et al., 1988; Moraska et al., 2002; Moraska & Fleshner, 2001). Using this robust animal model of stress-induced immune suppression, Moraska et al. (2001) tested the hypothesis that regular physical activity protects appropriate immune responses from dysregulation in the face of immune-disruptive challenges. Adult male rats were allowed to live with running wheels for six weeks. Sedentary controls were housed with locked running wheels. After six weeks, rats were injected with KLH and exposed to an acute, traumatic stressor (90 min, intermittent unpredictable and inescapable electrical shocks to their tail) or remained in their home cages and served as unstressed controls. Blood samples were collected weekly for four weeks, and KLH-specific immunoglobulin (Ig). Specifically, stress reduced anti-KLH IgM, IgG, and IgG2a were measured using ELISA. The results showed that stress suppressed the generation of anti-KLH IgM, IgG, and IgG2a in rats that were sedentary, but not physically active. Physical activity had no impact on the *in vivo* antibody response in the absence of challenge.

Animal studies are required to unravel the mechanisms underlying the protective effect of physical activity on the stress-induced suppression of antibody responses. Mechanisms for stress-associated immune dysregulation are varied, depending on the nature of the stressor and the immune measure (i.e., inflammatory proteins, anti-viral immunity, or antibody responses). Indeed, the stress-induced suppression of the anti-KLH antibody responses reported earlier is a stress-induced reduction in the number of anti-KLH CD4+ T cells and anti-KLH antibody-secreting cells in the spleen (Fleshner et al., 1995; Gazda et al., 2003). Additional studies determined that exposure to an acute traumatic stressor excessively activates central sympathetic neurocircuitry leading to catecholamine depletion of the spleen. Furthermore, it was determined that splenic catecholamine depletion is both necessary and sufficient to reduce anti-KLH antibody responses (Greenwood et al., 2011; Greenwood, Kennedy et al., 2003; Kennedy et al., 2005). Stress-responsive neural circuits such as the central sympathetic nervous neurocircuit, are regions of the brain that are activated by stressor exposure, and drive the behavioural and physiological responses to stress (Herman et al., 2003), including SNS output and the fight/ flight response. Regular physical activity constrains stress-induced activation of central sympathetic neurocircuitry reported to innervate the spleen and prevent splenic catecholamine depletion (Greenwood, Kennedy et al., 2003). Taken together, these findings and others (Day et al., 2004; Fleshner et al., 2002; Slusher & Acevedo, 2023) support the hypothesis that physical activity can reduce the negative impact of stress on immunity by modulating stress-responsive neural circuitry and immune-modulatory peripheral output (Barman, 1990; Greenwood, Kennedy et al., 2003).

The Fleshner lab has also explored if the impact of regular physical activity on antibody suppression after immune-disruptive challenge is translatable to humans. In this case, however, the immune-disruptive challenge was ageing, not stress (see Chapter 8). Using a cross-sectional experimental design, Smith et al. (2004) tested the hypothesis that regular physical activity reduces the immunosuppressive impact of ageing (Smith, Kennedy, & Fleshner, 2004a). Participants were men who were 65–79 and 25–35 years old. Physically active men performed aerobic exercise at least three times per week, and sedentary men had performed no regular exercise for at least two years. Fitness levels were measured using a graded treadmill exercise test. All participants were immunized with KLH. Blood samples were collected on day 0, 7, 14, and 21 days after KLH. Both young and older physically active participants had higher cardiorespiratory fitness than sedentary participants. The results showed that physical activity had no impact on anti-KLH immunoglobulins in young participants. In contrast, physically active older participants compared with sedentary older participants had less age-associated immune decline, as shown by an anti-KLH IgG response that was nearly equal to the younger participants. Compared with all other groups, sedentary older men had the lowest levels of anti-KLH IgM, IgG, and anti-KLH memory T cell responses (measured on day 21 using the delayed sensitivity test). Because this was a cross-sectional study and physically active participants also may practice healthy habits like better sleep and wellness, and regular doctor's visits, it is possible that these behaviours and not physical activity *per se*, were responsible for the improved response observed. In 2008, the Woods lab replicated these findings using an intervention experimental design (Grant et al., 2008). Participants, older sedentary men (~69 years old), were randomly assigned to a 10-week aerobic or stretching-exercise intervention. After 10 weeks, subjects in the aerobic group had higher VO_2 max measures than the stretching group. All subjects were immunized with KLH, and blood samples were collected on weeks 0, 2, 3, and 6. The results were that compared with the stretching group, older subjects participating in the 10-week aerobic intervention had higher levels of anti-KLH IgG1.

There are also studies on humans testing the impact of stress and physical activity on antibody responses to vaccination. Although most studies reported that regular physical activity reduced the psychological impacts of stressor exposure, several studies failed to detect either stress or physical activity-associated changes in antibody responses to vaccination (Burns, Carroll, Ring, Harrison, & Drayson, 2002; Long et al., 2013; Miller et al., 2004; Segerstrom, Hardy, Evans, & Greenberg, 2012; Shayea, Alotaibi, Nadar, Alshemali, & Alhadlaq, 2022; Wong et al., 2013). The participants in these studies varied in age from 18 to 74 years and included males and females. Physical activity was less effective in impacting antibody responses to common vaccinations for influenza, tetanus, and COVID-19 (Long et al., 2013; Miller et al., 2004; Segerstrom et al., 2012; Shayea et al., 2022; Wong et al., 2013). It is unknown why some studies failed to detect an impact of stress or physical activity on antibody responses to vaccination; however, it is possible that the lack of effect is due to the confounding associated with vaccination history and previous infections influencing baseline immunity. Participants injected with a novel antigen such as KLH might help further understand the interaction between stress, physical activity, and vaccine responses because participants will not have any confounding immunological memory for KLH (Smith, Kennedy, & Fleshner, 2004). In addition, primary versus tertiary antibody responses differ in their cellular interactions, signalling, and sensitivity to hormonal regulation (Friedman & Irwin, 2001; Silberman, Wald, & Genaro, 2003).

Exercise and physical activity modulate neural circuitry

This chapter is not intended to comprehensively review all literature in the field. Instead, the studies described were selected because they provide clear experimental evidence from both human and animal work that regular exercise and physical activity exert benefits on immunity when immune function is suboptimal due to immune-disruptive challenges, such as excessive and chronic stress or ageing. The next section will focus on developing the hypothesis that exercise and physical activity reduce the negative impact of stress on immunity by modulating the stress-reactive neural circuitry.

Stress neural circuitry and immunity

One study has highlighted regions in the brain impacted by stressor exposure and associated with stress-induced changes in peripheral immune responses (Chan, Poller, Swirski, & Russo, 2023). The authors linked specific stress-responsive regions with specific stress-evoked changes in immunity. For example, using pseudorabies virus (PRV) trans-synaptic retrograde tracing, Chan et al. describe a "*bone marrow-to-brain neurocircuit*". Retrograde tracing is a technique that labels neurons with a marker that travels in a retrograde manner, (i.e., from the axon terminal toward the soma). A trans-synaptic tracer is one that can 'jump' the synapse, moving from the cell body into axon terminals that innervate it. So, by using this technique, researchers can observe not only the neurons that directly innervate bone marrow but also neural circuits upstream of those neurons. The bone marrow is a primary immune tissue and home to hematopoietic stem cells (HSC). HSCs are the source for every myeloid-lineage and lymphoid-lineage cell in the body. The bone marrow also harbors long-lived antibody-secreting B cells and serves as a reservoir for neutrophils. One hallmark of the acute stress response is a very rapid increase in circulating neutrophils or neutrophilia (Dahdah et al., 2022; Denes et al., 2005). It is logical, therefore, that brain regions that rapidly respond to stressor exposure also innervate the bone marrow. This *bone marrow-to-brain neurocircuit* includes ventral lateral medulla, nucleus of the solitary tract (NTS), locus coeruleus (LC), and paraventricular nucleus of the hypothalamus (PVN) (Dahdah et al., 2022; Denes et al., 2005).

Exercise, physical activity, and neural circuitry

A variety of brain regions, receptors, neurotransmitters, neuropeptides, and neural transporters are sensitive to physical activity (Figure 15.2). Given the multi-system nature of exercise and physical activity, it is not surprising that many brain regions are fundamentally impacted. One of the most robust observations first reported by Cotman and colleagues (Neeper, Gomez-Pinilla, Choi, & Cotman, 1996) is that regular exercise and physical activity increases hippocampal brain derived neurotrophic factor (BDNF) mRNA and protein (Loprinzi & Frith,

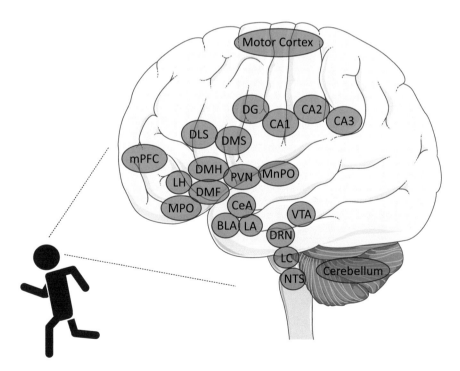

Figure 15.2 Brain regions sensitive to exercise, see Table 15.1 for citations.

Table 15.1 Brain regions sensitive to exercise paradigms

Region	Species	Sex	Type of exercise	Effect of exercise	Citation
PFC (medial prefrontal cortex)	Sprague Dawley rats	Male	VWR (12 days)	↑ Dendritic spine density, length, and expression of synaptic markers	(Brockett et al., 2015)
Cerebral Cortex	Human participants	Male	Bicycle ergometer 25W or 100W	↑ Blood flow to frontal, precentral, postcentral, parietal, temporal, occipital lobes	(Herholz et al., 1987)
Dorsal medial striatum	Fischer 344 rats	Male	VWR (6-7 weeks)	↑ DA (via microdialysis)	(Clark et al., 2015)
Dorsal striatum	Fischer 344 rats	Male	VWR (6 weeks)	↑ p-mTor protein in DLS -- p-mTor protein in DMS	(Lloyd et al., 2017)
VTA / SN; hippocampus (DG, CA3, CA1)	Long-Evans rats	Male	VWR (3 weeks)	↑ BDNF mRNA	(Van Hoomissen et al., 2003)
VTA	F344xBN rats	Male	VWR (2 days)	↑ Leptin sensitivity	(Scarpace et al., 2010)
Hippocampus (DG)	C57BL/6 mice	Female	VWR (12 or 29 days)	↑ Cell proliferation	(van Praag et al., 1999a)
Hippocampus (DG)	C57BL/6 mice	Female	VWR (10 days)	↑ Neurogenesis and long term potentiation	(van Praag et al., 1999b)
Hippocampus (DG, CA2, CA1)	Fischer 344 rats	Male	VWR (6 weeks)	↑ BDNF mRNA	(Greenwood et al., 2009)
Amygdala (LA, BLA, CeA)	Fischer 344 rats	Male	VWR (6 weeks)	↑ 5-HT2C receptor mRNA	(Greenwood et al., 2012)
Hypothalamus	Fischer 344 rats	Male	VWR (6 weeks)	↑ p-mTor protein in DMF, MPO, MnPO -- p-mTor protein in VMH, PVN	(Lloyd et al., 2017)
Hypothalamus	Fischer 344 rats	Male	Forced running wheel	↑ p-mTor protein in DMF, MPO, MnPO ↓ p-mTor protein in VMH	(Lloyd et al., 2017)
Hypothalamus (SON/PVN; PVN-NTS neurons)	Wistar rats	Male	Treadmill running (2x/ day, 5 days/week for 6 weeks)	↑ Neuronal excitability	(Jackson et al., 2005)
Hypothalamus (PVN, DMH, Arc)	Sprague Dawley rats	Male	VWR (7 days)	↑ DMH CRF mRNA ↑ Arc NPY mRNA ↑ DMH NPY mRNA	(Kawaguchi et al., 2005)
Hypothalamus (NTS, PVN)	Wistar-Kyoto rats	Male	Treadmill running 1 hour/day, 5 days/ week for 3 months	↑ Oxytocin mRNA	(Martins et al., 2005)

Region	Species	Sex	Type of exercise	Effect of exercise	Citation
Hypothalamus (NTS, PH)	Sprague Dawley rats	Male	VWR (50 days)	↓ Dendritic arborizations (PH and NTS) ↓ Avg. cell body diameter (PH)	(Nelson et al., 2010)
Cerebellum (vermis)	Sprague Dawley rats	Female	VWR (8 days)	↑ NA intervaricosity intervals ↑ NA varaicosity density	(Nedelescu et al., 2017)
Hypothalamus (PVN)	Sprague Dawley rats	Male	VWR (6 weeks)	↓ cFos mRNA after saline injection	(Campeau et al., 2010)
DRN	Sprague Dawley rats	Male	VWR (6 weeks)	↑ 5-HT1A autoreceptor mRNA	(Greenwood et al., 2003)
LC	HCR and LCR rats, SD rats	Male	VWR (21 days)	↑ Correlation between running distance and galanin mRNA -- TH mRNA	(Murray et al., 2010)
LC	Fischer 344 rats	Male	Treadmill running	↑ Prepro-galanin mRNA -- TH mRNA	(O'Neal et al., 2001)
NTS	Wistar-Kyoto rats	Male	Treadmill running 1 hour/day, 5 days/week for 3 months	↑ Oxytocin mRNA	(Martins et al., 2005)
NTS	Sprague Dawley rats	Male	VWR (50 days)	↓ Dendritic arborizations (PH and NTS) ↓ Avg. cell body diameter (PH)	(Nelson et al., 2010)
RVLM	Spontaneously hypertensive rats	Male	Treadmill running	↓ Glutamate concentration vGluT2 protein	(Zha et al., 2013)
Network connectivity	Human participants	Male and Female	Physical activity self-completed questionnaire	↑ Network connectivity	(Vinodh Kumar et al., 2023)

↑ Increase; ↓ decrease;--no change.
Abbreviations: Voluntary wheel running (VWR), ventral tegmental area (VTA), substantia nigra (SN), dentate gyrus (DG), lateral amygdala (LA), basolateral amygdala (BLA), central amygdala (CeA), supraoptic nucleus (SON), paraventricular nucleus of the hypothalamus (PVN), nucleus tractus solitarius (NTS), dorsomedial hypothalamus (DMH), arcuate nucleus (Arc), Corticotropin releasing factor (CRF), Neuropeptide Y (NPY), Posterior hypothalamus (PH), noradrenaline (NA), dorsal raphe nucleus (DRN), locus coeruleus (LC), high capacity rat strain (HCR), low capacity rat strain (LCR); rostral ventrolateral medulla (RVLM).

2019; Meijer et al., 2020; Zoladz & Pilc, 2010). Equally clear are the well-established impacts of exercise and physical activity on brain dopaminergic systems (Ruiz-Tejada, Neisewander, & Katsanos, 2022). Striatal dopamine, for example, is increased by exercise and physical activity (Bastioli et al., 2022; Fan, Kong, Liu, & Wu, 2022; Ruiz-Tejada et al., 2022). In addition, there is emerging evidence that the impact of striatal dopamine is associated with the innate rewarding properties of active behaviours (Tanner et al., 2022). More specifically, experimental results demonstrate that active behaviours begin as an 'action' and the transition from action to habit is largely dependent on the dorsal striatum and, within the dorsal striatum, activation of a subset of neurons that express the dopamine 1 (D1) receptor D (Nadel et al., 2021; Tanner et al., 2022). BDNF and dopamine pathways are two of many neural pathways reported to be sensitive to exercise and physical activity in a variety of brain regions. Table 15.1 provides the details.

Exercise, physical activity, and stress-reactive neural circuitry

There have been important advances in our understanding of how exercise and physical activity change the impact of stress on the brain in regions associated with depression and anxiety (Hwang et al., 2023; Wanjau et al., 2023). The focus of this chapter, however, is to present evidence that exercise and physical activity can reduce the negative impact of stress on immunity by modulating the stress-reactive neural circuitry and immune-relevant peripheral output. Using standard tools in neuroscience, it is possible to determine the intensity and connectivity of brain regions activated by stressor exposure and trace the neural outputs to peripheral tissues. Using PRV trans-synaptic retrograde tracing, brain regions that innervate the spleen have been shown to include the NTS, LC, PVN, dorsal motor nucleus of the vagus (DMV), and the bed nucleus of

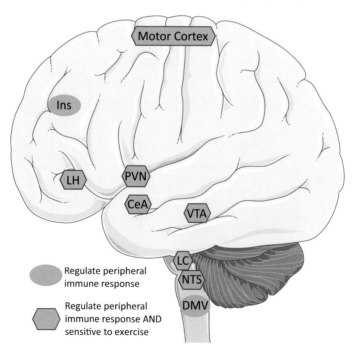

Figure 15.3 Brain regions at the intersection of stress, immune function, and regular exercise and physical activity.

the stria terminalis. All regions that are activated by stressor exposure and subserve behavioural and physiological features of the stress response, including endocrine and autonomic responses (Clark et al., 2015; Clark et al., 2014; Greenwood, Foley et al., 2003; Greenwood, Kennedy et al., 2003; Kennedy et al., 2005).

Summary

Acute short-term psychological stress induces a "fight or flight" response, similar to individual bouts of exercise that seem to enhance aspects of immunity, such as immune surveillance. However, exposure to excessive and sustained psychological stressors is associated with impaired immunity and increased risk of infections. While a physically active lifestyle is associated with lower disease risk and better immune function, evidence presented in this chapter supports the concept that regular exercise and physical activity protect immune function from so-called immune-disruptive challenges, such as excessive and chronic stress, by modulating stress-reactive neural circuitry and relevant immune outcomes associated with improved health. Thus, studies in exercise immunology should pay greater attention to potentially confounding, influencing, or underlying mechanistic effects of stress when interpreting the effects that exercise and physical activity have on aspects of immune function (Figure 15.3).

Key Points

- An appropriate immune response is important for countering infection with viruses, bacteria, parasites, and fungi, while also defending against tumors.
- Exposure to excessive and sustained stress is associated with impaired immunity, and increased risk of infections, therefore stress is considered to be an immune-disruptive challenge.
- Moderate-intensity exercise and physical activity protect the immune system from stress-induced dysregulation.
- If immune function is suboptimal due to immune-disruptive challenges, moderate-intensity exercise and physical activity have beneficial effects.
- Exercise and physical activity impact stress-sensitive brain regions, many of which innervate the spleen or bone marrow and are involved in the immune response.
- Evidence reviewed in this chapter supports the hypothesis that regular moderate-intensity exercise and physical activity reduce the negative impact of stress on immunity by modulating a stress-reactive neural circuitry, influencing immune function.

References

Adraskela, K., Veisaki, E., Koutsilieris, M., & Philippou, A. (2017). Physical exercise positively influences breast cancer evolution. *Clin Breast Cancer, 17*(6), 408–417.

Afnan, Saleem, A., Akhtar, M. F., Sharif, A., Akhtar, B., Siddique, R., … Alharthy, S. A. (2022). Anticancer, cardio-protective and anti-inflammatory potential of natural-sources-derived phenolic acids. *Molecules, 27*(21), 7286.

Afonina, I. S., Zhong, Z., Karin, M., & Beyaert, R. (2017). Limiting inflammation-the negative regulation of NF-kappaB and the NLRP3 inflammasome. *Nat Immunol, 18*(8), 861–869.

Aich, P., Potter, A. A., & Griebel, P. J. (2009). Modern approaches to understanding stress and disease susceptibility: A review with special emphasis on respiratory disease. *Int J Gen Med, 2*, 19–32.

Allen, A. P., Kennedy, P. J., Dockray, S., Cryan, J. F., Dinan, T. G., & Clarke, G. (2017). The trier social stress test: Principles and practice. *Neurobiol Stress, 6*, 113–126.

Ashcraft K.A, Bonneau R.H. (2008). Psychological stress exacerbates primary vaginal herpes simplex virus type 1 (HSV-1) infection by impairing both innate and adaptive immune responses. *Brain Behav Immun. 22(8),* 1231–1240.

Barman, S. M. (1990). Descending projections of hypothalamic neurons with sympathetic nerve-related activity. *J Neurophysiol, 64*(3), 1019–1032.

Bastioli, G., Arnold, J. C., Mancini, M., Mar, A. C., Gamallo-Lana, B., Saadipour, K., … Rice, M. E. (2022). Voluntary exercise boosts striatal dopamine release: Evidence for the necessary and sufficient role of BDNF. *J Neurosci, 42*(23), 4725–4736.

Bennett, J. M., Glaser, R., Malarkey, W. B., Beversdorf, D. Q., Peng, J., & Kiecolt-Glaser, J. K. (2012). Inflammation and reactivation of latent herpesviruses in older adults. *Brain Behav Immun, 26*(5), 739–746.

Bonneau R.H. (1996). Stress-induced effects on integral immune components involved in herpes simplex virus (HSV)-specific memory cytotoxic T lymphocyte activation. *Brain Behav Immun. 10(2),* 139–163.

Brown, J. D., & Siegel, J. M. (1988). Exercise as a buffer of life stress: A prospective study of adolescent health. *Health Psychology: Official Journal of the Division of Health Psychology, American Psychological Association, 7*(4), 341–353.

Burns, V. E., Carroll, D., Ring, C., Harrison, L. K., & Drayson, M. (2002). Stress, coping, and hepatitis B antibody status. *Psychosom Med, 64*(2), 287–293.

Chan, K. L., Poller, W. C., Swirski, F. K., & Russo, S. J. (2023). Central regulation of stress-evoked peripheral immune responses. *Nat Rev Neurosci, 24*(10), 591–604.

Chen, J. D. (1995). Benefits of physical activity on nutrition and health status: studies in China. *Asia Pac J Clin Nutr, 4 Suppl 1,* 29–33.

Chen, S., Deng, Y., & Pan, D. (2022). MicroRNA regulation of human herpesvirus latency. *Viruses, 14*(6), 1215.

Clark, P. J., Amat, J., McConnell, S. O., Ghasem, P. R., Greenwood, B. N., Maier, S. F., & Fleshner, M. (2015). Running reduces uncontrollable stress-evoked serotonin and potentiates stress-evoked dopamine concentrations in the rat dorsal striatum. *PLoS One, 10*(11), e0141898.

Clark, P. J., Ghasem, P. R., Mika, A., Day, H. E., Herrera, J. J., Greenwood, B. N., & Fleshner, M. (2014). Wheel running alters patterns of uncontrollable stress-induced cfos mRNA expression in rat dorsal striatum direct and indirect pathways: A possible role for plasticity in adenosine receptors. *Behav Brain Res, 272C,* 252–263.

Cohen, S., Janicki-Deverts, D., Doyle, W. J., Miller, G. E., Frank, E., Rabin, B. S., & Turner, R. B. (2012). Chronic stress, glucocorticoid receptor resistance, inflammation, and disease risk. *Proc Natl Acad Sci U S A, 109*(16), 5995–5999.

Dahdah, A., Johnson, J., Gopalkrishna, S., Jaggers, R. M., Webb, D., Murphy, A. J., … Nagareddy, P. R. (2022). Neutrophil migratory patterns: Implications for cardiovascular disease. *Front Cell Dev Biol, 10,* 795784.

Day, H. E., Greenwood, B. N., Hammack, S. E., Watkins, L. R., Fleshner, M., Maier, S. F., & Campeau, S. (2004). Differential expression of 5HT-1A, alpha 1b adrenergic, CRF-R1, and CRF-R2 receptor mRNA in serotonergic, gamma-aminobutyric acidergic, and catecholaminergic cells of the rat dorsal raphe nucleus. *J Comp Neurol, 474*(3), 364–378.

Denes, A., Boldogkoi, Z., Uhereczky, G., Hornyak, A., Rusvai, M., Palkovits, M., & Kovacs, K. J. (2005). Central autonomic control of the bone marrow: Multisynaptic tract tracing by recombinant pseudorabies virus. *Neuroscience, 134*(3), 947–963.

Ding, Y., & Xu, X. (2023). Dose-response relationship between leisure-time physical activity and biomarkers of inflammation and oxidative stress in overweight/obese populations. *J Sci Med Sport. 26*(11), 616–621.

Dinges, D. F., Douglas, S. D., Zaugg, L., Campbell, D. E., McMann, J. M., Whitehouse, W. G., … Orne, M. T. (1994). Leukocytosis and natural killer cell function parallel neurobehavioral fatigue induced by 64 hours of sleep deprivation. *J Clin Invest, 93*(5), 1930–1939.

Edwards J.P., Walsh N.P., Diment P.C., Roberts R. (2018) Anxiety and perceived psychological stress play an important role in the immune response after exercise. *Exerc Immunol Rev, 24,* 26–34.

Ellis, J. O., Lee, B. J., & Turner, J. E. (2021). One night of sleep fragmentation does not affect exercise-induced leukocyte trafficking or mitogen-stimulated leukocyte oxidative burst in healthy men. *Physiol Behav, 239*, 113506.

Fan, Y., Kong, X., Liu, K., & Wu, H. (2022). Exercise on striatal dopamine level and anxiety-like behavior in male rats after 2-VO cerebral ischemia. *Behav Neurol, 2022*, 2243717.

Fleshner, M. (2000). Exercise and neuroendocrine regulation of antibody production: Protective effect of physical activity on stress-induced suppression of the specific antibody response. *Int J Sports Med, 21(*Suppl 1), S14–19.

Fleshner, M. (2013). Stress-evoked sterile inflammation, danger associated molecular patterns (DAMPs), microbial associated molecular patterns (MAMPs) and the inflammasome. *Brain Behav Immun, 27*(1), 1–7.

Fleshner, M., Bellgrau, D., Laudenslager, M. L., Watkins, L. R., & Maier, S. F. (1995). Stress-induced changes in the mixed lymphocyte reaction is dependent on macrophages but not on shifts in phenotypes. *J Neuroimmunol, 56*, 45–52.

Fleshner, M., Brennan, F. X., Nguyen, K., Watkins, L. R., & Maier, S. F. (1996). RU-486 blocks differentially suppressive effect of stress on in vivo anti-KLH immunoglobulin response. *Am J Physiol, 271*(5 Pt 2), R1344–1352.

Fleshner, M., Brohm, M. M., Laudenslager, M. L., Watkins, L. R., & Maier, S. F. (1993). Modulation of the in vivo antibody response by a benzodiazepine inverse agonist (DMCM) administered centrally or peripherally. *Physiol Behav, 54*(6), 1149–1154.

Fleshner, M., Campisi, J., Deak, T., Greenwood, B. N., Kintzel, J. A., Leem, T. H., … Sorensen, B. (2002). Acute stressor exposure facilitates innate immunity more in physically active than sedentary rats. *Am J Physiol Regul Integr Comp Physiol, 282*(6), R1680–1686.

Fleshner, M., Deak, T., Nguyen, K. T., Watkins, L. R., & Maier, S. F. (2001). Endogenous glucocorticoids play a positive regulatory role in the anti-keyhole limpet hemocyanin in vivo antibody response. *J immunol, 166*(6), 3813–3819.

Fleshner, M., Hermann, J., Lockwood, L. L., Laudenslager, M. L., Watkins, L. R., & Maier, S. F. (1995). Stressed rats fail to expand the CD45RC+CD4+ (Th1-like) T cell subset in response to KLH: Possible involvement of IFN-gamma. *Brain Behav Immun, 9*(2), 101–112.

Fleshner, M., Laudenslager, M. L., Simons, L., & Maier, S. F. (1989). Reduced serum antibodies associated with social defeat in rats. *Physiol Behav, 45*(6), 1183–1187.

Fleshner, M., Nguyen, K. T., Cotter, C. S., Watkins, L. R., & Maier, S. F. (1998). Acute stressor exposure both suppresses acquired immunity and potentiates innate immunity. *Am J Physiol, 275*(3 Pt 2), R870–878.

Freeman M.L., Sheridan B.S., Bonneau R.H., Hendricks R.L. (2007). Psychological stress compromises CD8+ T cell control of latent herpes simplex virus type 1 infections. *J Immunol. 179(1)*, 322–328.

Friedman, E. M., & Irwin, M. (2001). Central CRH suppresses specific antibody responses: Effects of beta-adrenoceptor antagonism and adrenalectomy. *Brain Behav Immun, 15*(1), 65–77.

Gajewski, P. D., Getzmann, S., Brode, P., Burke, M., Cadenas, C., Capellino, S., … Wascher, E. (2022). Impact of biological and lifestyle factors on cognitive aging and work ability in the dortmund vital study: Protocol of an interdisciplinary, cross-sectional, and longitudinal study. *JMIR Res Protoc, 11*(3), e32352.

Galovski, T. E., Blain, L. M., Chappuis, C., & Fletcher, T. (2013). Sex differences in recovery from PTSD in male and female interpersonal assault survivors. *Behav Res Ther, 51*(6), 247–255.

Gazda, L. S., Smith, T., Watkins, L. R., Maier, S. F., & Fleshner, M. (2003). Stressor exposure produces long-term reductions in antigen-specific T and B cell responses. *Stress, 6*(4), 259–267.

Grant, R. W., Mariani, R. A., Vieira, V. J., Fleshner, M., Smith, T. P., Keylock, K. T., … Woods, J. A. (2008). Cardiovascular exercise intervention improves the primary antibody response to keyhole limpet hemocyanin (KLH) in previously sedentary older adults. *Brain Behav Immun, 22*(6), 923–932.

Greenwood, B. N., Foley, T. E., Day, H. E., Campisi, J., Hammack, S. H., Campeau, S., … Fleshner, M. (2003). Freewheel running prevents learned helplessness/behavioral depression: Role of dorsal raphe serotonergic neurons. *J Neurosci, 23*(7), 2889–2898.

Greenwood, B. N., Foley, T. E., Le, T. V., Strong, P. V., Loughridge, A. B., Day, H. E., & Fleshner, M. (2011). Long-term voluntary wheel running is rewarding and produces plasticity in the mesolimbic reward pathway. *Behav Brain Res, 217*(2), 354–362.

Greenwood, B. N., Kennedy, S., Smith, T. P., Campeau, S., Day, H. E., & Fleshner, M. (2003). Voluntary freewheel running selectively modulates catecholamine content in peripheral tissue and c-Fos expression in the central sympathetic circuit following exposure to uncontrollable stress in rats. *Neuroscience, 120*(1), 269–281.

Hatch-McChesney A., Radcliffe P.N., Pitts K.P., Karis A.J., O'Brien R.P., …, Karl J.P. (2023). Changes in Immune Function during Initial Military Training. *Med Sci Sports Exerc, 55(3),* 548–557.

Herman, J. P., Figueiredo, H., Mueller, N. K., Ulrich-Lai, Y., Ostrander, M. M., Choi, D. C., & Cullinan, W. E. (2003). Central mechanisms of stress integration: Hierarchical circuitry controlling hypothalamo-pituitary-adrenocortical responsiveness. *Front Neuroendocrinol, 24*(3), 151–180.

Houston, S. (2023). STAT3 and autoimmunity. *Nat Immunol, 24*(1), 1.

Hwang, D. J., Koo, J. H., Kim, T. K., Jang, Y. C., Hyun, A. H., Yook, J. S., … Cho, J. Y. (2023). Exercise as an antidepressant: exploring its therapeutic potential. *Front Psychiatry, 14*, 1259711.

Ingram, L. A., Simpson, R. J., Malone, E., & Florida-James, G. D. (2015). Sleep disruption and its effect on lymphocyte redeployment following an acute bout of exercise. *Brain Behav Immun, 47*, 100–108.

Ives, A. M., & Bertke, A. S. (2017). Stress hormones epinephrine and corticosterone selectively modulate herpes Simplex Virus 1 (HSV-1) and HSV-2 Productive infections in adult sympathetic, but not sensory, neurons. *J Virol, 91*(13), e00582-17.

Jasnow, A. M., Drazen, D. L., Huhman, K. L., Nelson, R. J., & Demas, G. E. (2001). Acute and chronic social defeat suppresses humoral immunity of male Syrian hamsters (Mesocricetus auratus). *Hormones and Behav, 40*(3), 428–433.

Joisten, N., Rademacher, A., Bloch, W., Schenk, A., Oberste, M., Dalgas, U., … Bansi, J. (2019). Influence of different rehabilitative aerobic exercise programs on (anti-) inflammatory immune signalling, cognitive and functional capacity in persons with MS - study protocol of a randomized controlled trial. *BMC Neurol, 19*(1), 37.

Kennedy, S. L., Nickerson, M., Campisi, J., Johnson, J. D., Smith, T. P., Sharkey, C., & Fleshner, M. (2005). Splenic norepinephrine depletion following acute stress suppresses in vivo antibody response. *J Neuroimmunol, 165*(1–2), 150–160.

Kiecolt-Glaser, J. K., Loving, T. J., Stowell, J. R., Malarkey, W. B., Lemeshow, S., Dickinson, S. L., & Glaser, R. (2005). Hostile marital interactions, proinflammatory cytokine production, and wound healing. *Arch Gen Psychiatry, 62*(12), 1377–1384.

Klopack, E. T. (2023). Chronic stress and latent virus reactivation: Effects on immune aging, chronic disease morbidity, and mortality. *The J Geront. Series B, Psychol Sci Soc Sci, 78*(10), 1707–1716.

Knechtle, B., & Nikolaidis, P. T. (2018). Physiology and pathophysiology in ultra-marathon running. *Front Physiol, 9*, 634.

Knight, J. M., Rizzo, J. D., Logan, B. R., Wang, T., Arevalo, J. M., Ma, J., & Cole, S. W. (2016). Low socioeconomic status, adverse gene expression profiles, and clinical outcomes in hematopoietic stem cell transplant recipients. *Clin Cancer Res, 22*(1), 69–78.

Kucuk, O. (2022). Walk more, eat less, don't stress. *Cancer Epidemiol Biomarkers Prev, 31*(9), 1673–1674.

Kunz, H. E., Spielmann, G., Agha, N. H., O'Connor, D. P., Bollard, C. M., & Simpson, R. J. (2018). A single exercise bout augments adenovirus-specific T-cell mobilization and function. *Physiol Behav, 194*, 56–65.

Laudenslager, M. L., Fleshner, M., Hofstadter, P., Held, P. E., Simons, L., & Maier, S. F. (1988). Suppression of specific antibody production by inescapable shock: stability under varying conditions. *Brain Behav Immun, 2*(2), 92–101.

Lin, L., Chen, S., Russell, D. S., Lohr, C. V., Milston-Clements, R., Song, T., … Jin, L. (2017). Analysis of stress factors associated with KHV reactivation and pathological effects from KHV reactivation. *Virus Res, 240*, 200–206.

Long, J. E., Ring, C., Bosch, J. A., Eves, F., Drayson, M. T., Calver, R., … Burns, V. E. (2013). A life-style physical activity intervention and the antibody response to pneumococcal vaccination in women. *Psychosom Med, 75*(8), 774–782.

Loprinzi, P. D., & Frith, E. (2019). A brief primer on the mediational role of BDNF in the exercise-memory link. *Clin Physiol Funct Imaging, 39*(1), 9–14.

Malecki, K. M. C., Nikodemova, M., Schultz, A. A., LeCaire, T. J., Bersch, A. J., Cadmus-Bertram, L., … Peppard, P. E. (2021). The Survey of the Health of Wisconsin (SHOW) program: An infrastructure for advancing population health sciences. *Front Public Health, 10,* 818777.

Marsland, A. L., Walsh, C., Lockwood, K., & John-Henderson, N. A. (2017). The effects of acute psychological stress on circulating and stimulated inflammatory markers: A systematic review and meta-analysis. *Brain Behav Immun, 64,* 208–219.

Masini M.A., Bonetto V., Manfredi M., Pastò A., Barberis E., …, Marengo E. (2022). Prolonged exposure to simulated microgravity promotes stemness impairing morphological, metabolic and migratory profile of pancreatic cancer cells: a comprehensive proteomic, lipidomic and transcriptomic analysis. *Cell Mol Life Sci, 79*(5), 226.

Matundan, H. H., Wang, S., Jaggi, U., Yu, J., & Ghiasi, H. (2021). Suppression of CD80 expression by ICP22 affects herpes simplex virus type 1 replication and CD8(+)IFN-gamma(+) infiltrates in the eyes of infected mice but not latency reactivation. *J Virol, 95*(19), e0103621.

Mehta S.K., Bloom D.C., Plante I., Stowe R., Feiveson A.H., … , Pierson DL.(2018). Reactivation of Latent Epstein-Barr Virus: A Comparison after Exposure to Gamma, Proton, Carbon, and Iron Radiation. *Int J Mol Sci, 19(10),* 2961.

Mehta S.K., Cohrs R.J., Forghani B., Zerbe G., Gilden D.H., Pierson DL. (2004). Stress-induced subclinical reactivation of varicella zoster virus in astronauts. *J Med Virol, 72*(1), 174–179.

Mehta S.K., Laudenslager M.L., Stowe R.P., Crucian B.E., Sams C.F., Pierson D.L. (2014). Multiple latent viruses reactivate in astronauts during Space Shuttle missions. *Brain Behav Immun, 41,* 210–217.

Meijer, A., Konigs, M., Vermeulen, G. T., Visscher, C., Bosker, R. J., Hartman, E., & Oosterlaan, J. (2020). The effects of physical activity on brain structure and neurophysiological functioning in children: A systematic review and meta-analysis. *Dev Cogn Neurosci, 45,* 100828.

Mika, A., & Fleshner, M. (2016). Early-life exercise may promote lasting brain and metabolic health through gut bacterial metabolites. *Immunol Cell Biol, 94*(2), 151–157.

Mika, A., Rumian, N., Loughridge, A. B., & Fleshner, M. (2016). Exercise and prebiotics produce stress resistance: Converging impacts on stress-protective and butyrate-producing gut bacteria. *Int. Rev. of Neurobiol, 131,* 165–191.

Mika, A., Van Treuren, W., Gonzalez, A., Herrera, J. J., Knight, R., & Fleshner, M. (2015). Exercise is more effective at altering gut microbial composition and producing stable changes in lean mass in juvenile versus adult male F344 rats. *PLoS One, 10*(5), e0125889.

Miller, G. E., Cohen, S., Pressman, S., Barkin, A., Rabin, B. S., & Treanor, J. J. (2004). Psychological stress and antibody response to influenza vaccination: When is the critical period for stress, and how does it get inside the body? *Psychosom Med, 66*(2), 215–223.

Moraska, A., Campisi, J., Nguyen, K. T., Maier, S. F., Watkins, L. R., & Fleshner, M. (2002). Elevated IL-1beta contributes to antibody suppression produced by stress. *J Appl Physiol (1985), 93*(1), 207–215.

Moraska, A., & Fleshner, M. (2001). Voluntary physical activity prevents stress-induced behavioral depression and anti-KLH antibody suppression. *Am J Physiol Regul Integr Comp Physiol, 281*(2), R484–489.

Morey, J. N., Boggero, I. A., Scott, A. B., & Segerstrom, S. C. (2015). Current directions in stress and human immune function. *Curr Opin Psychol, 5,* 13–17.

Nadel, J. A., Pawelko, S. S., Scott, J. R., McLaughlin, R., Fox, M., Ghanem, M., … Howard, C. D. (2021). Optogenetic stimulation of striatal patches modifies habit formation and inhibits dopamine release. *Sci Rep, 11*(1), 19847.

Nakayasu, E. S., Gritsenko, M. A., Kim, Y. M., Kyle, J. E., Stratton, K. G., Nicora, C. D., … Burnum-Johnson, K. E. (2023). Elucidating regulatory processes of intense physical activity by multi-omics analysis. *Mil Med Res, 10*(1), 48.

Neeper, S. A., Gomez-Pinilla, F., Choi, J., & Cotman, C. W. (1996). Physical activity increases mRNA for brain-derived neurotrophic factor and nerve growth factor in rat brain. *Brain Res, 726*(1–2), 49–56.

Niemiro, G. M., Coletta, A. M., Agha, N. H., Mylabathula, P. L., Baker, F. L., Brewster, A. M., … Simpson, R. J. (2022). Salutary effects of moderate but not high intensity aerobic exercise training on the frequency of peripheral T-cells associated with immunosenescence in older women at high risk of breast cancer: A randomized controlled trial. *Immun Ageing, 19*(1), 17.

Noce, A., Tranchita, E., Marrone, G., Grazioli, E., Di Lauro, M., Murri, A., … Cerulli, C. (2023). The possible role of physical activity in the modulation of gut microbiota in chronic kidney disease and its impact on cardiovascular risk: A narrative review. *Eur Rev Med Pharmacol Sci, 27*(8), 3733–3746.

Ortiz, G. C., Sheridan, J. F., & Marucha, P. T. (2003). Stress-induced changes in pathophysiology and interferon gene expression during primary HSV-1 infection. *Brain Behav Immun, 17*(5), 329–338.

Padilha, C. S., Von Ah Morano, A. E., Kruger, K., Rosa-Neto, J. C., & Lira, F. S. (2022). The growing field of immunometabolism and exercise: Key findings in the last 5 years. *J Cell Physiol, 237*(11), 4001–4020.

Park, J. H., Prochnow, T., Chang, J., & Kim, S. J. (2023). Health-related behaviors and psychological status of adolescent patients with atopic dermatitis: The 2019 Korea youth risk behavior web-based survey. *Patient Prefer Adherence, 17*, 739–747.

Peters E.M., Bateman E.D. (1983). Ultramarathon running and upper respiratory tract infections. An epidemiological survey. *S Afr Med J, 64*(15), 582–584.

Pierson, D. L., Stowe, R. P., Phillips, T. M., Lugg, D. J., & Mehta, S. K. (2005). Epstein-Barr virus shedding by astronauts during space flight. *Brain Behav Immun, 19*(3), 235–242.

Poirier, S., Gendron, P., Houle, J., & Trudeau, F. (2023). Physical activity, occupational stress, and cardiovascular risk factors in law enforcement officers: A cross-sectional study. *J Occup Environ Med, 65*(11), e688–694.

Popa, D. E., Pirlog, M. C., Alexandru, D. O., & Gheonea, D. I. (2022). The influence of the inflammatory bowel diseases on the perceived stress and quality of life in a sample of the South-Western Romanian population. *Curr Health Sci J, 48*(1), 5–17.

Pruimboom, L., Raison, C. L., & Muskiet, F. A. (2015). Physical activity protects the human brain against metabolic stress induced by a postprandial and chronic inflammation. *Behav Neurol, 2015*, 569869.

Rosa-Neto, J. C., Lira, F. S., Little, J. P., Landells, G., Islam, H., Chazaud, B., Pyne, D. B., Teixeira, A. M., Batatinha, H., Moura Antunes, B., Guerra Minuzzi, L., Palmowski, J., Simpson, R. J., & Kruger, K. (2022). Immunometabolism-fit: How exercise and training can modify T cell and macrophage metabolism in health and disease. *Exerc Immunol Rev, 28*, 29–46.

Ruiz-Tejada, A., Neisewander, J., & Katsanos, C. S. (2022). Regulation of voluntary physical activity behavior: A review of evidence involving dopaminergic pathways in the brain. *Brain Sci, 12*(3) 333.

Santos Rocha, C. L. A., Hirao, M. G. Weber, G. Mendez-Lagares, G. W. L. W. Chang, G. Jiang, J., … Dandekar, S. (2018). Subclinical cytomegalovirus infection is associated with altered host immunity, gut microbiota, and vaccine responses. *J Virol, 92*(13) e00167–18.

Santos, V. C., Ostler, J. B., Harrison, K. S., & Jones, C. (2023). Slug, a stress-induced transcription factor, stimulates herpes simplex virus 1 replication and transactivates a cis-regulatory module within the VP16 promoter. *J Virol, 97*(4), e0007323.

Schneider, K. M., Blank, N., Alvarez, Y., Thum, K., Lundgren, P., Litichevskiy, L., … Thaiss, C. A. (2023). The enteric nervous system relays psychological stress to intestinal inflammation. *Cell, 186*(13), 2823–2838, e2820.

Schoech L., Allie K., Salvador P., Martinez M., Rivas E. (2021) Sex Differences in Thermal Comfort, Perception, Feeling, Stress and Focus During Exercise Hyperthermia. *Percept Mot Skills. 128*(3), 969–987.

Segerstrom, S. C., Hardy, J. K., Evans, D. R., & Greenberg, R. N. (2012). Vulnerability, distress, and immune response to vaccination in older adults. *Brain Behav Immun, 26*(5), 747–753.

Segerstrom S.C., Miller G.E. (2004) Psychological stress and the human immune system: a meta-analytic study of 30 years of inquiry. *Psychol Bull, 130*(4), 601–630.

Seyle, H. (1978). *The Stress of Life*. New York, McGraw-Hill.

Shayea, A. M. F., Alotaibi, N. M., Nadar, M. S., Alshemali, K., & Alhadlaq, H. W. (2022). Effect of physical activity and exercise on the level of COVID-19 antibodies and lifestyle-related factors among vaccinated health science center (HSC) students: A pilot randomized trial. *Vaccines (Basel), 10*(12) 2171.

Silberman, D. M., Wald, M. R., & Genaro, A. M. (2003). Acute and chronic stress exert opposing effects on antibody responses associated with changes in stress hormone regulation of T-lymphocyte reactivity. *J Neuroimmunol, 144*(1–2), 53–60.

Silveira, S. L., Jeng, B., Cutter, G., & Motl, R. W. (2023). Diet, physical activity, and stress among wheelchair users with multiple sclerosis: Examining individual and co-occurring behavioral risk factors. *Arch Phys Med Rehabil, 104*(4), 590–596, e591.

Simpson, R. J., Bosslau, T. K., Weyh, C., Niemiro, G. M., Batatinha, H., Smith, K. A., & Kruger, K. (2021). Exercise and adrenergic regulation of immunity. *Brain Behav Immun, 97*, 303–318.

Simpson R.J., Campbell J.P., Gleeson M., Krüger K., Nieman D.C., …, Walsh NP. (2020) Can exercise affect immune function to increase susceptibility to infection? *Exerc Immunol Rev, 26*, 8–22.

Slusher, A. L., & Acevedo, E. O. (2023). Stress induced proinflammatory adaptations: Plausible mechanisms for the link between stress and cardiovascular disease. *Front Physiol, 14*, 1124121.

Smith T.J., Wilson M., Whitney C., Fagnant H., Neumeier W.H., …, Karl J.P. (2022). Supplemental Protein and a Multinutrient Beverage Speed Wound Healing after Acute Sleep Restriction in Healthy Adults. *J Nutr, 152*(6), 1560–1573.

Smith T.J., Wilson M.A., Karl J.P., Orr J., Smith C.D., ..., Montain S.J. (2018). Impact of sleep restriction on local immune response and skin barrier restoration with and without "multinutrient" nutrition intervention. *J Appl Physiol, 124*(1), 190–200.

Smith, T. P., Kennedy, S. L., & Fleshner, M. (2004). Influence of age and physical activity on the primary in vivo antibody and T cell-mediated responses in men. *J Appl Physiol, 97*(2), 491–498.

Straub R.H. (2014) Rheumatoid arthritis: Stress in RA: a trigger of proinflammatory pathways? *Nat Rev Rheumatol. 10*(9), 516–518.

Strewe C., Moser D., Buchheim J.I., Gunga H.C., Stahn A., ..., Feuerecker M. (2019). Sex differences in stress and immune responses during confinement in Antarctica. *Biol Sex Differ. 10(1)*, 20.

Sun, Y. I., Pence, B. D., Wang, S. S., & Woods, J. A. (2019). Effects of exercise on stress-induced attenuation of vaccination responses in mice. *Med Sci Sports Exerc, 51*(8), 1635–1641.

Tanner, M. K., Davis, J. K. P., Jaime, J., Moya, N. A., Hohorst, A. A., Bonar, K.,... Greenwood, B. N. (2022). Duration- and sex-dependent neural circuit control of voluntary physical activity. *Psychopharmacology (Berl), 239*(11), 3697–3709.

Tashiro, R., Ogawa, Y., & Tominaga, T. (2017). Rapid deterioration of latent HBV hepatitis during cushing disease and posttraumatic stress disorder after earthquake. *J Neurol Surg A Cent Eur Neurosurg, 78*(4), 407–411.

Thaiss, C. A. (2023). A microbiome exercise. *Science, 381*(6653), 38.

Theall B., Wang H., Kuremsky C.A., Cho E., Hardin K., …, Spielmann G. (2020). Allostatic stress load and CMV serostatus impact immune response to maximal exercise in collegiate swimmers. *J Appl Physiol, 128*(1), 178–188.

Turner J.E., Bennett S.J., Bosch J.A., Griffiths H.R., Aldred S. (2014). Ultra-endurance exercise: unanswered questions in redox biology and immunology. *Biochem Soc Trans., 42*(4), 989–995.

Uchakin, P. N., Parish, D. C., Dane, F. C., Uchakina, O. N., Scheetz, A. P., Agarwal, N. K., & Smith, B. E. (2011). Fatigue in medical residents leads to reactivation of herpes virus latency. *Interdiscip Perspect Infect Dis, 2011*, 571340.

Walsh N.P. (2018). Recommendations to maintain immune health in athletes. *Eur J Sport Sci, 18*(6), 820–831.

Wang, R., Cai, Y., Lu, W., Zhang, R., Shao, R., Yau, S. Y., … Lin, K. (2023). Exercise effect on the gut microbiota in young adolescents with subthreshold depression: A randomized psychoeducation-controlled Trial. *Psychiatry Res, 319*, 115005.

Wanjau, M. N., Moller, H., Haigh, F., Milat, A., Hayek, R., Lucas, P., & Veerman, J. L. (2023). Physical activity and depression and anxiety disorders: A systematic review of reviews and assessment of causality. *AJPM Focus 2*(2), 100074.

Warreman, E. B., Nooteboom, L. A., Terry, M. B., Hoek, H. W., Leenen, P., van Rossum, E., Ramlal, D., … Ester, W. A. (2023). Psychological, behavioural and biological factors associated with gastrointestinal symptoms in autistic adults and adults with autistic traits. *Autism, 27*(7), 2173–2186.

Weinstock M., Razin M., Schorer-Apelbaum D., Men D., McCarty R. (1998). Gender differences in sympathoadrenal activity in rats at rest and in response to footshock stress. *Int J Dev Neurosci. 16*(3-4), 289–295.

Wilson S.J., Padin A.C., Bailey B.E., Laskowski B., Andridge R., …, Kiecolt-Glaser J.K. (2020) Spousal bereavement after dementia caregiving: A turning point for immune health. *Psychoneuroendocrinology. 118,* 104717.

Wong, S. Y., Wong, C. K., Chan, F. W., Chan, P. K., Ngai, K., Mercer, S., & Woo, J. (2013). Chronic psychosocial stress: Does it modulate immunity to the influenza vaccine in Hong Kong Chinese elderly caregivers? *Age (Dordr) 35*(4), 1479–1493.

Yang, C., Gao, J., Du, J., Yang, X., & Jiang, J. (2017). Altered neuroendocrine immune responses, a two-sword weapon against traumatic inflammation. *Int J Biol Sci, 13*(11), 1409–1419.

Yazgan, I., Bartlett, V., Romain, G., Cleman, J., Petersen-Crair, P., Spertus, J. A., … Smolderen, K. G. (2023). Longitudinal pathways between physical activity, depression, and perceived stress in peripheral artery disease. *Circ Cardiovasc Qual Outcomes, 16*(8), 544–553.

Yu, W., Geng, S., Suo, Y., Wei, X., Cai, Q., Wu, B., X. … Wang, B. (2018). Critical role of regulatory T cells in the latency and stress-induced reactivation of HSV-1. *Cell Rep, 25*(9), 2379–2389, e2373.

Zavala J.K., Fernandez A.A., Gosselink K.L. (2011) Female responses to acute and repeated restraint stress differ from those in males. *Physiol Behav. 104*(2), 215–221.

Zoladz, J. A., & Pilc, A. (2010). The effect of physical activity on the brain derived neurotrophic factor: From animal to human studies. *J Physiol Pharmacol, 61*(5), 533–541.

Index

Note: **Bold** page numbers refer to tables and *italic* page numbers refer to figures.

For Product Safety Concerns and Information please contact our
EU representative GPSR@taylorandfrancis.com Taylor & Francis
Verlag GmbH, Kaufingerstraße 24, 80331 München, Germany